TRADITIONAL AND RELIGIOUS PLANTS OF WEST AFRICA

Mr. Daniel Abbiw

TRADITIONAL AND RELIGIOUS PLANTS OF WEST AFRICA

ISBN: 9988-600-87-9

First published in 2014

Smartline Limited
C3 Coastal Estates DTD
Baatsona, Spintex Road-Accra
Ghana-West Africa
Website: www.smartlinepublishers.com

4, Oloto Close, Norwood Residential Area, Agbogba, North Legon, Accra. Ghana.
Mobile: 024-4671485
E-mail: danielabbiw@yahoo.com

SMARTLINE

TRADITIONAL AND RELIGIOUS PLANTS OF WEST AFRICA

Plant uses based on:

- beliefs
- symbols
- signs and
- traditional values

Cover photo: *Adansonia digitata* Linn. Baobab. Bombacaceae (symbol of fertility)

Dedication

To
Mother: Mrs. Elizabeth Attuquaye
Brothers & Sister: Emmanuel Kow Abbiw, Isaac Nana Otu Abbiw &
Alice Beatrice Abbiw
Daughters: Dinah Ama Edufuwa Abbiw & Emilia Efuwa Botsewa Abbiw
And Old Achimotans – especially 1959 Year Group

For their encouragement and best wishes

CONTENTS AT A GLANCE

CONTENTS

FOREWORD

And he spake of trees, from the cedar tree that is in Lebanon
even unto the hyssop that springeth out of the wall;
he spake also of beasts, and of fowl,
and of creeping things and of fishes.
1 Kings 4:33

The author of this book is already well known in botanical and ethno botanical circles for his impressive work *Useful Plants of Ghana* (Intermediate Technology & Royal Botanic Gardens, Kew, 1990). He is now following it with one that has a narrower yet wider scope! Narrower in that, he has selected plants with traditional and religious uses - subjects that will appeal to a more extensive readership than botanists - and wider because he covers the whole of West Africa instead of one country, Ghana. In fact, readers will be fascinated to see that from time to time the author could not resist including details of similar traditions for plants in countries as far apart as America, Europe and Asia, as well as other parts of Africa.

Nothing like this has been attempted before for the Region since the information has been scattered in books, papers and among traditional men and women themselves. Much of this local information is in danger of being lost as traditional rural life gives way to urban living with its detachment from the countryside. This book will not only be a constant source of reference, but the text in its own right will intrigue readers. It is prepared for the general reader as well as the specialist; and for the traditionalist as well as the scientist. Hence, technical terms are kept to a minimum or where used are fully explained.

A word or two of warning, however, may be in order. Because the uses, myths, traditions and religious connotations are cited, it does not indicate that the author holds with or endorses them. He simply quotes what others do or what they believe. Not only that, but ill-informed experimentation with such prescriptions or self-medications could be dangerous or even fatal.

That said, as one who has spent all my working life at Kew studying the plant taxonomy of tropical Africa, and West Africa in particular, it gives me great pleasure to compliment Daniel Abbiw on his industry in compiling this work and to commend it to the reader.

F. NIGEL HEPPER
25A Montague Road

Richmond.

ACKNOWLEDGMENTS

In everything give thanks.
1 Thessalonians 5:18a

In addition to cited references, unpublished first hand information has been obtained from people from all walks of life. The informants are mainly from the rural communities, but a fair proportion of these contributions have also come from the urban centres. In some instances, the information was volunteered in a free chat, though the topic had neither been raised, nor any indication given that I was compiling the uses of plants that are based on beliefs, symbols, signs and traditional values. This could be an indication that the traditional and religious aspects of plants form part and parcel of everyday life. I am grateful to all these casual informants for this informal contribution.

In most instances however, chiefs, elders, medicine- and juju-men, herbalists, native doctors, witch-doctors and fetish priests who were recommended to me were consulted on a specific section of the subject, or for a more general view and consensus. A few of such persons were quite helpful outright, despite the secrecy that usually surrounds traditional rituals, sacrificial offerings, purification ceremonies, witch-hunting, ordeal trials, fetishism and juju practices. Others obliged only after receiving the customary 'drink'. Others would not say a word though - denying all knowledge on the subject; while yet others simply dishonoured all appointments. Among those whose contribution is worth special mentioning are the late Alhaji Asiedu Biney of CSRPM, Mampong; the late Mr. A.N.K. Mensah, a retired Headmaster of Keta; the late Dr. G.K. Akpabla, F.L.S., a retired Botanist; the late Dr. G.K. Noamesi, a traditional doctor at Hohoe; the late Moro Ibrahim, a herbalist at Legon; the late Mr. Ankoma-Ayew, an elder; the late Mr. S.K. Avumatsodo, F.L.S., a retired Botanical Artist; the late Nana Asifu Yao Okoamankran, the Apagyahene of Abiriw-Akwapim - known in private life as Opanyin Yao Odei; and the late Mr. M. Gamadi, an elder - all from Ghana.

I am equally grateful for the contributions of Mr. J.K. Eshun, a native doctor at Akyempim; Dr. Akligo-Zomadi, a solar scientist; Mr. Sallam Abdul-Wahab; Mr. G.O. Cofie, an elder; Mr. E.K. Siriboe, an elder - all from Ghana; Dr. L Aké Assi, a Botanist from Ivory Coast; Prof. J.C. Okafor, a Forester and Dr. Ayodele Edwards both from Nigeria; Mr. P. Klugah, an

elder from Togo; Mr. Seidu Bah from Sierra Leone; Miss Christiana Devi, a student from Liberia and Dr. F. Krôger, a German anthropologist.

My sincere thanks to the staff of the Africana Section of the Balme Library at the University of Ghana, Legon, for permission to borrow rare literature from their archives. I wish also to thank Prof. Kofi Asare Poku, Prof. Nana Apt van Ham both of Legon; Ms Oda Yukiyo, a PhD. Research Student in the Institute of African Studies, Legon from the Koyota University of Japan; Ms Julia Falconer, Ms Susanne Schaffer, a Forestry Consultant from Britain and a landscape student from Germany, respectively; for the provision of relevant references; and Miss Ruby Odzaba for her assistance in eventually leading me - after ten years search - to meet her relative, Miss Ama Doris Borbor, the lady allegedly whisked away by dwarfs at the age of three years at Gbadzeme, in the Avatime traditional area of the Volta Region in Ghana (see DWARFS AND FAIRIES: DWARFS: Abduction). I am thankful to Mr. David Daramani of the Department of Game and Wildlife in Ghana, a bird spotter of international repute and Prof. Yaa Ntiamoah-Baidu of the Zoology Department, Legon for the scientific names of the birds; to Prof. Dan Attuquayefio, Head of Zoology Department, University of Ghana, Legon, Mr. Francis Gbogbo of the same Department and Mr. Samuel K. Nyame from Department of Game and Wildlife for the scientific names of the animals. I am also thankful to Mr. Ayaa K. Armah of the Department of Oceanography and Fisheries, to Prof. Christopher Gordon and Dr. George Asare Darpaah, both of the Volta Basin Research Project (VBRP) at Legon for the scientific names of fish; and to Dr. David Wilson, Miss Millicent A. Cobblah both of Zoology Department and Prof. Kwame Afreh-Nuamah of the University of Ghana Agricultural Research Station, Kade for the scientific names of insects. Dr. Ted Annang, Department of Botany, University of Ghana, Legon provided information on the Teshie and Nungua designer or fancy, but expensive coffins.

Taxonomists Dr. Aaron Davis, Dr. Felix Forest, Dr. Alan Paton and Dr. Iain Darbyshire at the Royal Botanic Gardens, Kew; and Dr. Erin Tripp at the Rancho Santa Ana Botanical Gardens, USA helped with plant identification. Despite ill health, Prof. Nigel Hepper, formerly of the Royal Botanic Gardens, Kew, agreed to read through the entire first print of the Manuscript and made corrections and useful comments for improvement. In addition, Nigel also contributed to the selection of a suitable title for the book and undertook to write the Foreword. I am most grateful to Nigel for this assistance. Sadly, Nigel passed away in

the early hours of Thursday 16th May 2013 after a short illness at age 84. MAY HIS SOUL REST IN PERFECT PEACE. Prof. Emeritus E. Laing read through the Manuscript and commented freely. Miss Emilia Efuwa Botsewa Abbiw, my daughter, assisted in the conversion of the original Manuscript in Word Perfect on diskettes to Word on CD. In addition, Emilia sent letters and pictures by E-mail to specialists both locally and abroad. I would finally thank my wife, Beatrice, for her encouragement, contribution, assistance in the collection of data and advice throughout the book project; and all friends and colleagues, both in Ghana and elsewhere, for their patience in waiting for a book project that takes over two decades to complete.

INTRODUCTION

One generation passeth away,
and another generation cometh:
but the earth abideth for ever.
Ecclesiastes 1:4

Man has depended on plants and plant products for his various needs since prehistoric times. Directly, plants serve as food for man and fodder for animals; as material for housing, constructional work and furnishing; as items of trade or cash crops; as source of fuel for heating, lighting and cooking; and as implements or tools for farming and hunting. Plants are useful for household utensils, clothing and craft work; as medicine; as poisons and antidotes; as dyes and tannin; as avenue or shade trees; for decoration as house plants or for their aesthetic values; and also as manure and a source of nutrients for crops. Indirectly, plants influence the local climate (rainfall, relative humidity and wind); protect watersheds and catchment areas, animals, crops and the soil; prevent erosion; and assist in the drainage of swampy or inundated areas.

Some plants are used as food because they contain proteins to build the body, or carbohydrates to give energy or fats and oils to provide heat; but there are other uses of plants that are not necessarily based on any easily recognizable property, but rather appear to depend almost entirely on beliefs, values and social symbolic roles. For instance, the petals of *Rosa* species Rose, a widely cultivated thorny shrub in the plant family Rosaceae, is source of a valuable essential oil of commerce. Traditionally, however, the flower serves as garland for ceremonies; a symbol of love and charity, and of discretional secret; a badge of heraldry; a charm to influence court decisions; a decoration for shrines; and an emblem of martyrs (the five petals representing the five wounds of Christ). The Rose is also the National Flower of at least ten countries and states throughout the world. (For a full list of National, State or Emblematic Plants see APPENDIX - Table IV).

A combination of belief systems, historical events, environmental factors and the traditional value attributed to the plant have contributed to above uses and symbols in the culture of the people. Changes resulting either from development or from inventions or especially from civilization, invariably leads to a corresponding change or modification in the way of life. Old and outmoded local laws or customs, rules and regulations either die away naturally, or are banned by new legislation, or are modified. The refined end result is tradition.

TRADITION. Tradition is the oral transmission of custom, culture and beliefs from generation to generation, or the practice of such belief or custom. It portrays a well preserved lifestyle many, many centuries old. Some of the highlights include religious practices as rituals, purification and cleansing ceremonies, festivals, sacrificial offerings to the gods and ancestors and ancestral worship. Others are taboos, spirit possession, witch-hunting, casting out spells and exorcism, fortune-telling and divination. These practices and beliefs neither lend themselves to verification by scientific evaluation nor are yet fully comprehensible. The Doctrine of Signatures or sympathetic magic often explains away a section of these practices. However, by far the commonest single appellation for some of the traditional practices and beliefs, and more often synonymous with it, is superstition.

SUPERSTITION. Superstition is defined as a belief or practice resulting from ignorance, fear of the unknown and trust in magic or chance. It is suggestive of insecurity coupled with uncertainty. It may be classified into three main groups:

- Where a perfectly natural occurrence, such as an eclipse or a whirlwind or a hurricane or a volcano or lightning and thunder, through ignorance, is then mysteriously interpreted or believed to be the work of supernatural forces, and consequently deified or worshipped
- The fear of an entirely non-existing or baseless phenomenon, like the spirit of the ancestors, and the belief that they influence people and events. Finally the most serious is
- The involvement or indulgence in supernatural practices like witchcraft, magic or juju, the spirit world and occultism for services and favour with its inherent consequences.

Superstitious practices and beliefs transcend all races and social classes.

The subject is as diverse as it is universal. One researcher is reported to have found more than four hundred thousand different kinds. The habit of avoiding the number thirteen especially with rooms and hotel floors (many cultures believe that Friday 13th is a bad day) or walking under a ladder; the fear of a calamity if a chameleon, *Chamaeleon vulgaris*, or a snake, or even a black cat crosses ones path, or bird flies into the house, or a cock crows before midnight, or a dog makes an unusual racket at night; the use of a horseshoe, or a rabbit foot, or a monkey paw, or a wooden whistle, or a talisman, or *gris-gris* for protection; or the habit of checking the horoscope in the morning papers, are some of the few day to day examples - consciously or unconsciously. Africa has its fair share of superstitious practices and beliefs. The level of ignorance is a contributory factor; but unfortunately, the situation is seriously compounded by the lies, trickery, greed and dubious activities of some fetish-, juju-, or medicine-men and witch-doctors. On one hand, these men prophesy ominous events, impending disasters, spells and curses to their clients; and on the other, offer to ward off the catastrophe. This is usually on condition that an expensive and lavish sacrifice like sheep, *Ovis* species; goat, *Capra* species; fowl, *Gallus domestica*; tubers of yam; even cow, *Bos* species; and cash in addition to bottles of schnapps, is offered ostensibly to pacify the gods for mediation. The activities of some of the large numbers of spiritual churches and sects are equally to blame for this corruption.

There are superstitious beliefs about virtually all that influence human life - heavenly bodies like the sun, the moon and the stars; days of the week; numbers; disease; the dead; plants and animals. The editors of *The Illustrated Encyclopedia of Plants*, Michael Bisacre and others (1984) state that the mystical and symbolic properties once ascribed to plants often stemmed from improperly understood scientific facts. True as it is, some of these values are integrated with, and in fact, inseparable from the tradition of the people. A study of tradition invariably reflects the symbolic aspects of some plants and *vice versa*. As a result the works of plant collectors, ethnobotanists, historians, archaeologists and anthropologists alike show such uses of plants.

THE TEXT *Traditional and Religious Plants of West Africa* is a collection of plant uses that appear to be based on beliefs, symbols, signs and values (or is it natural uses ? See MEDICINE: SYMPATHETIC MAGIC). The information has been compiled from literature, general observation and from original, reliably authoritative and undeniably genuine knowledge

extracted from exclusive interviews with elders, medicine-men, fetish priests and witch-doctors - many though illiterate and belonging to the dying breed. However, since the information has been transmitted orally and over centuries, distortions are naturally expected, and do occur.

The text is not a 'recipe' to be copied or practiced through curiosity for several reasons. Besides possible failure and disappointment, it might be dangerous, or could result in serious repercussions, because the practices are usually preceded by incantations and rituals. There is also the danger of collecting the wrong plant to start with. In addition, sketchy prescriptive information, or only a few necessary ingredients are usually disclosed by the elders to maintain their reputation. This strategy of shelving knowledge on the subject, or reluctance in divulging the same, has been a major setback.

Some of the traditional uses of plants, no doubt, are mysterious. In a few cases there is an attempt at an explanation where there is reason to justify one; otherwise it has been left to the imagination and judgement of the reader. Equally left to open debate is the issue of whether the ancestral uses of plants as passed down from generations are, partly or wholly superstitious, or are natural uses; and whether they work, and have a meaning and a purpose.

There are fifty-two headings arranged alphabetically. A relevant verse from the King James Version (1611) of the *Holy Bible*, or a hymn or other verse, a simile, a proverb, a saying, or a slogan precedes each heading. Since many of the headings are interrelated and not separate concepts, there are several cross references. The chapter DISCUSSION sums up the advantages and disadvantages of traditional practices and beliefs; and the extent to which civilization has influenced these practices and beliefs.

There is usually a short description of each plant and its habitat immediately after the name. The binomial system of naming plants has been followed. The nomenclature (for indigenous plants) is as in the *Flora of West Tropical Africa* by Hutchinson, J. and Dalziel, J.M. (1954-72) (Revised by R.W.J. Keay and F.N. Hepper). Any name changes and revisions, since the appearance of this publication, have been put into parenthesis. Common names are given (immediately after the scientific name) where available, and local names where the meaning either helps to make the explanation clearer, or where it is the only alternative. With

the animal world, the scientific name rather follows the common name of the organism.

The emphasis is on West Africa, otherwise referred to as the 'Region', but free references have been made to many examples from other parts of Africa, and from lands far and near.

ILLUSTRATION The temperate plants were culled from the internet. The remaining photographs were all taken by the author.

the [illegible] nomenclature follows the [illegible] of [illegible].

[illegible] references have been made to many examples from [illegible] of Africa and [illegible].

[illegible] The remaining photographs were all taken by the author.

ANIMALS

Wildlife is life. Protect wildlife to enrich life.
Worldwide Fund for Nature

The interaction of plants and animals in folklore is common. There are many supernatural tales about plants and about animals, but not many of these stories involve both together. The few examples are about wild animals like the leopard, *Panthera pardus*; the lion, *Panthera leo*; the elephant, *Loxodonta africana*; and the hyena, *Hyaena hyaena*. Some plants are also named after these animals.

SPIRITUAL SIGNIFICANCE

Animals are traditionally believed to possess souls. (For animals with vindictive spirits, see HUNTING: THE HUNTED PROTECTED: Animals with Vindictive Spirits). For this reason, some animals are, by tradition, forbidden to be killed for food by certain tribal groups; in the belief that these tribes, in turn, enjoy the guardian protection of these animals, who may not kill and eat them up. For example, among the Bantu tribe of N Nigeria, it is deemed highly sacrilegious to kill a leopard, *Panthera pardus*. The superstition is that on any man who kills a leopard will fall a curse or evil disease, curable only by ruinously expensive process of three weeks' duration under the direction of Ukuku. So the natives allow the greatest depredation and ravages until their sheep, *Ovis* species; goats, *Capra* species; and dogs, *Canis* species, are swept away, and are roused to self defense only when a human being becomes the victim of the daring beast (Nassau fide 9). In S Nigeria, the Igbo at Aguka have an unexplained mystic

Sheep grazing

use for the fronds of *Nephrolepis undulata*, a fern, to tie over the eyes of a dead leopard (3c). In Ghana, the Nikwei family who have charge of the Teshie worship of Klan swear by the hyena, *Hyaena hyaena*, when they require a binding oath, and are not required to kill the animal (7).

Similarly, among the *Twidanfo*, a family division of the Akans of Ghana and Ivory Coast, the leopard is not only sacred, but members also abstain from the flesh of all feline animals. Should a member of this family chance to come in contact with a dead leopard, he must, as a sign of respect and sorrow, scatter shreds of white cloth upon it, and anoint the muzzle with palm-oil (5a). (One method of killing a cat, *Felis domestica*, is to coax it into a sack, tie up the opening, before hitting the cat's head with a club. The operation is completed by the ritual anointing of the cat's mouth with palm-oil. There is a firm unshaken belief that eating a cat's head ensures that one never dies away from home). Superstition in Senegal and Guinea evokes a saying: it is not done to beat a cat with a whip of Raphia for it will die (3b).

A second reason why some animals are protected by traditional groups is that these animals are believed to harbour the spirit of dead relatives. For instance, from observation, 'a doctor told me he once knew a man whose plantations were devastated by an elephant, *Loxodontha africana*. He advised that the beast should be shot, but the man said he dare not because the spirit of his dead father had passed into the elephant'(9). Writing about the Yoruba-speaking tribe of S Nigeria, another author adds: 'The animals in which human souls are most commonly reborn is the hyena, *Hyaena hyaena*, whose half-human laugh may perhaps account for the belief. Human souls are also re-born in different kinds of monkeys - species of *Cercocebus, Colobus, Cercopithecus, Erythrocebus, Papio, Piliocolobus* and *Procolobus* - but chiefly in the solitary yellow monkey called *oloyo;* and in these cases the human appearance and characteristics of monkeys, no doubt, furnishes the key to the belief' (5b). (See SLAVERY: LORE). (For animal-worship, see RELIGION: NATURE AND DEITIES).

Emblem

Some animals are used as emblem for traditional groups or clans; and for some countries. For example, the lion, *Panthera leo*, reckoned king of beasts, was the emblem of Ethiopia. The Red Lion signifies the Lion of Scotland and the Golden Lion that of England (10). The elephant,

Loxodontha africana, or its tusk is the emblem of Ivory Coast; and together with *Elaeis guineensis* Oil palm, the elephant served as the emblem on the common currency notes and coins of The Gambia, Sierra Leone, the then Gold Coast (now Ghana) and Nigeria - all former British colonies; as well as the motif of the W African Regiment. (For plants used as the emblem of some other countries worldwide, see SACRED PLANTS: NATIONAL AND STATE PLANTS. See also APPENDIX - Table IV). The traditional practice of attributing such properties as sacred, possessing a soul, harbouring the spirit of dead relatives, or serving as emblem to certain animals are, in effect, means of protecting these creatures from over-exploitation and ultimate extinction. The practice is therefore conservation of biodiversity for sustained yield. It is worth encouraging.

LORE, BELIEFS AND PRACTICES

The red dye of *Baphia nitida* Camwood, a shrub legume or small tree, used to mark the face or the forehead of a man during a particular ceremonial dance indicates that he has killed a man or a leopard (Thomas fide 8). In Liberia, the dried and pulverized leaves of *Aidia genipiflora*, an under-storey forest tree, are thrown into the air at a spot where a leopard has been seen and it will come (Cooper fide 4) - that is materialize. On the contrary, if *Sclerocarya birrea*, a savanna tree with edible fruits, is planted around a compound it keeps away leopards (3c). In S Nigeria, the Yoruba claim that drinking coconut milk makes men like monkeys. Perhaps this is a simianpromorphic fantasy arising from the three 'eyes' of the husked coconut, representing two eyes and the snout of an ape, *Pan troglodytes* (3b). The word *cocos* taken by Linnaeus for the generic name is, indeed, the Portuguese word *coco* meaning an ape (3b). Another plant is named after the Latin for the ape. The etymology of the generic name of *Mimulus gracilis*, a herb, is derived from Latin: *mimo*, an ape, from the supposed resemblance of the flowers to a mask, or monkey's face, hence the English name 'monkey flower' (3c).

In Ghana, the unopened leaves of *Annona senegalensis* ssp. *senegalensis* Wild custard apple are treated ceremoniously to tame dogs, *Canis* species, and wild beasts that may constitute a threat to man. Like the new leaves, the creature cannot open its jaws to bite - an instance of sympathetic magic. The leaves of *Datura innoxia*, a herbaceous annual, collected with rites, mashed and applied to a dog-bite serves the double purpose of healing the wound and killing the dog (12). In Egypt, *Epilobium hirsutum* Willow herb is used as a charm against poisonous animals;

and in Zambia, the root and bark of *Trichilia emetica* ssp. *suberosa (roka)*, a savanna tree, are used to tame the wild dove, *Streptopelia semitorquata*, a use that seems to depend on superstition rather than pharmacological effects (16). In Liberia, the leaves of *Gaertnera cooperi*, a small forest tree, are used to camouflage the traps used to catch monkeys on farms; the leaves are pulverized and scattered in the wind while calling the monkeys (Cooper fide 4). In Senegal, it is believed that when toads, *Bufo* species, die they become *Ceropegia deightonii*, a slender twiner of wet grassland. This is a fantasy based on the plant sap resembling the blood and secretion of a toad (6).

In E Africa, chips of wood of *Balanites aegyptiaca* Desert date or Soap berry are placed in elephant's dung, and with other pieces the remaining dung is heaped up; the procedure is a charm against the ferocity of the elephant (4). In N Nigeria, the bark of *Piliostigma thonningii*, a legume shrub or small tree in savanna region, is sacred to the bush cat, *Viverra civetta*, in the folklore of the Benue people; and strips are deposited beside the symbol of the cult (11). In Congo (Brazzaville), *Aspilia kotschyi*, an erect herbaceous composite, has superstitious application, and canoe-men claim the plant can protect them from crocodiles, *Crocodylus* species (2). In S Africa, the Wemba use the root of *Clematis* species, a climbing plant, probably *Clematis kirkii*, to frighten off the lion, *Panthera leo*; and in Kenya, the bark of *Ozoroa (Heeria) pulcherrima*, an erect savanna plant, is used by the Masai as a charm to ward off nocturnal attack by animals, by chewing a portion and spitting it out in all directions, saying at the same time *usho-usho* (16). In Europe, *Lunaria annua*, Honesty, a temperate plant, was once credited with such supernatural powers as possessing the ability to cast the shoe of horses, *Equus* species, which walked on it (14).

In the folklore of many countries, bats, species of *Cynopterus, Eidolon, Epomophorus, Glossophaga* and *Pteropus*; are believed to be sinister animals because of their nocturnal flight and screeching. Therefore the trees which they visit to pollinate are equally associated in the people's mind with ill omen. Bats are also agents of dispersal. (For trees associated with bats, see GHOSTS AND SPIRIT WORLD: INFLUENCE OF BATS: Tables 1 and 2).

ASSOCIATED PLANT NAMES

Some plants are named after wild animals - most likely in reference to, or comparison with the strength or ferocity of these animals. For instance, in Sierra Leone, the Susu name of *Combretum fragrans*, a medium-sized savanna tree, is 'samataga-na' from 'samana', meaning an elephant (3a). The elephant cannot break this tree, as the branches bend but do not break. The Akan name of *Stereospermum kunthianum* and that of *S. acuminatissimum*, trees of the savanna woodland and forest respectively, is 'esono-tokwa-kofo', meaning elephant fighter (8). In Sierra Leone, *Ipomoea asarifolia* and *I. mauritiana* are known as 'samajewo' in Mandinka, meaning elephant's pumpkin, and 'hale gbokpo' in Mende, meaning elephant's convolvulus (3a). The plants are known in Fula as 'layre ngabu' meaning creeper of the hippopotamus, *Hippopotamus amphibius* (3a). In S Nigeria, the Igbos call *Sorghum arundinaceum* Kamerun grass, a wild grass with edible grains, 'akpakpa enyi miri', meaning hippo corn (4). The name also applies to *S. lanceolatum* and to *S. vogelianum*.

Argyreia nervosa Elephant creeper

Pennisetum purpureum Elephant grass (associated with traditional religion)

In Nigeria, *Allanblackia floribunda* Tallow tree is called 'izeni' in Edo, meaning elephant rice; 'ala-enyi' in Igbo, meaning elephant's breast; and 'orogbo erin' in Yoruba, meaning elephant's bitter kola (4); and in Ghana, the tree is called 'sunkyi' in Anyi-Sefwi, meaning elephant's back (8). The Akan and the Diola names of *Microglossa afzelii* and that of *M. pyrifolia*, shrubby composites, are 'esono-mbabaa', meaning an elephant's sticks, and 'babum enab', meaning medicine of the elephant, respectively (3a). *Argyreia nervosa* Elephant creeper or Woolly morning glory is a strong creeper often cultivated as an ornamental and for shading terraces and verandas; and *Pennisetum purpureum* Elephant grass is a robust perennial often growing near the banks of streams. In N Nigeria, the Hausa call *Dalbergia*

saxatilis, a straggling leguminous shrub, 'runhun zaki', meaning a lion's shelter (4). In Ghana and Ivory Coast, the Akan name of both *Dialium dinklagei*, a leguminous forest tree, and *Rourea (Byrsocarpus) coccinea*, a savanna shrub, is 'awendade', meaning the lion (8). *Anopyxis klaineana* White oak, a large forest tree, is called 'kokoti' in Akan, meaning the bush pig or the red river hog, *Potamochoerus porcus* (8). The Ga name of *Anchomanes difformis* and *A. welwitschii*, araceous moist forest and savanna herbs respectively, whose tubers may be eaten as famine food, is 'batafoia kani', meaning a pig's cocoyam (4, 3a).

In S Nigeria, *Eragrostis ciliaris, E. gangetica, E. pilosa and E. tremula*, annual grasses are called 'irugbe-fon' in Yoruba - 'efon' meaning the buffalo, *Syncerus caffer (4)*. In Ghana, the Krobo name of *Cymbopogon citratus* Lemon grass and *C. giganteus*, a loosely tufted tall perennial grass, is 'gbetenga', meaning grass of the wolf, *Lupus* species (4); and *Acanthospermum hispidum* Star burr, a weedy composite with prickly fruits, is called 'peteku-nsoe' in Fante, meaning thorns of the hyena, *Hyaena hyaena* (3a) In N Nigeria, *Cucumis figarei (pustulatus)* and *C. metuliferus*, annual trailing cucurbits, are known collectively in Hausa as 'gunar kuuraa', meaning the hyena's watermelon, and 'noonon karse' or 'noonon kuuraa', meaning the hyena's milk (3a).

Acanthospermum hispidum Star bur.

A number of plants have reference to the leopard, *Panthera pardus*. In S Nigeria, the Yoruba name of *Hibiscus asper*, a fibrous savanna herb, and *Acanthus montanus* False thistle, a high forest plant; and the Igbo name of *H. surattensis*, a herbaceous prickly weed, 'anon ekun' and 'ile ago' respectively, both mean a leopard's tongue (3b). False thistle is also called 'anon ekun dudu' in Yoruba, and means a leopard's black tongue (3a). The plant is called 'edule imemein' in Ijo, meaning a leopard's claw (3a). In The Gambia and in Sierra Leone, the Mandinka and the Mende names of *Machaerium (Drepanocarpus) lunatum*, a thorny shrub in salt marshes and along river banks, 'solonoringo' and 'koli-ngengowi' respectively, have the same meaning - leopard's claw (4).

In S Nigeria, *Balanites wilsoniana*, a copal-producing forest tree, is called 'egungun-ekun' in Yoruba, and means bone of the leopard (4).

In Ghana, the Krobo name of *Sarcocephalus (Nauclea) latifolius* African peach, 'dikabi-awotso', means the leopard's fetish tree (8); and the Nzima call *Lagenaria (Adenopus) breviflora* 'aboa-ngatsee', meaning the leopard's groundnut - from the blotched fruit (4). In Sierra Leone, the Krio call *Ruthalicia (Physedra) eglandulosa*, a climbing cucurbit with bright red yellow-spotted fruits, 'koli-gojo', meaning the leopard's blow-blow trumpet (3a). In S Nigeria, the Igbo name of *Sansevieria liberica, S. senegambica*, and *S. trifasciata* African bowstring hemp, 'ebube-ago', means power of the leopard (4). The plant is commonly seen at leopard shrines, the speckled leaves being an ingredient in the magical werewolf prescription to effect change into the leopard form (15). Europeans and Native Americans referred to *Arnica montana* or *A. chamissonis* ssp. *foliosa* Arnica, herbaceous composites, as mountain tobacco and leopard's bane, and used the plants for sprains, bruises and wounds (1). In S Nigeria, the Yoruba names for the hemp or fibre of the plant 'ola ikoko' or 'pashan koriko' or 'oja koriko' means a hyena's girdle or whip.

Some flowers bear resemblance to the tiger, *Felis tigris*, and are accordingly named. For example, *Stanhopea devoniensis (tigrina)* is a Mexican orchid of great beauty. It is tiger-flecked with purplish markings (12). *Lilium lancifolium (tigrinum)* Tiger lily, which has been grown from centuries in the Far East for its edible bulbs, has rich fiery gold flowers with carved pointed petals and dotted with purple (12). *Tigridia pavonia* Tiger flower, is so called because of its flowers. The three large outer segments, come in vivid scarlet, orange, deep yellow and violet-mauve and are spotted inside near the base with contrasting or deeper colours (12). The genus name of *Equisetum arvense* Horsetail is derived from the Latin *equus* meaning horse and *seta* meaning bristle (1). The species name of *Humulus lupulus* Hops is derived from the Latin *lupus* (wolf), because it overgrows the gardens where it grows, as if it were a *Lupus* (the Latin word for wolf) (13). It could also mean that when it grows among osiers (willows), it strangles them by its light, climbing and embracing, as a wolf, *Lupus* species, does to a sheep, *Ovis* species (1). In the Greek language the genus name of *Crataegus oxyacantha* Hawthorn means 'strong goats' because when goats, *Capra* species, fed on the seeds they grew up agile and strong (13). (See also CATTLE, HUNTING, and RELIGION).

References.
1. Blumenthal & al., 2000; 2. Bouquet, 1969; 3a. Burkill, 1985; b. idem, 1997; c. idem, 2000; 4. Dalziel, 1937; 5a. Ellis, 1887; b. idem, 1894; 6. Ferry & al. 1974; 7. Field, 1937; 8. Irvine, 1961; 9. Kingsley, 1897; 10. Matson, 1970; 11. Meek, 1931; 12. Mensah, (pers. comm.); 13. Pamplona-Roger, 2001; 14. Perry & Greenwood, 1973; 15. Talbot, 1932; 16. Watt & Breyer-Brandwijk, 1962.

2 BEES AND HONEY

As busy as a bee

Some plants are frequented by bees, *Apis mellifera*, either because of the sweet flavour of the flowers or the fruits, or for their pollen and nectar. The colour of the flower may also attract bees. Blue, violet, purple, yellow and white flowers are all attractive to bees (13). The authors add that bees cannot perceive red, which explains the relative rarity of this colour in flowers. Other authors confirm this. The bee's colour spectrum is somewhat different from ours, it can see ultraviolet, which is invisible to us, but it cannot distinguish red, which appears black to a bee (11). The plants visited by bees are also associated with honey production. The trees in the Region (1, 3a, b, c, d, e, 4, 6, and 7) are listed in Table 1 below:

Species	Common Name	Plant Family	Colour of Flower
Acacia hockii	Shittim wood	Mimosaceae	Yellow
Albizia lebbeck	Lebbeck	Mimosaceae	White
Anacardium occidentale	Cashew nut	Anacardiaceae	Yellowish-pink
Asclepias curassavica	Blood flower, Red head	Asclepiadaceae	Red-orange
Blighia unijugata	-	Sapindaceae	Whitish
Burkea africana	-	Caesalpiniaceae	Cream
Cedrela odorata	Cedrela	Meliaceae	Yellowish
Ceiba pentandra	Silk cotton tree	Bombacaceae	White
Cocos nucifera	Coconut palm	Palmaceae	Cream
Crescentia cujete	Calabash nutmeg	Bignoniaceae	Yellow, with red or purple vein
Dialium guineense	Velvet tamarind	Caesalpiniaceae	White or pinkish
Diospyros mespiliformis	W Africa ebony tree	Ebenaceae	White
Drypetes floribunda	-	Euphorbiaceae	Cream
Elaeis guineensis	Oil palm	Palmaceae	Whitish
Eugenia jambos	Rose apple	Myrtaceae	Greenish-white

E. uniflora	Pitanga cherry	Myrtaceae	White
Haematoxylon campechianum	Logwood	Papilionaceae	Creamy-white or pale-yellow
Lantana camara	Wild sage	Verbenaceae	White, yellow, red or orange
Lophira alata	Ironwood	Ochnaceae	White
Mangifera indica	Mango	Anacardiaceae	Pinkish-white
Manihot glaziovii	Ceara rubber	Euphorbiaceae	Greenish-white or greenish-yellow
Melia azedarach	Persian lilac	Meliaceae	Lilac
Moringa oleifera	Horse-radish tree	Moringaceae	White or red
Octoknema borealis	-	Octoknemataceae	Yellowish-green
Parkia biglobosa (clappertoniana)	W African locust bean	Mimosaceae	Red or orange
Persea americana	Avocado pear	Lauraceae	Pale-yellow
Pithecellobium dulce	Madras thorn	Mimosaceae	White
Pouteria altissima	Aningeria	Sapotaceae	White
Psydrax (Canthium) venosa	-	Rubiaceae	Whitish
Pterocarpus erinaceus	Senegal rose wood tree	Papilionaceae	Golden-yellow
P. santalinoides	-	Papilionaceae	Brilliant-yellow
Sacoglottis gabonensis	Bitter bark tree	Humiriaceae	Yellow
Senna (Cassia) alata	Ringworm shrub	Caesalpiniaceae	Yellow
Tamarindus indica	Indian tamarind	Caesalpiniaceae	Yellow
Tetracera affinis	-	Dilleniaceae	White
Vitellaria (Butyrospermum) paradoxa	Shea butter tree	Sapotaceae	White
Vitex doniana	Black plum	Verbenaceae	Pale-yellow or pale-green

Table 1. Trees Visited by Bees

In addition to trees, other plants like the sedge, some grasses and orchids are also attracted by bees. Table 2 lists some of these plants.

Cyperaceae	Gramineae	Orchidaceae
Cyperus esculentus	*Andropogon tectorum*	*Disa racemosa*
	Heteropogon contortus	*Eulophia horsfalli*
	Zea mays	*Ophrys apifera*

Table 2. Other Plants, Besides Trees, Visited by Bees

Although bees - whether they are wild or reared - are primarily for honey production, the ability to sting could also be used (as appropriate technology or with traditional rites) depending on the ingenuity of the owner, for other purposes. These include preventing theft like the

Moringa oleifera Horse-radish tree

stealing of palm-wine (see preamble to BURGLARY), or ejecting a tenant from his or her room. In the latter instance, the bees first settle in a corner of the room as roommates, but if after twenty-four hours the tenant does not leave voluntarily, then he or she is forced out (9). Bees could also be used to chase away chain saw operators who intend to cut down sacred trees (see TABOO: FELLING TREES); or for warfare, for instance in Nigeria, an unidentified poisonous tree with bitter leaves locally called 'achineku' or 'agbu' in Tiv is part of a concoction for spilling over an enemy to attract bees to sting him (3e). The Ntumu tribal people of Cameroon have a proverb -'homeless bees never sting' (12).

CHARM PRESCRIPTIONS

Honey collectors may apply a magical touch to enhance their luck. In Senegal, the seeds of *Desmodium gangeticum*, an erect under-shrub of secondary forest, are considered by the Tenda to have magical properties. They are burnt on beehives, or put into an ointment to smear on the hive to ensure a good crop of honey (3c). In Nigeria, the seed of an unidentified plant locally called 'ikehegh' is put in a preparation inside a gourd to attract swarming bees to settle in it as a hive. The use seems to be basically magical as it is used to touch the gourd while pointing west (Ayoande fide 3e). Collectors may also smear their hands with insect repellents to protect themselves from bee stings. The practice

Desmodium gangeticum

could sometimes be a charm. In S Nigeria, Yoruba hunters of wild honey rub the leaf paste of *Baphia nitida* Camwood, a shrub or small tree legume, (and probably *B. pubescens*) on the body to prevent bee stings, but to eat any of the honey before completing the work is supposed to render the charm ineffective (4).

In S Nigeria, the leaves of *Allophylus africanus*, a shrub or small tree, are put in the entrance of bee-hives as a narcotic to stupefy bees (6); and among the Hausa of N Nigeria and the Zarma and the Fula of Niger the pounded leaves of *Annona senegalensis* var. *senegalensis* Wild custard apple is smeared on the hands by honey collectors to prevent sting (9) - probably superstitiously. In many farming villages the leaves of *Manihot esculenta* Cassava are often waved about by honey collectors during the operation to stupefy or calm down the bees (8). In Gabon, hives are fumed in smoke from burning *Lippia multiflora* Gambian tea-bush, a savanna annual; or the pulped foliage of *L. chevalieri* Tea bush is used to smooth the inside of the hives before placing them in trees. The practice attracts swarming bees to settle (3e). In Sierra Leone, honey-gatherers use the leaves of *Marantochloa leucantha,* a straight-stemmed undershrub, folded cone-shaped for collecting honey - hence the Mende name meaning 'honey leaf'(3d); and in The Gambia, the Diolas call *Momordica balsamina* Balsam apple 'ede hinde', meaning honey. The reason is uncertain. The genus name of *Melissa officinalis* Lemon balm, a perennial subshrub of the eastern Mediterranean region and western Asia, is from the Greek word for 'bee', referring to the bee's attraction to its flower and the quality of the honey produced from it (2). Honey bees also like the exquisite aroma of *Lavandula angustifolia* Lavender, a woody shrub labiate, and with the nectar of its flowers they produce a delicious honey (10).

Cassava (white-petioled)

In E Africa, Chagga beekeepers have to observe many ceremonies in connection with their beehives. The implement for making their boxes must be specially fashioned; the tree in which the hive is hung

is exhorted and threatened to secure its co-operation; and finally the collection of the honey is the occasion for ceremony, prayer and thanks (5). (For the relationship between beehives and witchcraft, see WITCHCRAFT: ACQUISITION: Beehives).

References.
1. Abbiw, 1990; 2. Blumenthal & al., 2000; 3a. Burkill, 1985; b. idem, 1994; c. idem, 1995; d. idem, 1997; e. idem. 2000; 4. Dalziel, 1937; 5. Dundas, 1910; 6. Irvine, 1961; 7. Keay, Onochie & Stanfield, 1964; 8. Mensah (pers. comm.); 9. Moro Ibrahim (pers. comm.); 10. Pamplona-Roger, 2001; 11. Raven, Evert & Curtis, 1976; 12. Sheppherd, 1988; 13. Steward & Hennessy, 1981.

3 BURGLARY (OR THEFT)

But lay up for yourselves treasure in heaven, where neither moth nor rust doth corrupt, and where thieves do not break through nor steal. For where your treasure is, there will your heart be also.
Matthew 6:20,21.

The use of charms, 'medicine' or juju - sometimes openly displayed and tied to a tree, farm paling or similar object; but usually hidden - in the belief of preventing theft, is practiced where pilfering of farm produce, trapped animals or fish, palm-wine and others is common. The punishment is varied, but often said to be instant. A snake might appear to frighten off the thief, or a bee, *Apis mellifera,* to sting the lower lip (in palm-wine pilfering). Alternatively, an implement provided for the purpose is used voluntarily by the would-be thief to hoe the farm non-stop until the owner arrives on the scene.

There are, in addition, charms that are concealed or buried in or around the home to protect the household and other personal effects from burglary; charms for the return of stolen items; and even counter-charms that aid the thief in his activities and protect him from possible maledictions.

PROTECTION OF FARMS AND FARM PRODUCE

Platycerium elephantotis
Stag-horn fern

In Sierra Leone, the Temne of the Port Loko area use *Ruthalicia (Physedra) eglandulosa*, a climbing cucurbit with yellow-spotted bright fruits, in a superstitious manner: a section of a leafy stem of the climber bearing a fruit is cut and attached with suitable curse to crops in the field to prevent thieving - as a punishment, the hands of the thief will swell up (6). In S Nigeria, whole plants of ferns such as *Platycerium elephantotis,*

P. stemaria Stag-horn fern and *Selaginella myosorus* enter into charms among the Igbo tribe to protect farms from thieves (16); and in Ghana, *Microgramma owariensis*, an epiphytic forest fern, has an unspecified but probably similar usage in Akropong Akwapim (14).

In S Nigeria, *Elaeis guineensis* Oil palm is either planted in *Cola acuminata* Commercial cola nut tree and *C. nitida* Bitter cola farms, among the Igbo tribe or the male or female inflorescence is hung around the kola tree to deter people intending to climb and harvest the kola nuts; and planting Oil palm in a farm or garden is also believed to discourage people from using the garden as a lavatory or a dumping ground for refuse (16). Other plants used by the Igbo tribe to protect farms include the leaves and twigs of *Adenia cissampeloides*, a climber in forest regrowth; the stem and leaves of *Baillonella toxisperma*, a very large forest tree with milky latex; and *Rhektophyllum mirabile*, a stout climbing aroid in forest.

In The Gambia, both *Scadoxus (Haemanthus) cinnabarinus* and *S. multiflorus* Blood flower or Fire-ball lily are planted as juju to protect farms from thieving, or planted in villages for other superstitious reasons (6); and in Sierra Leone, *Commelina diffusa* var. *diffusa*, an annual or perennial herb in open and wet places feature in a ceremony to protect rice on the farm (19). In S Nigeria, this herb is also used as a protecting charm on farms by the Igbo tribe (16). In Ghana, *Costus afer* Ginger lily, a forest perennial, is placed on cultivated field or on a path or the entrance to a house for protection. Other species like *C. deistelii, C. dubius, C. lucanusianus and C. schlechteri* are all similarly planted for protection.

In S Nigeria, the twigs and the pods of *Erythrophleum ivorense*, a forest tree legume; the dead branch of *Detarium senegalense* Tallow tree; the fruit of *Tetrapleura teptraptera*, a medium-sized forest tree legume; the twigs and leaves of *Newbouldia laevis*, a tree popularly believed to be fetish; and the bulb and leaves of *Crinum distichum* and *C. natans*, aquatic lilies, are hung on or around crops in the farm by the Igbo tribe to discourage theft (16). Among many of the tribal groups in the Region, it is the belief that *Gardenia ternifolia* ssp. *jovis-tonantis*, a savanna shrub popularly known to be protective against lightning, when planted in a farm protects the crops from thieves (2); and *Caralluma dalzielii*

Charm against theft

and *C. decaisneana*, succulent asclepiads with milky latex in savanna, are planted as fetish in yam fields, probably to discourage theft (6).

'Medicine' against theft of crops among the Bulsa tribe of N Ghana is described as follows - 'by burning certain herbs (unspecified) the owner of a farm produces black ash. With this he draws a black cross on a calabash and fixes it to a staff of *Pennisetum glaucum (americanum)* Pearl, Bulrush or Spiked millet or *Hibiscus cannabinus* Kenaf. If somebody who wants to steal crops from the field sees the calabash with the 'medicine' he will give up his wished plan, because he fears serious consequence (disease)' (13). While trekking from Yammin to Mankesim in Ghana, Ellis (7) observes that his team of thirsty men came upon a grove of palm trees with earthen pots containing palm-wine, but adds that 'there was a little stick, with a piece of rag tied on to it, that was stuck in the ground close to the pots. It was a fetish to prevent anyone stealing the palm-wine'.

A medicine against theft could also be in the form of a hedge. In the Region, *Euphorbia lateriflora,* a low shrub with poisonous latex, is a common hedge planted as a 'charm' by farmers to protect their crops (12). Many tree euphorbias are traditionally planted for protection against evil (3) - this could well explain its application here.

Snake Scaring

The practice of using 'spirit snakes' to frighten off would-be farm thieves is strongly believed to occur among farming communities in some parts of the Region. An egg with *Imperata cylindrica* Lalang grass, a notorious perennial weed with underground rhizomes, collected with rites, woven into a rope, then tied with a band of red cloth (likely a sign of danger) and supported on two forked sticks with a spell, is a charm for the spontaneous invocation of a snake to scare away farm thieves (18). Lalang grass appears to be spiritually associated with snake invocation. For instance in N Togo, a prescription drank by the Wangara tribe to protect oneself from bad juju in which the originator dies from snake-bite, is simply a tot of *akpeteshi* or *ogogoro* (local gin) taken from a bottle containing assorted poisonous snakes and a blade of the grass pickled with the gin - surely an example of sympathetic magic.

The woven, unopened leaflets of *Elaeis guineensis* Oil palm, collected with the necessary ceremony, and tied into a loop, with an egg among other ingredients, enter into similar charms to

invoke a snake to scare away crop thieves. The leaf of *Ananas comosus* Pineapple with the head of a poisonous snake as one of the ingredients is a snake-juju placed in farms with rituals and a spell to frighten prospective thieves away (14) - likely another instance of sympathetic magic. If the charm is prepared with the leaves of Pineapple with less prickly spines, and unspecified herbs as ingredients, the snake only chases away the thief; however, if the thorny and more prickly rhizomes of *Dioscorea praehensilis* Wild yam are used instead, a fatal bite is ensured (4).

A juju preparation of assorted ingredients including seven pieces each of roots that cross a path, and roots that run along a path, the bark of *Milicia (Chlorophora) excelsa* or *M. regia* Iroko; the stem of *Spiropetalum heterophyllum*, locally called 'homakyem' in Akan; the head of a mamba, *Dendroaspis viridis*; or any deadly snake, ground together (as one holds his breath, for it is forbidden to breathe while grinding the ingredients), then mixed with *akpeteshi* or *ogogoro* (local gin) into a poultice, 'dufa', and mounted directly on a leaf of *Baphia nitida* Camwood, which sits on a bamboo stump in a farm or stored corn, is invoked with libation in an incantation for a snake to materialize instantly - and depending on the degree of invocation - either to scare away or to bite to death any would-be thief (8) - another instance of sympathetic magic.

PROTECTION OF OTHER PROPERTIES

The use of plants to protect property, other than farms and farm produce, is recorded in the Region. In Sierra Leone, the Temne put the bark of *Craterispermum laurinum*, a shrub or small tree; or the fruit of *Cassia sieberiana* African laburnum in a superstitious application known as *wanka*, which is magic against theft (19). In this country, the leaf of *Bixa orellana* Anatto, a small tree commercially important as a food dye, is used as a remedy for pains in the ribs caused by *wanka*. The bark of *Milicia (Chlorophora) excelsa* or *M. regia* Iroko pounded and rolled into balls, then buried on the compound of a house, with rites, is used as a charm against thieving (14). The stone in the heartwood of Iroko (excrescences of calcium malate) buried on the compound of a house with the seed of *Cola verticillata*, a forest tree, and the hair of a fairy (said to be sold in medicine markets - for how it may be obtained, see DWARFS AND FAIRIES: FAIRIES: Nature and Characteristics), with rituals, serves as a preventive charm against burglary (1).

Among the plants traditionally used in prescriptions for preventing theft among other properties is *Calotropis procera Sodom* apple. The Hausa of N Nigeria, the Zarma of Mali and Niger and the Fula believe that when Sodom apple is planted around the house with the necessary ceremony, it is preventive of witchcraft and burglary during the night. The plant is said to materialize into a cobra, *Naja nigricollis*, or a frightening creature or a weird caricature to scare away any would-be thief. Alternatively, seven cut pieces of the stem may be hung around the house for the same purpose (15). Small pieces of twigs,(presumably of any plant) tied up to resemble miniature bundles of brushwood and suspended either over a door or window and invoked for the purpose, prevents a thief entering the house (7). In S Nigeria, a leaf of an unidentified plant locally called 'inunurin' in Yoruba, meaning 'sweet belly laughter' or 'happy laughter' is invoked to catch a thief (21).

To baffle thieves, feathers of a species of Shrike (a bird) found dozing along bush paths is wrapped in a leaf of Sodom apple and buried with rituals within the compound of one's house - the thieves cannot trace the house to execute their plan (14). (For a list of Shrike species in the Region, see WITCHCRAFT Table 1). Alternatively, if Sodom apple is planted with the necessary rites in the compound of a house simultaneously as a gang of seven armed men in uniform aim to shoot, the plant is endowed with power to prevent theft. An apparition of the armed men shooting re-appears to scare away any daring or would-be thief (1). To prevent theft in a household, the leaves of *Pergularia daemia*, a scandent asclepiad popularly believed to be fetish by many ethnic groups and sometimes tended around villages, are collected with rites and used to wrap up dried faecal matter. The parcel is covered up in a polythene bag and enclosed in an earthenware pot. Seven such pots are buried around the compound as a charm, amid incantations and a spell, invoking the contents of the pot to arrest any daring thief and to induce the culprit to defecate non-stop until daybreak (14) - surely an instance of sympathetic magic hinged on the buried faecal matter.

Another anti-burglary charm based on sympathetic magic records - 'a branch cut from a tree on which a thief has hanged himself, is used to carve any crude figure of a man. This is displayed in a corner of the house. The figure may thereafter be invoked in an incantation to prevent theft in the household, or failing that, command the figure in a sinister spell to compel any daring thief to hang himself on the nearest tree outside the premises (1). The charm is said to be very effective both as a watch-dog and as an executioner. Traditionally, a tree on which one hangs himself or herself is a bad omen and is cut down.

In N Nigeria, the leaves of *Lophira alata* Red ironwood, a big forest tree suitable for heavy-duty timber, are used among the semi-Bantu tribes as a charm to protect property from harm and theft; and pagan tribes lay the stem of *Stereospermum kunthianum*, a tree of savanna woodland, across the entrance to the hut as a charm to prevent thieving (6). In S Nigeria, *Triplochiton scleroxylon* African whitewood, a merchantable timber tree, is invoked by the Yoruba to stop thieves; *Tacca leontopetaloides* African arrowroot, a perennial herb in savanna, is credited with power to prevent theft, such as protecting property left by the wayside; and a length of the spiny vines of *Smilax anceps (kraussiana)* West African sarsparilla, a prickly forest climber, around a hut physically discourages robbers (5c). In its distribution area in the Region, the fronds of *Elaeis guineensis* Oil palm enter into prescriptions to protect goods against theft (5b).

In Botswana, *Psydrax (Canthium) ciliata*, a shrub, is used alone or with *Gymnosporia buxifolius* Staff tree as a charm to prevent anyone disturbing a body in a grave (23). This is likely a means of checking vandals who loot ornaments and jewels from corpses.

STOLEN GOODS RETURNED.

In addition to charms for the protection of property from theft, there are also charms in the Region and beyond for the return of stolen items. The full-length inflorescence of *Musa sapientum* Banana or that of *M. sapientum* var. *paradisiaca* Plantain collected with rites and buried amid incantations and offering of libation in front of the house in which the burglary took place, is a charm for the immediate and compulsory return of the stolen items or in default, face the consequences. If the items are not returned, and the culprit is a male, the penis would elongate like the inflorescence; and if a female, likewise the breast would similarly elongate (11). This is an instance of sympathetic magic. There is also the possibility that the stolen items are returned to avoid the vengeance of any juju to

Male inflorescence of Plantain

that effect. It is common knowledge that tribesmen openly threaten to curse or even kill thieves through juju unless stolen items are returned - a threat which often works. An example from Central African Republic records - 'most Azande can give instances of stolen property having been returned after magic was made to avenge the theft, and I have observed that this sometimes happens' (9).

Many other charm prescriptions for the return of stolen goods are recorded in the *Egyptian Secrets.* These include incantations under *Pyrus communis* Pear tree before sunrise with three horseshoe nails; or with nails of a hoof which have never been used; or alternatively, incantations under a Juniper tree at sunrise with the skull of a criminal - there is reference to sympathetic magic here. The trees are *Juniperus communis, J. drupaccea and J. excelsa.*

COUNTER CHARMS

An interesting use of plants as counter charms by thieves to further their nefarious activities is also practiced. *Biophytum petersianum* African sensitive plant, a weed of cultivated land, like *Schrankia leptocarpa* and *Mimosa pudica* both called Sensitive plant have probably entirely superstitious uses. For example, they are used by burglars as a charm to cause a deep sleep in their victims (6) - an instance of sympathetic magic. The plants are sensitive both to weather and to touch, which causes the leaves to close up or to 'sleep'. Other plants with sensitive leaves are *Neptunia oleracea, Mimosa pigra* and *M. rubicaulis.* A black powder prepared from the dried and charred whole plant of *Commelina lagoscensis,* a perennial weed of cultivated land, with a black cat, *Felis domestica,* and a mangrove crab, *Cardiosoma armata, Goniopsis cruentata* and species of *Sersama,* as ingredients, in a sinister incantation, is used by burglars to induce deep sleep in their victims (14).

In Lesotho, a Sotho thief uses the powdered or charred root of *Osteopermum acutifolium,* a composite, as a charm to protect himself and make him capable of acting unobserved. The preparation is applied to scarifications and some swallowed (23). In Ubangi, Central African Republic, a Mistletoe growing on *Clausena anisata* Mosquito plant is considered the talisman of robbers (5b).

A paraphernalia of second-hand charms and *gris-gris* worn by professional housebreakers, and attributed with similar properties is recorded - 'one, when placed near a house, caused all the inmates to fall into a deep and supernatural sleep; another contained powder that

destroyed the sight of all who looked upon it; and a third caused all doors to fly open, despite locks and bars, that came in contact with it' (7).

'Charms are also made to give the wearer the power of making himself invisible, and these are particularly useful to thieves - for the priests have no hesitation in taking fees from whatever quarters they are offered. A policeman friend of mine was covered with them as I discovered when I at last found him out and put him in prison, and his nickname in Jemaan Daroro was 'King of Door-blinds', because (I was told) he could pass his body into a house without disturbing even that flimsy protection' (20). *Datura metel* Metel or Hairy thorn-apple and *D. stramonium* Trumpet stramonium, semi-woody weeds often tended around villages, contain the hallucogenic alkaloid 'daturine' said to have been used by thieves to stupefy their victims (10). In N Nigeria, the Fula call *Leonotis* species , herbaceous labiates, 'hare gujjo', and the Hausa call *Leucas martinicensis*, another herbaceous labiate, 'kam barawo', both meaning thief's head, because the flower-heads are almost springy; and the term probably refers to the night thief's device of putting prickly and cutting things in his hair (5a). During the Middle Ages, supposedly, burglars put *Hyoscymus niger* Henbane, a shrub in the Garden egg family, on the coals that heated public baths in order to drowse the clients and then pick their pockets (17).

In Gabon, the leaves of *Greenwayodendron (Polyalthia) suaveolens* var. *gabonica*, a forest tree, boiled up, with the leaves of other plants, are used in ritual baths to cleanse maledictions brought by witch-doctors against crop thieves (22). (See also CATTLE, FARMING, and HUNTING).

References.
1. Akligo-Zomadi, (pers. comm.); 2. Ameyaw, (pers. comm.); 3. Ashieboye-Mensah, (pers. comm.); 4. Beloved, (pers. comm.); 5a. Burkill, 1985; b. idem, 1997; c. idem, 2000; 6. Dalziel, 1937; 7. Ellis, 1881; 8. Eshun, (pers. comm.); 9. Evans-Pritchard, 1937; 10. Everald & Morley, 1970; 11. Gamadi, (pers. comm.); 12. Irvine, 1961; 13. Kroger, (pers. comm.); 14. Mensah, (pers. comm.); 15. Moro Ibrahim, (pers. comm.); 16. Okigbo, 1980; 17. Pamplona-Roger, 2001; 18. Tete, (pers. comm.); 19. Thomas, 1916; 20. Tremearne, 1913; 21. Verger, 1967; 22. Walker & Sillans, 1961; 23. Watt & Breyer-Brandwijk, 1962

CARVINGS & IMAGES

What profiteth the graven image that the maker whereof hath graven it; the molten image, and a teacher of lies, that the maker of his work trusteth therein, to make dumb idols?
Habakkuk 2:18.

By tradition, wooden objects for rituals, ceremonies and fetish purposes have been carved from specific trees. These wooden objects include ceremonial staff, walking sticks, idols or figures, images, devil masks, ceremonial masks, stools, devil drums, fetish figures and fetish emblems (2, 11). Ordinary traditional carvings, on the other hand, consist of stools, drums, mortars, canoes and paddles; bowls, plates, utensils and cutlery; hair comb, gun-butts or gun stocks and tool handles (1).

Carver

The reasons why specific trees are used to carve wooden objects could either be symbolic, availability, suitability of texture, or a combination of these factors. Some of the preferred trees normally used in traditional carving such as *Holarrhena floribunda* False rubber tree, *Nesogordonia kabingaensis (papaverifera)* Redwood, a forest tree and *Diospyros mespiliformis* West African ebony tree are not known to be used for carving these wooden objects. This appears to suggest that the principal factor might be symbolism, and that the trees selected for carving these wooden objects are believed to be credited with spiritual attributes.

Images

THE EVIDENCE

There are many references that confirm the incidence of spirits dwelling within trees used for carving wooden objects; and the traditional rites required to be performed prior to felling these trees. Certain trees, the wood of which is used in carving, are believed to possess some sacred spirit; therefore they may not be felled without certain religious rites (2). In Sierra Leone, the Dan and the Mende masks used in ceremonies throughout the Region are often made with the wood of trees valued for their spiritual or mystical attributes (9). Many trees which play a role in traditional medicine, like *Milicia (Chlorophora) excelsa* and *M. regia* Iroko, locally called 'odum' in Akan, which is used for fabricating *akonnua* (stool) and *atwene* (drum), belong to the group of *sasammoa* (strong vindictive spirit); therefore their *sasa* (vindictive spirit) have to be pacified with libation and a food sacrifice (eggs and fowls, *Gallus domestica)* before human beings can make use of them (7).

Trees upon which the carver chiefly depends such as *Cordia platythyrsa* Drum tree, *Alstonia boonei* Pagoda tree, *Funtumia africana* and *F. elastica* False rubber tree and West African rubber tree respectively, are all trees with potentially vindictive spirits; thus the woodcarver strives to propitiate the spirit of the tree upon which he is about to ply his axe, by placing offerings before it (10). The tools for carving also have a rite performed over them, and also the blood of a fowl, *Gallus domestica*, with the customary prayer for assistance and freedom from accidents caused by the tools slipping and cutting the worker (10). (See also MARRIAGE: FAITHFULNESS IN MARRIAGE). A wood-carver might not go to work leaving unsettled any quarrel or grievance with his parents - the author concludes.

SYMBOLIC TREES

A list of the traditional trees used in carving wooden objects is given in Table 1 below:

Species	Common Name	Plant Family	Wooden Object/Objects
Albizia adianthifolia	West African albizia	Mimosaceae	Ogo - a club or image of the devil
Alstonia boonei	Pagoda tree	Apocynaceae	Devil masks or fetish emblems
Chrysophyllum delevoyii	African star apple	Apocynaceae	Fetish images
Cola verticillata	-	Sterculiaceae	Fetish masks and idol figures

Coula edulis	African walnut or	Olacaceae	Fetish figures
Gaboon nut			
Diospyros iturensis	-	Ebenaceae	Devil masks or fetishes
Discoglypremna caloneura	-	Euphorbiaceae	Ceremonial masks, devil masks
Milicia (Chlorophora) excelsa	Iroko	Moraceae	Ikenga - household god of Igbos
M. (Chlorophora) regia	Iroko	Moraceae	Ikenga - household god of Igbos
Psydrax parviflora	Common canthium	Rubiaceae	Ceremonial staff, walking sticks
(Canthium vulgare)			
Ricinodendron heudelotii	African wood-oil-nut tree	Euphorbiaceae	Fetish images
Trichilia monadelpha	-	Meliaceae	Devil masks
(heudelotii)			
Trichoscypha arborea	-	Anacardiaceae	Fetish masks
T. yapoensis	-	Anacardiaceae	Fetish masks
Vitex phaeotricha	-	Verbenaceae	Devil drums
Vitex doniana	Black plum	Verbenaceae	Pale-yellow or pale-green

Table 1. Symbolic Trees for Carving Wooden Objects.

Both *Psydrax parviflora (Canthium vulgare)* Common canthium, a tree beside rivers, and *P. subcordata (C. glabriflorum)*, a medium-sized myrmedophorous forest tree (meaning that the tree usually harbours black ants), have ritualistic uses particularly in providing the wood for which ceremonial staff and walking sticks are fashioned (8). In S Nigeria, the household god of the Igbo tribe, called Ikenga, is always carved from a solid block of *Milicia (Chlorophora) excelsa* or *M. regia* Iroko (3). In Liberia, the wood of *Trichoscypha arborea* and *T. yapoensis*, forest trees, are used to make carved fetish masks; and *Trichilia monadelpha (heudelotii)*, an under-storey rain forest tree, is preferred for devil masks (6). In Sierra Leone, *Discoglypremna caloneura*, a forest tree, is used to carve ceremonial or Devil masks (6); and in Congo (Brazzaville), the stems of *Diospyros iturensis*, a forest tree, are used to make carvings for masks and fetishes (4).

Carver

Devil masks serve as a screen to hide the identity of participants, and are an important regalia in invocation and initiation fetish ceremonies. The satanic look coupled with the gnashing teeth adds fear and horror to the ceremony. Devil masks may also be used as wall hangings - just for decoration and curiosity.

In S Nigeria, the Yoruba name of *Albizia adianthifolia* West African albizia, 'ayinre ogo' a tree legume of high altitudes, reflects the use of the wood to carve the ceremonial 'ogo' club or the mask of the 'Devil' carrying one (5); and in Liberia, *Vitex phaeotricha*, a forest tree, is reported to be used to make small 'Devil drums' (6). The wood of *Coula edulis* African walnut or Gaboon nut, a forest tree with edible walnut-like fruits, is used to make certain fetish figures (6). The native uses of *Alstonia boonei*, Pagoda tree, a forest tree with whorled leaves and branches, include carving 'Devil masks', fetish emblems and other objects (6, 11). In S Nigeria, the white hard wood of *Cola verticillata*, a forest tree, is used in carving fetish masks and idol figures. Other trees traditionally used in carving fetish images include *Ricinodendron heudelotii* African wood-oil-nut tree, a forest tree especially in secondary regrowth, and *Chrysophyllum delevoyii* African star apple, a tree with edible fruits, often cultivated or tended in and around farms and homes for the fruits.

The paste of *Baphia nitida* Camwood, a shrub or small tree legume, is used as paint for images and objects of superstitious practice (6). Camwood is symbolic of intelligence and wise counselling. (For a list of other symbolic plants, see APPENDIX - Table 1.) (See also DEVIL; FETISH and TABOO)

References.
1. Abbiw, 1990; 2. Antobam, 1963; 3. Basden, 1921; 4. Bouquet, 1972; 5. Burkill, 1995; 6. Dalziel. 1937; 7. Fink, 1989; 8. Gledhill 1972; 9. Jedrej, 1986; 10. Rattray, 1927; 11. Taylor, 1960.

5 CATTLE

And Abram was very rich in cattle,
in silver, and in gold.
Genesis 13:2

For increase in the number of herds, in milk production, and the protection of the herd, cattle-men may adopt traditional methods - including the use of plant prescriptions. There are also charms to drive away disease in cattle.

INCREASE IN HERDS

In N Nigeria, the Fula give both fruits and leaves of *Ficus sur (capensis)*, a common tree often with abundant cluster of figs on the trunk, (sometimes along with the tuber of *Trochomeria macrocarpa (macroura)*, a trailing cucurbit), to their cattle to bring about increase in the herds and to increase the milk in cows, *Bos* species (2). This fig tree is always superstitiously linked with abundance of yield or fertility because of the numerous fruits (see FERTILITY: SYMBOLS OF FERTILITY: Sympathetic Magic). The Fula also feed the milky latex of *Euphorbia balsamifera*, a tree, to their cattle, *Ovis* species, and other stock to promote fertility and to increase their milk (1b) - an instance of sympathetic magic.

Cattle grazing

MILK PRODUCTION

In N Nigeria, the Fula feed *Cissus populnea*, a shrub in savanna, to their cattle ostensibly to increase milk production, and sometimes with

Caesalpinia bonduc Bonduc or Gray nickernut 1/5 natural size

Pergularia tomentosa, a milky juice plant, perhaps on the Doctrine of Signatures (1d). An infusion of the leaves of *Guiera senegalensis* Moshi medicine, a small shrub in sandy wastes, is used in milking by women to ensure abundance of yield or richness in cream (2). In the Scottish Hebridean Islands, *Caesalpinia bonduc* Bonduc or Gray nickernut, a tropical drift dissemule (see RELIGION: SYMBOLIC PLANTS : Sea-Beans), known as the white Indian nut, is believed to have magical powers of cleansing up blood in cow's milk. When a nickernut was placed in a pail of milk, the milk was supposed to clear up almost immediately (Martini fide 4). In comparatively recent times, the English Worcestershire farmers would feed *Viscum album* Mistletoe, an epiphytic parasite, to the first cow calving in the New Year, to ensure good luck in the dairy (6). In Niger, *Tapinanthus globiferus* Mistletoe, an epiphytic parasite, growing on *Grewia bicolor*, a shrub in savanna thickets, is valued by herdsmen as fodder for cattle (1d).

PROTECTION FROM THEFT

As a preventive of cattle theft, plant charms may be used. Cattle owners use the fruits of *Opilia celtidifolia*, a climbing shrub in savanna woodland, to prevent cows, *Bos* species, from straying or being stolen - thus it may be burned along with the dung of the bush cow, *Syncerus caffer*, to cause smoke which is supposed to have this effect, or the fruit may be administered internally along with other medicines, which will cause the cattle to gore an intruder (2). As protection at night, Fula cattle-men may sprinkle wood-ash with rites and a spell in a circle round the herd of cattle - a practice which is reported to keep the cattle together, and within the circle overnight.

In Kenya, witch-doctors of the Turkana people apply *Caralluma russelliana (retrospiciens)*, an erect succulent asclepiad with white latex - often cultivated - in magic to protect cattle from theft by other tribesmen (Mwangangi fide 1a, 7). In S Africa, the Zulu use the powdered root of *Ipomoea crassipes* Wild patata as a charm against a person who wishes to

Scoparia dulcis Sweet broomweed

interfere with cattle. The powder is administered by the mouth to the animal and the hand of the person administering the powder is then passed along the backbone of the animal from head to tail (7). On the contrary, in S Nigeria *Scoparia dulcis* Sweet broomweed is used by the Igbo tribe under the epithet *aiya* to 'drive a cow', *Bos* species (1d).

PROTECTION FROM DISEASE

Plants may also be used with milk in a ritual against disease in cattle, or as a tranquilizer. To reduce high temperatures in calves among some cattle herdsmen in the Region, the dung is besmeared on the forelegs of the mother for the young to rest on (5). In Kenya, the Masai put *Commelina africana*, a prostrate herb, into a milk-container with water, milk and honey; and sprinkle the liquid around the cattle enclosure as a charm to drive away disease (Glover fide 3); or use the bulb of *Albuca abyssinica*, a lily, as a charm, waving it around the cattle-pond to prevent their stock getting malignant catarrh when the wildebeest, either *Connochaetes gnou* or *C. taurinus*, is calving (1c).

In S Africa, *Popowia caffra*, a climbing shrub, is placed in front of huts and at kraal gates in Tembuland for the cattle to jump over when they are diseased (7). In Ethiopia, culms of *Cyperus alternifolius*, a robust leafless cultivated sedge, are laid across the kraal to rid people and cattle of illness - its green and cool appearance is held to cool the malady (Strecker fide 1a). If a cow behaves strangely and climbs into the bush like a goat, *Capra* species, a switch of the sedge is used to splash a mixture of milk and water over it, and then some of the sedge is laid down by the side of the cow (Strecker fide 1a). (See also FERTILITY: CATTLE).

References
1a. Burkill, 1985; b. idem, 1994; c. idem, 1995; d. idem, 2000; 2. Dalziel, 1937; 3. Guillarmod, 1971; 4. Gunn & Dennis, 1979; 5. Mensah, (pers. comm.); 6. Perry & Greenwood, 1973; 7. Watt & Breyer-Brandwijk, 1962.

6 CEREMONIAL OCCASIONS, PLANTS FOR

...and thou shalt rejoice in thy feast, thou, and thy son, and thy daughter, and thy manservant, and thy maidservant, and the Levite, and the stranger, and the fatherless, and the widow, that are within thy gates.
Ecclesiastes 5:15

A ceremony is usually an occasion of joy and merry-making, but also a solemn one. It is performed by an experienced elder, but involves all those gathered. The event could be either on the family level, like child-naming; on tribal or ethnic level, like the traditional annual festivals; or on the national level, such as celebrations marking the independence from colonial rule. The occasions may be divided into ritual ceremonies, purification and sacrificial ceremonies, festivals, feasting and merry-making, and festive occasions. Each of these may also be sub-divided. Plants may feature in all these occasions.

RITUAL CEREMONIES

Among many traditions in Africa and elsewhere, some specified activities are usually preceded with customary rites to seek the approval and blessing of the spiritual guardians. The objective of these rites is to create the necessary atmosphere for the function. The rites may take the form of a prayer or an utterance, an offering or libation, self-denial, the correct timing of the performance, or an appropriate combination of these rituals. It is the belief that failure to perform this sacred duty is an offence against the spiritual guardians, and cause of any disappointments, mishaps, failure or even accidents and death.

In The Gambia, the Mandinka and the Susu people make a sort of bread or baked cake with rice flour and honey, used also in ceremonial ritual; and the sap of *Rothmannia megalostigma*, a shrub or tree in swamp forest, is rubbed into cuts as ritual (9). In N Nigeria, the bulb

Adowa dance

of *Crinum* species is probably the sacred bulb *gadali* used for ceremonial ritual by the various Benue tribes (9). In Gabon, the sap of *Costus lucanusianus*, a tall forest herb in the ginger family, and the nut of *Coula edulis* African walnut are used in benediction rituals (32, 8d); and in S Africa, *Senecio (Crassocephalum) latifolius*, a composite, is of great ritual significance to the Mpondo (33). In Burkina Faso, *Harungana madagascariensis* Dragon's blood tree, a shrub or small tree, and *Terminalia ivorensis*, a forest timber tree, enter into rituals for supplication (8a, e). In the Region, *Calotropis procera* Sodom apple has both ritualistic and medicinal uses (18); and *Excoecaria (Sapium) grahamii*, a semi-woody herb of savanna and fringing forest, is used for ritual scarifications (9).

In Ghana, *Cola acuminata* Commercial cola nut tree and *C. nitida* Bitter cola have social and ceremonial uses in the southern and northern parts, respectively (19b); and in Ivory Coast and Mali, the nuts of Bitter cola have a place in all social ceremonies, baptism, marriages, funeral and fetish sacrifices (8e). (See also LOVE: HOSPITALITY AND FRIENDLINESS). However, the colour of the dried nut (whether red, pink or white) determines their popularity and ritualistic use (18). In traditional practice, the colour of kola-nut is important, because use of the wrong colour type is believed rather to have the opposite effect. (See MARKETING: CHARMS; LOVE: CHARM PRESCRIPTIONS; and WAR: PEACE for other instances).

In Botswana, *Cymbopogon dieterlenii*, locally called 'lebatjana', is used by the Southern Sotho, with another grass *Elyonurus argenteus* Lemon-scented grass, locally called 'loko', as a medicine for wounds in treatment of 'modikana' (an eruption which affects people who have not undergone certain rites) (33). In S Nigeria, *Dialium guineense* Velvet tamarind has an important place in rituals among the Igbo society (Iwu fide 8c). In Guinea, the leaf of *Daniellia oliveri* African copaiba balsam tree is used in Maninka rituals; and in Senegal, the resin of *D. ogea* Gum copal tree is sometimes burnt in magic ceremonies (14). In this latter country, the leaves of *Hibiscus sterculiifolius*, a small forest tree, are used by the Tenda to make costumes for ritual dances (14). In W Cameroon, *Dichrocephala integrifolia,* a composite, has use in several different rituals and masquerades in the Mambila area; and in Brazil, *Aeollanthus suaveolens,* a short-lived perennial labiate, is an important voodoo plant used in ritual baths among the Afro-Brazilian population (8c)

Libation and Prayers

Normally traditional functions start with libation and a prayer. 'Libation or prayer, used to communicate with the ancestors has its origin in the belief that there are relations between *Onyame* (God) and the ancestors. The libation is a means by which wishes are directed to God through the ancestors. It is the typical function of prayer to establish contact between God, the ancestors, the *abosom* (local deities) and other spiritual beings and man, and to explain the reasons for this contact and to state requests '(16).

In many parts of the Region, palm-wine - the wine tapped from *Elaeis guineensis* Oil palm or from *Raphia* species Wine palm - either fresh or distilled into local gin *(akpeteshi or ogogoro),* is important in social functions for libation or prayers to the gods and ancestral spirits and for heralding and announcing ceremonial occasions. In Ghana, the people of Akropong-Akwapim use the wine from *Raphia* species for drinking, but never for libation (17). Among the Ga tribe, however, corn-wine is used only in religious ceremony - palm-wine and white man's wine is the common cheer (15). In addition to palm-wine, other parts of the Oil palm may have ceremonial significance.

In S Nigeria, for instance, during some ceremonies among the Igbo tribe, priests or those who are not expected to talk, hold the young leaves or branches of Oil palm between their lips (24). In the islands of South Pacific, *Piper methysticum* Kava kava, a herb, is used as a ritual beverage for ceremonial purposes, including welcoming of important guests (5). Pope John Paul II, Queen Elizabeth II, President Lyndon B. Johnson, Lady Bird Johnson and Hillary Rodham Clinton are all known to have drunk Kava kava upon being welcomed to Fiji (the Pope) and Samoa (the others) (5).

General Ceremonies

In Gabon, Fang 'orators' masticate some leaves of *Ocimum basilicum*

Sweet basil on ceremonial occasions before speaking at palavas in the belief that this gives them ideas and aplomb (32); and in Tanzania, *Hoslundia opposita*, an erect or scrambling shrub, has ceremonial uses (33). In SE Nigeria, those who have taken *tooru* tie *Eleusine indica*, a tufted annual grass, in their hair during the *uzii* ceremony among the Ijo so as not to 'disturb' others (8b). In SE Senegal, Tenda singers chew the raw root of *Acridocarpus spectabilis*, a savanna under-shrub, on ceremonial occasions in the belief that this strengthens the voice for singing; or for the same purpose the root is pounded, dried, mixed with salt and eaten. On the contrary, a mistletoe found growing on an unidentified species of *Combretum* is said to be detrimental to singers' voice (14). In Ivory Coast, Guéré and 'Kru' witch-doctors colour themselves with the wood powder of *Baphia nitida* Camwood for ceremonial occasions (20). This is also applied to the hair of young dancers of the Blon cult in the Tai region (29). In S Nigeria, the Igbo tribe prepare a wash from the leaves of *Cordia aurantiaca*, a shrub or small tree, which is used during dance ceremonies (31a).

In Sudan, throwing-sticks carved from *Balanites aegyptiaca* Desert date or Soap berry, a savanna tree with edible yellow fruits, that have been used in warfare, are now for hunting and ceremonial occasions (2). In Nigeria, the leaves, bark and flowers of *Maerua crassifolia*, a shrub, are used in ceremonies; and in Ethiopia, the fruit of *Solanum incanum* Egg plant or Garden egg has unspecified ceremonial use (8e). In Sierra Leone, the leaves of *Cissus diffusiflora*, a slender forest climber, are used in superstitious practice in connection with Poro; and *Dodonaea viscosa* Switch sorrel, a shrub or small tree of sandy coastal areas, enters ceremonial uses (Deighton fide 9). In this country, the leaves of *Berlinia confusa*, a forest tree legume, is said to be an ingredient of ceremonial sauces (28)

Initiation Ceremonies

In the Region, *Ceiba pentandra* Silk cotton tree is planted at the entrance to sacred groves during initiation ceremonies (29). In Sierra Leone, the tree is the centre of a ceremony to pray for long life, wealth, good harvest, prosperity and the well-being of the population (8a). Similarly, among the Ada people of Ghana, *Adansonia digitata* Baobab is believed to be a god to which prayers are said and sacrificial offerings presented on ceremonial occasions for general prosperity, protection and guidance, fertility of the women folk and good behaviour of the youth (1). In Ubangi, Central African Republic, bark-cloth beaten from *Sterculia setigera* Karaya gum tree is used in initiation ceremonies (8e).

In Kenya, Kikuyu elders who are candidates for the senior grades which enable them to try cases, are given a bunch of the leaves of *Clausena anisata* Mosquito plant at the time of their initiation (33). In Botswana, a decoction of *Myosotis afropalustris* Forget-me-not with *Cissus* species, a climber in the grapes family; *Galium wittebergensis*, a herb; and *Clematis* species, a climbing plant; are used in the initiation of the Southern Sotho 'witch-doctor' to develop the memory and make the initiate mentally fit for his work (33). In Gabon, *Lapportea aestuans*, and *L. ovarifolia*, irritating herbs of waste places, are a fetish of the masonic rites of the 'Ndjembe' and the 'Nyemba' societies. During initiation the young novitiates are rubbed with leaves and stems and are then put in a bath which causes intolerable pain (32).

In E Senegal, the stems of *Grewia lasiodiscus*, a savanna shrub, enter into Basari ritual ceremonies in the form of wands used by initiates; and the closely-related Bedik people use them to play jokes on initiates (14). In SE Senegal, *Combretum collinum* and *C. lecardii*, shrub or tree and scandent shrub respectively, play a role in the initiation ceremony of the Tenda (14). In this country, the tribe consider Mistletoe growing on the latter plant to have a powerful spirit so one does not utter its name (14). As part of the initiation ceremony of herbalists in the Region, vines of *Momordica balsamina* Balsam apple or *M. charantia* African cucumber, climbing cucurbits traditionally believed to be fetish, are placed round their necks (3). Among the Ga tribe of Ghana, the women and the children wear wreaths and girdles of fresh green leaves, chiefly of African cucumber vine which is the chief herb in all *Kple* ceremonies (15). (For the initiation of fetish priests, see FETISH: INITIATION)

Circumcision Ceremonies

An occasion like circumcision is associated with traditional rites throughout Africa. (This sub-heading deals with only male circumcision. Female circumcision or female genital mutilation (FGM) is illegal in a number of African countries). In Sierra Leone, the bark of *Parinari excelsa* Guinea plum, a forest tree with edible

fruits, is pounded to be used as a dressing after ritual circumcision (18); and in its distribution area, the spiny flower-heads of *Centaurea perrottetii* or *C. praecox*, erect composite herbs, are said to be carried by boys while enduring the circumcision wound, so as to drive off others who approach too near (9). In Sierra Leone, the Mende allow the resin from the cut stems of *Vismia guineensis*, a small tree in secondary forest, to drip onto the freshly-made wound in the circumcision ceremony. This is followed by the operator chewing the seeds of *Aframomum melegueta* Guinea grains or Melegueta and blowing the cud onto the wound (8b).

In S Africa, the pole (mokogoro) erected by the Tlhaping at the circumcision lodge is usually a trunk of *Acacia giraffae* Giraffe thorn; and the leaf fibre of *Sansevieria thyrsiflora* Pile root, which is extremely strong, is used among others, for making the ceremonial garb used during circumcision ceremonies in this country (33). In Botswana, *Phygelius capensis*, locally called 'mafifi-matso', is one of the ingredients in the medicine horn used at the circumcision ceremonies among the Southern Sotho (33). In Gabon, an ointment prepared from the wood of *Pterocapus soyauxii* African coral wood has ceremonial use for rubbing on newly-married persons, or newly-delivered women, or women on becoming widowed, or the newly circumcised, or the followers of the Bwiti Secret Society either on dance days or on days of initiation (32).

In S Africa, a small piece of meat, roasted in a smoking fire made of the wood of *Pteroxylon obliquum* Sneezewood; *Rhus mucronata* var. *villosa* and *Capparis citrifolia* Cape caper, wrapped in a covering of the thorny twigs of the *Capparis*, is thrown to each boy in turn during the Xhosa circumcision ceremony (33). It must be caught in mid-air to give the initiate courage and strength to resist evil influences (33). In Botswana, the Southern Sotho rub the dry crushed root of *Asparagus africanus* Wild asparagus into scarifications on boys undergoing the circumcision rites, and believe that this makes them brave and strong (33). In W Africa, *Ritchiea capparoides* var. *capparoides*, a scandent shrub or liane, enters circumcision rites in The Gambia; and in both Ivory Coast and Senegal, the leaf sap or white latex of *Strophanthus sarmentosus,* a stout climber, is used on sores and wounds to give rapid healing (8a). For this reason in some villages of Casamance in Senegal, the climber is considered a fetish

of the circumcision ceremony (8a). In N Nigeria, *Blepharis linariifolia*, a herbaceous acanth, is known as 'daudar-magusaawaa' in Hausa, meaning protection for boys who have been circumcised (8a).

Funeral Ceremonies.

Occasions like funerals or mourning the dead are equally associated with rituals. In Ashanti, the seeds of *Cardiospermum grandiflorum* and *C. halicacabum* Heart-seed or Balloon vine, herbaceous climbers, are used to make chaplets at the funeral of important chiefs and other ceremonial occasions (27a, b). Ritual uses of the climbers are also recorded in S Nigeria (31b). In Tanzania, *Commiphora charteri*, locally called 'ubani', on account of its pleasant aroma, is burned at funeral rites (33). In Mozambique, the Thonga celebrate the feast of the first fruits by pouring the fresh juice of the fruit of *Sclerocarya caffra* Cider tree on the tomb of deceased chiefs in the sacred wood. Branches of the tree are also used in the funeral rites of the tribe (33). In Havar and Yemen, *Catha edulis* Abyssinian tea or Chirinda redwood is an important item in social intercourse, especially at birth, circumcision, marriage and funeral services (33). (For rituals connected with the New Yam Festival, see FESTIVAL below).

Hallucinogen Influenced.

Cannabis sativa Indian hemp or Marijuana, an annual herb, and a native of Asia introduced to the Region, is the source of the drugs *ganja, charas* and *bhang*, and has extensive ritual uses in many countries (12). The active principles are *cannabidiol, tetrahydrocannabidol* and *cannabinol*; and the effect is to produce a sort of intoxication with hallucinations and a feeling of unreality and complete detachment, sometimes with delirium (8a). Similarly, *Lophophora williamsii* Peyote, a bulbous cactus from N America, contains the drug *mesculine* (trimethoxyphenephylamine) which is a well-known intoxicant first used for ritual purposes by the Aztecs (26). In Mexico, *Solandra* species, the god-plant of mythology, is recorded used by the Huichol Indians for ritual purposes and for its hallucinogenic effects (21). These Indians use *Nicotiana rustica* Wild tobacco in parts of W Africa and the Sierra Madre mountains of Mexico, ceremoniously (30) - most likely for its narcotic properties. (For the effects of other hallucinogenic plants, see GHOST AND SPIRIT WORLD: EXORCISM: Accident victims;

FORTUNE-TELLING AND DIVINATION: GENERAL CASES: Hallucinogenic Induced Divination; and RELIGION: RELIGIO-HALLUCINOGENIC PLANTS)

PURIFICATION AND SACRIFICE

Invariably, ceremonial occasions are characterized by purification or cleansing and sacrificial offerings to the gods and ancestral spirits.

Purification

In Congo (Brazzaville), *Senecio (Crassocephalum) biafrae*, a climbing herbaceous forest composite - sometimes eaten as spinach - is used to wash initiates at sect ceremonies (6). In Ghana, the climber is also reported by the Apagyahene of Abiriw to be used for purification in Abiriw-Akwapim. Similarly in Congo (Brazzaville), *Acanthus montanus* False thistle, an erect plant in high forest, mixed with other plants is used in ceremonies of purification (6). In Ghana, the use of the red powdered bark of *Baphia nitida* Camwood for ceremonial purposes probably has the significance of cleansing and purification, because it is sometimes used in washing where soap is unobtainable to cleanse and soften the skin (15). In this country, this red dye was smeared over the *odwira suman* (god of *odwira*) during the Odwira Ceremony in Ashanti. During this ceremony the Ashanti king replaced his gorgeous robes with bark cloth - the garb of the poorest slave in the realm - made of beaten bark of *Antiaris toxicaria* Bark cloth tree (27b). The reason is that it would not be fitting for an inferior to go into the presence of his superior - the spirits - in fine attire (27a). The fibre from the bark of *Sterculia setigera* Karaya gum tree, a savanna tree, is also used to make ritual costumes (18b). The Odwira Ceremony is, in effect, a cleansing or purification festival in which prayers are offered, asking for prosperity for the nation, freedom from sickness, plentiful crops, and many children.

Among the Akans of Ghana and Ivory Coast, a clay pot filled with water and assorted herbs, *nyankonsuo* (divine water), is placed in the fork of *Alstonia boonei* Pagoda tree, a forest tree with whorled branches, locally called 'nyame-dua'. 'The divine water had the function of symbolically purifying the state, the dwellers of a house, visitors, and women returning home after the end of menstruation' (16). Some of the tribal groups within its distribution area utilize

Eragrostis tremula, a tufted annual savanna grass, in a ritual for self-purification after sin - among others. The grass is immersed with a dried chameleon, *Chamaeleon vulgaris*, and a piece of a butcher's log, in water for bath on Mondays, Tuesdays and Wednesdays only. After bathing, the body is not wiped but allowed to dry. The ritual is also for healing powers, for marketing and for prospects in business (22).

In Ghana, the Ga folk use the leaves of *Adenia rumicifolia* var. *miegei (lobata)*, a tall woody climber with yellow fruits, in ceremonial bathing (19b). In N Nigeria, a small pear-shaped *Lagenaria siceraria* Calabash or Bottle gourd, called 'jallo' or 'dan jallo', is used to carry water for ceremonial ablutions or on a journey; and in S Nigeria, *Dissotis rotundifolia*, a scandent or prostrate herb, enters Yoruba ritual in the consecration of idols, purification ceremonies and cleansing of brass (9). In Botswana, the Sotho-Tswana group use *Aloe saponaria* Soap aloe, a lily, for ritual purification (33).

Sacrifice

In W Africa, *Combretum grandiflorum*, a scandent shrub or forest liane with brilliant red flowers, has some superstitious and ceremonial uses. In Sierra Leone, it is used in 'satka' by the Temne (9). The green, nearly ripe pods of *Acacia tortilis* ssp. *raddiana*, a savanna tree legume, are commonly sold in the markets bordering the Sahara at the beginning of the dry season as fodder especially to fatten sheep, *Ovis* species, for sacrificial purposes (9). In Ghana, *Ocimum canum* American basil enters into fetish practices and is generally mixed with the sacrificial fowl, *Gallus domestica*, before it is eaten (19a). In SE Senegal, Tenda men make annual ritual sacrifices at the base of *Lannea microcarpa*, a savanna tree with edible fruits (14); and in S Nigeria, *Physalis angulata*, an annual herb, is used by the Igbo tribe in the Onu ceremony for sacrificial purposes (8e).

In the northern Peruvian Andes the seeds of *Thevetia peruviana* Milk bush or Exile oil plant or Yellow oleander and *Myristica fragrans* Nutmeg are some of the plants put in holy water and talcum powder with the seeds of *Citrus aurantiifolia* Lime and *Zea mays* Maize or Corn and used in ritual offerings to propitiate the spirits (10). The smoked leaves of *Nicotiana tabacum* Tobacco or *N. paniculata* serve as offerings to the spirits (10). In S Africa, the Xhosa burn twigs of *Halleria lucida* White olive when offering a sacrifice to the ancestral

spirits; and in Tembuland when a sacrifice is offered by a Zulu, the wood of *Popowia caffra*, locally called 'isidwaba', is sometimes used instead of *Ptaeroxylon obliquum* Sneezewood or Olive twigs on which to place the meat of the slaughtered animal (33).

Human Sacrifice

In addition to animals, human sacrifice was practiced. One of these rituals is described among the Ashantis about a century ago. 'It is considered necessary that blood should be poured into the hole where the new yam is dug; and for this purpose, slaves are sacrificed on the spot so that their blood may be mixed with the herbs and animal matter which at this custom, they stew up and eat as a powerful fetish, when the chiefs and captains renew their oaths of fealty to the king' (11). Similarly, in S Nigeria the celebration of New Yam Festival in ancient times among the Igbo tribe was associated with human sacrifice to some gods of the land and oracles, but in more recent times these offerings are made with goats, *Capra* species; sheep, *Ovis* species; and fowls, *Gallus domestica* (24). In the state of Mali, the execution of a criminal (by beheading) in the yam-field on a king's order is said to be essential ingredient in the annual custom (24). (See FETISH: APPEASEMENT SACRIFICES: Human Sacrifice; and FISHING: RITUAL OFFERINGS for other instances of human sacrifices).

FESTIVALS

Traditional festivals symbolize the spiritual bond that binds the community together. These festivals portray the belief in life after death and the nearness of dead ancestors to their living descendants; who remember their past leaders and ask for their help and protection (25). The occasion is also used to purify the whole state so that the people can enter the New Year with confidence and hope (25). In S Nigeria, the Ijo hold a festival to the plant called 'odufi' or eating the cocoyam (34). Similarly, the New Yam Festival is celebrated throughout the whole of the Yam Zone of W Africa - which extends from central Ivory Coast to the Cameroon mountains, spanning both forest and the southern (more humid) parts of the savanna (4). The great amount of ritual, ceremony and superstition which surrounds almost every aspect of yam cultivation and utilization is also indicative of the antiquity of their use in W Africa (4).

Treculia africana var. *africana* African breadfruit (germinating seeds)

The Ashantis believe that the time for the Adae Festival is marked by the falling of the fruit of *Treculia africana* var. *africana* African breadfruit, which takes place in about twenty one days from its appearance - all the birds and beasts in the neighborhood crying out simultaneously (7). (The African breadfruit fruits from February-April). Like the Warsaws, also of Ghana, they believe that from the many seeds of the fruit spring various kinds of vegetables. In N Ghana, *Tacca leontopetaloides* African arrowroot, a perennial savanna herb with tuberous rhizome, is used on special occasions such as the Bugum (fire) and Damba Festivals around Nalerigu area. In Ivory Coast, the Baoulé people of the Tiébissou region use young *Zanthoxylum senegalense (xanthoxyloides)* Candle wood, particularly the thorns, frequently as a flame on festival occasions (20). In the Far East, the blossom of *Firmiana colorata*, a tree from India, Ceylon, Thailand and Sumatra, is used to decorate the horns of cattle, *Bos* species, during certain festivals (26).

Feasting and Merry-making

Festivals are characterized by feasting and merry-making. In Ghana, the *kpekpei* of the Ga people, a *Homowo* festival meal, is prepared from corn dough, kneaded with red palm oil and served with palm soup usually prepared with sea bream, *Pagrus pagrus*, locally called 'tsire', a large reddish fish. The corn is the old-year corn, and the idea is that none of this old corn must be left over into the New Year. The palm-nuts, in particular, always come from a special tree (15). In S Nigeria, a meal prepared with *Treculia africana* var. *africana* African breadfruit is served as a dominant meal at a number of festivals and ceremonies. For instance, during a festival staged by young girls to intimate their friends and relatives of their impending departure from their respective homes to their husbands' homes, breadfruit meal is the centre of attraction (see also CHILDBIRTH: POST-NATAL); and it is also served as the main meal during the burial ceremony of a woman, when several large bowls of breadfruit are presented to the female relatives of the deceased

as their entertainment (23). As observed above, African breadfruit is also of traditional significance among the Ashantis of Ghana.

In N Nigeria, the flowers and young leaves of *Balanites aegyptiaca* Desert date or Soap berry may be added to *daudawa*, a food prepared from *Parkia biglobosa (clappertoniana)* African locust bean for ceremonial consumption. At a child's naming ceremony in this country, *Ampelocissus africanus (grantii)*, a savanna climbing herb, is used along with *Waltheria indica*, an erect shrubby weed, (a common children's medicine) as a decoction to be drunk (9). In Ghana, *Ocimum canum* American basil enters into fetish practices and is generally mixed with the sacrificial fowl, *Gallus domestica*, before it is eaten (19a). The cooked, sliced and fermented seeds of *Pentaclethra macrophylla* Oil-bean tree mixed with stockfish and/or the fruits of *Solanum* species, features conspicuously in children's naming ceremonies (see also CHILDBIRTH: CASUALTIES), and girl's outing ceremony (to herald maturity) (23).

Waltheria indica

In parts of Senegal, oil extracted from the seeds of *Adansonia digitata* Baobab is used in certain festive dish; and a drink prepared from the root-fiber of *Alchornia floribunda*, a shrub or tree in forest undergrowth, macerated for some days in palm- or banana-wine is drunk to stimulate energy during festive dances (19b). In E Africa, *Commelina imberbis*, a prostrate or straggling herb, is held to have magical powers by the Masai of Kenya. The tribe drape it round a pot containing milk from which milk and water is drunk on ceremonial occasions (Glover & Samuel fide 8a). In N America, *Amaranthus hybridus* ssp. *incurvatus* Love-lies-bleeding or Inca wheat was the Achita, the Jataco or the Quihuicha of the Aztecs, who used its seeds for food and for ceremonial purposes (26). In the annual ceremony of the Swati in S Africa, the black bull, *Bos indicus* and *B. taurus*, which is slaughtered for the occasion, is first struck on the back by the Paramount Chief with a switch of *Carissa bispinosa* Yum yum, locally called 'umvusankunzi', which is said to have

aphrodisiacal properties - it is supposed to make the bull fierce so that it will not fall easy victim to the warriors who have to overcome it with their bare hands (32).

THE INFLUENCE OF CHRISTIANITY

With the introduction of Christianity to the Region, Christmas, which has evolved from a pagan festival, and the Passion have become important occasions for family reunion.

Christmas

The Celts and the Goths paid great attention to *Viscum album* Mistletoe, an epiphytic parasite, as its fruiting coincided with their winter solstice festivals when they used it as a decoration during rituals (13). With the spread of Christianity the Church found the old pagan rites difficult to suppress, and although the Mistletoe was widely banned because of its heathen association, its decorative use was transferred to the home at Christmas (13). In W Africa and other parts of the world, plants like *Ilex aquifolium* English holly and *Hedera helix* English ivy whose blossom coincide with the Christmas season are accordingly named after this season. Table 1 below lists some of these plants.

Species	Plant Family	Distribution	Features	Common Name/s
Aechmea racinae	Bromeliaceae	Espirito Santo	Orange-red inflorescence	Christmas jewels
Alchornea cordifolia	Euphorbiaceae	Trop Africa	Shrub or small tree	Christmas bush
Blanfordia punicea	Liliaceae	Tasmania	Yellow or scarlet flowers	Christmas-bells
Cattleya trianae	Orchidaceae	Colombia	Purple-crimson petals	Christmas orchid
Ceratopetalum gummiferum	Cunoniaceae	New Guinea to E Australia	Flower for dec at Xmas	Christmas bush
Combretum racemosum	Combretaceae	Trop Africa	Dark red inflorescence	Christmas rose
Erica melanthera	Ericaceae	S Africa	Pale rose flowers	Christmas heather
Euphorbia pulcherrima	Euphorbiaceae	Mexico	Bright red leaves	Poinsettia, Christmas star or Christmas flower

Helleborus niger	Ranunculaceae	C & S Europe	Saucer-shaped flowers with golden stamens	Christmas rose
Hildegardia barteri	Sterculiaceae	Trop Africa	Scarlet flowers	Christmas tree
Metrosideros tomentosa	Myrtaceae	New Zealand	Flowers with red stamens	Christmas tree
Morinda lucida	Rubiaceae	Trop Africa	White flowers	Christmas tree
Polystichum acrostichoides	Aspidiaceae	Nova Scotia to Texas	Green fronds for Xmas decoration	Christmas-fern
Prostanthera lasianthos	Labiatae	Australia, South Wales and Tasmania	New Blooms from Sept-Jan	Victorian Christmas bush
Ruellia paniculata	Acanthaceae	Jamaica	Blue purple flowers	Christmas pride of Jamaica
Sandersonia aurantiaca	Liliaceae	Natal, S Africa	Bright orange-yellow inflated flowers	Christmas-bells
Schlumbergera truncata (Zygocactus truncatus)	Cactaceae	Brazil	Blooms at Xmas	Christmas cactus
Senna (Cassia) alata	Caesalpiniaceae	Trop America	Yellow inflorescence	Christmas candle
Veitchia merrillii	Palmaceae	Philippines	Bright red fruits around Xmas	Christmas palm

Table 1. Plants Named After Christmas

The Passion

Senna alata Christmas candle.

In W Africa, the fronds of *Elaeis guineensis* Oil palm and other palm trees are used to symbolize Palm Sunday and Easter Sunday activities. In Europe, *Buxus sempervirens* Box was once used instead of *Salix babylonica* Weeping willow on Palm Sunday; and *Anemone pulsatilla (Pulsatilla vulgaris)* Pasque flower, is so called because it blooms around Easter time. For the same reasons *Pulsatilla hirsutissima* is called American pasque flower; and *Lilium longiflorus* var. *eximum* is known as Easter lily - an indication of the respective traditional links with these Christian festivals.

References.
1. Adibuer, (pers comm); 2. Arkell, 1939; 3. Ayensu, 1981; 4. Ayensu & Coursey, 1972; 5. Blumenthal & al. 2000; 6. Bouquet, 1969; 7. Bowdich, 1819; 8a. Burkill, 1985; b. idem, 1994; c. idem, 1995; d. idem, 1997; e. idem, 2000; 9. Dalziel, 1937; 10. De Feo, 1992; 11. Ellis, 1881; 12. Emboden, 1972; 13. Everard & Morley, 1970; 14. Ferry & all, 1974; 15. Field, 1937; 16. Fink, 1989; 17. Gilbert, 1989; 18. Gledhill, 1972; 19a. Irvine, 1930; b. idem, 1961; 20. Kerharo & Bouquet, 1950; 21. Knab, 1977; 22. Mensah (per comm); 23. Okafor, 1979; 24. Okigbo, 1980; 25. Opoku, 1970; 26. Perry & Greenwood, 1973; 27a. Rattray, 1923; b. idem, 1927; 28. Savill & Fox, 1967; 29. Schnell, 1950; 30. Siegle, Collins & Diaz, 1977; 31a. Thomas, 1910; b. idem, 1913-14; 32. Walker & Sillans, 1961; 33. Watt & Breyer-Brandwijk, 1962; 34. Williamson, 1970.

CHARM, GENERAL

...he took bread, and blessed it, and broke it, and gave it to them. And their eyes were opened, and they knew him; and he vanished out of their sight.
Luke 30b:31

In traditional Africa and elsewhere, persons who have course to, do use charms for various purposes - general or specific. These charms are made either with plants or some other materials, alone or with ingredients. A charm is a magic formula or spell. The object may be worn as a ring, amulet, armlet, bracelet, girdle, talisman; or applied as powder to incisions cut on the body, or it may be taken - either in food or drink. Depending on the purpose for which it is intended, a charm may be acquired secretly or bought openly on the market from medicine-men, fetish- or juju-men, witch-doctors and malams, or Moorish hawkers (popularly called *kramo* or *eduro-eduro*) or from snake charmers. Some charms may be inherited as a legacy.

PLANT PRESCRIPTIONS.

In the Region, the roots of *Securidaca longepedunculata* Rhode's violet, a savanna shrub, collected with rites, are sold as medicine and a charm; and in Sudan, the seeds of *Mucuna sloanei* Horse-eye bean or True sea-bean are sold for use as a charm (2). (See also DISEASE: ELEPHANTIASIS; and MEDICINE: CHARMS AND TALISMAN). The stem bark of *Securidaca* is similarly sold as a charm. In N Nigeria, *Annona senegalensis* var. *senegalensis* Wild custard apple is held in superstitious esteem by the Moslems and used as a charm; and pieces of the bark or the red heartwood or the flower of *Erythrina senegalensis* Coral flower or Bead tree are used as a charm (2). In S Nigeria, Yoruba herbalists use *Coprinus ephemerus*, an agaric mushroom, in charms; and *Lentinus tuber-regium,* a large rigid toadstool with a cup-like cap, has magical potential to counter harmful charms invoked by enemies (6).

In S Africa, *Portulaca oleracea* Purslane or Pigweed; or *Albuca cooperi,*

Erythrina senegalensis Coral flower or Bead tree (symbol of reconciliation)

a lily; or *A. major*; or the reed of *Phragmites communis* Common reed are used as a charm by the Southern Sotho (8). The leaf fibre of *Scilla saturata*, a lily, is used by the tribe to fasten horns and charms round the neck of a child (8). In this country, *Kalanchoe thyrsiflora* White lady is used as a charm to smooth away difficulties; and *Euryops annae*, a composite, is largely used by the tribe in the preparation of charms (8). The Xhosa of this country use the wood of *Hippobromus pauciflorus* Horse-wood as a charm (8).

In N Nigeria, *Ludwigia octovalvis* ssp. *brevisepala* Prime-rose-willow, a hairy herb with yellow flowers usually growing in damp places, locally called 'sha shatan', is also the name of a malam's charm in Hausa; and in S Nigeria, the Igbo-speaking people use *Chassalia kolly*, a soft-stemmed forest shrub, to make a child's charm (7). In Angola, the seed of *Entada gigas* or *E. rheedei (pursaetha)* Sea bean or Sea heart is used as a charm (3). The Angola species *Lycopodiella (Lycopodium) cernum*, a fern, is used in Surinam in a herb bath and as a charm; and in Surinam, *Lantana camara* Wild sage or Lantana, a notorious weed of cultivated land, is similarly used (8). In S Africa, the Kuba use pieces of *Asclepias* species as an amulet; and the Karanga use any species of *Aloe*, except *A. dichotoma* for a variety of medicinal and charm uses (8). In this country, *Verbena officinalis* Vervain has been used in part as a charm (8). In Kenya, a Masai charm is made from some part of *Cordia ovalis*, locally called 'olokora'; and in Botswana, *Heteromorpha arborescens* Kraaibos is planted with charm intent in every 'lekhotla' (8).

In W Africa, the seed of *Melia azedarach* Persian lilac are used for rosaries, and they are supposed to act as a charm worn by women round the waist (2). In N Nigeria, the Hausa appear to use *Holarrhena floribunda* False rubber tree or similar plants popularly as a charm; and women thread the ripe berries of *Physalis angulata* Wildcape gooseberry,

an erect annual herb, as a chaplet for ornament or as a charm (2) (see also FERTILITY: FERTILITY IN HUMANS: Sympathetic Magic). In this country, the black pigment from the seeds of *Vitex doniana* Black plum is used by the Moslems to prepare an ink for the *marabouts* - a charm (5). The Nupe name of *Lagenaria siceraria* Calabash or Bottle gourd, 'bingi', means a small bottle-gourd exclusively associated with charm (1a). In S Nigeria, *Phallus aurantiacus*, a gasteromycete stinkhorn toadstool, is used to prepare bad charms, to make people mad, to make harmful charms on other persons, and hence protection for oneself (6). In Nigeria, the fruit of *Tetrapleura tetraptera*, a tree legume, is worn as a charm round the neck (1b).

In Botswana, the Sotho use *Cynodon dactylon* Bermuda grass, alone or with other plants; or *Cephalaria zeyheriana*, locally called 'thswene', as a charm (8). In S Africa, the Xhosa and the Mpondo make necklaces of the wood of *Spirostachys africanus* African sandalwood as a charm; or the Xhosa witch-doctor uses either the fruit juice or a decoction of the leaf or the root of *Cucumis africanus*, a cucurbit, as an emetic charm (8). In this country, the Xhosa and the Zulu put *Ptaeroxylon obliquum*, a tree, to charm; and *Bulbine asphodeloides* Balsam, a lily, is used by the Nguni, the Xhosa and the Sotho-Tswana group as a charm (8). In W Africa, the seeds of *Afzelia africana*, a tree legume in forest or fringing forests in savanna regions, are used as a charm to dispel evil; and *Platostoma africana*, a slender labiate herb in damp places, locally called 'asirisiri' in Akan, is popularly used as a charm by fetish-men (4). The leaves are mixed with shea-butter and rubbed on the face, which enables the fetish-man to charm people and he does not fear anyone (4). The Akan name means to laugh, because he can then laugh with impunity (4). Similarly in Zimbabwe, the Manyika herbalist uses an infusion of the root of *Berkheya radula*, a composite, with which to wash once a month with some charm intent (8).

A charm may not necessarily be for personal protection - it could also cover a whole household. For instance, in E Sudan and E Africa, the fruit of *Kigelia africana* Sausage tree is hung as a charm outside huts (2). In Congo (Brazzaville), *Sansevieria cylindrica* Africa bowstring hemp is cultivated by the African as a charm in Ubangi-Shari, evidently by planting it in village surroundings (8). In addition to the general charms listed above, there are charms for specific purposes. (For specific charms, see BURGLARY (OR THEFT); COURT CASE; FARMING; FERTILITY; FETISH; HUNTING; JUJU OR MAGIC; KINGSHIP; LIGHTNING

AND THUNDER; LOVE; LUCK; MARKETING; POPULARITY AND SUCCESS; PROTECTION; SNAKES AND SNAKE-BITE; and TALISMAN).

References.
Ia. Burkill, 1985; b. idem, 1995; 2. Dalziel, 1937; 3. Holland, 1922; 4. Irvine, 1930; 5. Kerharo & Bouquet, 1950; 6. Oso, 1975; 7. Thomas, 1913-14; 8. Watt & Breyer-Brandwijk, 1962.

8 CHILDBIRTH

Lo, children are a heritage of the Lord;
and the fruit of the womb is his reward.
Psalm 127:3

Precautions are normally taken for the safe delivery of expectant mothers in traditional medicinal practice. In addition, when complications are suspected or are imminent - particularly those attributable to evil forces - plant prescriptions are administered to counteract and neutralize these forces. The prescriptions may be divided into three main groups: those used at the later stages of pregnancy, those used during parturition, and those used after the birth of the child. There are also plants that enter into ceremonies or rituals associated with unsuccessful delivery or the death of the mother during labour.

PRE-NATAL

Among the Ga tribe of Ghana, prenatal cleansing of the mother, known as *Ayowie ceremony*, consists of bathing the woman in the sea (8). Steps are also taken to find out the wishes of the coming child. If it asks for money, a threepence coin is waved three times round the mother's head and then put in a basin with the leaves of *Kalanchoe integra* var. *crenata* Never die, a herb, and either *Momordica balsamina* Balsam apple or *M. charantia* African cucumber and washing sponge; the whole is then shot on the rubbish heap and the coming child's unseen friends pick the money (8).

Rauvolfia vomitoria Swizzle-stick

In many parts of the Region, a palm-wine decoction with the roots of *Rauvolfia vomitoria* Swizzle-stick, a shrub with milky latex, dug up with the necessary ceremony, is drunk at intervals during early pregnancy when there are fears that the foetus is failing to develop, to induce its rapid growth (15). In S Nigeria, the Yoruba invoke *Setaria verticillata,* a loosely tufted annual grass, for its adhesive property, and *Trilepisium madagascariense,* a forest tree, in an incantation to keep a miscarried baby on earth (25). In N Nigeria, Hausa boys use *Excoecaria (Sapium) grahamii,* a savanna plant with milky latex, to raise blisters on the skin and say 'eliminate prenatal evil influences' (5). In Congo (Brazzaville) a pregnant woman threatened with a miscarriage will tie a length of *Synclisia scabrida,* a slender forest menisperm, as a girdle around her waist (1).

In W Africa, special amulets are worn by Ashanti women during this period to protect them and the unborn child against witchcraft, to the influence of which both seem, at this stage, particularly susceptible (19b). There is also a purification ritual performed by the Ashanti tribe with *Portulaca oleracea* Pigweed or Purslane and common salt 'to allow an infant to come out peacefully', and this consists of gift exchange by husband and wife and a ceremonial meal (19a). In S Nigeria, a Yoruba woman washes once every two weeks with soap prepared with *Hybanthus enneaspermus,* an annual herb, about the seventh month of pregnancy; a month later she uses it weekly; and when she reaches the ninth month she uses it daily for four consecutive days - the practice is believed to afford easy and painless delivery (23). The plant also features in a Yoruba incantation for the good delivery of a pregnant woman under the name *'abiwéré',* meaning 'give birth gently'; or alternatively under an even more cryptic name *'ewé loko lepon'* meaning 'owner of penis and testicles' (25).

In Ivory Coast the bark, or more especially the seed, of *Pleioceras barteri,* an erect or climbing shrub with white latex, is an emmenagogue (13). In S Nigeria, it is taken by Yoruba women once or twice a month during pregnancy albeit with care to avoid abortion, and because the medication is held to induce movement of the foetus (5). For this reason it is given shortly before full-term to ensure a head-presentation, and explains too the significance, apparently medico-magical, for the Yoruba name 'pari omoda' or 'pa ori omo da', meaning cause child's head to turn (5). The prescription requires half a fruit crushed and cooked with a fish called 'aro' or 'ejaro', a silurid which is known in connection with fetish worship, and has the habit of turning over on the surface of the water. Sometimes the brown rat, *Cricetomys gambianus,* is substituted for the fish (5).

In S Africa, the Mfengu tribe uses a cord of the runners of an *Ipomoea* species worn round the lower abdomen as a charm against abortion, and to relieve uterine pain, and to calm foetal movement (27). Should a barren Southern Sotho woman become pregnant after drinking a root decoction of *Aloe kraussii*, a lily, she bathes herself with the decoction throughout the gestation period, at the confinement, and during the puerperium (27). When a Southern Sotho mother becomes pregnant soon after the birth of a child, a decoction of the root of *Nidorella anomala*, a composite, is administered to both child and mother as a charm remedy (27). Generally African women who are pregnant make periodic prophylactic use of the strong purging action of the root of *Phytolacca heptandra* Umbra tree in order to prevent their children being born with birth marks (27).

In S Nigeria, the fruit of *Colocynthis lanatus (vulgaris)* Water melon is invoked by the Yoruba in an incantation for the safe delivery of a pregnant woman, doubtless on an anthropomorphic basis, because the plant normally fruits abundantly and is the 'owner of many children' (25). In this country, *Amorphophallus dracontioides*, a herbaceous aroid in savanna with tubers 'the size of a child's head' and flattened above, locally called 'ere otoro', is given to a woman during pregnancy to create a feeling of well-being till the time of her delivery (3a). The corm is peeled and cooked with corn or plantain and fish with pepper and salt to taste, the woman will urinate off the 'otoro' and feel at ease (3a) - the sympathetic magic probably referred to the tubers.

In S Africa, the Zulu administer the powdered root-stock of *Gloriosa simplex* Climbing lily to determine the birth of a child of the desired sex; and the Venda administer the powdered bark of *Sclerocarya caffra* Cider tree to an expectant mother to regulate the sex of the child (bark from a male tree for a boy, and bark from a female tree for a girl) (27). In this country, the bark of *Osyris compressa* African sandalwood is one of the ingredients in the Sotho *horn of fertility* which is used to assist in changing the sex of the next child, when only male or female children are being born to a couple (27). In

Gloriosa simplex Climbing lily

W Africa, *Hybanthus enneaspermus*, an erect savanna bush, is put into a formula by some Ivorean midwives to confer upon their customers the ability to have a boy or a girl baby at choice. The prescription is sold in medicine markets in S Nigeria and it is added to food for pregnant and parturient women in order to strengthen the child and to prevent undesirable after-effects (2, 13). (For birth of the desired sex among the Gas of Ghana, the Guéré, the Kru and the Shien of Ivory Coast and other tribal groups elsewhere, see FERTILITY: FERTILITY IN HUMANS: Predicting the Sex of Child).

PARTURITION

The prescription of plant medicine at the commencement of parturition is aimed at hastening it, keeping away the evil one while summoning the beneficial forces for the safe delivery of the mother and protection of both mother and child thereafter. Writing about the ethno-phytotherapy of Yoruba medicinal herbs and preparations, Tremearne (23) observed that, 'difficult births' are managed with the aid of incantations, coupled with the use of the roots of male *Carica papaya* Pawpaw - chewed with seven seeds of *Aframomum melegueta* Guinea grains or Melegueta during labour. In addition, a powdered bark of *Blighia sapida* Akee apple previously mixed with locally made black soap, is used in bathing throughout the period of pregnancy, to ensure easy labour for pregnant women (23). If, however, labour is still difficult despite all precautions, and it becomes evident that evil machinations are the root-cause, the problem is combated by the use of kernels which have been previously swallowed by a vulture, *Necrosyrtes monachus*, and recovered from its faeces - the application being accompanied by psychic recitations (22).

In Ghana, *Ocimum gratissimum* Fever plant and *O. canum* American basil, erect strongly aromatic labiates often planted around villages, are used in fumigation at birth in Ashanti to drive away evil spirits; and in Sierra Leone, both *Platycerium elephantotis* and *P. stemaria* Stag-horn fern, called 'sasabonsam kyew' by the Akan tribe of Ghana and Ivory Coast, meaning Devil's hat is reported to be used in child-birth (12a) - probably to give supernatural protection and safe delivery. In Kenya, *Oxystelma bornouense*, an attractive climber with white latex, has magical use in case of delayed childbirth when the patient ties a stem around the middle (3a). In S Nigeria, an unidentified plant locally called 'sadii' in Yoruba, meaning 'run to bottom', is used by the tribe for easy

Abutilon mauritianum Bush mallow

delivery of babies (25). In Gabon, the mucilaginous quality of the leaves of *Abutilon mauritianum* Bush mallow is used in poultices to help in the extraction of foreign bodies, and in sympathy with this effective property the women of some races carry a cord made of the fibre of this plant around the waist during their first pregnancy in the belief it will ease delivery (25).

Traditionally tropical drift dissemules (see RELIGION: SYMBOLIC PLANTS : Sea-Beans) are believed to possess magical powers. For example, among the Norse, a woman in childbirth could seek relief from pain by drinking a strong brew or ale from a cup made from the seed coat of *Entada gigas* or *E. rheedei (pursaetha)* Sea bean or Sea heart (10). In the Hebrides, a woman in labour was assured an easy delivery, if at the proper time she clenched a seed of *Merremia discoidesperma* Mary's-bean, another drift seed, in her hand. Seeds were handed down from mother to daughter as treasured keepsakes (10). In the temperate regions, *Aristolochia clematis* Birthwort, an introduced climbing decorative from southern Europe and Caucasus, was once widely cultivated as an aid to childbirth (7), because the flowers are similar to the female genital organs (both external and internal) (17). The use of Birthworth is likely an instance of sympathetic magic. The corolla resembles the womb of the female body, and consequently used in female diseases, especially if connected with childbirth (4). Corms of *Cyclamen repandum* Common sowbread, a Mediterranean plant, was reputed to have most excellent properties in assisting childbirth (7); and it has been observed (17) that *Artemisia vulgaris* Mugwort, a composite, is named after the Goddess Artemis, also called Diana, since like the goddess, the plant helps women in labour, without ever failing.

In W Africa, the red stipular sheath of *Musanga cecropioides* Umbrella tree, a quick growing soft-wooded tree in forest clearings, attracts attention, in part at least, on the Doctrine of Signatures, for treatment of gynaecological conditions. To hasten childbirth, the whole sheath boiled in soup is used by the Ashantis of Ghana as a powerful emmenagogue (5, 12b). The *asuman* (fetish) that carry names and come close in

importance to *abosom* (deities) among the Dormaa people of Ghana include 'Anyinsem'. This is worn by pregnant women and helps them to deliver securely and without pain (9).

In Ghana, the Ashantis either squeeze the leaf-juice of *Acacia kamerunensis,* a common prickly climbing shrub, over the woman's breast and abdomen in the event of a difficult labour; or a mixture of the leaf-juice of *Platostoma africanum,* a labiate herb, and *Ocimum gratissimum* Fever plant (18b). A decoction of the leaves of *Hyptis pectinata,* a labiate, is also administered as a drink (18b). In this country, if a woman's labour shows signs of being difficult among the Ga tribe, one of the first remedies tried by the midwife is fetching a broom and giving her a good sweeping all over to clear away evil influences (8). (For the traditional significance of the broom, see TABOO: GENERAL FOR ALL AND SUNDRY: Sweeping Brooms). Alternatively, the leaves of either *Momordica balsamina* Balsam apple or *M. charantia* African cucumber (previously used in bathing rituals during pregnancy and kept for the purpose) are soaked in water and given her to drink (8). The leaves of *Holarrhena floribunda* False rubber tree are similarly used. In Nigeria, the leaves of *Sarcocephalus (Nauclea) latifolius* African peach are placed over a stool on which a newly-delivered mother in Ingalaland sits after birth (3c).

It is traditionally believed that difficult labour, among other causes, could also be attributed either to unfaithfulness during pregnancy or even to conception with a paramour. In an unmarried woman, it could also be deliberate concealment, and therefore protection from responsibility, of the biological father of the unborn child - where more than one man may be involved. In all instances, however, the only prescription to safe delivery is confession - otherwise death could result (21).

In W Africa, the crude fibre of *Sanseviera liberica, S. senegambica* or *S. trifasciata* African bowstring hemp is used for tying the umbilical cord at birth (5) - this is probably to keep the evil one away. In Ghana, the Akyem people however take the contents of the fruit of *Coccinea barteri,* a trailing cucurbit, mixed with fresh lime juice and apply it with a feather to the umbilical cord of a new-born baby, then tied near the body with a strand of *Ananas comosus* Pineapple leaf-fibre or *Musa sapientum* var. *paradisiaca* Plantain fibre (3a). The application is repeated till the cord falls off. Both the placenta and the umbilical cord could be used in juju prescriptions against mother and child. For this reason, it is buried in the family bath-house by the grandmother or an elderly woman. In S

Nigeria, it is the customary practice among the Igbo tribe to bury the naval cord that falls off from a new-born child underneath the tap root of *Dacryodes edulis* Butter tree or Native pear (11). In Ghana, the Ga tribe bury the placenta under *Cocos nucifera* Coconut palm or other tree, which will then be expected to bear a great abundance of fruit (8).

Among many of the traditional groups in the Region, the village or town where the umbilical cord is buried is also where the body would eventually be buried. The practice appears to date back to biblical times. For instance, in Genesis 47:30 Jacob requested from his son Joseph:

> *But I will lie with my fathers, and thou shalt carry me out of Egypt, and bury me in their burying place. And he said, I will do as thou has said.*

In S Nigeria, it is the belief among the Igbo tribe that *Pentaclethra macrophylla* Oil-bean tree, a forest tree legume sometimes cultivated, and *Milicia (Chlorophora) excelsa* or *M. regia* Iroko, timber trees, are all sacred trees credited with furnishing souls for the new-born (20). (For a list of other sacred trees, see SACRED PLANTS; and APPENDIX: Tables I-V).

POST-NATAL

Post-natal rites, involving naming ceremonies, could also be for spiritual protection. In Ghana, immediately a child is born anklets and armlets of *Musa sapientum* var. *paradisiaca* Plantain fibre are bound round its limbs by the Ashantis (18 a, b) - for protection from evil influences. In Tanzania, a necklace of the rhizomes of *Cyperus articulatus*, a sedge cultivated for the aromatic rhizomes, is worn by the Sukuma mother after childbirth as protection for herself and the baby against evil influences (3a). If the child gives cause for concern, the mother chews the root and the child is quietened (3a). In Congo (Brazzaville), the sap of *Tetracera alnifolia* and *T. potatoria* Sierra Leone water tree, forest lianes whose cut stems yield clear water, is used to 'purify' mother and child immediately after birth; it must be given as it is a sort of colostrum to a baby for its first suckle (1). In Gabon, the leaves of *Amaranthus spinosus* Prickly amaranth are prepared into a ceremonial bath for cleansing new mothers one month after delivery (26); and in Ivory Coast, Kyama women bathe a baby in a root decoction of *Panda oleosa*, an under-storey forest tree, to protect it from evil spirits (2).

Among many of the coastal tribes in Ghana, the sweeping broom placed by the side of a sleeping baby during the absence of the mother is believed to protect it from evil spirits. The broom is normally made from the leaflets of *Elaeis guineensis* Oil palm. (For the traditional values, see TABOO: GENERAL FOR ALL AND SUNDRY: Sweeping Brooms). In Ghana, the root of *Mimosa pudica* or *Schrankia leptocarpa* both called Sensitive plant together with a corn sheath, tied up into seven knots (representing all the days of the week), is attached to the shell of a tortoise, *Kinixys* species, and used to cut seven incisions on each side of the ribs, and three each at all the joints of an infant, with incantations shortly after birth. The ritual is an immunity against all evil eyes (15). This is likely an instance of sympathetic magic hinged on the tortoise shell. (For the use of Sensitive plant and tortoise blood with rites for protection, see EVIL: SPIRITUAL PROTECTION: Sympathetic Magic). In Sierra Leone, the bark of *Erythrophleum ivorense* Sasswood tree and that of E. *suaveolens* Ordeal tree is placed outside a house where there is a new-born baby to keep away evil spirits (3b).

In Central African Republic, the Oubangui plant a tree for a new-born child (26). For female children a fast-growing profuse fruiter is planted and the child's development is linked to the growth of the tree. Thus if the tree's growth declines, people fear for the health of the child, and a healer is called upon. When the child is sick, it is brought to the tree for treatment; and when the tree begins to fruit, the time will have come for the child to marry (26). In S Africa, the Xhosa plant a young plant of *Euphorbia tetragona* Map tree in the precincts of the kraal when a child is born, with the superstitious belief that the health and death of the tree and that of the child coincide (27). In W Africa, several scores of *Borassus aethiopum* Fan palm are planted at a child's birth in parts of Senegal and they serve to some extent as a dowry (5). In Japan, *Paulownia tomentosa (imperialis)* a quick growing tree, was traditionally planted at the birth of a baby girl, so that it might be cut down to make a chest of drawers for her dowry (19).

In S Nigeria, the wood of *Treculia africana* var. *africana* African breadfruit is burnt by the Igbo tribe when a male child is born (16) - the significance of the ritual is not clear. In Ghana, *Solenostemon monostachyus*, a herbaceous labiate with a long inflorescence, is used at birth ceremonies by the Ashantis (18 a, b). In Sudan, *Evolvulus alsinoides*, a spreading hairy savanna herb, is an ingredient along with other herbs as a charm to exorcize evil; as such women will burn the plant to fumigate

the hut during the puerperium, and use the warm infusion as a wash during the forty days purification (5). In Colombia and Ecuador, the Tunebo Indians wrap their new-born babies in the leaves of *Canna indica* Indian shot (17).

Child-naming Ceremony

The child-naming ceremony takes place on the eighth day after birth among many tribes in the Region. The grace period is to determine whether the child will live or die. (See CASUALTIES below). During this grace period only the close family is allowed to see the child. Family friends and visitors are normally prevented from seeing the mother and child. The precaution is to avoid any possible acts of evil or witchcraft - to which the child is prone at that tender age. In N Nigeria, a decoction of *Ampelocissus africanus (grantii)*, a herbaceous savanna perennial, is drunk at a child's naming ceremony along with *Waltheria indica*, a savanna shrub which is commonly incorporated in children's medicine (5). In Ivory Coast, *Milicia (Chlorophora) excelsa* and *M. regia* Iroko are known as 'die die' by the Bété and the Shien people, and any child born to a mother who has had medicine prepared from the tree will be given that name (2). A similar practice is recorded in Guinea should sacrifices to the tree to have children result in success: the Kono: 'gué', and the Mano: 'ghei' (3c).

Child-naming

CASUALTIES.

Should the mother die during childbirth, which is considered a great disgrace among the Ashantis of Ghana, all the expectant mothers in the village point the shoots of a budding plantain leaf at the corpse shouting and wailing in sympathy, while they plead on their own behalf, for a successful labour (18b). She must on no account be buried with the child in her womb; consequently, just before interment the body is cut open and the child removed. It may then be buried with her. The non-observance of this custom would be what is known as a 'red taboo', that is one the violation of which would be deemed to affect adversely the whole nation (18b). Similarly, among the Ga tribe of Ghana, such a death is unspeakably disgraceful and she is buried secretly without mourning, weeping, wailing, or any ordinary rites in *Ko tsa*, the accursed grove, outside the town (8).

In S Nigeria, care is taken by the Niger Delta tribes when a woman dies and leaves a child over six months old. For fear that the spirit of the mother will return to fetch the child, the corpse is tricked into believing that the baby has been placed beside her, by bringing in the crying child; then the child is hastily smuggled out of the hut while a bunch of *Musa sapientum* var. *paradisiaca* Plantain is put in with the body of the woman and bound up with the funeral binding clothes (14). In Ghana, should the mother die in childbirth and the child itself be born alive among the Twi-speaking tribe, it is customary to bury it with the mother... The idea seems to be that the child belongs to the mother, and it is sent to accompany her to *samanadzi* (realm of the dead), so that her *saman* (ghost) may not grieve for it (6). In the past it was a custom among the Dormaa of Ghana to bury childless women without rituals, and to drive thorns into the soles of their feet. Such a miserable departure from earth was to prevent the *okra* (the soul) from being reborn in a different shape but with the same fate (9). In S Africa, the Southern Sotho use *Cotyledon orbiculata* Hondeoor as a remedy for many diseases, and for making a charm for an orphan child (27).

According to Ga custom if the child dies before the eighth day, it is considered as having never been born and has no name (8). Should the infant die before the eighth day among the Ashantis, the body is whipped (sometimes it is mutilated by having a finger cut off), wrapped in sharp-cutting spear-grass or *Pennisetum purpureum* Elephant grass and buried (19b); without ceremony (9). The punishment is to discourage a repetition of premature infant death, while the mutilation serves as a positive identification - it is believed - in event of a rebirth. The parents are neither allowed to moan nor morn for the child is regarded as only a temporary visitor which was sent by God for exactly this purpose (9). In Yoruba language, 'emeres' means children who incarnate into this world as spirits, and are also known to die very early in life. These spirit-children are believed to be 'wanderers' with an unusual short life-span, shuttling between the earth and the world beyond (23). Among the Akan-speaking in Ghana and Ivory Coast, and some of the coastal

tribes in the Region generally, a decoction prepared with cut stems of *Gouania longipetala,* a scandent shrub or liane in forest believed to be fetish, with three other plant ingredients is used by the mother for bath in rituals associated with *begyina* (successive infant deaths). The ceremony is usually preceded by the sacrificial offering of a hen, *Gallus domestica* and libation (22).

'When several children of one mother have died in infancy, means must be taken to avert a similar fate in case of those born subsequently' (24). 'First a special name is given, *Ajuji* being a favourite in case of both males and females. Next a special charm is worn on neck and waist until the child is grown up; and sometimes the hair will be shaved or dressed in a special way. The mother, too, may partake in the last; if three children have died she will shave one side of the head; if four, the whole. Very often in the case of other peoples, an opprobrious name is chosen for a child born after the death of others, so as to depreciate it and make the evil influence less likely to be exerted against it'(24).

In successive infant deaths *(fea)*, among the Akwapims of Ghana and some other tribes in the Region, like accident victims *(atofo)*, the corpse is wrapped with the dried leaves of the brown-stemmed type of *Musa sapientum* Banana, locally called 'kwadu-pa', and *Ricinus communis* Castor oil plant instead of a shroud. *Atofo* include all forms of sudden death through unnatural means - that is except through old age or sickness; namely by hanging, by drowning, by lightning or electrocution, through lorry accident, snake-bite, tree-fall, and so on. (Accident victims are to be distinguished from'difficult bodies'). That is either a person suspected of *bayie* (witchcraft) who dies as a result of its own mischief, or is killed by the gods, or a person who commits suicide, or died childless, or drowned (9).

'Nevertheless they are both buried in the same section of the cemetery - (see GRAVES AND CEMETERIES : SUPERSTITIOUS VALUES). The spirits of such victims are traditionally believed to take delight in haunting, as such the burial ritual might be a form of exorcism. (See also INFANTS).

References.
1. Bouquet, 1969; 2. Bouquet & Debray, 1974; 3a. Burkill, 1985; b. idem. 1997; 4. Crow, 1969; 5. Dalziel, 1937; 6. Ellis, 1887; 7. Everard & Morley, 1970; 8. Field, 1937; 9. Fink, 1989; 10. Gunn & Dennis, 1979; 11. Ilogu, 1974; 12a. Irvine, Herb. GC; b. idem. 1961; 13. Kerharo & Bouquet, 1950; 14. Kingsley, 1897; 15. Lordzisode, (pers. comm.); 16. Okigbo, 1980; 17. Perry & Greenwood, 1973; 18a. Rattray, 1923; b. idem, 1927; 19. Richards & Kaneko, 1996; 20. Talbot, 1932; 21. Teiko, (pers comm.); 22. Thomas, 1988; 23. Tremearne, 1913; 24. Verger, 1969; 25. Vergiat, 1969; 26. Walker & Sillans, 1961; 27. Watt & Breyer-Brandwijk, 1962.

9 COURT CASE AND FAVOUR

Judge not, that ye be not judged, For with what judgement ye judge, ye shall be judged: and with what measure ye mete, it shall be measured to you again.
Matthew 7:1, 2

Defendants in court may seek the assistance of medicine-men, fetish priests, witch-doctors or some other forces either to win a case, get acquittal, obtain favour or lenient sentence, or for a case to be declared a foolish one or indefinitely postponed. The prescription is normally a charm of some sort, and it is either worn, applied as a paste, taken in food or drink or otherwise said to be effected through the occult. A witness may also be given a prescription to test the veracity of his or her evidence.

OATH-TAKING

Witnesses in court are normally required to take an oath, as a pledge to speak the truth, before giving evidence. Depending on one's religious beliefs, the oath may take the form of swearing by the Bible, the Koran, a fetish, a juju, a local deity, a river, a hill or an ancestral spirit. One such native oath taken by a witness in S Nigeria is described : 'The oath taken was usually the heathen *mbiam*. For this were needed a (human) skull and a vile concoction (of unspecified herbs) in a bottle, that was kept outside the Court House on account of the smell. After a witness had promised to speak the truth, one of the members of the Court would take some of the stuff and draw it across his tongue and over his face, and touch his legs and arms. It was believed that if he spoke falsely, he would die. After Miss Slessor took up her duties, a heathen native, who had clearly borne false witness, dropped down dead on leaving the Court, with the result that *mbiam* was in high repute for a time in the district' (14).

In N Nigeria, oaths used to be taken on iron of old, and even now many of the less civilized Hausa people are tested with this metal, a bayonet being passed across their throats, and then between their legs (18). 'I found an even better method. A cartridge was put in a calabash of water, and the witness had to drink some. The rifle was rested upon the head for a moment, and then pointed at his heart, and he was told that it would thus know where to find the child (the cartridge being supposed to have communicated its properties to the water) if the swallower told an untruth. I have known this method to break up a case that had looked quite hopeless a few minutes previously' (19). In Liberia, *Scottellia coriacea.* a forest tree, is an ingredient in legal oath-taking (Cooper & Record fide 7).

An oath may be taken for various other purposes such as keeping a secret (see SECRET) or claiming a debt from the estate of the deceased. To make it binding, something that pertains to a deity, is usually eaten or drunk. For instance, a person swearing by *samantan* (a fairy) eats small pieces of the bark of *Ceiba pentandra* Silk cotton tree in which a deity of this kind is believed to reside (12).

CHARM PRESCRIPTIONS.

The leaves of *Vitex doniana* Black plum, a tree of savanna woodland often tended for the edible fruits, collected with ceremony, with assorted ingredients, including a black and white powdered stuff, then tied together and immersed in a container of seawater with a prayer, before going to court, is a charm prescription by medicine-men for the indefinite postponement of the case (2). In W Africa, the whole plant of *Sida linifolia*, an erect little malvaceous weed, called 'wodoewogbugbo' by the Ewe-speaking tribe, meaning 'they planned but retreated', placed under the pillow with cowry shells while praying, is used as a charm for favour at a trial, or for the case to be declared a foolish one. The practice is reported to be common and popular among the tribe (11). The weed chewed with sugar just before going to court; or alternatively *Phyllanthus niruri* var. *amarus*, a small weed of cultivation, chewed with salt, have the same effects on court decisions (16).

In Ghana, *Gardenia ternifolia* ssp. *jovis-tonantis*, a savanna shrub traditionally believed to be protective against lightning, is approached by the Ada people early in the morning with prayers. Some milled corn, contained in a new calabash, is sprinkled around the plant. The ritual is for favour in an impending court case (3). Among many tribal

groups in the Region, the flower of the red variety of *Rosa* species Rose ground with a drop of lavender or Olive oil is rubbed on the skin in an incantation to win any case even before going to court (16). However, the charm must not be attempted if the case is heavily against the individual, else the consequences would be even more drastic (16). A piece of the skin of a wolf, *Lupus* species, that has been ground together with *Amaranthus lividus* or *A. viridis*, annual weedy herbs of cultivation, and local black soap and used for bathing, is a charm to win court cases (16). On the other hand, if favour is required then the person's name is mentioned while one is bathing (16).

Amaranthus viridis

In Congo (Brazzaville), it is believed that by eating some leaves of *Senecio (Crassocephalum) biafrae*, a climbing composite forest herb sometimes eaten as spinach, someone pleading before a tribunal will put forward a good case; and *Annona senegalensis* ssp. *onlotricha*, a small tree, is held to have magical properties to advance a case before a tribunal (5). In the Democratic Republic of the Congo (Zaire), *Microdesmis puberula*, a forest shrub, is put into various concoctions to improve one's luck and to advance a case before a tribunal (6); and in Congo (Brazzaville), a litigant going before a magistrate will add some leaves of *Selaginella myosorus*, a herbaceous forest fern, to a perfume which is applied on the forehead to ensure winning the case (6). In Gabon, *Eclipta alba*, a common annual composite herb with whitish flowers that grows in damp places, is put in a lotion with other plants to anoint a supplicant for special favour from influential persons (21).

In N Nigeria, *Scoparia dulcis* Sweet broomweed or *Heliotropium bacciferum*, *H. subulatum* or *H. ovalifolium*, erect or sub-erect or prostrate herbs with woody root-stock, and *Evolvulus alsinoides*, a spreading hairy herb, are used to secure favour of officials or to seek court favour (10). A defendant in court wears the preparation, made up with other medicines as a charm or waist girdle, or sucks a piece along with natron (native sesquicarbonate of soda). The Hausa name of these plants 'ruma fada', 'rauma fada', or 'roma fada', implies seeking favour (10). In S Nigeria, *Gossypium barbadense* Cotton enters a Yoruba Odu incantation in a mystical way to win a court case, that is, for the cotton to tie up one's adversary and tether him or her like an animal (20). In Nigeria, *Ludwigia octovalvis* ssp. *brevisepala* Primrose-willow, an erect herb with yellow flowers - often growing in marshy areas, is used among both the Hausa and the Yoruba to ensure favour and to escape punishment. *L. stenorraphe*, a herb or shrubby plant, has superstitious uses as well. The Yoruba also invoke *Dioscorea rotundata* White yam or Guinea yam to make a policeman forget a case or to make someone generous (20).

In Ghana, *Ritchiea reflexa*, a scrambling savanna shrub, is approached with incantations and invoked to influence court decisions in one's favour (16). In the Region, the charms prescribed by medicine-men for favour in court include a leaf of *Bryophyllum pinnatum* Resurrection plant with that of *Phyllanthus niruri* var. *amarus*, a common herbaceous weed, concealed in the mouth (8). Alternatively, three leaves of Resurrection plant on which the chewed slimy nuts of *Cola verticillata*, a forest tree, have been sprayed are buried in the house with the necessary ceremony before going to court (4). The egg-like seed of *Okoubaka aubrevillei*, a rare forest tree occurring mainly in Ivory Coast and Ghana, believed to be fetish and sold as such, when placed in water, may also be used in a ritual bath for favour in court or for a more lenient sentence (9). The leaf of *Sida acuta* Broomweed, a common weed of cultivated land, placed under the tongue in court is a charm to win a case (18). Broomweed also enters into charms to prevent the other party from appearing in court altogether, or to lose interest in the case (16).

In N Nigeria, the root of *Cyperus articulatus*, a sedge, is chewed by a defendant in court as a charm to secure acquittal (15). In Ghana, *Paullinia pinnata*, a climber with tendrils, and *Momordica balsamina* Balsam apple or *M. charantia* African cucumber, trailing cucurbits traditionally believed to be fetish, are invoked in juju practice to stop or 'kill' court cases (16). In the Region, fetish priests use *Vernonia cinerea* Little ironweed, a weedy composite, in prescriptions as charms for court cases and for favour. The plant is visited at 1.00 pm. After offering it powdered roasted corn as a symbol of payment or compensation, some of the leaves are collected and ground with a type of salt called gelo, amid prayers and incantations. The preparation is rolled up into pellets or poultice and dried. In the event of a court case, one pellet is placed under the

Solandra species (legendary and mythical plant).

tongue with a bit of white kola-nut (5). The ground leaves of Little ironweed in lavender applied to the skin as cosmetic also enters into prescriptions for favour (5).

In Botswana, *Berkheya alba*, a composite, is used as a charm for procuring things wished for (22). In this country, the Southern Sotho chews *Trifolium africanum* Cape clover as a charm when going to court over difficult litigation, and this is thought to turn the jury in his favour (22). In S Africa, a Zulu litigant often places a few leaves of *Acridocarpus natalitius*, a climbing shrub, under his tongue in order to ensure success by causing his opponent's witness to become tongue-tied (22). In Mozambique, a decoction of the leaf and root of *Lonchocarpus capassa* Lance tree or Mbandu is administered among the Tonga to disputants who appear before a witch-doctor for settlement of an argument (22). In Mexico, people seeking certain favours such as better singing ability, skill in embroidery or weaving and so forth among the Huichol Indians, go to the nearest important god-plant, *Solandra* species, a tree on rocky cliffs, bearing various offerings (13).

Sympathetic Magic

Some of the prescriptions appear to be based on sympathetic magic. For instance, it has been observed that the unopened young leaves of *Piliostigma thonningii*, a common savanna shrub or small tree legume, collected early in the morning before talking to anyone, are used as a charm for obtaining favour in court among the Hausa of N Nigeria, the Zarma of Mali and Niger and the Fula. The inference here is that like the unopened leaves, one's adversary cannot open his or her mouth in court (17) - an instance of sympathetic magic. (The ritual is also used both in seduction, and in preventing the creditor from demanding repayment from the debtor. For the details, see MARRIAGE: FAITHFULNESS IN MARRIAGE: Adultery: Seduction and WEALTH: CHARM PRESCRIPTIONS, respectively). In these countries, the scrapings from the bark where two branches of a tree rub against each other (provided the tree is non-poisonous), with that

of *Ficus platyphylla* Gutta percha tree is powdered and taken in food and also used in bath water (on three consecutive Fridays in men, and four in women) as a charm to win cases, especially chieftaincy disputes (17). In S Nigeria, a display of the scarlet flowers of *Alstonia congensis*, a forest tree, in the course of a dispute, is a sign of challenge (1). This is likely an instance of sympathetic magic hinged on the red colour.

Ninety-nine cut stems each of *Sida acuta* Broomweed and *Boerhaavia* species Hogweed with seven coins placed in a blazing fire with prayers, and immediately quenched with consecrated water, is a charm to win a court case, or to render the case less serious, as the water quenches the fire (5). This is clearly an instance of sympathetic magic. Similarly, *Portulaca quadrifida* Ten o'clock plant, a trailing fleshy weed, is employed in charms to suspend court cases indefinitely. Some of the plant is collected and then placed under other plants of the same species in an incantation to sit on the case (16). Yet another instance of sympathetic magic, this time hinged on one plant directly 'sitting' on another. Some sensitive plants also enter into charms to win court cases. For instance, either *Mimosa pudica* or *Schrankia leptocarpa*, both called Sensitive plant, or *Neptunia oleracea* is visited at dawn with an egg as an offering, and amid prayers the plant is invoked to shut up the case with the closing of the leaves in the evening. The procedure is repeated later in the evening at the same plant (5). Surely this is another example of sympathetic magic hinged on the closing leaves. (See also CHARMS, GENERAL; and LOVE: CHARM PRESCRIPTIONS).

References.
1. Adams, 1943; 2. Addo (pers. comm.); 3. Adibuer (pers. comm.); 4. Ashieboye-Mensah (pers. comm.); 5. Blankson (pers. comm.); 6. Bouquet, 1969; 7. Burkill, 1994; 8. Cofie (pers. comm.); 9. Commeh-Sowah (pers. comm.); 10. Dalziel, 1937; 11. Dokosi (pers. comm.); 12. Ellis, 1887; 13. Knab, 1977; 14. Livingstone, 1914; 15. Meek, 1931; 16. Mensah (pers. comm.); 17. Moro Ibrahim (pers. comm.); 18. Noamesi (pers. comm.); 19. Tremearne, 1913; 20. Verger, 1967; 21. Walker & Sillans, 1961; 22. Watt & Breyer-Brandwijk, 1962.

DEATH AND DYING

...all flesh shall perish together,
and man shall turn again into dust.
Job 34:15

Death is interpreted in many ways in African folklore, varying from region to region, and among the different ethnic groups. However, in general either human or supernatural malice, or an evil force is believed to be responsible, and accordingly blamed, for the loss of loved ones - hardly any deaths occurring from natural causes; except of course old age. The general belief is that sickness was unnatural, and that death never occurred except from extreme old age; so when a man became ill or died, sorcery would be alleged (31). The belief in witchcraft is the cause of more African deaths than the slave-trade; for at almost every death a suspicion of witchcraft arises (29). In many cultures worldwide, plant nomenclature sometimes reflects life and death. While some plant names suggest life, others are ominous and symbolize death.

PLANT NAMES AND LIFE

In S Nigeria, the Yoruba in calling *Thonningia sanguinea* Crown of the earth, 'edele' or 'ade-ile', invoke the plant for longevity and to avoid death; and in dubbing *Dioscorea cayenensis* Yellow guinea yam, 'riddle', the tribe invoke the plant in a repetitive incantation against death; similarly, as 'prostitute' or 'beautiful woman', imply a similar desire for biological continuity (49). Two other plants locally called 'apongbe' and 'etichoro' in Yoruba, meaning 'put on back and carry' and 'rabbit's ear', are used by the tribe in an invocation to keep a dead baby on earth and to send death away, respectively (49). Similarly, *Piptadeniastrum africanum* African greenheart, a forest tree, is imprecated in an Odu incantation to grant a long life by making one invisible from death (49). In Benin Republic, *Elaeis guineensis* var. *idolatrica* King oil palm, or Palmier fétiche, is known as 'iviromila', meaning palm of everlasting life; and in N

Nigeria, the Hausa name of *Commiphora africana* var. *africana* African bdellium, 'daashi mai-yawan rai', means 'daashi' with seven lives - a reference to immortality. In S Nigeria, *Termitomyces microcarpus*, an agaric mushroom, is put into propitiation to the gods to prevent mortality and to increase the population (38). Some composites that retain the flowers such as *Helichrysum bracteatum*, an Australian perennial, and *H. plicatum*, that grows in the high mountains of Lebanon, are sometimes referred to as 'Everlasting' in English, 'Immortelle' in French and 'Khalidah' in Arabic. (41). Ancient Chinese tradition has it that dew collected from the flowers of *Chrysanthemum* species Chrysanthemum preserves and restores vitality in man; and certain individuals even hoped to become immortal by eating Chrysanthemum petals regularly (17). (See also MANHOOD: LONG LIFE). Among Christians, immortality was well shown by *Amaranthus* species Amaranth (11).

In N Africa where it is indigenous, *Phoenix dactylifera* Date palm is known as the 'Tree of Life' and is much prized in the desert region as the staple food, diet and source of wealth from antiquity (24). Similarly in Gabon, *Tabernanthe iboga* Iboga, an apocynaceous forest shrub with hallucinogenic properties, is called the 'Tree of Life' (9). The common name of *Guaiacum officinale* Lignum vitae, an introduced decorative tree with hard wood, means 'Tree of Life'. From the wood is also obtained the medicinal resin *guaiacum*. There are five references to the 'Tree of Life' in *The Holy Bible*: two of these are in the first book - Genesis 2:9 and 3:22; two in the last book - Revelation 2:7 and 22:19; and one about mid-way - in Proverbs 3:18. The deciduousness of a tree gives it an ambiguous image which reflects the tree's power to give life and rebirth, as well as to bring about death (18).

PLANT NAMES AND DEATH

Some other plant names and uses, however, are ominous or suggestive of death. By tradition, certain occurrences are also signs of impending death. In many traditions worldwide, plants feature in the various death ceremonies - from rituals performed at the last minutes of life, through burial, purification after burial, sacrifices to the dead, communication with the dead to remembrance of the dead. Indeed, from the cradle to the grave, man finds use for plants in one way or the other. In S Nigeria, the Edo name of *Pararistolochia goldieana*, 'ugbogiorinmwin', means grave of the King of the Dead - probably on account of the smell of decaying

flesh of the mature flowers, or that the unopened flower is the haunt of a small venomous viper, species of *Bitis* or *Echis*, that uses the smell to attract its prey (25). In this country, the Igbo tribe associate *Phallus aurantiacus*, a gasteromycete stink-horn toadstool with the Devil and with death - an Yoruba name 'oga-egungun' specifying it as the barrier between life and death (38). In The Gambia, the Mandinka name of *Celosia leptostachya*, an amaranth herb, 'furai nyamo', meaning 'death grass' refers perhaps either to the use of the plant in death ceremonies, or on account of its smell (13). In Ivory Coast, 'okuo' means death, and 'baka' means tree among the Anyi tribe. Therefore *Okoubaka aubrevillei*, the rare forest tree popularly believed to be fetish, literally means 'tree of death' (5). Similarly, in S Nigeria, *Jatropha curcas* Physic nut is known by the Efik as 'eto-mkpa', meaning 'tree of death' - probably due to the poisonous seeds. In Ghana, *Mareya micrantha* "Number one", a forest shrub or small tree, is called 'odubrafo' in Twi, meaning 'executioner' - in direct reference to the poisonous leaves.

In one area of its distribution *Sarcocephalus (Nauclea) latifolius* African peach is called 'tsoiatsuru', which means 'if father dead, eat half; if orphaned, eat all' (17). In Ghana, the Akim name of *Dicranolepis persei*, a forest shrub, is 'owudako', meaning to die one day; and in S Nigeria, the Igbo name of *Cleome rutidosperma (ciliata)*, an annual weed in waste places, is 'akidimmo', meaning 'beans of the dead'. However, the reasons for both names are not clear. In Ghana, the false fruits of *Coix lacryma-jobi* Job's tears, a perennial grass by rivers and wet places, is referred to in Fanti as 'owu-amma- mannka-m'asem, meaning death makes me mute, and is worn as a necklace to denote mourning for children (13). In the olden days it was customary to place garlands of *Vinca minor* Lesser periwinkle, a temperate plant, on the biers of dead children, which probably accounts for the Italians knowing the plant as Flower of Death (41). Similarly, in the Middle Ages, Lesser periwinkle was traditionally associated with death, and garlands were made of the flowers to adorn corpses and condemned criminals (17).

Taxus baccata Yew (symbol of death)

In Europe, a superstition that the shade of *Taxus baccata* Yew was deadly to those who slept beneath it seems to have originated by Andreas and to have persisted into the eighteenth century. In Tanzania, the exhalation from a species of *Oldfieldia* was said to be fatally poisonous - the African believing that if a dog, *Canis* species, goes to sleep under the tree in flower, it will be killed by the perfume (53). In S Africa, the Zulu and Swati believe that to touch *Monadenium lugardae*, a succulent euphorb, or even lie in its shadow brings certain and sudden violent death (53). In Japan, the strange fact has been noted that when some types of bamboo flower, all members of that species in the country - and even further afield - also flower within two seasons, amounting to a mass suicide of bamboos. The bamboos are species of *Arundinaria, Bambusa, Dendrocalamus, Oxytenanthera, Phyllostachys, Sasa* and *Shibataea*. Consequently, the 'flowering of the bamboo' has come to be a portent of disaster (43). *Cupressus sempervirens* Cypress is an almost sinister tree (40). Firm and solemn next to graveyard gates, it points up towards the sky with its crown, and to the tombs with its elongated shadow, seeming to remind humans of the deadly destiny awaiting us (40). Cypress is the tree that best symbolizes death (40).

Cupressus sempervirens (symbol of death)

In W Africa, it is the belief and fear among many tribal groups that those who deliberately break a taboo and enter a sacred or a fetish grove on an unauthorized day would get lost and never return home. As such wherever *Amorphophallus* species, tuberous aroids in forest and savanna areas, is seen growing in a grove, it is said to indicate where such an adventurer had died (46). The species in the Region include *A. abyssinicus, A. aphyllus, A. dracontioides, A. flavovirens*, and *A. johnsonii* Johnson's arum. In Sumatra, *A. titanus,*

which has one of the world's largest inflorescence - about three metres tall - is locally called 'bunga bangkai', meaning corpse flower. It is the belief in its distribution area that whenever a dog is observed chewing *Pennisetum subangustum* or *P. pedicellatum*, common grass weeds, it is an indication that the animal is about to die (34). In Nigeria, *Calopogonium mucunoides*, an annual or perennial legume, is known as 'duwei iyau' in Ijo-izon (Kolokuma) language, meaning 'deadman's yam-bean'.

In Ghana, there is a superstition that a sick person near death will express a great desire to eat *Ananas comosus* Pineapple (26); and that those near death express the desire to drink water; or are given water to help them on the journey through the spirit world (3). In this country, 'the Ashantis declare that in order to reach the *samanpow* (place of ghosts) a steep hill must be climbed. They see the dying man panting for breadth, and think of his soul struggling up some steep incline, and this draught of water is to speed him on his journey' (42). This water is also given at the grave-side (21). (See BURIAL below).

THE LAST MOMENTS.

Traditional medicine-men use some specific plants to bring about a peaceful, gentle and easy death - referred to in western medicine as euthanasia - in case of an incurable and painful disease. For instance, in S Nigeria, the leaves of *Croton zambesicus*, a small tree planted in villages and towns, is drawn gently over the face of the dying by the Ekoi people to cause the spirit to pass painlessly (13); and the Yoruba invoke *Dichrocephala integrifolia*, a composite, in an Odu incantation to produce calm - to kill gently (49). Among some of the ethnic groups in the Region, a leaf macerate of *Newbouldia laevis*, a medium-sized tree popularly believed to be fetish, in a container, which the wooden staff for preparing corn meal has been used, in a ritual, to make a sign of the cross at the opening the previous evening, is used to wash the sick the following morning to induce a peaceful death (34). In Sierra Leone, the Mende name of *Newbouldia*, 'poma magbe', meaning

Croton zambesicus (symbolizes Ekoi people of Nigeria)

'corpse drive on' derives from the use of leafy branches of the tree to fan a corpse to help its spirit on its way (44). However, it is reported that the fanning is rather to keep away the flies (13).

In Congo (Brazzaville), *Senecio (Crassocephalum) biafrae*, a climbing composite forest herb, is used to assure rest to the spirit of a departed sect member (7a); and *Kolobopetalum chevalieri*, a menisperm, is a medication of final departure when all else has failed. The patient's arms and hands are scarified and the sap is applied. If there is no favourable response, there is nothing else to be done but to leave this patient to die (7b). However, in Kenya, Masai witch-doctors prepare an infusion of *Polygonum pulchrum*, a perennial herb, with which to wash a person on the point of death to effect a magical recovery (8d). In Senegal, the 'Socé' use a decoction of the root of *Gardenia ternifolia* ssp. *jovis-tonantis*, a shrub or savanna tree, as a medicament of last resort against undiagnosed illnesses when all else has failed 'so long as life remains' (26b). In Congo (Brazzaville), *Cissus aralioides*, a forest liane, may be attached to a moribund person, delaying death, and permitting the person's final wishes to be stated (7a). The leaves of *Newbouldia* placed under the pillow of a dying person enables him or her to hold on to life for at least an hour to await any arrivals to whom a message is to be given - irrespective of the pain and suffering - until the leaf is removed (2). In Ghana, the bark decoction is used among the Ga tribe to hold on the dying for three more days (30); and in Guinea-Bissau, a root-infusion of *Spondias mombin* Hog plum or Ashanti plum is considered to be an important tonic able to prolong the life of the dying by two or three days (23).

In N Nigeria, a lotion of *Ficus abutifolia*, a savanna tree on rocky hills, is used in a superstitious practice by persons present at a death, lest the spirit seizes a relative or some other person (13); but in Ghana, small pieces of the hair of the deceased are sometimes kept among the Ga tribe and worn round the knees of relatives as a protection against the dead man (20). In S Africa, *Strychnos spinosa* Kaffir orange is used in death ceremonies by certain tribes; and the Southern Sotho use *Eriocephalus punctulatus*, a composite, with *Metalasia muricata*, another composite, to fumigate a hut after a death (53). On Fouta Djallon, there is the belief that if one has some fig (8d) or a kola nut (8e) in one's mouth at the moment of death, one will enter Paradise. In Central African Republic where a tree is planted for a newborn child among the Oubangui tribe, and gifts are occasionally left for this tree throughout a person's life (see CHILDBIRTH: POST-NATAL), it is the belief that when someone dies

their spirit goes to reside in their personal birthright tree (50).

FUNERAL OBSEQUIES

This is a solemn, usually religious ceremony to dispose of the dead - either by burying or by burning. The preparations prior to this include the following: embalming the body, washing the body, laying it in state and ritual separation of the dead from the living relations.

Embalming the Body

The preparation of the dead for burial may be accompanied by the use of specific plants. In Sierra Leone, the reeds of *Costus afer* Ginger lily are laid under the bodies of the dead (13). By extension of the idea of protecting the skin, the leaves of *Hyptis suaveolens* Bush tea-bush or *H. spicigera* Black sesame, strongly aromatic labiates, are rubbed onto the skin of cadavers in Senegal during obsequies for embalming (27a, b) - hence the Diola names, 'kafulogay', meaning 'medicine of the dead'. In N Nigeria, *Boswellia dalzielii* Frankincense tree may be added to the juice berries of *Acacia nilotica* ssp. *tomentosa* Egyptian thorn, a thorny gum-yielding shrub legume in savanna, in ritual mummification practiced by various tribes (33).

In Burkina Faso, the spiny branches of *Acacia gourmaensis*, a savanna shrub, enter into funeral rituals in the Moore country - perhaps the tenacity of the hooked spines presents some symbolism (39). The oil prepared from *Lawsonia inermis* Henna is known in Arabic as *dunk-ul-fagiya,* and has been used from ancient times in Egypt for embalming and religious practices (8c). In the Niger Republic, the butter from *Vitellaria (Butyrospermum) paradoxa* Shea nut tree has ceremonial funerary use for embrocating a dead person (8e). In the Middle East, *Commiphora molmol* Myrrh was used as an embalming agent by the Ancient Egyptians; and *Thymus vulgaris* Thyme essence was used in perfumes and embalming oils (6). In the Niger Delta, *Marantochloa purpurea,* an erect herb of swampy forest with a red under-side leaf, is used to cover the corpse of anyone dying by suicide or by accident (8d).

Washing the Body.

In Burkina Faso, *Combretum glutinosum*, a savanna tree, has funerary use: two items fastened in a cross are placed in a hole in which a newly dead corpse is laid for shaving and washing (39). In N Nigeria, the Fulfulde use pounded leaves of *Ocimum basilicum* Sweet basil in water to wash the dead (8c). In Ghana, the dead are washed, among the Ashantis, the Fantis, the Gas and many of the southern tribes, exclusively with sponge prepared from *Momordica angustisepala,* a stout cucurbit forest climber often cultivated for the spongy fibres, locally called 'sapowpa' or 'ahensaw' in Akan. This is probably a sign of last respect, since the local names signify superiority and kingship.

'After the washing the sponge is coiled into a great many little rings. Three days after the burial of the body, these rings together with the nails and hair, are put into a small box and buried in the dead person's father's house, under the floor of his father's sleeping-room. If a goat, *Capra* species, is killed to purify the house of bad things which caused the death, the fat of the goat is mixed with shea-nut butter and added to the sponge, hair and nails in the box'(20). In S Africa, an ancient custom of the Muslim at the Cape is to make a saponaceous infusion of the bark of *Polygala myrtifolia* Myrtle-leaved milkwort with which to wash the dead before interment (53). The Swahili wash the dead with *Cinnamomum camphora* Camphor water and insert the solid into the natural orifices of the deceased (53). Indian Muslims in funeral obsequies use a leaf-decoction of *Ziziphus spina-christi* Christ's thorn to wash a corpse on account of the veneration with which the tree is held (8d).

Laying in State

It is common to scent shrouds of the dead either by rubbing the cloth with *Ocimum basilicum* Sweet basil or by fumigation (8c). In Ubangi, bark-cloth beaten from *Sterculia setigera* Karaya gum tree is used as shroud for the dead (8e). In SE Senegal, the leaf of *Icacina oliviformis* False yam, an erect woody shrub, enters into Tenda ritual in a burial ceremony: a head-pad is placed under the head of the corpse (19). The tribe also have a ritual use of *Andropogon tectorum,* a robust perennial grass, during funeral (19). In Ghana, the people of Techiman-Bono believe that pillows stuffed with cotton are taboo at *asaman* (ghosts), so pillows stuffed with soft grass are used, and later buried with the corpse (52). The practice is economically sound, because burying bodies with expensive jewelry, valuable gold ornaments and costly coffins encourages grave looting . For

example, among the Dormaa people of Ghana it has been observed: 'Apart from food, drinks and beads, the deceased are provided by their families with jewelry, cloth, gold and money for their journey to *asamandu*' (21).

Separating the Dead from the Living

Various ceremonies or rituals are performed before, during or after burial to separate the spirit of the dead from the living family and friends. For instance, among the Ga tribe of Ghana, each takes a kola-nut or lime, cuts it in halves, and throws the two halves in the air. 'If they fall with one cut surface upwards and one downwards, it is taken as a sign that the dead man is willing to part company with the living friends. If both cut surfaces fall the same way, the dead man has not consented to dissolve the friendship, and the spirit may remain to trouble the survivor. In this case the kola-nut is thrown again and again until the dead man consents to depart' (21). The practice is, in fact, a form of divination. (See FORTUNE-TELLING AND DIVINATION for plants used in the art).

In Gabon, in the event of death of a husband or a wife, the spouse must undergo a lustral cleansing with a leaf-macerate of *Leea guineensis*, an erect or sub-erect shrub; and an Nkomi widow is required to wear a necklace of *Abrus precatorius* Prayer beads seeds during the mourning period (51). In this country, *Lygodium microphyllum*, a climbing fern, is used as a head-crown worn by Fang women one or two days after the death of their husband (51). In Ivory Coast, a deceased person's hut is sprinkled with a decoction of *Hyptis pectinata*, an aromatic herb, to cleanse it (28); and in Uganda, it is appropriate to sprinkle a preparation of *Withania somnifera*, Winter cherry, a semi-woody perennial herb, over people at a wedding or after a death (8e). In S Africa, the Kgalagadi use *Grewia flava* Raisin tree in their death rites. They tie a piece of the inner bark around the right arm of the

Traditional funeral

deceased and around the right arm of each of the children, beginning with the eldest and continuing in sequence to the youngest (53).

The Coffin

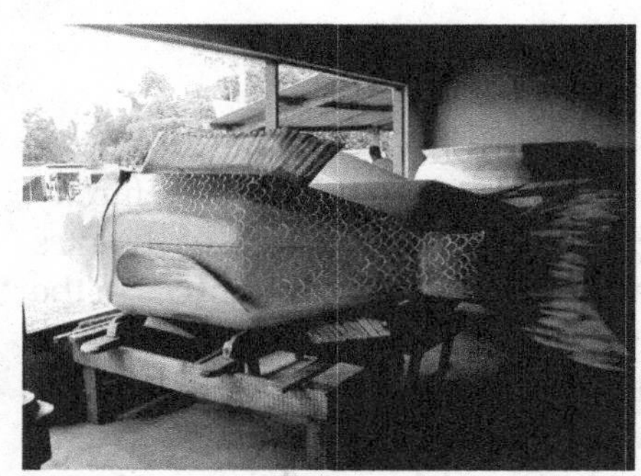

Fish-shaped coffin for a fisherman

Several decades ago coffins used in Igboland were special baskets made from the petioles of *Elaeis guineensis* Oil palm. These are usually woven in the form of coarse mat, *ute ekwere,* which is rolled round the body and then draped with cloth before burial underground or above ground on branches of trees in thick, sacred bushes or forests (37). The young leaflets of Oil palm are used to strike a note of danger or alarm; for example, vehicles conveying corpses are tagged with them (36). In N Nigeria, freshly-picked leaves of *Ocimum basilicum* Sweet basil are used to scent coffins in Bornu, a practice similarly known in Kordofan (8c).

Hen- or a Cock-shaped coffin for a Poultry farmer

Nowadays, coffins are mostly wooden caskets. Some are very expensive, and may be designed to conform with the profession or status of the dead body - a book or a pen for a scholar; an okro or okra, pepper or cocoa pod for a farmer; a fish, a lobster, species of *Panulirus* or *Scyllarides,* or canoe for a fisherman; a bag of flour for a flour dealer or a baker; a cobra, *Naja nigricollis,* for a snake charmer; a lion, *Panthera leo,* for a hunter; a mammy truck for a driver or a transport owner; a petrol tanker for a tanker driver;

Stool-shaped coffin, for a royal

and a saloon car for a private car owner. Others include a gun for a warrior; an airplane for a pilot; a ship for a sea-man; a sewing machine for a tailor; a drum for a drummer; a cow for a cattle owner; or a microphone for a singer. Teshi and Nungua, suburbs of Accra in Ghana, are famous for these designer or fancy, but expensive coffins. The practice of burying the dead in expensive coffins, further adds to grave looting - (see Laying in State above).

Burial Ceremony

One of the characteristics that distinguishes the human race from the lower animals is that it buries its dead - as a last respect for the dead, fear of invoking the wrath of the *sasa*, and to enable the spirit to rest in peace. It is also a hygienic means of disposing a corpse. Even when the body cannot be found as in death by drowning or through plane crash or by complete burning or even if the relation is presumed dead, a coffin containing sand or an effigy representing the dead is accordingly buried. Acting in a socially acceptable manner, as burying the deceased in the proper way, brings reward (32).

In Ghana and Ivory Coast, a sacrifice has to be made to *Asase Yaa* (mother earth) among the Akan-speaking, before digging a grave - only then may 'her child' be buried in her fold (21). 'At a traditional burial, the dead is given the funeral meal - *eto* (mashed unsalted yam) and a glass of water - at the grave. The repast, which must be prepared by an elderly woman of the matrilineage of the dead, has two functions. Firstly, it serves as provisions for the long journey to *asamandu* (realm of the dead); and secondly, the *nsamanfo* (ancestral spirits) are invited to participate in the burial' (21). Just before burial of a corpse in Ashanti, the head of the deceased's family (that is blood) steps forward, holding in either hand a branch of *Costus afer* Ginger lily; touching the coffin with each branch alternately in an incantation to separate the soul of the dead from the family. One of the branches is laid on the coffin and buried with it, and the other is placed at the head of the sleeping-place of the person who performed the rite (42). In Nigeria, the Ebina place branches of *Sclerocarya birrea,* a savanna tree with edible fruits, over a newly-used grave to prevent hyenas, *Hyaena hyaena,* digging up the body. Should they do so, it is thought that the deceased was a witch (8e). (See GRAVES AND CEMETERIES: SUPERSTITIOUS VALUES for the burial of accident

victims and 'difficult bodies').

Ginger lily also serves the dual purpose of 'retrieving' the spirit of the dead. Among the Nzima people of Ghana and the Akan-speaking of both Ghana and Ivory Coast, Ginger lily or other *Costus* species, is dragged along the ground from a fatal accident spot to the house in a ritual ceremony to 'bring home' the spirit of the victim (11). The plants include *C. afer, C. deistelii, C. dubius, C. lucanusianus, C. schlechteri* and *C. spectabilis*. Failure to perform this solemn ritual is traditionally believed to be the direct cause of frequent motor accidents at former fatal accident spots. It is traditionally believed that such spots are haunted by the ghosts of the accident victims.

During the pyramid age in Egypt (ca. 2614-2287 B.C.) the wood of *Cedrus libani* Cedar of Lebanon with its acclaimed imputrescibility was especially desired in preparation for the after life of the entombed. Since it was thought that the spirit might like to visit the body to provide for the needs and desires of the united spirit and body. Many of these objects as well as some sarcophagi were fashioned from cedar wood (10). Egyptian Pharaohs reportedly ordered a bouquet of *Rosmarinus officinalis* Rosemary put on their tombs to perfume their trip to the world of death (40). *Lawsonia inermis* Henna has been known from antiquity for funerary cloths, and the nails of mummies of Ancient Egypt have been found henna-tinted (8c). *Nymphaea lotus* Water-lily is figured in Ancient Egyptian and Assyrian tombs (8d); and *Cymbopogon schoenanthus*, a tufted perennial grass, has been found amongst the funerary materials of Ancient Egyptian tombs; and is said to have been used in the toilet and burial preparations of the Prophet Mohammed (8b). Cloth woven from *Linum usitatissimum* Flax, an annual herb, have been found in Egyptian tombs (6); and remains of *Punica granatum* Pomegranate have been found in Egyptian tombs more than 4,000 years old (40).

In N Nigeria, *Euphorbia balsamifera* Balsam spurge is normally used for hedges and field boundaries, but in the northerners' cemetery on the Accra Plains in Ghana, one plant was grown for each person buried (26b). The link of Balsam spurge with burial is not clear, but it could probably be for superstitious reasons. In S Africa, a similar use of tree euphorbias is reported (17) (see TWINS: CEREMONIES: Birth). In Gabon in olden times, *Musanga cecropioides* Umbrella tree

provided the fuel for anyone condemned to death by burning (51).

Exceptions.

There are, however, a few instances when a body may not be buried. Among the Akan-speaking tribe of Ghana and Ivory Coast, 'when a stranger dies in a village, the inhabitants seldom bury him. The body is placed on a raised platform of wattles outside the town, with any property belonging to the deceased, and is there left. If the relatives are known to the people of the village, they send to inform them of the death; but if not, the body is allowed to decay, or be devoured by vultures, *Necrosyrtes monachus* (14b). Among the people of Benin in S Nigeria, 'it is unlawful for any person who has been killed by lightning to be buried, and it is commonly believed on the Slave Coast that the bodies of those who have met their death in this manner are cut up and eaten by the priests of So, the god of thunder and lightning' (14a). (See LIGHTNING AND THUNDER: GOD OF THUNDER). 'If the relations of the deceased offer a sufficient payment, the priests usually allow the corpse to be redeemed and buried' (14a). In some of the traditional areas, a barren woman is denied a normal burial, and men without an issue are similarly treated. (Should a body be buried in defiance of these traditional norms, nature reacts accordingly. For the consequences of non-compliance, see <u>Cremation</u> below).

Cremation.

With a few exceptions - as in the case of a murderer, whether by physical or by spiritual means (that is bad juju, evil medicine, sorcery or witchcraft) - the burning of dead bodies was not practiced in traditional Africa. Presently, cremation does occur, but of some expatriates and few indigenous people, consequently, some cemeteries do have a crematorium. One such example of burning a dead body was practiced among the Ewe tribe of Ghana, Togo and the Benin Republic in former times.

'When a person died, the spirit was summoned to find out whether the dead person ever killed anyone whilst he or she was alive. If so, the body was burnt' (1, 22). 'Before the ceremony, the corpse was whipped and dragged naked by a rope through the village to the outskirts. The fuel-wood for burning the body was provided by the kinsmen of the deceased. Any pieces of the body torn off during the

whipping and dragging were swept, and together with the personal effects, burnt with the body. At the highlight of this ceremony, there were incantations and libation by the medicine-men and fetish priests present to bind the screaming and yelling spirit of the dead, so that it was burnt with the corpse to prevent this spirit from running adrift to wreak vengeance on its captors. The smoke from the burning body either drifted straight up or spiraled upwards. All the residue ashes were collected' (1, 22). The practice is now illegal - like trial by ordeal. The objective of the punishment and open disgrace of the corpse, followed by burning, was to pacify and compensate the spirits of the victims killed by this murderer; so that these souls might rest in peace in full satisfaction that their murderer has been exposed and annihilated.

The story is also told of the punishment meted out by victims of one such murderer buried at 'Do me a bra' (meaning, 'Come, all that love me') - a cemetery at Cape Coast in Ghana. 'At about midnight, intermittent lashing with tree branches, and screaming disturbed the quiet night for about thirty minutes - from the direction of the cemetery. In the morning, news spread through the town that a body buried the previous day had been exhumed, removed from the coffin and severely battered by whipping with tree branches which still littered the cemetery and adjoining street' (15). A parallel incident is recorded from N Nigeria. 'It is related to one chief that he used to kill not only everyone who displeased him, but that he would even cut open living women with child so that he could see the stages of development. On his death a grave was dug, and he was put in it, but the earth threw him out again. A second time he was put in, but once more he was ejected, and a hut had to be built for the corpse' (47).

These incidences are traditionally interpreted not only as instant disapproval of burying the corpse in the first place, or at that particular cemetery, or even of denying these corpses a final resting place, or an exposure of their true nature; but rather a lesson to the living that as mortals, it does not pay to resort to any evil methods to shorten the lives of others - for eventually we shall be held accountable for our deeds.

Newbouldia laevis (symbol of authority and deities; purification after death and reconciliation; forbidden to be used as fuel-wood or cut)

Purification after Burial.

Purification of the body after a burial ceremony forms part of the funeral rites among many ethnic groups in traditional Africa. It serves as an additional ritual to separate the dead from the living. The leaf of *Newbouldia laevis*, a medium-sized tree believed to be fetish, is used in the purification of the body after a burial ceremony by many ethnic groups throughout its distribution area (2). The leafy shoots of *Spondias mombin* Hog plum or Ashanti plum are similarly used (1). *Newbouldia* also enters into rites to cleanse the body after digging a grave. Purification of the body after burial may also be effected by washing the hands and face in water or salted water or sea-water. Among the Akan-speaking people of Ghana and Ivory Coast, it has been observed that: 'after interment, they proceed in procession to the nearest well or brook, and sprinkle themselves with water, which is the ordinary native mode of purification' (14b). Alternatively, water in which *Momordica balsamina* Balsam apple or *M. charantia* African cucumber or *Ocimum canum* American basil or *Portulaca oleracea* Purslane or Pigweed has been immersed may be used for the purification.

In Zambia, a piece of the bark of *Strychnos innocua* Kaffir orange with the bark from other trees is used by mourners to dab the head and upper parts of their body after a funeral; and Kaffir orange enters into the Tonga death rites (53). In Mozambique, the Tonga employ *Acridocarpus natalitius*, a shrub, in purification rites after a death (53). A decoction of the bark of *Sclerocarya caffra* Cider tree is used by some Africans for steaming, and is taken internally to remove defilement arising from eating food in the house of relatives where there has been a death without the performance of the necessary purification rites (53). In S Africa, the Southern Sotho give a decoction of the root of *Malva parviflora* Mallow to a person who has lost a near relative - he or she must not mix with others before taking the medicine, as otherwise terrible sores will break out on the body (53). In Botswana, *Melolobium alpinum*, a legume, is used by the 'witch-doctor' as a sedative for comforting a person who has passed through a great sorrow or who is depressed and sad, thus it is administered to a person after the death of a near relative (53).

THE AFTER DEATH

As part of the culture and beliefs in the Region and elsewhere, the living communicate with the dead and the spirit world, and offer sacrifices to these ancestors on occasions. The practice is usually preceded with prayers, rituals and libation.

Communicating with the Dead

In Liberia, the Basa name of *Eugenia whytei,* a shrub or small tree by the sea-shore or in forest, 'bloe', means 'fearful' or 'trembling heart', the tree being credited with magical powers for the initiated, and used to communicate with spirits of the dead (13). In its distribution area, the leaves of *Newbouldia laevis,* a popular fetish tree, are used in an invocation ritual on the grave by some fetish priests to communicate with the dead (2). 'Amid libation, the name of the deceased is called in an incantation as the grave is stumped three times with the foot, and then hit with the leaves until a groan is heard from within - then you speak' (2). In Ghana and Ivory Coast, *Dracaena arborea,* another popular fetish tree commonly planted around shrines, is similarly used among the Akan-speaking (16).

Some plants are said to enter into rituals to resurrect the dead. For instance, in S Nigeria, *Dioscorea rotundata* White yam or Guinea yam is used by the Yoruba in an invocation to arouse dead people (49). In Congo (Brazzaville), *Aspilia kotschyi,* a weedy composite, is said to be used superstitiously in nasal instillation to revive the dead (7a); and among some tribesmen in the Region, the leaf-juice of *Heliotropium indicum* Indian heliotrope or Cock's comb is instilled in the nostrils and mouth in a ritual to wake up the dead (35) - as a prelude to communicating with these spirits. There is a strong suspicion that some Secret Societies do practice these rituals. The instances above sound like the magico-religious practice in Haiti where the dead are allegedly raised to life as Zombies (the living dead) - in a mid-night voodoo ceremony which has its ancestral roots traced to Dahomey (now Benin Republic) - to serve as slaves or cheap labour on plantations. *Sauromatum venosum (guttatum)* Voodoo lily

is an aroid which grows in the Democratic Republic of the Congo (Zaire), Congo (Brazzaville) and Central Africa - probably so named because of its association with the cult. (For other plants associated with the cult or used to neutralize its effects, see WITCHCRAFT: NATURE AND CONCEPT).

Sacrifice and Feeding of the Dead

It is the custom before meals, to place a bit of the food on the ground for the ancestral spirits. This tradition is practiced by many tribes in the Region. In Ghana, no food made especially for the dead ever contains salt and pepper among the Ga tribe (20); and the Akan-speaking people of Ghana and Ivory Coast and many of the coastal tribes in the Region regularly offer boiled eggs to departed ones on occasions. In S Nigeria, the Ijo of the Niger Delta use the corm of *Anchomanes difformis* and the berries of *Lasimorpha senegalensis (Cyrtosperma senegalense)* Swamp arum, both aroids, in making sacrifice to the dead (8a). An account is given of another method in Old Calabar, S Nigeria, about a century ago. 'A curious local custom is that called 'Feeding the Dead'. When they bury their dead, the relations, before the earth is filled into the grave, place a tube, formed of bamboo, or pithy wood with the pith extracted, and sufficiently long to protrude from the earth heaped up over the body, into the mouth of the deceased. Down this they pour, from time to time, palm-wine, water, palm-oil, and so on. They appear to imagine that the dead men do not require solid food at all, and as they pour the liquids down two or three times a month, are not very thirsty souls' (14a).

Another feeding custom among the Ashantis in Ghana is recorded (42) in which ghost husbands or skeleton spouses or royal skeletons were fed about 11 am. and were served with palm-wine about 4 pm. at the royal mausoleum at Bantama by the living wives. The writer even records instances when food is served to the body while lying in state. 'The food generally consists of a fowl, *Gallus domestica*; eggs and mashed yams, and water, which are placed beside the body, which has been laid on its side purposely to leave the right arm and hand free for eating' (42). Among the Twi-speaking people, some food is even buried with the corpse. 'In the grave with the corpse are placed food and drink, tobacco, pipes, gold-dust, trinket and cloths, according to the wealth and position of the deceased. The first two

are for use during the journey to *samanadzi* (realm of the dead), and the remainder on arrival there' (14b).

The tube (mentioned earlier) in reality, served a double function. 'They believed that after death the deceased suffers from the same ailments as he did in life, and sometimes very filial natives will go to the doctor, and simulate the complaints from which the paternal or maternal ancestor suffered, in order that they may obtain the requisite medicine to pour down the grave' (14b). Undoubtedly, this belief is not unique to the people of Old Calabar alone. Commenting on rum and money and gold and soap buried with the dead among the Ga-speaking tribe of Ghana, it has been observed, 'this money is not only for the ferry-fare into the land of the dead, but to help the dead man to pay to be cured of the sickness of which he died' (20). The practice is also recorded among the Peruvian Indians. 'The burial of the dead with food and drink, jewels and textiles, and even arms, continued up to Colombian contact among the common people, who kept their dead relatives happy by renewing their food and clothes and drink' (48).

REMEMBRANCE

Remembrance of the dead occurs on occasions like funerals, festivals, family gatherings, child-naming or out-dooring ceremonies and anniversaries. At such functions, prayers are said and libation poured to the ancestral spirits as gratitude to the dead for the continued welfare of the living, and for the success during the deliberations and ever thereafter.

Wreaths

It has become the tradition to lay wreaths on the grave side at burial ceremonies; and sometimes on subsequent anniversaries of the death. In Ghana, the Christian Council once ruled against the practice of

Species	Common Name	Plant Family	Part Used
Casuarina equisetifolia	Whistling pine	Casuarinaceae	Leaves
Codiaeum variegatum	Garden croton	Euphorbiaceae	Leaves
Cupressus species	Cypress	Cupressaceae	Leaves
Cycas circinalis	Cycas	Cycadaceae	Leaves

C. revoluta	Sago palm	Cycadaceae	Leaves
Encephalartos barteri	Ghost palm	Zamiaceae	Fronds
Gomphrena globosa	Bachelor's button	Amaranthaceae	Flowers
Panax species	Panax	Araliaceae	Leaves
Pedilanthus tithymaloides	Red-bird cactus Slipper-flower Jew-bush	Euphorbiaceae	Leafy shoots
Polyscias species	Polyscias	Araliaceae	Leaves
Sabal mexicana	Sabal	Palmaceae	Flowers
Thuja species	Arbor-vitae	Cupressaceae	Leaves

Table 1. Plants Used for Long-term Wreaths

laying wreaths at burial, mainly because of the cost involved and the rampant pilfering of same immediately afterwards. Nevertheless, the practice still persists. Tradition dies hard. Wreaths may be made either with artificial flowers or with natural ones. Plants popularly

Species	Common Name	Plant Family	Part Used
Bougainvillaea glabra	Bougainvillaea	Nyctaginaceae	Flowers
B. spectabilis	Bougainvillaea	Nyctaginaceae	Flowers
Combretum grandiflorum	-	Combretaceae	Flowers
Delonix regia	Flamboyante, Flame tree	Caesalpiniaceae	Flowers
Ixora duffii	Ixora	Rubiaceae	Flowers
Petrea volubilis	Queen's wreath	Verbenaceae	Inflorescence
Plumeria rubra var. acutifolia	Frangipani, Temple flower 'Forget-me-not'	Apocynaceae	Flowers
Portulaca grandiflora	Rose moss	Portulacaceae	Flowers
Sun plant of Brazil	Polyscias	Araliaceae	Leaves
Rosa species	Rose	Rosaceae	Flowers
Setcreasea purpurea	Purple heath	Commelinaceae	Leafy shoots
Tagetes species	Marigold	Compositae	Flowers

Table 2. Plants Used for Short-term Wreaths

Combretum grandiflorum

Petraea volubilis Queen's wreath

Setcreasea purpurea Purple heath

used for making Long-term and Short-term wreaths are listed in Tables 1 and 2 respectively, above:

The fronds of *Elaeis guineensis* Oil palm serve as wreath frames, bordered by leaves of *Syndapsus aureus*, a lofty aroid, *Pandanus* species Screw pine and *Ficus elastica* India-rubber fig.

Tagetes species Marigold, a native to the New World, has been used in religious ceremonies for the dead in Mexico and Guatemala dating back to pre-Colombian times (45). Many centuries ago, the Ancient Egyptians used flowers of *Narcissus tazetta*, a lily, for funeral wreaths, and even after 3,000 years the preserved remains are recognizable (41). In Ancient Egypt *Nymphaea lotus* Water Lily was used for the funeral wreaths of priests and Pharaohs (41).

Rosmarinus officinalis Rosemary (legendary and mythical plant; symbol of fidelity and remembrance)

In Ancient Greece, *Carum petroselinum* Parsley was a ceremonial plant dedicated to the dead, and tombs were festooned with wreaths of Parsley (5). In Europe, the flowers of *Buxus sempervirens* Box was once used for funeral sprays; and *Rosmarinus officinalis* Rosemary, a labiate, were worn in olden times at funerals. Rosemary was once regarded as a symbol of fidelity and remembrance (41). 'Rosemary was believed to stimulate the brain and help the memory and so it was associated with remembrance. The old custom of leaving Rosemary at the grave-side and for handing a bunch to those bereaved is carried on to this day' (5). (For remembrance of the war-dead, see WAR: THE WAR DEAD).

Ruellia simplex

MAN AS A FLOWER

In reference to time and mortality, the glory of bright blooming flowers, shortly fading, wilting and dying has been compared with man. In the morning it flourisheth, and groweth up; in the evening it is cut down, and withereth (Psalm 90:6). For instance, the flowers of *Ruellia simplex*, a perennial ornamental Acanth from S America introduced to Ghana, open at dawn, bloom during the morning, then drop and wither by late afternoon. In Sierra Leone, *Neomarica gracilis*, an erect or reclining exotic herb, is now being cultivated as an ornamental. The mature plant is floriferous, flowering continuously with 1-day flowers opening at dawn and fading before dark (8b). In Central America, the beautiful flowers of *Cereus grandiflorus*, a cactus, have very short lives. In the same night they grow, they produce their aroma and die (40). Similarly, *Convolvulus tricolor*, an attractive annual from Spain and Portugal, was known in the 18th century as the Life of Man, because it has flower buds in the morning, which will be full bloom by noon and withered up before night (41). (See also GHOST AND SPIRIT WORLD; GRAVES AND CEMETERIES; SACRED PLANTS; and WAR).

References.

I. Agbordzi, (pers. comm.); 2. Akpabla, (pers. comm.); 3. Antobam, 1963; 4. Aubréville, 1959; 5. Back, 1987; 6. Blumenthal, 2000; 7a. Bouquet, 1969; b. idem, 1972; 8a. Burkill, 1985; b. idem, 1994; c. idem, 1995; d. idem, 1997; e. idem; 2000; 9. Cantor, 1990; 10. Chaney & Basbous, 1978; 11. Crow, 1969; 12. Cudjoe, (pers. comm.); 13. Dalziel, 1937; 14a. Ellis, 1883; b. idem, 1887; c. idem, 1894; 15. Entsuah, (pers. comm.); 16. Eshun, (pers. comm.); 17. Everald & Morley, 1970; 18. Falconer, 1990; 19. Ferry & al., 1974; 20. Field, 1937; 21. Fink, 1989; 22. Gamadi, (pers. comm.); 23. Gomes e Sousa, 1930; 24. Graf, 1978; 25. Green, 1951; 26a. Irvine, 1930; b. idem, 1961; 27a. Kerharo & Adam, 1962; b. idem, 1974; 28. Kerharo & Bouquet, 1950; 29. Kingsley, 1897; 30. Lamptey, (pers. comm.); 31. Livingstone, 1914; 32. Martin, 1995; 33. Meek, 1931; 34. Mensah, (pers. comm.); 35. Nkwantabisa, (pers. comm.); 36. Okafor, 1979; 37. Okigbo, 1980; 38. Oso, 1977; 39. Ouedraogo, 1950; 40. Pamplona-Roger, 2001; 41. Perry & Greenwood, 1973; 42. Rattray, 1927; 43. Richards & Kaneko, 1988; 44. Savill & Fox, 1967; 45. Siegel, Collings & Diaz, 1977; 46. Teiko Tackie, (pers. comm.); 47. Tremearne, 1913; 48. Van Sertima, 1976; 49. Verger, 1967; 50. Vergiat, 1969; 51. Walker & Sillans, 1961; 52. Warren, 1974; 53. Watt & Breyer-Brandwijk, 1962.

DEVIL, THE

And he laid hold on the dragon, that old serpent, which is the Devil, and Satan, and bound him a thousand years.
Revelations 20:2

The Devil is a legendary figure in the folklore of many countries throughout the world, who is believed to be responsible for the misfortunes of man. It is synonymous with the mythical *sasabonsam* in Ghanaian folklore, or the *ombuiri* of the M'pongwe of Gabon. Recounting an encounter with this figure, it has been observed: 'what we object to in this spirit is that one side of him is rotting and putrefying, the other sound and healthy, and it all depends on which side of him you track whether you see dawn again or no' (13). Another description reads: 'the *sasabonsam* of the Gold Coast and Ashanti is a monster which is said to inhabit parts of the dense virgin forests. It is covered with long hair, has large blood-shot eyes, long legs, and feet pointing both ways' (17).

Anything sinister is linked to the Devil. Traditionally, it is believed to be a killer and the force or power behind evil activities, bad juju, sorcery, witchcraft, sinister curses and spells - a no mean enemy of mankind. *Sasabonsam* are evil by nature and always associated with *abayifo* (witches), bringing misfortune to the community (8).

PLANTS NAMED AFTER THE DEVIL

Some plants are named after the Devil either because of the origin, colour, structure, destructive activities, or some other superstitious reasons. For instance, in S Nigeria, the Efik call *Desmodium adscendens*, a straggling forest perennial, 'groundnut of the demon'; and the Igbo name of *Hoslundia opposita*, 'uto mmo', meaning 'devil plant', is perhaps also significant (4b). The English name of *Afrotrilepis (Catagyna) pilosa* Devil grass implies magical attributes (4a); and in the West Indies and elsewhere *Cynodon dactylon* Bermuda grass becomes a pest in plantations and is called Devil's grass (5) - (but see SACRED PLANTS: NATIONAL

AND STATE PLANTS). *Argemone mexicana* Mexican or Prickly poppy is also known as Devil's fig; and *Succisa pratensis* Devil's bit scabious, an European plant, earned its name because the Devil was envious of its virtues and bit the black-root so far to have destroyed it (15, 16).

Argemone mexicana Devil's fig

The large oval orange-yellow pulpy fruits of *Mandragora officinarum* Mandrake are sometimes known as Devil's apple, probably because it was long believed that Mandrake dwelt in dark places of the earth and thrived under the shadow of the gallows, being nourished by the flesh of criminals on the gibbet (16). The name of *Stachytarpheta angustifolia, S. cayennensis and S. indica* Devil's coach whip suggests superstitious application; and *Datura stramonium* Devil's apple or Devil's trumpet stramonium connotes magical use. (*Datura* species in general do have hallucinogenic properties). For the same reasons *Amorphophallus* and *Sansevieria* species are called Devil's tongue; and *Epipremnum aureum*, a decorative variegated yellow and green introduction from the Solomon Islands, is called Devil's ivy. In Indo-Malaya, *Alstonia scholaris* is called 'Devil tree'.

Among the many names of *Tradescantia virginiana*, a trailing decorative with grassy leaves and showy blue, purple, red or white flowers is Devil in the pulpit or Widow's tears (16). *Plumbago scandens*, from West Indies is Devil's herb - probably because juices from the roots and leaves irritate the skin and cause blisters, a circumstance exploited by beggars who create sores to excise pity (16). In California, USA, the arid desert slopes of 11,485ft Mt. Gorgonio which is planted with a multitude of cacti, being primarily wickedly armed *Opuntia* species Cholla and *Ferocactus acanthodes*, cylindrical and as tall as a man in maturity, is known as Devil's garden (9).

In W Africa, the common name of *Crotalaria retusa* Devil bean, a weedy legume on cultivated land, associates the plant or the fruit with the Devil; and in Senegal, *Spathodea campanulata* African tulip tree, a very decorative forest tree, often cultivated, is called 'baton du sorcier' (4). In Ghana and Ivory Coast, the Akan name of *Abrus precatorius* Prayer

beads, 'anyen-enyiwa', means Devil's eyes - in reference to the red and black colour of the seeds. The local name could equally mean 'witch's eye'. In these countries, the Akans call the red variety of *Zea mays* Maize, 'abonsam aburow', meaning Devil's corn; and in Sierra Leone, *Oryza barthii* Wild rice is called 'pa-kin-kin', in Sherbro meaning Devil's rice. The decorative epiphytic fern *Platycerium elephantotis* and *P. stemaria* Stag-horn fern are called 'sasabonsam-kyew' by the Akans, meaning Devil's hat. The leaves of the fern tied round the head is supposed to give supernatural powers (10) - (see CHILDBIRTH: PARTURITION).

Plants named after the Devil are listed in Table 1 below:

Species	Plant Family	Common Name	Habit
Abroma angusta	Sterculiaceae	Devil's cotton	Tree
Achyranthes aspera	Amaranthaceae	Devil's horsewhip	Herb
Afrotrilepis pilosa	Gramineae	Devil grass	Herb
Alstonia scholaris	Apocynaceae	Devil tree	Tree
Amorphophallus species	Araceae	Devil's tongue	Herb
Argemone mexicana	Papaveraceae	Devil's fig	Herb
Cassytha ciliolata	Lauraceae	Devil's tresses	Parasitic Herb
Crotalaria retusa	Papilionaceae	Devil bean	Herb
Cynodon dactylon	Gramineae	Devil grass	Herb
Datura metel	Solanaceae	Devil's trumpet	Herb
D. stramonium	Solanaceae	Devil's trumpet straminium	Herb
Dicerocaryum zanguebaricum	Pedaliaceae	Devil's thorn	Herb
Distemonanthus benthamianus	Caesalpiniaceae	Devil's tree	Tree
Emex australis	Polygonaceae	Devil's thorn	Shrub
Epipremnum aureum	Araceae	Devil's ivy	Herb
Harpagophytum procumbens	Pedaliaceae	Devil's thorn	Herb
Mandragora officinarum	Solanaceae	Devil's apple	Herb
Platycerium elephantotis	Polypodiaceae	Devil's hat	Herb
P. stemaria	Polypodiaceae	Devil's hat	Herb
Plumbago scandens	Plumbaginaceae	Devil's herb	Herb
Sansevieria liberica	Agavaceae	Devil's tongue	Herb
S. senegambica	Agavaceae	Devil's tongue	Herb
S. trifasciata	Agavaceae	Devil's tongue	Herb
Smeathmannia laevigata	Passifloraceae	Devil bush	Shrub
Solanum aculeastrum	Solanaceae	Devil's apple	Shrub
S. hispidum	Solanaceae	Devil's fig	Shrub
Starchytarpheta angustifolia`	Labiatae	Devil's coach whip	Herb
S. cayennensis	Labiatae	Devil's coach whip	Herb
S. indica	Labiatae	Devil's coach whip	Herb

Succisa pratensis	Dipsacaceae	Devil's bit scabious	Shrub
Tacca species	Taccaceae	Devil's plant	Herb
Tradescantia virginiana	Commelinaceae	Devil in the pulpit	Herb
Tribulus terrestris	Zygophyllaceae	Devil's thorn	Herb

Table 1. Plants Named after the Devil.

Crotalaria retusa Devil bean

Tribulus terrestris Devil's thorn

Distemonanthus benthamianus African satinwood (legendary and mythical plant; symbol of evil; forbidden to be used as fuel-wood or cut)

PLANTS ASSOCIATED WITH THE DEVIL.

Other plants are associated with the Devil for one reason or the other. In Tanzania, witch-doctors use the leaves of *Oldenlandia herbacea,* an annual, for 'sheitani' (devil-worship) (14). The monstrous appearance of *Adansonia digitata* Baobab, with its swollen, bottle-shaped trunk and short dumpy branches sticking up in the air like thick roots, gives a picturesque view of the belief in Kenya that the Devil planted this tree upside down (16). Another huge tree associated in African mythology with the Devil is *Ceiba pentandra* Silk cotton tree. Describing the link between the tree and this mythical figure, it has been observed: 'He lives in the forest, in or under those great silk-cotton trees around the roots of which the earth is red. The coloured earth, identifies a silk-cotton tree as being the residence of a *sasabonsam,* as its colour is held to arise from the blood it whips off him as he goes to the underworld home after a night's carnage. All silk-cotton trees are suspected because they are held to be roots of Duppies. But the red earth ones are feared with great fear, and no one makes a path by them, or camps near them at night' (13a, b).

Ceiba pentandra Silk cotton tree (symbol of maleness; legendary tree; associated with traditional religion; forbidden to be cut; National Plant of Nicaragua).

The association of *sasabonsam* with the Silk cotton tree is confirmed. 'It sits on high branches of *Milicia (Chlorophora) excelsa* or *M. regia* Iroko, 'odum', or *Ceiba pentandra* Silk cotton tree, 'onyina', and dangles its legs, with which at times it hooks up the unwary hunter' (17). In folklore, this *sasabonsam* is said to be married to *samantan,* a female deity - also believed to be associated with *Ceiba*. '*Samantan* always lives among the huge silk-cotton trees, which rear their enormous trunks far above the surrounding trees of the forest' (6). The author explains why the two deities are associated with *Ceiba*. 'The belief in *samantan* and *sasabonsam*

who reside among the Bombacaceae, is due to the frequent loss of life caused by falling off such trees. The Silk cotton trees seem to be especially attacked by white ants, *Macrotermes* species, and towering far above the other trees of the forest, and offering a prominent object, are frequently struck by lightning. In either case the tree is killed, but the tall gaunt trunk remains standing until it falls through a gradual process of decay, or, half-decayed, is prostrated by a tornado' (6).

Among the curiosities and mysteries of the plant kingdom is the problematic *Welwitschia mirabilis* Welwitschia, a plant that is neither Gymnosperm nor Angiosperm, but betwixt and between (18). The specific name is the Latin for 'wonderful'. The plants are so different from any other plant that scientists classify them as a unique family and genus made up of only one species. Welwitschia grows in the Namib Desert near the Walvis Bay in Namibia. It was first collected in 1860 by the Australian botanist, Friedrich Welwitsch who admits he felt afraid to touch it at first in case it vanished. It is one of the most weird plants ever discovered (1); and was described by Joseph Hooker as 'the most wonderful plant ever brought to this country (England) and the very ugliest' (18).

When *Coffea arabica* Coffee plant of commerce was introduced from the Abyssinian mountains (now Ethiopia) by traders and explorers, it was superstitiously believed to be satanic - probably because of its origin; so not until the Pope had tasted it and given the papal approval that the beverage was accepted for cultivation as a stimulating drink. Other cultivated species include *C. canephora* Rio Nunez coffee, *C. liberica* Liberian or Monrovian coffee and *C. stenophylla* Sierra Leone (Upland) or Narrow-leaved coffee. Similarly, *Solanum tuberosum* Potato, a crop whose output exceeds all others in both volume and value, did not achieve this supremacy without first overcoming considerable public opposition. Acceptance of the root crop was held back because other known members of the potato family , Solanaceae, were associated in the public mind with witchcraft and murder (7). For example, *Mandragora officinarum* Mandrake was supposed to have a man-shaped root with mysterious sexual powers and shriek as it was dragged out of the earth (7). *Hyoscyamus niger* Henbane and *Atropa bella-donna* Deadly nightshade were known to contain poisonous drugs; and at the same time there was a lingering distrust of *Nicotiana tabacum* Tobacco, introduced from the Americas with the potato (7).

PLANTS AGAINST THE DEVIL

However, some other plants are used in the belief that they possess protective powers against the forces of the Devil. For instance, *Ocimum gratissimum* Fever plant is believed to have magical attributes throughout the Region, and a fetish for exorcizing spirits and demons (3). In Ivory Coast, *Ocimum basilicum* Sweet basil is believed to exorcize Devils (3, 12). In this country, the frothing of the root and the stem of *Carpolobia lutea*, a tree, in water leads to medico-magical uses of the plant, so the froth is used to wash persons who have drunk human blood or eaten human flesh, and to exorcize demons and evil spirits (3). In Sierra Leone, the strong aromatic smell of *Hyptis spicigera* Black sesame, a labiate, evokes belief in protective powers, so the leaves are rubbed on the body to keep the Devil away (4b). In The Gambia, the seed of *Abrus precatorius* Prayer beads are buried on the site of a proposed new building; should they disappear overnight, a new site must be found since their removal is indication of a Devil occupying that position. The beans are deemed to confer protection against spirits and Devils who fly through the air like arrows to pierce the skin and cause smelly sores (4b).

In Congo (Brazzaville), *Thomandersia hensii (laurifolia)*, a shrubby forest acanth or small tree, is planted by a hut entrance to keep away demons (2). In this country, a piece of *Cissus rubiginosa*, a trailing savanna herb, is held to have power to repel demons, and is hung over a door to confer protection in a house (2). A few drops of sap from the crushed leaves of *Acanthus montanus* False thistle, a prickly semi-woody herb, on the eyebrows are believed to protect one from Devils, and with other plants are used for exorcizing (2) (see EVIL: SPIRITUAL PROTECTION). In Gabon, *Adenia rumicifolia* var. *miegei (lobata)*, a huge forest climber, is ascribed with magical power of exorcism of the Devil from persons possessed (4b); and in Senegal, the Diola in Fogny consider *Acridocarpus plagiopterus*, a scandent woody savanna shrub, a fetish to exorcize Devils (11). In Tanzania, the fruit of *Vepris lanceolata* White ironwood is used 'to throw out the Devil' by casting it on to a fire and making the 'possessed' person inhale the smoke (20).

In Europe, there was a belief that every *Punica granatum* Pomegranate fruit contains one seed which has come from Paradise - thus it was once thought that a whipping with Pomegranate branches could drive out the Devil (14). In Mexico, it is the belief among the Huichol Indians that a cross of *Tagetes lucida* Marigold, a decorative composite, is protection

against the Devil, and prevents him from entering one's house (19). On the contrary, in Tanzania, the bark of *Stereospermum kunthianum*, a slender savanna tree, is used for Devil worship (4a) (but see, TABOO: FUEL-WOOD; and FELLING TREES). (See also CARVINGS AND IMAGES; and GHOST AND SPIRIT WORLD).

References.

1. Benson, 1959; 2. Bouquet, 1969; 3. Bouquet & Debray, 1974; 4a. Burkill, 1985; b. idem, 1995; 5. Dalziel, 1937; 6. Ellis, 1887; 7. Everard & Morley, 1970; 8. Fink, 1989; 9. Graf, 1978; 10. Irvine, Herb. GC; 11. Kerharo & Adam, 1962; 12. Kerharo & Bouquet, 1950; 13a. Kingsley, 1897; b. idem, 1901; 14. Kokwaro, 1976; 15. Pamplona-Roger, 2001; 16. Perry & Greenwood, 1973; 17. Rattray, 1927; 18. Rowley, 1972; 19. Siegel, Collings & Diaz, 1977; 20. Watt & Breyer-Brandwijk, 1962.

12 DISEASE

(He) Himself took our infirmities, and bare our sickness.
Matthew 8:17b

It is the belief in African tradition that the causes of diseases (including spread of germs and viruses) are either physical or spiritual. Spiritual causes are usually effected through the occult or a curse or a spell; and is either a result of some wrong doing or sacrilege on the victims part, or just ill-will. Every plant, every mountain, river or lake has a spirit that can bring about all kinds of illnesses either for reasons beyond man's control or through want of respect for it (13). 'If a disease is diagnosed in African medicine as having a 'natural' cause then the treatment falls within the area of responsibility of a herbalist. However, should a 'supernatural' origin be suspected, a priest-healer is to be sought. Included under 'supernatural' causes of disease are: the influence of the gods, dead ancestors, spirits and witchcraft' (20). Where evil forces, juju, sorcery, witchcraft or other human or spiritual malice are believed to be the cause of affliction of the ailment, the traditional methods used in healing or in prevention or in protection are equally based on superstitious faith. In S Nigeria, *Phyllanthus niruri* var. *amarus (amarus),* an euphorb herb, features in an incantation against disease (47).

AMENORRHOEA

(The absence of menstruation). In Ghana, *Olyra latifolia,* a forest grass, is used as a male abortifacient in Kwabenya, a village on the Accra Plains. If a man swallows three fruits and then has sex with a girl who has missed her period, it will come (21); and in Senegal, the root of *Cochlospermum tinctorium,* a bushy savanna plant with golden yellow flowers, is used with other drug plants for amenorrhoea (25). The leaves of *Mucuna pruriens* var. *pruriens* Cow itch, a climbing legume, contain a red sap when fresh. In the Ivory Coast, this is used on the Doctrine of Signatures as an emmenagogue (promoting menstruation) (8a). (See JAUNDICE below)

ASTHMA

(Chronic chest illness causing wheezing and difficulty in breathing). In Liberia, *Ruthalicia (Physedra) eglandulosa,* a climbing cucurbit, has medico-magical application by the Mano people to treat shortness of breath - the leaves are rubbed on the chest and then on to a stick, whereby the malady is 'transferred' to the stick (Harley fide 12). In Ivory Coast, a decoction made of the leaves of *Parinari curatellifolia,* a savanna tree, from the edge of a path, that is, those brushed against by passers-by, is reported as effective against asthma if one washes with it at night while the world sleeps (1).

BOILS

(Infected swelling under the skin, producing pus). In Liberia, *Microdesmis puberula* a forest shrub to small tree, is regarded as preventive of boils - for each whole fruit swallowed one acquires a year's immunity (12). In Ghana and Ivory Coast, the ground seeds of *Omphalocarpum ahia, O. elatum* and *O. procerum,* forest trees with cauliferous fruits on the stem like boils, and appropriately locally called 'duapompo' in Akan, meaning 'tree with boils', is applied to boils to bring to a head. This is an instance of sympathetic magic. The Bulsa of NW Ghana apply a red-brown poultice prepared from the bark of *Khaya senegalensis* Dry-zone mahogany, that of *Acacia* species, and the bark taken from the living side of a tree with one side dead for the treatment of boils and any swollen parts of the body (30).

In N Nigeria, the Hausa and the Fula will not use *Amblygonocarpus andongensis,* a tree in savanna woodland, for firewood in the belief that whoever does so will develop incurable boils (8c). In S Sudan, a species of *Euphorbia* which some men plant in their homesteads is among the medicines given offerings by the Azande, as the milky sap which oozes from the stem, furnished them with an arrow poison. 'The owner of one of these plants occasionally places a crab, *Cardiosoma* species, on it as an offering, because it is believed that unless it receives offerings from time to time it will cause boils and ulcers to form in members of his household' (16).

BURNS

(Destroy, injure or mark skin by fire, heat or acid). In Senegal, the Niominka use a decoction of the whole plant of *Cassytha filiformis* Dodder, a parasitic plant, for third degree burns. The treatment must be accompanied by incantations (26b, c).

COLLAR CRACK (OF COCOA).

(A fungal disease of the crop). There is a native belief that an outbreak of this disease will occur if *Ricinodendron heudelotii* African wood-oil-nut tree is felled on a farm of *Theobroma cacao* Cocoa (Bunting fide12).

COUGH

(Illness, infection that causes a person to send out air from the lungs violently and noisily). In S Nigeria, *Cleistopholis patens* Salt and oil tree, a soft-wooded forest under-storey, features in a Yoruba Odu incantation for curing cough, hence the Yoruba name 'apako', meaning 'killing cough' (47). In this country, birds appear to be able to eat the fruits of *Spondianthus preussi*, a poisonous forest tree in swampy areas, with impunity. This fact may have some significance in a curious Yoruba Odu incantation involving erotic rats, *Cricetomys gambianus;* birds and this plant for the cure of cough (47). In Zimbabwe, *Becium obovatum* var. *hians*, a labiate, is used by the male Manyika for a cough which results from intercourse with a woman other than his wife or with a menstruating woman (49). (For cure of convulsive coughing, see MEDICINE: SYMPATHETIC MAGIC).

CRAWCRAW

(A form of filariasis, occurring especially in W Africa and characterized by the formation of papules or pustules). In S Nigeria, an unspecified part of *Strophanthus gratus*, an apocynaceous liane with milky latex, is used with incantations by the Yoruba to cure crawcraw (47). The tribe also use *Aspilia africana* Haemorrhage plant, a weedy composite, in a similar manner. The Yoruba name of the weed 'yun yun', means scratch.

DIARRHOEA & DYSENTERY

(Condition that causes waste matter to be emptied from the bowels frequently, and in a watery form). Fula medicine-men follow a complicated prescription in which the bark of *Annona senegalensis* var. *senegalensis* Wild custard apple is taken only from the left fork of a V-shaped branch on the east side of a tree, that is, the side facing the rising sun (27). The practice probably suggests sun-worship. In a medico-magical application also suggestive of sun-worship, the Tandanké of Senegal take the bark of *Anogeissus leiocarpus*, a graceful savanna tree, in the morning from the east and the west sides of the trunk for infantile diarrhoea (26b, c). In Ivory Coast and Burkina Faso, the application of *Momordica balsamina* Balsam apple or *M. charantia* African cucumber as both purgative and anti-dysenteric, while seemingly in opposition, is based on the precept that dysentery is a sign of 'beasts in the abdomen' and the best treatment is to charge them out with a good purge (27). In S Africa, *Eriocephalus punctulatus*, a composite, is used by the Southern Sotho with *Metalasia muricata* Blombos, another composite, to fumigate the hut of a person suffering from cold or diarrhoea during the illness (49). *Piper nigrum* Black pepper has been optimistically used as preventive against cholera and dysentery. (See also WITCHCRAFT: SORCERY PLANTS: Plants Against The Craft).

EARACHE

(Pain in the ear-drum). In Senegal, a herbalist is reported to cure the disease by prescribing the patient to wear a collar of *Sansevieria senegambica* African bow-string hemp (26d). In Nigeria, *Euphorbia kamerunica*, a cactiform tree, may be parasitized, perhaps by a loranth, which is used by the Ebina of Adamawa to cure deafness, probably ascribing magical attributes (8b).

ELEPHANTIASIS (LYMPHATIC FILARIASIS)

(A tropical disease in which the limbs or the scrotum become abnormally enlarged and the skin thickens). The traditional methods of treating elephantiasis appear to be based on sympathetic magic. For instance, healers in the Region prescribe a calabash-full of palm-wine - stored in the hollowed-out fruits of *Picralima nitida*, a small forest tree, and allowed to ferment in the sun for a few hours - to be taken as a cure for elephantiasis of the scrotum. This is an instance of sympathetic magic

Picralima nitida 1/5 natural size

since the fruit has resemblance to the ailment. In Ivory Coast, a decoction of the whole fruit of *Allanblackia floribunda* Tallow tree, an evergreen tree of the high forest, is used to relieve elephantiasis of the scrotum. This may simply be based on the Doctrine of Signatures, because of the size and shape of the fruit (7).

In S Sudan, the fruit of the *danga* resembles the human scrotum and it is burnt so that the ashes may serve as remedy for scrotal hernia and elephantiasis (16). This is yet another instance of sympathetic magic. Similarly, a cure for elephantiasis of the scrotum, by the Azande of this country is an application to the scrotum of the pulp of the fruit of *Kigelia africana* Sausage tree (16).

Elephantiasis of the limb

In Mali, *Mucuna sloanei* Horse-eye bean or True sea-bean is sold in the market as a charm; and one native belief is that the person wearing the seed will not beget a child to suffer elephantiasis (6, 12). The implication or connection of the seeds to the ailment is uncertain.

EPILEPSY

(Disease of the nervous system that causes a person to fall unconscious; often with violent uncontrolled movement of the body). In Senegal, *Caralluma dalzielii* and *C. decaisneana,* succulent asclepiads in savanna, are used in a medico-magical treatment of epilepsy (26c). In Ivory Coast and Burkina Faso, a preparation of leaves of *Melanthera scandens,* a weedy composite, (and probably also *Aspilia africana*

Haemorhage plant) is rubbed on the patient's head for epilepsy (27). This is probably a superstitious use. In Ivory Coast, *Vicoa leptoclada*, an erect herbaceous annual composite, is held to have magical properties, and is thus used in a prescription for treating hysteria and epilepsy (27). In this country and Ghana, the Brong give a decoction of *Secamone afzelii*, a scrambling shrub with milky latex, by draught or in baths to children in the treatment of 'epilepsy' caused by the passage overhead of certain birds such as hornbill, species of *Tochus* or *Ceratogymna*; or crowned crane, *Balearica pavonina* (27).

Early N American settlers used the flower infusion of *Caltha palustris* Kingcups in the belief that it cured fits in children; and bead necklaces made from the dried root of *Paeonia officinalis* Common paeony, an introduction from southern Europe, strung on leather were worn by infants to prevent convulsion (39). In Britain, the flower of *Cardamine pratensis* Cuckoo flower, a wild plant of the Northern Hemisphere, at one time were esteemed for their properties in alleviating hysteria and epilepsy - the blooms being roasted on pewter dishes over a fire, and the resultant powder stored in bottles stoppered with leather, never with a cork for some reason (39). In Europe, the leaves of *Anchusa officinalis* Bugloss, a herbaceous introduction to the Region, were recommended as specific against melancholia and epilepsy (17).

In W Africa, the bark of *Ekebergia senegalensis*, a medium-sized tree in dry forest and savanna regions, is used in Senegal for epilepsy, but it must be detached from the tree with the help of wood not iron (Sébire fide 8a). In this country, the Tenda people crush the root of *Cochlospermum tinctorium*, a savanna plant, in water which is used for washing the body over several days for epilepsy. It is necessary to drink with the right hand three times, then undress and wash, and on return to the village, not to leave (18). In Mali, the smoke of the burnt fruit of *Calotropis procera* Sodom apple is used to drive away the evil spirit that causes epilepsy (10). Mothers in its distribution area in the Region place a piece of the stem of Sodom apple under the pillow to stop cases of children 'shaking' in bed or convulsion (35).

Among the Ewe-speaking tribe of Ghana, Togo and Benin Republic, two or three dried nuts (depending on the age of the patient) of *Thevetia peruviana* Exile oil plant or Milk bush or Yellow oleander collected with rituals and ground with a few seeds of *Piper guineense* West African black pepper enters into nasal instillation to clear the phlegm instantly from the breathing tract of convulsive or epileptic patients - especially cases suspected to be of evil causes (31). With the Akan-speaking tribe of Ghana and Ivory Coast, a clay pot which is filled with water and herbs, *nyankonsuo* (divine water), is placed in the fork of *Alstonia boonei* Pagoda tree, a forest tree with whorled branches, locally called 'nyame-dua'. 'The water is of particular importance in the treatment of children who have convulsion. At the beginning of the treatment these children are washed with 'divine water' in order to cool their hot bodies' (20).

In S Nigeria, the Yoruba community believe that convulsion may be prevented by immunizing the unborn child while still inside the womb by giving the mother - at 5 month's pregnancy - a herbal soup or 'agbo' (a herbal concoction made from *Ehretia cymosa,* a soft-wooded shrub; *Picralima nitida,* a small forest tree; *Aframomum sceptrum,* a ginger-like herb; and *Abrus precatorius* Prayer beads, a climbing legume) (44). In this country, convulsion is also considered to be a 'folk illness' or illness caused by supernatural powers - such as the 'Evil Eye'. In this situation herbal remedies, exorcism and recitation or singing of incantations are employed in its control, and curative rites are directed to the body and the mind of the convulsive patient (44). In Congo (Brazzaville), *Renealmia africana,* a herb in the ginger family, is believed to hold magical power. The leaf-sap is used to wash the face of epileptics (6). In Tanzania, *Lantana viburnoides,* an aromatic shrub, is used clinically in cases of 'sheitani', resulting in attempts to exorcize possession of the evil eye, or the Devil, perhaps manifested by attacks of epilepsy (8d).

EYE COMPLAINTS

(Various ailments of the eye). In E Africa, charcoal made from the roots of *Annona chrysophylla* Wild soursop, which for magical purposes must be made by burning on a hoe, is rubbed by the Lobedu around the eye to relieve twitching; and the Shambala use the leaf of *Phyllanthus stuhlmannii,* an euphorb, as a charm for corneal opacity (49). In Tanzania, there is a magical use for *Asparagus flagellaris,* a scandent shrub in the lily family. Seven fruits eaten will ensure good eyesight for seven years (8c). In S Nigeria, *Hexalobus crispiflorus,* a forest tree, enters into a Yoruba Odu incantation against eye disease under the name *apara,* meaning a joke (47). In a cryptic way the Yoruba invoke the spirit of *Andropogon gayanus,* a robust grass, to come dancing as a cure against bleeding eyes (47). In its distribution area throughout the Region, traditional healers use *Alternanthera pungens (repens)* Khaki bur or weed, a prostrate

prickly amaranth, in juju both to treat an eye trouble supposedly caused by witches and apoplexy (34). A magical usage of *Crotalaria pallida (mucronata)*, an erect legume under-shrub, as an eye medicine is reported - either as a lotion or sometimes rubbed in the palm of the hands, which are then passed over the eye (12).

In Europe, *Ruta graveolens* Common rue or Herb of grace was a famous medicinal and magical herb, said to be good for eyesight (17). In Greece, *Agrimonia eupatoria* Agrimony was believed to have magical powers, including that of healing the eye (5); and the white stain on the leaves of *Trifolium pratense* Red clover was the reason why people defending the doctrine of signs said the plant was good for cataract (38). Commenting on the local diet in Africa, it has been observed that manioc *(Manihot esculenta)* meal is the staple food, the bread equivalent, all along the coast...fou-fou on the Leeward, kank on the Windward, m'vala in Corisco; and ogoma in the Ogowe; but acquaintance with it demonstrates that it is all the same - manioc (28). On its effects on the eye, the natives themselves say that a diet too exclusively maniocan produces dimness of vision, ending in blindness if the food is not varied (28).

FEVER

(Abnormally high body temperature, in humans above 98.4 Fahrenheit). In the Region, a waist-band of the bark of *Bauhinia rufescens*, a legume shrub or small tree on river banks in savanna country and sometimes planted, is worn as a charm against fever (12). In S Nigeria, *Pergularia daemia*, a semi-woody climber with milky latex of forest edges and savanna, serves as protection against 'insult fever' among the Igbo tribe; that is, a fever resulting from insult by a boy of a big man, who, if he suffers such should immediately wash in a preparation of the plant (43). In this country, the Yoruba imprecate *Artocarpus altilis* Bread fruit against fever (47). In N Nigeria, the Hausa believe that there is a fever which breaks up when *Sorghum bicolor* Guinea corn is ripe, and the only way of avoiding it is to give presents of *Zea mays* Corn to the poor (45). In SE Senegal, a Mistletoe growing on *Ficus cordata*, a fig tree, is steeped in water which is used by the Tenda to wash children who are always getting fever (18). In Kenya, the Masai give their babies a piece of the bark of *Zanthoxylum* species, shrubs or trees in the orange family, to chew as a preventive against fever, as they think that 'the fever is afraid of this tree' (49). (For the treatment of fever influenced by evil, see EVIL: INFLUENCE ON DISEASE).

HEADACHE (MIGRAINE).

(Continuous pain in the head). The Hausa of N Nigeria, the Zarma of Mali and Niger and the Fula prescribe scrapings of the root-bark of *Capparis erythrocarpos*, a scandent thorny shrub in savanna, tied up in a piece of cloth as an inhalant for the instant cure for headache. However, it is only by using a coin to scrape the root would the medication be effective (35). In S Nigeria, the Yoruba invoke *Nicotiana rustica* Wild tobacco to send its smoke to cure headache (47). In Yoruba folklore, Eji-Ogbe, the premier disciple of Ifa, the God of Divination, a messenger of Orunmila, the Supreme deity representing God on earth, ordered that *Lentinus tuber-regium*, an agaric, a large rigid toadstool with a cup-like cap, be used for curing ailments such as headaches, stomach pains, fever and colds. The toadstool is ground with pepper and onions in palm-oil (37). In Uganda, *Panicum maximum* Guinea grass is tied round the head for the relief of headache (49). Similarly in Europe, a circle of *Viola odorata* Sweet violet used to be worn round the head to cure dizziness or headache (5).

HEART DISEASE.

(An ailment of the heart). According to folklore it was through sympathetic magic that the healing properties of digitalin was discovered - a decoction of the heart-shaped leaves curing a patient. *Digitalis purpurea* Common foxglove is the source of the drug and poison digitalin, widely used in the treatment of heart complaints. Its introduction into legitimate medicine around 1780, following the investigations of Dr. William Withering is a classic example of the incorporation of a folk cure into official pharmacopoeia (17). Similarly, the leaves of *Citrus sinensis* Sweet orange have a little heart on their base, and have thus been recommended for the treatment of heart dysfunction (38). In S Nigeria, *Musanga cecropioides* Umbrella tree is implicated in a Yoruba incantation against heart disease (47). In Congo (Brazzaville), the leaves of *Impatiens niamniamensis*, a herb in montane regions, is eaten as a vegetable for heart-troubles due to evil spirits (6).

HICCUPS.

(Sudden involuntary stopping of the breath). In Ivory Coast, a friction with crushed leaves of *Lygodium microphyllum,* a climbing or scandent fern, on the left side of the body from feet to head is practiced for hiccups (7). This seems to be an act of magic (8d).

HUMP

(Deformity on a person's back or front, from an abnormal curvature of the spine). Traditional healers in the Region prescribe the pulverized roots of *Rourea (Byrsocarpus) coccinea,* a shrub or climber in secondary forest or savanna thickets, to be taken internally and applied externally for humps, especially those cases resulting from, or attributed to the effects of lightning (34).

ITCH

(An irritating sensation in the skin relieved by scratching). In S Nigeria, the Yoruba invoke *Commelina erecta,* a herbaceous weed, in an incantation 'to give an itch' to someone (47). In S Africa, a decoction of the burnt and crushed root of *Asparagus scandens,* a climbing lily, is taken by the Southern Sotho to cure a rash which is thought to appear after a snake has been seen (49).

JAUNDICE/YELLOW FEVER

(Disease caused by an excess of bile in the blood which makes the skin and the whites of the eye become abnormally yellow). Since the external symptom of jaundice is usually yellow eyes, plants or plant parts with this colour are usually used to effect treatment. In effect some of the traditional methods of treating the disease appear to be based almost entirely on sympathetic magic. For instance, the leaves of *Crateva adansonii (religiosa),* a savanna tree, are used in fumigation for the disease, perhaps on the Doctrine of Signatures as the wood is yellow in colour (26a, b). Similarly, a yellow or brownish-yellow dye is obtained from *Cochlospermum tinctorium,* a flos-producing savanna plant, with the result that in Senegal the root, on the same theory, is used with or without other drug plants, for jaundice and liverish fevers (26a, b). The powdered root of the plant in water or millet-beer is supposed to cure jaundice by sympathetic magic (23b). Traditional healers in the

Region prescribe a copious infusion of the bark and roots of *Sarcocephalus (Nauclea) latifolius* African peach, collected and prepared with rites, to be drunk to cure jaundice (34). This is probably on the same theory, as the plant parts yield a yellow colour. The thickened rhizome of *Stylochiton lancifolius*, an aroid in savanna woodland which is eaten as a famine food, is bright yellow in colour and jaundice is treated by wearing beads made from the plant (12). In Sierra Leone, Ivory Coast and Nigeria, the yellow colour of the gum of *Harungana madagascariensis* Dragon's blood tree, a shrub or small tree, on the Doctrine of Signatures, evokes usage of the bark, roots or the gum itself for jaundice (7, 8b, 27).

In Europe, people supporting the doctrine of signs recommended *Calendula officinalis* Calendula, a herbaceous composite, for jaundice and gall bladder disorders because of the bile-like colour of the flowers (38). In Ivory Coast, *Chrysophyllum delevoyi* African star apple is used medicinally for jaundice - probably a superstitious usage (27). In this country and in Burkina Faso, the leafy twigs of *Cussonia arborea (barteri)*, a deciduous savanna tree, are used with magical rites for yellow fever (27). However, if the jaundice is suspected to be influenced by witchcraft, then the healers tend to recommend a bath with a macerate of either *Momordica balsamina* Balsam apple or *M. charantia* African cucumber in water (34).

LEPROSY

(Infectious disease affecting the skin and nerves, causing disfigurement and deformity). In Burkina Faso, the roots of *Guiera senegalensis* Moshi medicine, a small shrub in sandy wastes, is put into a medico-magical prescription for leprosy by the Moore (27); and in Senegal, *Capparis tomentosa*, a thorny climbing savanna shrub, is used in a medico-magical treatment of leprosy (26b, d). From the likeness of defoliated *Cussonia arborea (barteri)*, a deciduous savanna tree, to the deformed limbs of a leper, the plant is used on the Doctrine of Signatures in the treatment of

Cussonia arborea

leprosy (8a). In S Nigeria, the Yoruba name of *Euphorbia poisonii* and *E. unispinosa*, erect euphorbs with candelabriform branching, 'oro-adete', has reference to a leper (12) - probably for the same reason. In the Region, *Piper nigrum* Black pepper has been optimistically used as preventive against leprosy. In E Sudan, a remedy for leprosy is attained by writing a verse of holy scripts one thousand times on the pods of *Acacia nilotic*a ssp. *tomentosa* Egyptian thorn, a thorny savanna tree legume, after which a decoction of these is drunk copiously and rubbed over the body (Anderson fide12).

Many of the uses of *Phragmanthera* and *Tapinanthus* species Mistletoe, epiphytic parasites, are based on superstition and vary according to the nature of the host. The epiphytes indigenous to the Region are listed in Table 1 below:

Species	Plant Family	Common Name
Phragmanthera incana	Loranthaceae	Mistletoe
P. kamerunensis	Loranthaceae	Mistletoe
P. lapathifolia	Loranthaceae	Mistletoe
P. leonensis	Loranthaceae	Mistletoe
P. nigritiana	Loranthaceae	Mistletoe
Tapinanthus bangwensis	Loranthaceae	Mistletoe
T. belvisii	Loranthaceae	Mistletoe
T. buntingii	Loranthaceae	Mistletoe
T. farinari	Loranthaceae	Mistletoe
T. globiferus	Loranthaceae	Mistletoe
T. heteromorphus	Loranthaceae	Mistletoe

Table 1. Mistletoes in the Region

For instance in N Nigeria, the berries of the parasites found on *Vitellaria (Butyrospermum) paradoxa* Shea butter tree or on *Vitex doniana* Black plum are used in prescriptions among the Hausa for leprosy along with washings of the Koran to be drunk (12). In this country, the leaves and roots of *Excoecaria (Sapium) grahamii,* an under-shrub euphorb with milky latex in savanna, are ingredients in a prescription for leprosy with the leaves of Black plum, Mistletoe on Shea butter tree and various other additions in a superstitious ceremony (12). In this country, the smoke of *Stereospermum kunthianum*, a tree of savanna woodland, is believed to conduce to leprosy in Sokoto (12). In Burkina Faso, some thirty-six plants are added to *Tamarindus indica* Indian tamarind to treat lepers brought to the fetish called Ydtaba in the Kaya region (27).

In S Sudan, it has been observed that 'at a certain time of their growth the stems of the creeper *araka* lose their leaves. These are replaced by a double row of bands, joined to the stalks, which little by little dry, split, and fall in pieces just like the extremities of the hands and feet disappear in *'la lepre mutilante'*. This creeper is highly thought of as furnishing treatment for this kind of leprosy among the Azande tribe. Likewise, the *kunga* tree furnishes a remedy against cutaneous leprosy. The reddish patches which are conspicuous on its trunk resemble the patches which appear in the early stages of this disease' (16).
In Europe, *Solanum tuberosum* Potato had an unfavourable reputation for many years in France, where people believed it to be responsible for leprosy and other ailments (39); and *Fritillaria meleagris* Leper's lily is so called because the bell-shaped flower represents the leper's bell once carried by the diseased, and the colouring represents the disease itself (17). *Hyssopus officinalis* Hyssop was highly esteemed by Jewish priests as a purge and to cleanse themselves after contact with lepers (40).

LEUCORRHOEA

(A discharge of whitish mucus and pus from the female genitals). In S Nigeria, the Yoruba grind *Calvatia cyathiformis* with *Daldina concentrica*, both fungi, mixed with African black soap as a remedy for leucorrhoea. The soap compound is used by the patient for washing locally at prescribed intervals (37). Among the Akan and other tribal groups in Ghana and Ivory Coast, fetish priests prescribe a decoction of *Indigofera hirsuta* Hairy indigo, an annual herbaceous legume, taken in draught on the patient's day of birth on two consecutive weeks (34).

LUMBAGO

(Backache in the region of the loin). In Liberia, the pulverized bark of *Homalium smythei*, a small tree with a very hard wood, or its ash, is mixed with palm-oil and rubbed on the bark for lumbago; but in rubbing it in, the fingers must not be used, only by employing the toe to apply the salve will relief be soon effected (9). In N Nigeria, a strip of bark of *Acacia senegal*, a savanna tree legume with thorns, twisted and worn as a waist girdle is supposed to prevent lumbago and kidney troubles (12). In this country, *Aloe buettneri*, *A. macrocarpa* var. *major* and *A. schweinfurthii*, perennial lilies, are similarly worn in the form of a waistband, along with the leaves of *Calotropis procera* Sodom apple, the tail feathers of a

kite, *Elanus riocourii* or *E. caeruleus*, locally called 'shirwa', and others as a charm to cure lumbago (12).

Among the Ewe-speaking farming villages in Ghana, Togo and Benin Republic, it is the belief that waist pains experienced during hoeing on the farm could be mysteriously cured by wearing a leafy branch of *Clausena anisata* Mosquito plant as a waistband. In addition, the tribe deliberately tend *Sida linifolia*, a herbaceous weed popularly believed to be fetish, around their homestead and on their farms. It is the belief that deliberately weeding this plant causes waist pains (34). In Mozambique, the root of *Gymnosporia senegalensis* Senegal pendoring is boiled and applied by the Shangana as a poultice first to the opposite side of the chest and then to the painful part for intercostal pain. The decoction is then made into a porridge with a mealie meal and given to the patient to eat (49). In S Africa, the Southern Sotho use *Scilla lanceifolia* Wild squill, a lily, in the following way as a treatment for lumbago. 'The patient lies on his stomach, the native doctor puts his right foot into a bowl of water which has been boiled with some of the bulbs, he then places his foot on the portion of a red-hot hoe for a few seconds and rubs the small of the patient's back with the sole. This is repeated until relief is obtained' (49).

LYMPHATIC FILARIASIS

(See ELEPHANTIASIS above).

MENTAL DISORDERS

(State of being insane or insane behaviour). Writing on *Causes of mental illness in Yoruba ethno-medicine*, (44) lists among others:

1. 'Epe' or 'Ase' - a curse or evil wish by some wicked people as a revenge on somebody for a past offence involving the use of a special preparation in cursing others, or what is asked to happen to a person must happen as wished, or might possibly involve the use of juju or black magic.
2. Evil spirits that can cause insanity if one crosses their path.
3. The use of witchcraft by those who possess the power on victims at any time of the day, but preferably better used when the victim is asleep.
4. The influence of a family god 'Orisa Idile', and unless an

atonement is made to it, no amount of treatment can save the victim.

Before a traditional doctor can start treating a psychiatric patient, he must first try to establish the type of mental disorders involved. Among the methods of determining the causes of mental illness are 'Ifa oracle', the 'Sixteen cowries oracle', the 'Agbigba oracle' and the 'Kola-nut method of investigation' (44). In S Nigeria, the Yoruba invoke the fruit as well as the bark of *Rauvolfia vomitoria* Swizzle-stick, a shrub or small tree with milky latex, in separate incantations to cause mental disorders (47). The root decoction of this plant has been mentioned in traditional medicine (Shapara fide12) as a sedative to induce several hour's sleep in mad people. In Mali, the roots and bulbils of *Dioscorea bulbifera* Potato yam with ingredients prepared at dawn or at sunset, is an effective cure against mental disorders - this secret is only known to the fetish Sagara and his son (10). In Ivory Coast and Burkina Faso, the powdered roots of *Cnestis ferruginea*, a scrambling shrub with bright red fruits, in *Carica papaya* Pawpaw fruits is given to anyone suffering from mental disorders caused by ill fortune (27).

. In Senegal and The Gambia, *Tamarindus indica* Indian tamarind, a savanna tree, enters into a medico-magical treatment for mental disorders, erectal disorders and sterility. The removal of the parts of the tree required must be preceded by rituals and sacrifice (26b, c). In Senegal, the roots of *Detarium microcarpum*, a savanna shrub or small tree, enters into a medico-magical treatment, in the Cayor and the Ferlo areas, for mental conditions (26c). In this country, perhaps due to the strong smell released when bruised, *Ageratum conyzoides* Billy goat weed finds use in the treatment of mental disorders among other medico-magic (26b, c); and *Ekebergia senegalensis*, a savanna tree, is believed to hold magical power. It is used in exorcism treatments and for mental states (26c). Other plants used for mental troubles in this country are *Capparis tomentosa*, a thorny climbing shrub in savanna thickets, *Caralluma dalzielii* and *C. decaisneana*, succulent asclepiads in savanna (26b, c).

In Ivory Coast, *Tragia benthamii* Climbing nettle, an irritating twining plant, pulped with the heart of a young dog, *Canis domestica*, and the skin of an electric catfish, *Malapterurus electricus*, is said to have magical power to induce mental disorders according to the 'Kru' (27). In this country, the leaf of *Rhaphiostylis beninensis*, a scandent forest shrub, when touched by fire crackles violently, thus conjuring up magical use to chase away spirits, and to quieten fits of madness in possessed persons (27). Also in this country, *Carpolobia lutea*, a shrub or small tree, is used in medico-magical psychiatry to relieve those who, because they have sinned, have gone mad. A necessary prerequisite to the ablutionary bath being that the patient's fetishes have to be burnt (7). In Senegal, the Fula use *Khaya senegalensis* Dry-zone mahogany, a savanna tree, and the root of *Afraegle paniculata* Nigerian powder-flask fruit, a tree in the orange family, in medico-magical treatment for mental disturbance (26b, c).

In Botswana, the Sotho believe that the burning of a leafy branch of *Osteospermum auricatum*, a composite, in a hut cures insanity; and the Southern Sotho believe that a burning branch of *Chrysanthemoides monilifera* Bush tick berry in the hut of a madman will cure him (49). In Gabon, *Lipocarpa chinensis*, a tufted perennial sedge, is ascribed with magical properties to overcome crises in mental disorders - some of the plant is carried in the hair or on the wrist (48). In Congo (Brazzaville), a small spoonful of the root decoction of *Cogniauxia podolaena*, a herbaceous climbing cucurbit, is taken morning and evening to calm fits of insanity (6).

In W Africa, mad patients are forced to inhale a snuff prepared from *Erythrophleum suaveolens* Ordeal tree, especially at the early stages of the attack, to induce them to talk; or the patient is washed with the leaves of *Sida acuta* Broomweed, a common malvaceous weed, and the blood of a goat, *Capra* species - in a ritual bath (34). In Senegal, a healing village for mental patients is established in the Casamance region at the foot of *Ceiba pentandra* Silk cotton tree because of the symbolic image of the tree (46). In S Nigeria, the Yorubas have an incantation to *Cannabis sativa* Indian hemp or Marijuana for the smoke to make their enemies mad (47); and in Ghana, the powdered bark of *Milicia (Chlorophora) excelsa* or *M. regia* Iroko is believed to induce mental disorders. The bark, collected and ground with rituals, is rubbed into incisions on the wrist and anyone knocked by such a fist, goes mad instantly (34). In Cameroon, it is the belief among the Ntumu tribe that if a female eats one of the tabooed foods, the child will be deformed or mentally deficient (42). In Ghana, a drummer must on no account carry his own drums, lest he should become mad (41). (See TABOO: SPECIFIC FOR TARGET GROUPS).

Corchorus aestuans

The Ewe-speaking tribe of Ghana, Togo and Benin Republic invoke *Corchorus aestuans*, an annual or perennial semi-woody

plant often cultivated as a vegetable, in an incantation to cause mental disorders in an adulterous wife (4). The 'stuff' is buried with a spell at the entrance to the house, and when the suspect walks over the charm after her illicit affairs, it is triggered, and she goes mad instantly (4). There is however, a counter-charm or secret method of nullifying or neutralizing the effects of this charm. If an adulterous wife suspects a 'trap', she first removes her red under-cloth before entering the house - to 'fool' the charm (32). In addition, an adulterous wife may use other methods to nullify the effects of this charm. For instance, a maceration of the whole plant of *Hygrophila auriculata*, a spiny acanth herb commonly found around wet and inundated places, used as douche to wash the private parts is preventive of mental disorders arising from unfaithfulness to one's husband (34).

The flowers of *Buxus sempervirens* Box, an introduction from southern Europe and N Africa, yield honey which is so strong and noxious that in Tudor times it was believed to cause mental disorders (39). In Mexico, it is the belief among the Huichol Indians that sudden madness or death will certainly befall anyone who molests or disturbs the god-plant *Solandra* species (29). According to legend three mestizos who destroyed one of the god-plants while looking for gold at its base, became crazy and leapt to their deaths on the rocks below as a result of this desecration (29). *Helleborus niger* Christmas rose is an European plant with romantic history, and for centuries was thought to cure mental disorders and other ailments; *Lunaria annua* Honesty was believed to be efficacious in curing mental disorders; and the fruit of *Cassia fistula* Showers of gold, an introduced decorative tree from India, are said to 'help mad people sleep'(39). (For the treatment of mental disorders due to evil spirits, see EVIL: INFLUENCE ON DISEASE).

MUMPS.

(Epidemic Parotitis - acute infectious inflammation of the parotid gland near the ear, caused by a virus). Among many of the coastal tribes in Ghana, Togo and Benin Republic - the Fantes, the Nzimas, the Gas and the Ewes - the disease is treated with a thin smear of an aqueous paste of the seeds of *Piper guineense* West African black pepper, *Aframomum melegueta* Guinea grains or Melegueta, *Monodora myristica* Calabash nutmeg, *Xylopia aethiopica* Ethiopian pepper, *Zingiber officinale* Ginger and some *Capsicum frutescens* Chillies on both cheeks. On this hot, spicy base is conspicuously superimposed a multi-coloured pattern of red, white (both clay), blue (washing powder) and black (charcoal) spots.

It is these spots that attract attention, and to the embarrassment of the patient, he or she is hooted and jeered at by both the young and the old in a spontaneous and popular chorus. The belief is that it is the chorus that drives away the ailment. In past centuries, phytotherapists when observing the bladders of *Fucus vesiculosus* Fucus, an algae, filled with air (floats), thought that according to the Doctrine of Signatures, it could be useful against diseases such as mumps and scrofula (an inflammation of neck ganglion, often caused by tuberculosis) (38).

OEDEMA

(An excessive accumulation of fluid in the tissues). In Ivory Coast and Burkina Faso, the leafy twigs of *Cussonia arborea (barteri)*, a deciduous savanna tree, are used with magical rites for oedema (27); and in Ghana, Togo and Benin Republic, the fruit of *Kigelia africana* Sausage tree, collected with the necessary ceremony, is employed on the Doctrine of Signatures, in a juju practice by fetish priests to cure cancer of the breast and general oedema (34).

PARALYSIS

(The inability of muscles to perform, caused by injury to the nerve or nerve-cells). In Congo (Brazzaville), paralytics are laid on a bed of the young leaves of *Spondias mombin* Hog plum or Ashanti plum which are pounded to a pulp to the accompaniment of incantations, and the patients are massaged with this pulp. It is essential that in preparing the material it should froth (14). In S Africa, it is traditionally believed that when an older man marries a much younger woman - such as a University Professor marrying his student - the old don only ends up with either paralysis or stroke or both. In SE Senegal, the Tenda use Mistletoe growing on *Dichrostachys cinerea* Marabou thorn to treat a form of arthritis (?paralysis). A water extract to which palm leaf bases have been added is used to wash the patient who must be accommodated in an isolated house where he remains throughout the treatment. Anyone visiting him must also wash with the preparation before leaving (18).

In Ivory Coast and Burkina Faso, the leafy twigs of *Cussonia arborea (barteri)*, a deciduous savanna tree, are used with magical rites for paralysis (27). In these countries, witch-doctors treat paralysis by wrapping a cloak over the patient and placing him over a hole in the ground in which hunks of oil-palm and fronds of *Pteris atrovirens*, a ground forest fern, are set on fire to envelope him in the smoke (27). Among the personified diseases in Hausaland, N Nigeria, is *Dogua*, an evil spirit which injures *Tamarindus indica* Indian tamarind and *Adansonia digitata* Baobab, and causes paralysis and death to those eating the fruits (45). In Ghana, the leaf-juice of *Bryophyllum pinnatum* Resurrection plant, collected for the purpose and squeezed onto one's footprint mark with a spell, is a juju to inflict the victim with paralysis (31). The general practice in the Region is that trees struck by lightning are usually avoided until the traditional rituals to render these trees 'safe' again are performed - the belief and fear being that deliberate defaulters will be afflicted with paralysis or hunch, among other ailments (see LIGHTNING AND THUNDER: 'NEUTRALIZING THE CHARGES')

PILES (HAEMORRHOIDS)

(An enlarged vein of the anus engorged with blood). In Ivory Coast, the red flowers of *Thonningia sanguinea* Crown of the earth, a root parasite, perhaps on the Doctrine of Signatures, are crushed with pimento into a paste for use as an enema for haemorrhoids (8a). In Ghana, the people of the Upper West Region - mostly of the Dagarti, the Lobi, the Sisala and the Wala tribes - use a decoction of *Phragmanthera* or *Tapinanthus* species Mistletoe that is growing on *Parkia biglobosa (clappertoniana)* West African locust bean in the treatment of piles (36). This is probably an instance of sympathetic magic based on the red flowers of the epiphyte and that of the *Parkia*, with the bleeding piles. *Biophytum petersianum* African sensitive plant, a weed sensitive to weather and to touch, and reputed to be spiritually powerful, with ingredients also enter into prescriptions with rites for the treatment of the disease among this tribal group (36). In Peru, all one must do to cure haemorrhoids is to carry an amulet made of the leaves of *Ficus carica* Common fig tree or *Citrus aurantiifolia* Lime (13).

PLAGUE

(Any deadly infectious disease that kills many people). In Ghana, *Palisota hirsuta*, a robust forest herb, is regarded as deterrent to the spiritual influences caused by plague, and placed on paths during epidemics which the plague cannot pass (23a). In Ivory Coast, the black doctors of the Man region confess that in case of epidemics, they put the leaves of *Phyllanthus niruroides*, a weed of cultivated land and waste places, in

Palisota hirsuta

water for a week, wash their face with it and sprinkle the rest round the buildings where the sick are, to drive the epidemic away (27). In Senegal, Niominka medicine-men, at times of epidemic, prepare a beverage-offering at the foot of *Faidherbia (Acacia) albida*, a legume tree, which the people imbibe thrice daily to drive out the illness - ablutions are also necessary (26b).

In E Africa, particularly among the Kikuyu, appeasement sacrifices especially of rams, *Ovis* species; and male goats, *Capra* species; have been made to *Ficus* species from generation to generation to stop epidemics; and in S Africa, to stamp out an epidemic of colds (?influenza) all the inhabitants of a Sotho village must bathe *coram publico* in an infusion of *Aloe saponaria*, a lily (49). In Gabon, *Mammea africana* African mammee -apple is held to have magical powers for warding off epidemics (48).

In the temperate zone, *Gentiana cruciata*, an European plant, was at one time called *Ladislai Regis herba* because distressed by the suffering of his people, who were afflicted with plague, the King of Hungary prayed to the Almighty to guide a shot arrow towards some herb which might alleviate the misery - the arrow plunged straight into the heart of a Gentian plant, the root of which was immediately tried and found to possess remarkable properties (39). In the Middle Ages, the country folk in Europe would watch the budding of *Euonymus europaeus* Spindle tree with close concern, for if the flowers were scanty all would be well, but great apprehension was felt if there was a profusion of blossom for this meant that plague would sweep the land (39). In order to ward off such dire disaster, great quantities of the Spindle tree flowers (Nature's own antidote) were consumed, in fearsome concoctions which caused violent nausea and purging (39).

Again, when Europe was ravaged by plague in the Middle Ages, the archangel Gabriel apparently came to a monk in a vision, telling him *Angelica archangelica* Angelica would effect a cure - and ever since then, the plant has borne the name Angelica (5, 38). Similarly, the legend goes that Charlemagne was advised by an angel that *Carlina acaulis* Carline thistle, a thorny composite, was useful to prevent plague in his armies, hence he obliged his soldiers to eat, thus avoiding a terrible epidemic (38). In Europe, *Ruta graveolens* Common rue or Herb of grace was credited with anti-magical powers, and when plagues were abroad people carried sprigs of Rue to ward off infection (39). Similarly, *Allium sativum* Garlic was thought to guard the eater against the plague, and was eaten in wine; and *Centaurea cyanus* Cornflower was believed to give protection against the plague and other infections and dangerous diseases (5). In medieval England, floors of courts of justice were strewn with the leaves of *Acorus calamus* Sweet flag as protective of plague (40).

PNEUMONIA

(Inflammation of the lung). In Liberia, pneumonia in children is treated by the Mano people by rubbing the leaves extract of *Ageratum conyzoides* Billy goat weed on the chest and then 'transferring to a stick' (Harley fide12). In S Africa, the Venda mix the powdered root of *Capparis tomentosa*, a thorny savanna shrub, with dried hyena, *Hyaena hyaena;* the blood of an antelope, *Adenota kob kob;* and the fat of an ox, *Oryx* species; for the ritual treatment of pneumonia. For this purpose the 'doctor' makes three incisions on the right side of the patient's chest and two on his left side with a spear or the thorn of *Scolopia ecklonii* var. *engleri*, a shrub, to which the mixture has been applied. The spear or thorn is then thrust into the ground to bury the cause of the illness (49). In Europe, *Pulmonaria officinalis (Pulsatilla vulgaris)* Lungwort or Pulmonaria, a plant with irregularly spotted bristly leaves, drew the attention of herbalists in the past, and according to their Doctrine of Signatures, they decided that the spotted leaves would cure spots on the lungs and accordingly named the plant (39).

RABIES

(Fatal virus disease causing madness in dogs, *Canis* species, and other animals, transmitted to humans by a bite). In S Nigeria, the leaves of *Acalypha ornata*, an euphorb shrub, are compounded with the leaves of other drug-plants into a draught for children with rabies. This is perhaps on the extension of a widespread belief found in Tanzania of magical powers residing in the roots of the plant (8b).

RHEUMATISM

(Diseases of the nerves, bones, joints, muscle and tendons). In Congo (Brazzaville), *Tylophora sylvatica* , a herbaceous twiner with milky latex, is tied around the leg to prevent rheumatism (6). In S Nigeria, the Ijo put the leaves of *Emilia coccinea* Yellow taselflower, a composite herb, (and probably also *E. praetermissa*), with fish and cut-up plantains into a medico-magical preparation for treating painful nerves in the legs. The concoction must be prepared in an earthenware (not metal) pot used by a woman (Williamson fide 8a). In N Nigeria, the leaves of *Vernonia colorata,* a shrubby composite in savanna, is expressed in cold water for a bath, and the limbs smeared with the finely grated root in the treatment of the disease; but some fetish is connected with this operation (Thonning fide 22). As a cure for rheumatism by some native doctors in the Region, a chick, *Gallus domestica,* is offered to *Millettia thonningii,* a savanna tree legume, in a midnight ritual and the roots dug up, powdered and taken in an alcoholic drink (34). In Europe, *Arctium lappa* Burdock was a popular herb in folk medicine. The seeds, hung in a bag around the neck, were believed to protect the wearer against rheumatism (5).

RICKETS

(A Vitamin D deficiency disease of children marked by disordered bone formation). Homeopathically, a bark decoction of *Ceiba pentandra* Silk cotton tree is given to rickety children on the precept of the tree's rapid growth (8a); and in Ivory Coast, the magical property of Silk cotton tree is used in remedies for rickets in children, pregnancy and labour (23b).

SCABIES

(A contagious infestation of the skin caused by *Sarcoptes scabiei,* an insect). In Europe, the name Scabious was bestowed on *Scabiosa atropurpurea* Sweet scabious or Mournful widow because of the supposed efficacy of the herb and certain other species of the genus to cure skin diseases, such as scabies (39).

SLEEPING SICKNESS

(Trypanosomiasis - the disease is characterized by irregular fever, enlarged lymph-glands, and increasing somnolence). In Ivory Coast and Burkina Faso, the leafy twigs of *Cussonia arborea (barteri)*, a deciduous savanna tree, are used with magical rites for the disease (27). In Senegal, *Detarium senegalense* Tallow tree, a savanna tree, has medico-magical attributes in connection with sleep (26b). In Europe, on the contrary, Pliny recommends wrapping *Sedum acre* Stonecrop in bark cloth, and secreting this (unknowingly) beneath the pillow of sufferers to cure insomnia (sleeplessness) (39). Similarly, in S Nigeria, the Yoruba invoke *Iodes africana*, a liane in evergreen forest, in an Odu incantation against sleeplessness (8b). An unidentified plant locally called 'ase' in Yoruba, meaning 'power', is used in an invocation by the tribe against sleeplessness (47). In Peru, the leaves of *Juglans neotropica* Nogàl, put in bed, keep away the spirits that disturb one's sleep (13).

SMALLPOX

(A contagious disease characterized by sudden high fever, an irruption of pimples which increase in size all over the body, blister, form pus, crust, and leave a pock mark). In S Nigeria, the Yoruba invoke *Canarium schweinfurthii* Incense tree and the red berries of *Anchomanes difformis*, a large herbaceous aroid, as a cure for smallpox, and in an incantation for protection against saponna (smallpox), respectively (47). The Yoruba also call on *Homalium letestui*, a small forest tree, in an incantation against smallpox (46). In the Plateau state of this country, *Tacca leontopetaloides* African arrowroot is an important ingredient of the rite of *Karen* to gain protection from one's enemies, evil spirits and epidemics such as smallpox (Walu fide 8d). In Ivory Coast and Burkina Faso, *Phyllanthus niruroides*, a semi-woody herbaceous euphorb, is an ingredient of a medicinal nostrum sprinkled about houses during smallpox epidemics (27); and in Kenya, the leaves of *Blighia unijugata*, a medium-sized tree, with pock-marked galls are used for smallpox - perhaps this is homeopathic sympathetic magic (8d).

In S Nigeria, soup prepared with a snail, either *Achatina achatina* or species of *Archachatina* with its own fluid and the leaves of *Baphia nitida* Camwood, a shrub or small tree legume; the leaves of *Brillantaisia patula*, a robust forest acanth; the roots of *Marantochloa cuspidata*, a forest undergrowth; and the seeds of *Piper guineense* West African black pepper,

ground together in a pot with oil and salt, is eaten by each member of a Yoruba household to guarantee protection and prevention from a smallpox attack for at least a year (44). For continued immunity the practice is repeated yearly (44). *Piper nigrum* Black pepper has been optimistically used against smallpox. Among the precautions taken in historic times to stop the spread of smallpox from entering Nungua, a suburb of Accra in Ghana, (besides physical barricades across all the roads leading into the town), was purification with water in which leaves of either *Momordica balsamina* Balsam apple or *M. charantia* African cucumber has been immersed (19).

In N Nigeria, women and children in the Kordofan wear the yellow curled pods of *Faidherbia (Acacia) albida*, a large savanna tree, as a charm against smallpox (8c). In Ivory Coast, the Shien seek to protect their villages, in the event of a smallpox epidemic, by sprinkling around a decoction of the leaves of *Holarrhena floribunda* False rubber tree (27). Alternatively, to prevent the passage of the disease from village to village in this country, *Mikania cordata* var. *cordata* Climbing hemp-weed, or *Synedrella nodiflora* Nodeweed or Starwort, both composites, is macerated in palm-wine with *Senna (Cassia) occidentalis* Negro coffee, and the preparation is spread along tracks between villages - it is forbidden to pass the points treated (27). In S Africa, the ashes of *Ajuga ophrydis*, a labiate, mixed with fat in 'doctoring' pegs round villages prevents entry of smallpox by the Southern Sotho (49).

In S Nigeria, *Sida acuta* Broomweed, a perennial semi-woody herb, is invoked in a Yoruba incantation, under the first name - *agidi magbayin* - for the protection of smallpox (47). In this country, an unidentified plant locally called 'iboti' in Yoruba, meaning 'malt of guinea corn', is used by the tribe in a magical protection against small pox (47). In Ivory Coast, the Bété crush the leaves of *Vernonia amygdalina* Bitter leaf or *V. colorata*, shrubby composites in savanna, with white sand and spread this along tracks and at certain points in a village to confer protection from smallpox (27). In this country and Burkina Faso, the powdered roots of *Cnestis ferruginea*, a shrub or climber with bright red fruits,

Vernonia colorata

in a love-philtre sprinkled on roads, also enter magical treatment to prevent the spread of the disease during an epidemic (27). In these countries, *Costus lucanusianus*, a tall forest herb in the ginger family, has superstitious uses to confer protection in villages against epidemics, and to repel evil spirits (27).

These practices reflect the traditional belief that smallpox, like many other diseases, is a curse. The rituals are, therefore, not without libation, prayers and appeasement sacrifices as fowls, *Gallus domestica*; sheep, *Ovis* species; yams and bottles of schnapps to the gods and ancestors, and incantations to drive away the evil believed to cause and spread the disease.

While some plants are used in rituals to prevent the spread of the disease and protect the people from the epidemic, some other plants enter into rituals to infect persons with smallpox. For instance, fetish priests in the Region formulate a paste from four cut pieces of *Anchomanes difformis* or *A. welwitschii*, large herbaceous aroids, and ground each piece separately with seven seeds of *Aframomum melegueta* Guinea grains or Melegueta and some lavender. The four pastes are then applied separately with a spell to four incisions (two cut on the elbow, and two cut on the knees). The ritual is aimed at protecting oneself from any maledictions, and to return any evil intention to the originator; and to infest the same with smallpox (2). Among the jumble of herbalist's lore surrounding *Anchusa officinalis* Bugloss, a herbaceous Europeans plant, is the belief that the roots promote smallpox (17).

In S Nigeria, 'Shankpanna' is the god of smallpox in Yoruba mythology (15); and in N Nigeria, among the personified diseases in Hausaland is *'Dan Zanzanna'*, who gives smallpox to his enemies (45).

STOMACH COMPLAINT

(Ailments of the stomach). In a medico-magical treatment of an acute stomach ache among some of the coastal tribes in the Region, a macerate of seven leaves of the red-petioled variety of *Manihot esculenta* Cassava collected with the necessary ceremony, is drunk by the medicine-man (not the patient) as his left foot is placed on the patient's stomach with incantations to drive out the ailment (3). In Nigeria, the bark of *Lonchocarpus sericeus* Senegal lilac is supposed to be a remedy for the

Cassava (red-petioled)

disease if worn by young children round the waist (12). In Senegal, *Combretum lecardii*, a scandent shrub, is recorded with medico-magical properties for stomach complaints (26c); and in Ghana, *Solenostemon monostachyus*, a weedy herbaceous annual labiate, rubbed on the abdomen of the pregnant woman will prevent the child having stomach ache (Irvine fide 12). In Sierra Leone, if the bark of *Distemonanthus benthamianus* African satinwood, a forest tree with reddish bark, is cut between sunrise and sunset it is said to cure stomach diseases (Savill & Fox fide 8c). In the Mediterranean region, the fallen petals of *Punica granatum* Pomegranate are often threaded on necklaces hung round the necks to alleviate stomach ailments (39).

SWELLINGS

(Abnormal swollen place on the body). In N Nigeria, the Fula use an infusion of *Phragmanthera* or *Tapinanthus* species Mistletoe, epiphytic parasites growing on *Combretum glutinosum*, a savanna tree, to poultice swellings (24). This is perhaps a case of sympathetic medico-magic (8a). In Central African Republic, the root of *Curculigo pilosa*, a herbaceous plant with grass-like leaves, reduced to a pulp is applied topically to swellings held to be of fetish origin by the porcupine, *Artherurus africanus*, or by the aardvark, *Orycteropus afer* (8b).

SYPHILIS

(A venereal disease caused by *Treponema pallidum*, and usually transmitted by sexual contact). In N Nigeria, the roots of *Terminalia avicennioides*, a savanna tree, forms part of a ceremonious treatment of syphilis among the Jukun (33). It is said in native folklore that syphilis may be prevented by swallowing all the seeds in one pod of *Sesbania pachycarpa*, a herbaceous legume usually growing near water - probably introduced (12). In Tanzania, the root of *Ageratum conyzoides* Billy goat weed, a composite weed, is placed in a small shell and an adolescent boy, who must be wearing a copper armlet, urinates into it. The fluid is rubbed over a stone and the slimy mess is applied to a syphilitic sore with the feather of a cock, *Gallus domestica*. The application is repeated daily for a month by which time a cure is thought to have taken place (49). In S Sudan, it has been recorded among the Azande that 'in treating syphilis they take the root of an unidentified plant locally called 'bafuafu' or 'bamoru' and scrape and cook it with sweet potatoes. They mash the sweet potatoes with the scrapings, and the patient drinks an infusion from the mixture after he has had a long spell uttered over it. The medicine will then cause violent sickness and clear the liver, which is the place where syphilis and, indeed, most other diseases are considered to be localized' (16). (See also VENEREAL below)

TAPEWORM.

(Taenia - long, flat, segmented intestinal worms that are parasites on man and other animals). In W Africa, the red sap washed from the bark of *Harungana madagascariensis* Dragon's blood tree taken from the east and from the west sides of the tree (suggestive of sun-worship) is drunk as remedy for tapeworm (12).

THRUSH

(A children's disease caused by an yeast-like germ, *Candida albicans*). This is a fungal disease of the mucous membrane especially in the mouth of children. Native doctors in the Region collect, with the necessary ceremony, the whole plant of *Pergularia daemia*, a scrambling asclepiad often tended around villages and believed to be fetish, dry and char the whole plant into a black powder and then apply it locally as an important medico-magical cure for the disease (34).

TOOTHACHE

(Pain in a tooth or teeth). In N Nigeria, the roots and probably the stems of *Salvadora persica*, a small savanna tree, are widely used by the Moslems, in imitation of the Prophet, to clean the teeth or to relieve toothache (12) - probably an instance of sympathetic magic.

TUBERCULOSIS

(Infectious wasting disease in which growths appear on the body tissue, especially the lungs). From the sixteenth century onwards, people defending the doctrine of signs saw in the leaves of *Pulmonaria officinalis (Pulsatilla vulgaris)* Lungwort or Pulmonaria, the resemblance of an ill lung with tuberculosis knots; thus many people suffering from tuberculosis were treated with Lungwort, in some cases achieving success (38). In Sudan, the Dinka call *Crotalaria aculeata*, an annual spiny herb legume, 'dates of the spirits', and chew the roots for lung and throat complaints (8c). In E Africa, a Chagga mother allows the powdered root of *Polygala usambarensis,* a shrub, mixed with butter and held in the mouth, to melt into a child's mouth before the first breast feed, as a prophylactic against tuberculosis if the child's forefathers have suffered from the infection (49)

ULCER

(An excavated sore, the base of which is inflamed). In Sudan, there is a belief among the Azande that an ulcer may be transferred. 'If you want to rid yourself of those ulcers, go to the centre of a path and take some grass in each hand from either side of the path and tie them together. Then squeeze an ulcer and put some of its pus on the grass, saying: 'You are ulcer. I place you on the grass. I tie you up in the centre of the path. If a man passes by and breaks the knotted grass, then in breaking them, he will take the ulcer with him. May they not come back to me'. Some man who is walking in the place comes along the path, and breaks the knotted grasses.... This man thereupon gets the ulcer' (16).

'If ulcers greatly trouble a man, he takes a bean to a cross-roads, and there digs a shallow hole. He holds the bean with the pus on it in his hand and addresses it. 'You, bean, I put you with ulcers. May the bean spring up at this cross-roads and take away ulcers for ever. May ulcers never again trouble me''(16). (See also BOILS above).

VENEREAL

(Relating to, or caused by sexual intercourse). Among the tribal groups in the Region and elsewhere, there are various magical ways of infecting women (particularly married women) with venereal disease that may be contracted, as a punishment, by any man (other than the husband) who has sex with her. The curse could also take the form of a spell by a former husband over the woman, and to punish any would-be future partner. Naturally, there are also counter-charms against the malediction. For instance, it is the practice among fetish priests in the Region to sprinkle the leafy juice of *Pergularia daemia*, a trailing herbaceous asclepiad often tended around villages as a magic and fetish plant, on suspected objects to drive away any evil therein. Consequently, it is widely believed that when one collects a leaf of the asclepiad with rites, and chews it before or during intercourse with an 'infected' woman, one goes 'scot-free' (34).

In S Africa, Zulu men administer powdered *Cyanotis nodiflora* Wandering Jew, a commelinaceous herb, as a 'medicine' to their wives. This is supposed to cause disease about the genitalia of the woman if she subsequently have illicit intercourse (49). Similarly, the tribe apply a paste of the leaf of *Mikania capensis*, a composite, and other plants over the bladder (the skin being previously anointed) for a 'disease' of the urinary organs contracted from intercourse with the girl of another youth (49). In this country, a strong infusion of *Thunbergia dregeana*, an acanth, is taken by a Zulu man to cure problematical disease which he may acquire from intercourse with his wife if she has been 'doctored' with the intention of infecting him (49).

In S Sudan, there is a similar 'medicine' as above among the Azande. 'There is a medicine called *moti*, recently introduced from the Belgium Congo (now Congo (Brazzaville)), which causes a venereal chancre. It is a plant with purple flowers from which an infusion is made. A husband before intercourse puts drops on his member, uttering a spell against any other man who has intercourse with his wife. His wife is ignorant of his action, and neither he nor she are injured by the medicine, though a man will only use it after he has eaten the antidote. If any other man afterwards has adulterous relations with the wife his member swells and becomes a great pus-laden sore which usually proves fatal' (16). In Congo (Brazzaville), *Hyptis suaveolens* Bush tea-bush a strongly aromatic herbaceous labiate, provides medicine to be taken against illness caused by 'unclean' persons, especially illness resulting from recent sexual intercourse (6). (See also SYPHILIS above; and MARRIAGE: FAITHFULNESS IN MARRIAGE).

According to the doctrine of signs it was deduced from the pearly seeds of *Lithospermum officinale* Gromwell, a vivacious plant of the Boraginaceae family, the property of being able to dissolve kidney stones (38). Since the small bulbs at the stem base of *Saxifraga granulata*

Saxifrage, a herbaceous plant, resemble some kind of urinary calculi, from ancient times onwards it was supposed that the use of Saxifrage be recommended to heal the 'stone disease' or urinary lithiasis (38). The fact that *Spergularia rubra* Sand spurry, a creeping plant, grows in sandy soils, and also is successfully able to treat sand which forms in the urinary tract seems quite curious, as such those supporting the doctrine of signs took this plant as one more argument which demonstrated their doctrine (38). In Europe, *Utricularia* species Bladder-wort, water plants with submerged leaves upon which are born small bladders in which insects are caught, was used in diseases of the urinary bladder (11). This is an instance of sympathetic magic

VERMIFUGE

(A drug or agent that expels intestinal worms). In W Africa, women drink a decoction of the leaves of *Vernonia amygdalina* Bitter leaf or that of *V. colorata*, shrubby composites in savanna, during the puerperium to affect the milk and act as a vicarious preventive of worms in the infant (12). In Liberia, the sap washed out from the bark of *Harungana madagascariensis* Dragon's blood tree, a shrub or small tree, taken from the east and west sides of the trunk (a relic of sun-worship ?), is used as remedy for tapeworm (12)

YELLOW FEVER

See JAUNDICE above.

(For the treatment of other ailments, see CATTLE; EVIL; and MEDICINE).

References.

I. Adjanohoun & Aké Assi, 1972; 2. Alhaji Asiedu Biney, (pers. comm.); 3. Ameyaw, (pers. comm.); 4. Avumatsodo, (pers. comm.); 5. Back, 1987; 6. Bouquet, 1969; 7. Bouquet & Debray, 1974; 8a. Burkill, 1985; b. idem, 1994; c. idem, 1995; d. idem, 2000; 9. Cooper & Record, 1931; 10. Coppo, 1978; 11. Crow, 1969; 12. Dalziel, 1937; 13. De Feo, 1992; 14. De Wildeman, 1906; 15. Ellis, 1894; 16. Evans-Pritchard, 1937; 17. Everard & Morley, 1970; 18. Ferry & al. 1974; 19. Field, 1937; 20. Fink, 1989; 21. Hall, 1978; 22. Hepper, 1976; 23a. Irvine, 1930; b. idem, 1961; 24. Jackson, 1937; 25. Kerharo, 1967; 26a. Kerharo & Adam, 1963; b. idem, 1964; c. idem, 1974; 27. Kerharo & Bouquet, 1950; 28. Kingsley, 1897; 29. Knab, 1977; 30. Kroger, (pers. comm.); 31. Kulefianu, (pers. comm.); 32. Lordzisode, (pers. comm.); 33. Meek, 1931; 34. Mensah, (pers. comm.); 35. Moro Ibrahim, (pers. comm.); 36. Murrey, (pers. comm.); 37. Oso, 1977; 38. Pamplona-Roger, 2001; 39. Perry & Greenwood, 1973; 40. Pitkanen & Prevost, 1970; 41. Rattray, 1923; 42. Sheppherd, 1988; 43. N.W. Thomas, 1913-14; 44. Thomas, 1988; 45. Tremearne, 1913; 46. Trincaz, 1980; 47. Verger, 1967; 48. Walker & Sillans, 1961; 49. Watt & Breyer-Brandwijk,1962

13 DWARFS AND FAIRIES

...for thou shalt worship no other gods:
for the Lord, whose name is Jealous,
is a Jealous God.
Exodus 34:14

In folklore, dwarfs and fairies are said to be small human-like, spiritual beings with supernatural powers. Both are said to be invisible to the uninitiated, but are generally believed to be beneficial to mankind. Whether these mysterious beings do exist or not is still a matter of ever continuous and unresolved controversy. On one end, some people claim that these spirit beings are merely mythological and a figment of the imagination. 'If dwarfs really exist, why is it that they are not seen by anybody? Why do they prefer the tranquility of the forests to the hubbub of city life?' (Reynolds fide 11). On the other hand, others, medicine-men, fetish priests, juju-men and traditional practitioners maintain that these beings are real. 'I would emphatically say that dwarfs are spirits. In my area, they live in a cave, above which hang huge stones, supported strangely by ropes' (Herrick fide 11). It is the claim of the initiated that they have seen these spirit beings, communicated with them, and received advice, guidance, predictions, divination, healing prescriptions and healing powers from them on several occasions.

DWARFS

Dwarfs are said to be mythical being residing in the forest. In Ghana, *Milicia (Chlorophora) excelsa* and *M. regia* Iroko, forest timber trees, are believed to be dwellings for dwarfs in the Ho area of the Volta Region. As such underneath these trees, ritual sacrifices are performed (4).

Favourite Food

In the folklore of many of the people in the Region - especially in the forest zone - *Musa sapientum* Banana is believed to be a delicacy of dwarfs. Those believed to have been spirited away by these

Uvaria globosa 1/5 natural size

supernatural beings report that they were fed only on bananas and biscuits. The brown-stemmed type of banana, with shorter fruits, called 'kwadupa' in Akan, and the ripened fingers of *M. sapientum* var. *paradisiaca* Plantain are believed to be a delicacy of dwarfs. These fruits enter into rituals by medicine-men to entice dwarfs into sacred groves (8). Another delicacy of these spirit beings is said to be the fruits of *Uvaria chamae* Finger-root and *U. globosa*, scrambling savanna shrubs with edible fruits (3); or groundnuts, palm-kernels and sugar (16); or eggs (12). Yet another delicacy is *Saccharum officinarum* Sugarcane.

Game of Marbles

The seeds of *Dioclea reflexa* Marble vine, a climbing legume in forest, are used in a game of marbles by tribal groups in the plants distribution area and beyond. It is the belief in folklore that this game is also a favourite of dwarfs. 'When at times you are entering the cave, you will see marbles being played by unseen hands on cement paper on which sand is spread' (Herrick fide 11). The spot in the forest where dwarfs have been playing marbles is sometimes reported to be clearly marked; and elders always advise all citizens to avoid the area. It is traditionally believed that should one deliberately walk over this spot, he or she would lose the path back home, and get lost in the forest - to the amusement of the dwarfs. Other seeds used in the game of marbles are those of *Mucuna sloanei* Horse-eye bean or True sea-bean and *Canavalia ensiformis* Sword bean, products of wild and cultivated climbing legumes, respectively.

Dioclea reflexa Marble vine (seeds) 1/3 natural size

Nature and Characteristics.

Dwarfs grow a bushy hair over the face, call one another by whistling and their heels are transplanted (1). The whistling and transplanted heels are supported (16). 'The most characteristic feature of these Ashanti 'little folk' - the word *mmoatia* probably means 'the little animals' - is their feet, which point backwards. They are said to be about a foot in stature, and to be of three distinct varieties: black, red and white; and they converse by means of whistling' (16). More details on the nature and characteristics of these spirit beings is given (12). 'There are three types of *mmoatia,* having a dark, red and yellowish skin, respectively. Dark-skinned *mmoatia* are considered as good-natured harmless creatures, whereas those with a yellowish and red skin are capable of all kinds of mischief. The light-skinned *mmoatia* are thought to possess special herbal knowledge and ability to fabricate *asuman* (amulets) and wooden statues, which they exchange with humans, particularly priests, priestesses and herbalists, in what is called 'silent trade' (12).

Abduction

In folklore, there are sometimes stories of an incident in which a person was reported to have been abducted by dwarfs. More often, the victim returns by himself or herself, or is returned by the dwarfs after a certain period of time. In Ghana, an authenticated account of how a 3-year old girl, Ama Doris Borbor, was whisked away into the forest allegedly by these superhuman creatures in 1971 on one Saturday afternoon from Gbadzeme, a town in the Volta Region, is recorded in the May 1971 issue of the DRUM MAGAZINE. In this instance however, the town folk did not wait for her to be returned by her abductors, but rather organized a search party to look for her. With combined prayers and supplication by Christians of the Evangelical Presbyterian Church on one hand to God not to allow the Asafomen to suffer defeat at the hands of the invisible, invincible dwarfs; and libation with *akpeteshi* or *ogogoro* and palm-wine by an elder to the gods and ancestral spirits exhorting the deities to lead them into the

Gbazeme Town

Doris in dwarf land in 1971

battle - amid 'death to the dwarfs' ritual dance by the womenfolk of the town; an Asafo search party eventually found her sitting alone in the forest, after being abducted for one night and one day. In her right hand was a herb, and by her foot - one finger of banana. How the little toddler managed to reach the mountainous forest many miles from home by herself in the first instance, much more surviving all this period, is still a mystery. At the time of the first draft of this manuscript in 1992, Doris was an apprentice seamstress in Accra. The author met Doris with her mother, step father and son in a suburb of Accra called Fish pond on Sunday, 29th September, 2002.

Curative Powers

Doris today 2013

The herb is of interest, for dwarfs are reputed as source of some of the hidden curative secrets of plants (see MEDICINE: INSIGHT: Revelation). The traditional belief is that Doris would grow up to be a powerful, prosperous herbalist. (Doris though is presently a qualified seamstress). Some medicine-men, herbalists and fetish-priests and fetish-priestesses claim to have acquired their healing skill in this way. Such a healer often builds a shrine in which he or she consults the spirits before undertaking to cure any kind of disease (10).

FAIRIES

In the folklore of many countries, toadstools are associated with fairies (6). This connection with the supernatural may have arisen from the toadstool's sudden overnight appearance in the fields and woods, or

perhaps it stemmed from the effects of eating certain species (6). Since Medieval times, macrofungi have been associated with superstitions, myths and prejudices (18); and with fairies, witches, mythical beings and legends (15). In W Africa, toadstools include species of *Bolbitis, Calvatia, Chlorophyllum, Hygrocybe, Lentinus tuber-regium, Pleurotus and Pycnoporus.* The rest are species of *Schizophyllum, Termitomyces* and *Volvariella.*

'Fairy Rings'

As the name suggests, 'fairy rings' are traditionally associated with fairies and the supernatural. Explaining the occurrence of the rings, it has been observed that: 'the mycelium from which mushrooms are produced spreads underground, forming a ring which may grow as large as 30 metres in diameter. In an open area, the mycelium expands evenly in all directions, dying in the centre and fruiting at the outer edges, where it grows most actively because this is the area in which there is the most fresh nutritive material. As a consequence, the mushrooms appear in rings, and, as the mycelium grows, the rings become larger in diameter. Such circles of mushrooms are known in European folk legend as 'fairy rings' (17).

It has also been observed that: 'in the past, the regularity of shape of the rings together with the presence of a bare central area were explained in a number of ways. The supernatural was often invoked so that the circles became the abode or dancing floor of fairies or toads, and to step into one was to invite misfortune. Later, sceptics attributed them to the action of lightning or to large or small animals moving in circles and manuring the ground' (9).

Nature and Characteristics

Like dwarfs, fairies are said to grow a long hair. This hair is a vital ingredient in some traditional medicinal prescriptions (see BURGLARY: PROTECTION OF OTHER PROPERTIES). A lock of the fairy's hair is said to be obtained by 'tricking' this being to drink an excess of its favourite wine - the wine tapped from *Phoenix dactylifera* Date palm - until it is completely intoxicated and asleep (14). When in this situation, a lock of the hair may be cut off by medicine-men, fetish-priests or witch-doctors after performing the necessary rituals (14). It is to be expected that naturally, false ones would also be offered for sale in medicine markets by fraudulent traders. The false ones may be distinguished from the genuine ones which are said to coil around the red tail feather of a grey parrot, *Psittacus erithacus*, when the two are brought together (2).

In Ghana the Guang of Gonjaland say that fairies live in *Isoberlinia tomentosa*, a tree of the Guinean-savana forest (7). In this country, it is believed that the presence of *Platycerium elephantotis* or *P. stemaria* Stag-horn fern, decorative epiphytic plants in rain forest, on farms prevent fairies from entering and residing within (5). Where dwarfs and fairies are believed to occur, measures are taken to protect potential victims from being spirited away. For instance, at one time, prudent residents along the Western Coast of Ireland placed a tropical drift dissemule or sea-bean under the pillow at night as a charm against the nocturnal visits of fairies (13). (For the definition of tropical drift dissemule, see RELIGION: SYMBOLIC PLANTS: Sea-beans).

Psittacus erithacus Parrot.

References.
1. Adoboe, 1971; 2. Akligo-Zomadi, (pers. comm.); 3. Annan, (pers. comm.); 4. Asamoah, 1985; 5. Atoe, (pers. comm.); 6. Bisacre & al., 1984; 7. Burkill, 1995; 8. Cofie, (pers. comm.); 9. Cooke, 1977; 10. Dokosi, 1969; 11. Drum Magazine, August, 1971; 12. Fink, 1989; 13. Gunn & Dennis, 1979; 14. Mensah, (pers. comm.); 15. Oso, 1977; 16. Rattray, 1927; 17. Raven, Evert & Curtis, 1976; 18. Zoberi, 1973.

14 EVIL

For every one that doeth evil hateth the light,
neither cometh to the light,
lest his deeds should be reproved.
John 3: 20

In the folklore of Africa and many other countries throughout the world, the belief in evil - either real or imaginary - and its influence on man is ever imminent. This haunt calls for a variety of measures either to obstruct it, or to neutralize it or to drive it away. Different institutions adopt different methods of achieving this common objective. The Clergy use prayers and burn incense; the spiritual churches and sects burn candles and sprinkle Florida water in addition to the prayers; the occultist fights it on the astral level; the traditional medicine-man, fetish priest or witch-doctor by incantations and the invocation of the supernatural powers of plants.

Collectively, evil embodies anything but good - that is harmful, malicious, sinister, and above all devilish. Some of the medicines against these vices are either administered seven times or are made up of seven units of the constituents, or a multiple of this number. The number 'seven' is apparently, both anti-evil and anti-witchcraft; or favourable to the beneficial forces. It is said to be a mysterious number. For instance, the Moorish people *(kramo* or *eduro-eduro)* are very fond of mystical numbers, and often quote seven (13). On the contrary, evil influences may be borne by smoke, so it is inadvisable to destroy an evil object by burning it, since its potency is thereby distributed over the country-side (Larken fide 24). Such objects are rather said to be destroyed by immersing in water. This explains why bad juju is always protected from rain.

Traditional medicine-men use plants in various ways against the forces of evil - such as in disease, at childbirth and for general protection from demonic forces and bad juju. Some plants are believed to symbolize evil, while others are inimical to evil forces.

INFLUENCE ON DISEASE

The belief in the role of evil influences on the causes of disease is emphatically stressed in the folklore of many countries. It is, in fact, the rule rather than the exception. There are, however, also traditional plant prescriptions to counteract this - and even to reflect the ailment to the originator if need be. Medicinal plants have curative properties only because they possess a spirit and they exert their power by fighting the other powers ('encantos') that caused the illness (19).

General Sickness

In S Nigeria, *Scoparia dulcis* Sweet broomweed is taken by the root to drive out the evil spirit from a sick person (53). In Senegal, *Psorospermum senegalense,* a savanna shrub, enters Fula magical treatment to confer protection against evil: when the bark is burnt over live embers it makes much smoke in which the naked patient is fumigated (35b). In Ghana, the scent from burning leaves of *Ocimum canum* American basil put on a fire where there is a sick person is believed to drive evil spirits away (34a); and in Ivory Coast, *O. basilicum* Sweet basil is believed to ward off evil (12, 36). In Nigeria, the plant is burnt in a room where there is a sick person in order to drive out the evil sprit (18); and in Tanzania, the live plant is placed beneath a bed for the same purpose (60). In Sudan, the root of *Datura metel* Hairy thorn-apple or Metel is an ingredient in a remedy to exorcize an evil spirit causing disease (18); and in Congo (Brazzaville), a decoction of *Cissus rubiginosa,* a trailing savanna herb, will exorcize evil spirits from a sick person (11).

In S Nigeria, the Yoruba use the bark of *Erythrophleum suaveolens* Ordeal tree to fumigate a house in order to purge it of evil spirits, and therefore of sickness caused by them (18). In Ivory Coast and Burkina Faso, a macerate of the entire plant of *Pancratium trianthum* Pancratium lily used as a face-wash is considered protection against all illness (36). In Ivory Coast, fetish properties are attached to *Platycerium stemaria* Stag-horn fern, an epiphytic fern, and it is placed in the house of sick persons to protect them from evil spirits and sorcery (12). Among the Ewe-speaking tribe of Ghana, Togo and Benin Republic, a warm decoction of *Acanthospermum hispidum* Star burr, a weedy composite, is used by the tribe in a ritual bath of patients suspected to be victims of evil forces, while the leaves are chewed as protection against enemies (41). In S Africa, *Amphidoxa gnaphaloides*

is mixed with *Conyza pinnata*, both composites, and burnt in the hut of a sick person to drive away illness; and in Botswana, the Southern Sotho also burn the leaf of *Valeriana capensis* Cape valerian in the hut of a sick person to drive away the illness (60).

In Senegal, *Caralluma dalzielii* and *C. decaisneana*, succulent savanna asclepiads with milky latex, are ascribed with strong magical attribute by the Fula witch-doctors who use it for cleansing against illness, to confer protection against spells and to exorcize spirits (35a, b). The plants are also planted at praying grounds and mosques as a charm to keep away evil (18). In this country, the Fula and the Wolof prescribe the roots of *Diospyros mespiliformis* West African ebony tree in ritual invocation to chase away evil spirits, to effect cures, and to obtain good fortune (35b). In Congo (Brazzaville), the leaves of *Impatiens niamniamensis*, a herb in the Balsam family, are eaten as a vegetable for serious illnesses due to evil spirits; and *Adenia gracilis*, a sub-herbaceous forest climber, is used in prescriptions for persons with serious illnesses ascribed to evil spirits and fetishes (11). In this country, *Quassia (Simaba) africana*, a forest shrub, prevents spirits from entering a house, when hung from a door lintel; and when placed under a sick person's bed it will guard the patient from bad spirits and sorcerers (11).

In W Africa, *Costus* species, robust perennial herbs, are generally put in medico-magical formulations to protect people and villages from evil spirits and illnesses (14a). Thus, as part of the Apo ceremony of the people of Techiman, in Ghana, when the procession returned from the river, the head priest crossed the path after the last person with three long branches of the herb, and a handful of sand and clay on top (51). The significance of the ritual is that all evil of the past year was now behind, and this was a precaution against any of it finding its way back (51). The indigenous species in the Region include *Costus afer* Ginger lily, *C. deistelii*, *C. dubius*, *C. lucanusianus* and *C. schlechteri*. In the Region, when fainting is suspected to be caused by evil forces, the leaves of *Sida acuta* Broomweed, a common malvaceous weed, are collected by medicine-men with rites, tied in a piece of white cloth and attached to the wrist of the patient to revive him or her (41).

Reflecting Ailment To Originator In Ghana, a concoction of *Commelina* species, perennial or annual succulent herbs, and an old corn cob in an earthenware pot which, for a purpose, is supported on a bundle of the red variety of *Tragia* species, twiners with irritating

hairs, is used in a simulated bath as a general protection from disease; and to redirect any intended evil, with protracted illness, swollen stomach and eventually death on to the originator (6). The sympathetic magic appears to be the irritating plant. In this country, a decoction of *Phyllanthus niruri* var. *amarus (amarus)*, an annual weed of cultivated land, is used to rinse oneself after bath, during which some of the bitter decoction is sipped as protection against bad juju, evil influences and intentions. While washing the body the plant is invoked in an incantation to redirect same onto the originator (26). Similarly, in S Nigeria, *Alchornia cordifolia*, a shrub or forest understorey tree, enters into a Yoruba incantation to make 'bad medicine' rebound on the sender (56a). In N Nigeria, *Indigofera astragalina*, a herbaceous annual, enters Hausa folklore in wishing the return of evil that might befall one back upon the evil-doer causing it. The expression: *kaikayi koma kan mashe kiya* meaning 'o chaff, return to the winnower', is applied to this plant and others like it (18).

Fever

In Congo (Brazzaville), *Setaria megaphylla*, a course perennial grass, with *Cissus aralioides*, a forest liane, and *Selaginella* species, a herbaceous fern, is used to prepare a bath for someone with fever, especially if the fever is due to meeting an evil spirit (11). In this country, a massage with the leaf-pulp of *Ceiba pentandra* Silk cotton tree and baths in bark decoction are considered excellent for evening fevers, especially those deemed to arise from evil influence (11). Among some of the tribesmen in the Region, fever suspected to originate from evil spirits is treated by drinking a decoction of the leaves of *Jatropha gossypiifolia* Red physic nut with that of *Cymbopogon citratus* Lemon grass - collected and prepared with the necessary rites (26). (See DISEASE: FEVER for other instances of combating the ailment traditionally).

Bidens bipinnata Spanish needles

Mental Disorders

In the treatment of mentally deranged patients - especially cases suspected to be caused by evil forces, fetish priests prescribe a ritual bath with a decoction of *Bidens bipinnata* Spanish needles

on seven consecutive days as a cure (6). In Ivory Coast, the Anyi put the sap of *Laggera alata*, a strongly aromatic composite herb, into nasal instillation as a rapid cure for attacks of mental disorders caused by evil influence (36). In Ivory Coast and Burkina Faso, a decoction of twigs, leaves, bark and branches of *Ficus sur (capensis)*, a fig tree often with abundant cluster of figs on the trunk, is put into draughts and into baths to confer protection against evil spirits, in particular mental disorders provoked by them (36). In Tanzania, the indigenous population use the strongly perfumed inflorescence of a *Pandanus* species as a charm to drive out evil spirit from the insane (60). (For more prescriptions see DISEASE: MENTAL DISORDERS).

Children's Cases

Children's diseases are particularly believed to be influenced by evil spirits, and as such the necessary measures are taken to counteract this. For instance, small lumps of gum copal are hung around the neck of children who are vulnerable to these influences, or burned in the household to drive away evil spirits believed to be causing disease. A list of the copal trees in the Region is given in Table 1 below:

Species	Common Name	Plant Family	Remarks
Balanites wilsoniana	-	Balanitaceae	Forest tree
Boswellia dalzielii	Frankincense tree	Burseraceae	Savanna tree
Canarium schweinfurthii	Incense tree	Burseraceae	Forest tree
Commiphora africana var. africana	African bdellium	Burseraceae	Savanna shrub or small tree
C. dalzielii	-	Burseraceae	Savanna shrub or tree
C. pedunculata	-	Burseraceae	Savanna tree
Daniellia ogea	Gum copal tree	Caesalpiniaceae	Forest tree
D. oliveri	African copaiba balsam tree or Niger copal tree	Caesalpiniaceae	Savanna tree
Guibourtia copallifera	Sierra Leone gum copal	Caesalpiniaceae	Forest tree
G. ehie	Bubinga	Caesalpiniaceae	Forest tree
Hymenaea verrucosa (*Trachylobium verrucosum*)	East African copal	Caesalpiniaceae	Forest tree
Lannea kerstingii	-	Anacardiaceae	Savanna tree
Pellegriniodendron diphyllum	-	Caesalpiniaceae	Forest tree

Table 1. Indigenous Copal Trees in the Region.

The most famous incense *Commiphora molmol, C. myrrha* Myrrh, *C. gileadensis* Balm-of-Gilead, *Boswellia carteri, B. papyrifera* and *B. sacra* Frankincense are tapped from north-east Africa and southern Arabia. *Ervatamia coronaria* East Indian rosebay is an introduction - the wood is burnt as incense; and *Myroxylon balsamum* var. *balsamum* and var. *pereirae,* introductions from tropical America, yield the medicinal balsam of Tolu and balsam of Peru, respectively (61).

Allium sativum Garlic is widely believed to have magical powers and extensively used in many countries of its distribution to counteract evil forces (see TALISMAN: FUNCTION: Evil Spirits). In W Africa, the scent of Garlic is believed to be inimical to evil influences causing disease in children. In Africa, the Arab wears a piece of Garlic as an amulet against the evil eye; and *Allium cepa* Onion is similarly used as an amulet to drive out spirits (60). In S Nigeria, the Yoruba place the pounded leaf of *Sanseviera liberica, S. senegambica* or *S. trifasciata* African bowstring hemp with chalk and the leaf of *Baphia nitida* Camwood as a charm on the door where a child is sick, to keep evil away (18). In Ghana, *Boerhaavia* species Hogweed, enters into charms to drive away evil spirits or witchcraft believed to cause children to cry during the night. The leaves are collected with rites, knotted in a white cloth and tied to the left wrist of the child (41). The species in the Region are *B. coccinea, B. diffusa, B. erecta* and *B. repens.* The seeds of *Antrocaryon micraster,* a forest tree, as necklet serve the same purpose (41).

Antrocaryon micraster - 1/3 natural size

In its distribution area, *Okoubaka aubrevillei,* a rare forest tree, is believed to be such a powerful fetish that the egg-like nut is either placed in a child's bathing water, or a poultice from the ground scrapings with water is smeared on the child's body as a dressing to drive away all evil influences and disease. For adults, the nut is placed under the pillow for the same effects (46). In Ivory Coast, people often wash babies with the bark infusion of this tree believing that it protects them spiritually from disease (36). The nuts are commonly

sold in medicine markets for this and other purposes. The nuts are sold for a sum of money plus a coin or a pebble - likely a symbolic gesture (16). (Incidentally, the fruits of *Balanites wilsoniana*, a copal-producing forest tree, are sometimes inadvertently peddled as those of *Okoubaka aubrevillei* since the two seeds are very similar (46) - the former is a capsule and splits when dry, but the latter is a nut so does not split open when dry).

In Ghana, a decoction of the whole plant of *Hygrophila auriculata*, a thorny acanth in marshy places, is used by the Ga tribe to wash babies against the evil eye (15). The roots of *Ziziphus spina-christi* Christ's thorn are used by nursing mothers in a superstitious practice as necklace and girdles for children as protection against, or cure for diseases and misfortune (18). This is probably an instance of sympathetic magic, with reference to the Crown of thorns (see SACRED PLANTS: RELIGIOUS PLANTS: The Christ). In Congo (Brazzaville), *Asystasia gangetica*, a weedy herbaceous acanth, is considered a fetish protective of children, especially against illness (11). In Ivory Coast, a species of *Voacanga*, a savanna shrub or small tree, is recognized by the Kru as a fetish tree to protect children from illness and evil spirits. Protection is first acquired by the child on reaching the age of unaided walking by being bathed in a leaf-macerate each morning (36). In this country, the Gagou give the sap crushed from *Secamone afzelii*, a scrambling shrub with milky latex, or the powdered whole plant to young children as a protection against evil spirits (36); and *Aerva lanata*, a straggling amaranth herb in moist places, is similarly deemed to confer protection against evil spirits (36).

Animals

Among the Ewe-speaking tribe of Ghana, Togo and Benin Republic, when there is suspicion that evil forces are destroying livestock, the kraal or coop is fumigated with *Acanthospermum hispidum* Star burr, a weedy composite, to keep off the forces (41). In S Africa, the Ndebele use the twig of *Salix woodii* Wild willow in marking pegs which are 'medicated' and inserted in the ground in and around a kraal for protective purposes (60). In Europe, pieces of the root of *Helleborus niger* Christmas rose, a temperate plant, were inserted in a hole cut through the ear or dewlap of a sick animal with the idea of warding off the evil spell - on its removal twenty-four hours later,

the trouble was supposed to be cured. The plants are still seen near cottage doors, a link with the past when they were often set near the threshold to prevent evil spirits from entering (49).

SPIRITUAL PROTECTION

Some of the plant prescriptions in traditional medicine are believed to expel evil influences, spells, bad spirits, curses and bad juju; and to offer protection against these forces. In Ivory Coast, *Ocimum gratissimum* Fever plant as with other labiates, has power to keep people in good sorts and to ward off evil. A decoction made of leafy twigs to which *Citrus* leaves may be added is sprinkled on persons in need for protection (12). Like other strong-smelling plants, *Hyptis suaveolens* Bush tea-bush, *H. pectinata* and *H. spicigera* Black sesame, all strongly aromatic bushes, have 'protective' values. In this country, people believe the plants ward off evil spirits, and are taken in medication for the purpose (12). In Ghana, it is believed that if the inflorescence of *Stachytarpheta angustifolia* Devil's coach whip, a herbaceous annual of waste places, is chewed early in the morning and water in which the leaves have been rubbed is used in bathing it will render one immune from evil influence (21). In Senegal, Serer doctors make up washes with the macerated root of *Morus mesozygia,* a forest tree, to give protection against evil influences (35b).

In S Nigeria, a Yoruba medicine that prevents curses affecting a person includes the fruit of *Drypetes chevalieri*, a shrub, with the tuber of *Plectranthus* species, a labiate; and the seedless pod of *Xylopia aethiopica* Ethiopian pepper, 'pound with soap and use it each time you wash' (59). Another prescription consists of the leaves of an unidentified plant locally called 'monwokuro' with the head of a cat, *Felis domestica,* and an egg, 'cook to make an infusion and wash with it on a refuse tip in the night' (59). In Ghana, *Scoparia dulcis* Sweet broomweed, an annual or perennial weed, with ten other herbs, is used in a ritual bath to reverse a curse (5). In the Region, a mixture of the fruit juice of *Okoubaka aubrevillei,* a rare forest tree believed to be fetish, with that of *Raphia* species Wine palm in either water or palm-wine drank after narrating ones complaints is believed to offer spiritual protection (7).

In Gabon, culms of *Pennisetum purpureum* Elephant grass may be laid across a door threshold to prevent entry of evil spirits; or for the same purpose clumps of the grass are grown near to houses (58). In this country, the Fang believe *Clerodendrum melanocrater,* a small climbing

herb, to be a charm protective against evil spirits (58). In Tanzania, the sap of Elephant grass, on a similar line of thought, enters into formulations with a number of other drug-plants to be taken by mouth as a tranquilizer for someone possessed by spirits (14b). In S Nigeria, spiritous association is also found in the Efik name of Elephant grass, 'mbokok ekpo' - meaning 'sugar-cane of the demon' (14b). In Central African Republic, the root of *Phacelurus gabinensis*, a perennial grass, is worn by women in Ubangi as a perfume with the added satisfaction that it repels evil spirits (57). In Congo (Brazzaville), the leaves of *Lantana camara* Wild sage and *Olax latifolia*, a forest shrub, have a place in magic to wash away bad spirits (11). In this country, a preparation of the leafy stems of *Renealmia africana,* a perennial herb in the ginger family, crushed and left to stand in the sun will drive out evil spirits which make a person gabble at night (11).

In Ghana, the pulverized leaf of *Croton lobatus*, an annual euphorb weed, is burnt as incense in a ritual to fight and drive away any evil influences (41); and in N Nigeria, a person washing with a lotion prepared from the plant keeps evil at bay (18). In Ghana, a bath of *Launaea (Lactuca) taraxacifolia* Wild lettuce infusion is believed to give protection against evil spirits - the burnt plant as powder has the same effects (41). In its distribution area, *Leptadenia hastata*, an asclepiad twiner with light green latex in dry savanna - sometimes tended as a magic plant, is used in fumigations to drive away evil spirits (41). In Ivory Coast and Burkina Faso, the whole plant decoction of *Paullinia pinnata,* a sub-woody climber with tendrils, used as bath offers protection against evil spirits and from danger (36).

In S Nigeria, the Yoruba invoke *Urera mannii,* a shrubby climber, and *Cissus populnea,* a liane of woody savanna ascribed with fetish or magical power, in an incantation to send away evil and to keep calamity away from one's head, respectively (56a). In this country, *Bridelia micrantha,* a forest tree, has magical attribute as a protection against one's enemies (56b). In Senegal and Guinea, Mistletoe found growing on the plant if placed in a new field is held by the Tenda people to have protective powers (27). In Gabon, *Guibourtia tessmannii* Rosewood, a large forest tree confers protection against evil influences (14b); and the people powder up the seeds of *Sterculia oblonga,* a forest tree, to put into ritual ablutions for protection against the 'evil-eye' (58). In Ivory Coast and Burkina Faso, the odour of *Costus afer* Ginger lily and *C. lucanusianus,* tall forest herbs, are regarded as being inimical to evil influences (36). In Ivory Coast and Ghana, Ginger lily is placed in front of the door leading to the stool room (where all the stool regalia are housed) as preventive of any evil intention, bad eye or bad juju among the Nzima and the Akan-speaking people (17).

In Congo (Brazzaville), *Thomandersia hensii (laurifolia)*, an acanth shrub, is planted by a hut-entrance to keep away demons, or the sap as a drink will exorcize evil spirits (11); and a tisane of bark or leaf of *Psychotria venosa,* a forest shrub, taken by draught or put into a bath, is considered able to protect against spirits and malign influences (11). In this country, *Helixanthera mannii* Mistletoe, an epiphytic parasite, enters in magic practices for protection against evil spirits and to prevent bad dreams (11). In Ivory Coast, protection against evil spirits is sought by drinking and washing with a leaf decoction of *Parinari curatellifolia,* a savanna tree (3). Among the Ewe-speaking people of Ghana, Togo and Benin Republic, *Corchorus aestuans,* an annual or perennial semi-woody plant often cultivated as spinach, is buried in the house against evil (34b) - (but see DISEASE: MENTAL DISORDERS). The tribe places the leaf of *Newbouldia laevis*, a medium-sized tree often planted around fetish places, collected with rites, in a room as deterrent of evil things (5). The tree is also often planted as a hedge around a bathhouse in villages for spiritual protection, as it is traditionally believed that one is more vulnerable to evil forces when in the nude (9).

Another common hedge-plant in villages against evil spirits and their influences is described. 'The entrance to the long street-shaped villages are frequently closed with a fence of saplings, and this sapling fence you will see hung with fetish charms to prevent evil spirits from entering the village; and sometimes in addition to charms you will see the fence wreathed with leaves and flowers' (37). In many parts of W Africa, *Jatropha curcas* Physic nut is planted as a hedge or a live fence or planted in the vicinity of the household, as protection from evil forces and witchcraft. In Peru, Physic nut is among the plants used for their drastic and emetic activity to expel bad spirits ('contagios' or 'danos') from the body (19). Other plants used for the purpose are given in Table 1 below:

Species	Plant Family	Local Name
Phytolacca weberbaueri	Phytolaccaceae	Yumbi
Senecio elatus	Compositae	Hornamo amarillo,
Valeriana adscendens	Valerianaceae	Hornamo morado

Table 1. Plants Used to Expel Bad Spirits in Peru

In the Region, fetish priests prescribe the fruits of *Solanum americanum (nigrum)* Black nightshade, a herbaceous erect weed, with the flowers of *Ageratum conyzoides* Billy goat weed and that of *Boswellia dalzielii* Frankincense tree or *Daniellia thurifera* Niger copal tree to dispel evil forces and bring in good ones (38). In N Nigeria, *Monocymbium ceresiiforme*, a tufted perennial grass, is used against the 'evil eye' in Bornu (18); and in Tibesti, *Geigeria alata,* a composite herb with 3-winged stem, enters superstition, and is remedy against the 'evil eye' (42). In Ghana, *Clausena anisata* Mosquito plant is commonly hung in houses or put on fires to keep away evil spirits (34b). (See also GHOST AND SPIRIT WORLD: EXORCISM: Widowhood Rites). In Gabon, *Elytraria marginata,* a small rosette acanth, is used to ward off evil spirits (58).

In S Africa, a decoction of the bark of *Acacia karroo* Gum arabic tree is a Zulu emetic often used to expel any deleterious matter arising from the activity of a witch or other evil person; or the tribe burn the green twigs of *Rhamnus prinoides* Dogwood to smoke out evil influences from their kaffir corn lands (60). In this country, the herbalist uses *Strophanthus grandiflorus*, a climbing shrub with white latex, as a charm against evil; or sprinkle the powdered stem and leaf of *Adenia gummifera*, a climbing plant, about the kraal entrance to ward off evil influences (60). The herbalist may also sprinkle an infusion of the tendril roots of *Begonia* species on the path leading to the kraal to ward off evil spirits (60). In this country, the Xhosa and the Zulu use *Plumbago capensis* Plumbago as a charm to ward off evil; and the medicine-man of the former tribe uses it to confound an enemy in every sense of the word (60). Bantu evil doers also use the plant to confound an intended victim and his dog, *Canis familiaris* (60).

In Mozambique, an aspirant Tsonga medicine-man prescribes *Acridocarpus natalitius*, a shrub, as a protection against the evil charms of competitors; and in Botswana, the Southern Sotho witch-doctor uses *Ipomoea crassipes* var. *longepedunculata,* a creeper, as a charm to prevent harm befalling a village (60). In Tanzania, *Areca catechu* Betel with *Cinnamomum cassia* Chinese cassia, *Myristica fragrans* Nutmeg and *Elettaria cardamomum* Cardamon is chewed among the Africans and the Indians by a person supposed to be possessed by evil spirits (60). In this country, a root decoction of *Uvaria leptocladon* var. *holstii*, a climbing shrub, is used for possession by spirits (60).

In Congo (Brazzaville), *Brillantaisia patula*, a stout forest acanth, is held to have magical power to exorcize evil spirits; and *Annona senegalensis* var. *onlotricha (arenaria)*, a small tree, is held to have magical properties in sorcery to keep away evil spirits (11). In Ivory Coast, the Dyimini prepare an ointment of the pulped leaves of *Leptadenia hastata*, a twiner in dry savanna, which spread over the body confesses protection to a traveller from evil spirits (36). In this country, *Hybanthus enneaspermus*, a small erect plant, is used by some fetish to drive away evil spirits (36). In Ghana, *Ochna afzelii* and *O. membranaceum*, savanna shrubs, when used as a walking stick by the Ada people, is believed to protect the owner from bad spirits; and when this stick is placed at the entrance of a house, it is believed to prevent evil spirits from entering (2) (see also FETISH: FETISH PLANTS: Shrubs and Lianes).

The Ewe-speaking people of Ghana, Togo and Benin Republic burn the fallen leaves of *Croton zambesicus*, a small tree with silvery leaves believed to be fetish and juju and planted in villages, as a fumigant in the belief that it repels evil spirits (8). Among the Akan-speaking people of Ghana and Ivory Coast, a corn-cob, baked in fire, then bound with *edow* (a twine prepared from the leaf-sheaths of *Raphia* species Wine palm) and hung up in a doorway, is believed to prevent any evil entering the house (22). The Akans also believe that the seeds of *Dioclea reflexa* Marble vine, called by the natives *enteh,* pierced with a hole and hung up over the door of a house, prevents people talking scandal about the inmates (22). The bark of *Albizia adianthifolia* West African albizia exudes a clear gum - which is probably the *bongbo* used in the Region as a cosmetic, and to drive away evil spirits (Millson fide 18). In S Nigeria, the leaves and bark of *Barteria nigritana* ssp. *nigritana,* a small tree in coastal areas, have a superstitious use in Lagos as medicine to act at a distance or divert a charm against the doer (18).

In W Africa, many of the spiritual churches and sects believe that *Myrtus communis* Myrtle, an introduction from W Asia, is a religious tree and therefore inimical to evil influences, and accordingly use the leaves in prayers to drive away evil spirits (43). (The belief in the religious significance of the Myrtle extends beyond the Region). In Europe, *Ruta graveolens* Common rue or Herb of grace, a native of the Balkan Peninsula, Turkey and Crimea, and cultivated for at least four centuries, is famous as being a potent antidote against evil forces and witchcraft (25). In Sicily, *Rosmarinus officinalis* Rosemary, an evergreen aromatic labiate, is one of the several cures for evil-eye *(malocciu)* - bunches of the plant are placed around the victim and burned on the Island of Pantelleria; and Rosemary burned in the house will get rid of evil-eye as well (28).

Among the Ewe-speaking people of Ghana, Togo and Benin Republic, the whole plant of *Sida linifolia,* a little malvaceous weed of cultivated

land called 'wodoewogbubgo' in Ewe, meaning 'they planned but retreated', is used with the necessary rites by the tribe in various ways as a personal charm for an indefinite postponement of all adverse intentions (5). For instance, the powdered leaf mixed with an ointment and applied is preventive of evil spirits (45). Alternatively, either a coin or a cowry shell or a bottle of lavender is placed with rites near the plant as compensation, while a request for spiritual protection is made, before collecting a leaf into ones pocket. This is followed later by an egg offering when the request is successful (45).

In the Region *Sida acuta*, a common malvaceous weed, is also used for spiritual protection. Fetish priests prepare a medico-magical prescription with a leaf each of the plant picked with rituals from seven different plants, then tied in a corn husk and boiled with thunder rock, then ground in an earthenware (6). The preparation is licked while lying flat on the floor as preventive of all evil forces and spirits. The prescription could also be used in a knotted form when a chief is carried in a palanquin, for spiritual protection from evil forces (6). In the Region, the fruit of *Citrus aurantiifolia*, Lime is believed to counter-act evil intentions and bad juju, and for this reason it is also worn by chiefs for protection against such forces when they are carried in a palanquin (5). Among the Ga-speaking and the Ewe-speaking people, *Momordica balsamina* Balsam apple or *M. charantia* African cucumber is worn around the neck by chiefs at enstoolment ceremonies and festivals for protection from evil spirits (41).

In Botswana, a mixture of the crushed bulb of *Scadoxus (Haemanthus) hirsutus,* a lily, and water is sprinkled around huts and other places to ward off evil influences; or *Gomphrena celosioides,* a common weed, is used along with species of *Asclepias* and *Myosotis* to ward off evil spirits (60). The plants are said to have this effect because they have a disagreeable odour (60). In this country, the Southern Sotho rub the powdered ash of the root of *Melolobium microphyllum,* a tree legume, mixed with the gall of a black sheep, *Ovis* species, into incisions on the forehead to ward off evil spirits (60). *Albuca* species, lilies, are used in general to ward off evil spirits; and the wood of *Teclea utilis,* a tree in the orange family, is used to carve drums for driving away spirits (60).

In W Africa, the dried bark of *Elaephorbia drupifera* and *E. grandifolia,* euphorb trees with abundant caustic latex in savanna and forest respectively, collected with the necessary rites and hung in the house; or the dried, pulverized leaf, or a little of the caustic latex ingested is both inimical to witchcraft and preventive of evil influences generally

(54). In Gabon, as with fleshy euphorbias generally, the trees are also ascribed with fetish and magic properties on account of the bizarre form. They are planted in the streets and before houses as a fetish-protector to keep away evil influences and thunder-bolts; and twisted around arrows, spears and pointed stakes (14b). In this country, *Combretum racemosum,* a scandent shrub or forest liane, is frequently planted at village entrances, or a branch is hung over a house-door to keep away spells (14a). Sorcerers in this country put *Acrostichum aureum,* a large terrestrial fern in wet places, into fumigations for exorcizing evil spirits (58).

In Senegal, *Allophylus africanus,* a shrub of forest margins, is credited with magical attributes capable of displacing evil circumstances. A length of wire with small pieces of the wood tied at each end is stretched across a path being an entrance to a house, a concession or a rice field. Should the wire be displaced, an evil influence is deemed to have passed by, and thus immediate ablutions must be made with leaves of this plant pounded in water to render the individual invulnerable (35a). In this country, the Nyominka use *Calotropis procera* Sodom apple as a fetish to confer protection of their huts, by fastening branches over the door (35a). In Ivory Coast, various species of *Psychotria* are commonly used for magical preparations to drive away spirits and evil influences (12). In Gabon, the bark of *Bridelia grandis* ssp. *grandis,* a forest tree, in fumigations has superstitious use to drive away evil influences (58).

In S Nigeria, an unidentified plant, locally called 'olobgogin' in Yoruba, meaning 'yawning cat, *Felis domestica*', is the vegetable of spirits, and is invoked to drive evil away from home (56a). In Congo (Brazzaville), washing in a decoction of *Anonidium mannii,* a small tree, confers individual protection from evil forces (11). In Benin Republic, a mini scarecrow-like charm was used to repel evil. 'They place a ridiculous caricature of the human form, made of grass, old calabashes, or any rubbish, on the doorstep of their houses and on the gates of the enclosures, to keep evil spirits from entering therein' (22a).

In the Hebrides, *Caesalpinia bonduc* Bonduc or Grey nickernut, a popular tropical drift dissemule was worn by the Islanders as an amulet and was useful in warding off the Evil Eye - the seeds were supposed to turn black when harm was intended to the wearer (Martin fide 32). (For the origin of tropical drift dissemule, see RELIGION: SYMBOLIC PLANTS: <u>Sea Beans</u>). In Ghana, Bonduc seeds are used as fetish for hanging on small children (Thonning fide 33) - likely to expel evil influences causing disease. In Europe, the thick roots of *Mandragora*

officinarum Mandrake were valued by the Ancients to keep off evil spirits and render other notable services; and a small piece of the roots of *Paeonia officinalis* Common paeony, a native of southern Europe, worn round the neck as an amulet, was thought to protect the wearer against evil spirits (49). In the Mediterranean region, *Punica granatum* Pomegranate was thought to give protection against evil spirits (49); and in medieval days, *Hyssopus officinalis* Hyssop was considered a charm against the evil eye (50). In the Middle Ages, *Leonurus cardiaca* Motherwort was used to protect individuals from evil spirits (10).

In S Nigeria, the Yoruba say that *Heliotropium indicum* Indian heliotrope or Cock's comb can help turn away evil, for in calling it 'agogo igun', meaning 'beak of vulture', implies that when it bears fruit it turns the wrong way like the beak of a vulture, *Necrosyrtes monachus*, when feeding (56a). The tribe address *Luffa cylindric*a Loofah or Vegetable sponge in an incantation 'to wash badness out of the body', alluding to the way the plant creeps in the forest like a liane (56a). In Ghana, images used to be hung on *Morinda lucida* Brimstone tree, in a superstitious usage, and gifts of food offered to ward off evil influences, such as death of several members of the household (34b). The leaves of Brimstone tree were also burned in broken pots to drive off evil spirits (34b). In Ivory Coast, the smoke from the flame of *Zanthoxylum senegalense (xanthoxyloides)* Candle wood, a small savanna tree in the orange family, is always considered by the Baoulé people of the Tiébissou region to possess magic properties to drive away evil spirits (36).

Native doctors in the Region prescribe a bath with an infusion of *Tridax procumbens* Coat buttons, a weedy composite, as a precaution against evil spirits. As part of the rituals however, and for maximum efficiency, the plant has to be urinated upon before collection (41). A wash with a maceration of the leaves of *Asystasia calycina* or *A. gangetica*, scrambling weedy annual acanths, in water, in addition to a rub with the leaves on the whole body expels evil spirits (41). Alternatively, water in which the stone (excrescences of calcium malate) in the heartwood of *Milicia (Chlorophora) excelsa* or *M. regia* Iroko has been immersed is used in ritual cleansing at dawn for personal protection from evil forces (41). The seeds of Iroko may also be used against evil forces (29).

In Senegal, the Bedik, in a superstitious sense administer a bath of the macerate leafy twigs of *Guiera senegalensis* Moshi medicine, a small shrub in sandy wastes and semi-desert areas, to persons possessed by evil spirits (35a, b). Chewing the gummy exudate of *Piliostigma thonningii*, a savanna shrub or tree legume, is regarded as passing some of the tree's

own property of discouraging evil spirits to the person who chews it (30). Medicine-men in the Region sprinkle the leafy-juice extract of *Pergularia daemia,* a trailing asclepiad sometimes tended around villages as a magic plant, in a fetish ritual to drive away evil from suspected objects (41). The whole plant may also be buried with the necessary rites in front of the house; or the plant may be charred together with dried palm fruit husks, and the mixture sprinkled round the whole house for the same effects (29).

Among the Hausa of N Nigeria, the Zarma of Mali and Niger and the Fula, medicine-men prescribe the leaves of *Cassia sieberiana* African laburnum to be used to safeguard juju from destruction by evil forces (43). In parts of the Region, *Securidaca longepedunculata* Rhode's violet, a savanna shrub, is believed to be inimical to evil forces, as such a tincture of the roots or root-bitters in a pomade is applied to the skin for spiritual protection from these forces (45). In Congo (Brazzaville), *Strychnos icaja,* a large forest liane, has magical application by medicine-men to confer protection against sorcerers, spirits and fetishes (11). In Mozambique, a cold-water infusion of the powdered root of *Securidaca* is drunk by persons who are believed to be possessed of an evil spirit and is taken, together with an infusion of a plant which is probably *Ozoroa (Heeria) abyssinica,* a tree, as a purifying medicine after ceremonial defilement (60). In this country, the Chopi use *Securidaca* with *Sphedamnocarpus pruriens* Malpighian hair as a medicine for persons 'possessed' of evil spirits (60).

Sympathetic Magic

Some of the plant prescriptions for spiritual protection against evil appear to be based on sympathetic magic. In many parts of the Region, medicine-men sprinkle an infusion of *Bryophyllum pinnatum* Resurrection plant in places suspected to be infested with inimical forces to neutralize them (41). It is also the general belief that the leaves of Resurrection plant chewed with kola-nut protects one from any evil intention (20). The sympathetic magic is said to lie in the plant's natural ability to resist burning (see also JOURNEY: COMPULSORY RETURN HOME: Sympathetic Magic). In S Africa, water into which *Kalanchoe hirta,* a herbaceous perennial plant in the same plant family as Resurrection plant, has been thrown, is sprinkled by the Zulu around a kraal to ward off night visits by evil spirits known as *nkovu* (60).

In W Africa, the whole plant of *Schrankia leptocarpa* Sensitive plant, ground or macerated in water, sieved, then fortified with the blood of the tortoise, *Kinixys* species, lavender and common salt as ingredients in an earthenpot, is used for ritual bath on seven consecutive nights for spiritual protection. The pot with the remaining contends thereafter is crushed on a rubbish heap or at crossroads (7). The inclusion of tortoise blood in addition to 'bath on seven consecutive nights' appears to be an instance of sympathetic magic. In Ghana, the root of *Sophora occidentalis*, a maritime shrub with yellow flowers, collected early in the morning while naked and ground with a millipede, *Iulus terrestris;* a slug, *Limax* species; and forty-nine seeds (seven times seven) of *Aframomum melegueta* Guinea grains or Melegueta as ingredients, enter into charms preventive of all sinister and evil forces (6).

In Ghana, *Cissampelos mucronata* and *C. owariensis,* climbing shrubby menisperms in savanna and forest respectively, have similar uses. One plant is 'talked to', while another nearby is uprooted and, with an egg, used in bath water on seven consecutive days. The ritual is both for protection against evil forces and for purification (15). In the Region, medicine-men prescribe the young flowers of *Calotropis procera* Sodom apple collected with the necessary ceremony and chewed one a day for seven consecutive days as protection against evil forces (41). The Hausa of N Nigeria, the Zarma of Mali and Niger and the Fula believe that, a piece of the dried stem of the plant carried by any individual is believed to warn others, and to protect him or her from evil forces (43). In Gabon, the seeds of *Omphalocarpum procerum,* a forest tree, are put into a rattle with which to chase out evil spirits; and *Smilax anceps (kraussiana)* West African sarsparilla, a prickly forest climber, is ascribed with magic power - a tangle of spiny vines around a hut gives protection enough to turn away evil spirits (58). The sympathetic magic appears to be hinged on the noise of the rattle and the prickles of the climber, respectively. In its distribution areas of the Region, the dried seed of *Martynia annua,* a foetid decorative introduction which grows around villages, is either placed in pomade (46) or in lavender (1) and applied as a cosmetic to counteract evil influences. The clawed seed may also be worn as a necklet against evil - either naked or covered in a leather case like a talisman (1). The sympathetic magic is hinged on the clawed seeds.

The butter extracted from *Vitellaria (Butyrospermum) paradoxa* Shea butter tree is used domestically both for cooking and as a cosmetic among others. Spiritually, this butter with a supersaturation

Vitellaria paradoxa Shea butter tree

of common salt, is believed to be preventive of evil forces when it is licked as needed (23). In N Nigeria, salt seems to be very generally regarded as being particularly inimical to evil spirits, the idea being based probably on its power of preventing decay. This is likely an instance of sympathetic magic. In this country, the Hausa mother says that her baby's flesh is salt, so that the witches will not take it (55); and the practice of putting salt in coffins was both religious and utilitarian (55). The Hindus wave salt round the head of a bride and bridegroom and bury it near the house as a charm (55). The Roman Catholic priests still use salt in baptism (55).

There is a popular traditional belief among many of the ethnic groups in the Region that *Elaeis guineensis* Oil palm planted on the compound of a house, protects the whole household from evil spirits (6); and that an infusion of the roasted seeds of *Abelmoschus (Hibiscus) esculentus* Okro or Okra taken as tea with salt is protective against evil influences (4). In the Region, the spiritual uses of Madagascar periwinkle are based on the colour of the flowers. The white-flowered *Catharanthus roseus* Madagascar periwinkle chewed with salt is preventive against evil intentions, while the purple-flowered type chewed with salt strengthens one against adversaries (8).

Catharanthus roseus Madagascar periwinkle

Protection Against Poisoning

The fear of taking in food or drink - that is either deliberately or inadvertently poisoned cannot be ruled out. Persons who suspect foul play in an evil, unfriendly or hostile environment either abstain from any food or drink altogether, or carry their own food and drink along with them, or otherwise take precautionary measures to acquire

the necessary protective charms or talisman believed to counteract this bad juju.

In Liberia, the Mano people eat the fruit of *Pentadesma butyraceum* Tallow tree to act as a 'preventive against poisoning' when going among strangers (18); and in S Nigeria, the Igbo tribe at Asaba are reported using *Elytraria marginata*, a small acanth with rosette leaves, perhaps superstitiously, in medicine against poisoning caused by bats, species of *Cynopterus, Eidolon, Epomophorus, Glossophaga* and *Pteropus* (14a). In Ivory Coast, *Gloriosa simplex* and *G. superba* Climbing lily are credited with the ability to cause poisoning at a distance (12); but in S Nigeria, Climbing lily, locally called 'ewe-aje' in Lagos, are used in a charm to prevent one from being poisoned (Dawodu, Herb. Kew fide18). Traditional medicine-men in the Region prescribe the leaf of *Newbouldia laevis*, a popular fetish tree, to be used as a personal protection against poisoning (5); and in Ivory Coast, the chewed leaves of *Hoslundia opposita*, an erect odorous labiate, are said to give protection from poisoning and evil fate (36). In the Region, the fruit of *Citrus medica* Citron features in the folklore of western Soudan: 'if one carries one in one's pocket, all attempts to poison one would fail' (14c).

In Ivory Coast, a decoction of the whole plant of *Sporobolus pyramidalis* Rat's tail grass, is given in draught as a counter-poison in the Man region (36); and in Senegal, *Arachis hypogaea* Groundnut is considered by the Peul and the Tukulor people to be anti-dotal to poisoning by magic (35a). In Ghana, incisions cut on the wrists and applied with charred assorted herbs (botanical identities unspecified but including *Leptadenia hastata* and *Pergularia daemia* – both asclepiads) collected and prepared with rites, is an instant protective and preventive charm against any poisoned food or drink. The plate, bottle or drinking glass will crack immediately on contact. In folklore, *Indigofera astragalina*, an erect or straggling legume, *I. hirsuta* Hairy indigo and many species of *Tephrosia*, sub-woody legumes, are used in a similar manner. An infusion of the plant is used as a wash every morning in the belief that any intended evil, such as attempted poisoning to return to hurt the evil doer. In S Nigeria, a similar idea is found among the Yoruba with reference to *Argemone mexicana* Mexican or Prickly poppy (18).

In S Nigeria, medicine enabling one to eat poisoned food with impunity among the Yoruba, include the following: *Piper guineense*

West Africa black pepper with the meat of an elephant, *Loxodontha africana;* python, *Python regius;* and pigeon, *Columba unicincta* or *C. guinea* as ingredients, 'burn in palm kernel oil and lick a little at a time'; or *Ageratum conyzoides* Billy goat weed, a weedy composite, with onion and salt, 'grind in palm-oil and lick a little at a time'; or *Aframomum melegueta* Guinea grains or Melegueta with the meat of a horse, *Equus equus;* an elephant, and the head of a pigeon as ingredients, 'burn and lick a little at a time'; or *Abelmoschus (Hibiscus) esculentus* Okro or Okra, 'put the dried seeds into a chameleon, *Chamaeleon vulgaris,* plant it in a refuse tip, when they germinate and they are about one foot high, uproot them and burn with the seeds of Melegueta - this is to be licked (59). In Central African Republic, the irritating hairs of *Mucuna poggei* Buffalo bean have been used with evil intent in Ubangi to lace food causing great discomfort (57).

Columba species Pigeon

In Congo (Brazzaville), the leaves and roots of *Carpolobia lutea,* a forest shrub or small tree, enters various formulae to confer protection against evil influence, to combat repeated miscarriage and food-poisoning (11). In Zambia and Zimbabwe, the root of *Solanum linnaeanum (sodomeum)* Apple of Sodom is tied round the neck or carried in the pocket by the Manyika as a protection against poisoning by an 'umtakata' - the tribe also carry some of the root of a *Solanum* species as a protection against being poisoned (60). In S Africa, the Xhosa chews a small piece of the root of *Dicoma anomala* Swartstorm, a composite, as a charm so that if he receives any poisoned food, he may immediately vomit it (60). 'In Ancient times, the bitterness of *Ruta graveolens* Common rue or Herb of grace, was thought to be a sign of virtue, and was considered an antidote for poisoning. Many rulers of that period ate it daily as a possible remedy against plots to poison them' (50). (For physical protection, see PROTECTION, PHYSICAL).

PLANTS FOR EVIL

Some plants are used to evoke evil or to symbolize evil. In S Nigeria, *Dioscorea rotundata* White yam or Guinea yam is evoked by the Yoruba to give bad medicine to an enemy; and *Bombax buonopozense* Red-Flowered silk cotton tree, called 'poponla', meaning 'big post' is evoked by the tribe in the spirit form as a cudgel by which to send evil to one's enemy (56a). In N Nigeria, the Hausa regard *Andira inermis* Dog almond as evil and so the people are reticent as to its properties (18). In this country, the Hausa also ascribe some superstition to *Parkinsonia aculeata* Jerusalem thorn. They apply the epithet *ka-ki-zo-ila* to the plant, - *ila* meaning unexpected but predestined evil (18). (But see PLANTS AGAINST EVIL below). *Brachystelma togoense*, a perennial herb in lowland and montane situations with edible tubers is reported to have evil-smelling flowers (14a).

Parkinsonia aculeata
Jerusalem thorn

Throughout the Region, there is a general belief that *Erythrina crista-galli* var. *variegata* Cockspur coral-tree, a decorative introduction, is symbolic of evil or enhances the activities of these forces (44). In Ghana, the Akan name of *Ageratum conyzoides* Billy goat weed, a composite, is 'gu-ekura', which literally means 'to disband a village'. This suggests a belief that its appearance in excess is a bad omen (34a). Similarly, in this country, *Emilia sonchifolia,* another composite weed, has reputation as a bad omen if present in a village, because it is frequently found growing on ruins (Williamson fide14a). In Nigeria, *Combretum collinum* ssp. *binderianum*, and two other sub species - *geitonophyllum* and *hypopilinum*, all savanna trees, locally called 'shafa' in Hausa, enter superstitious uses (40). The Kilba of this country prepare a bundle of the leaves tied up with a piece of the tail of a lion, *Panthera leo*; or the hair of a hog, *Potamochoerus porcus;* a bug, *Clarigralla shadabi* or *Nezera viridula;* a louse, *Pediculus humanus*; an arrow shaft; the quill of a porcupine, *Artherurus africanus;* a stalk of grass and a stick used by grave diggers (40). These are supposed to symbolize the evil things that guilty persons stand for. The bundle is called 'shafa' upon which the Kilba take an oath (40).

In many traditions worldwide, trees on which a person hangs himself

or herself are said to symbolize evil - and parts of such trees enter into various charms. In Yoruba mythology when 'Shango', King of Oyo was dejected, lost in the forest with his slave, and hungry, he hanged himself in desperation on *Distemonanthus benthamianus* African satinwood. 'After waiting a long time, the slave, as his master did not appear, went in search of him, and before long found his corpse hanging by the neck from an *anyan*-tree' (22b). The tree is thus associated with evil - not only in Nigeria but also among the Akan-speaking people of Ghana and Ivory Coast. Similarly, *Cercis siliquastrum* Judas tree, from eastern Mediterranean, is traditionally believed to be the tree upon which Judas Iscariot hanged himself - the flowers having blushed with shame and retained their purple-rose colour ever since (25). Legend has it that before the crucifixion the flowers were white (52). It has been observed that confusion between Judas and Judea has led to the belief that this was the tree on which Judas hanged himself (39). It has equally been observed that the branches of an old Judas tree are spreading and quite tortured in shape, and this, along with the red flowers, gives rise to the story that this was the tree on which Judas hanged himself (52).

From the twisted sepals of *Nymphaea lotus* Water lily, Professor Goodyear in *Grammar of the Lotus* traces the Ionic capital, and from that the Greek fret or meander (49). This doubled becomes the swastika, earliest of all symbols, and according to the way it faces the representative of good or evil, male or female, light or darkness (49).

PLANTS AGAINST EVIL

Some other plants are traditionally used against evil, or to punish evil. These plants are also believed to symbolize perfection. Native doctors in the Region confess that *Pergularia daemia,* an asclepiad sometimes tended as a magic plant, and *Newbouldia laevis,* a tree popularly believed to be fetish, are two of the most powerful plants used by many tribal groups in ceremonies to destroy the evil effects of other herbs (41). In its distribution area, the stems of *Pennisetum purpureum* Elephant grass, a robust savanna grass, when used to hit an evil-disposed person is believed to neutralize his or her evil powers (44).

The stems of *Hypselodelphys poggeana* and *H. violacea,* marantaceous forest perennials with bamboo-like habit, also enter into rituals to neutralize evil (41). In N Nigeria, the Hausa saying *shuka hali,* means *Parkinsonia aculeata* Jerusalem thorn will not thrive if planted by an evil-disposed person (18). (But see PLANTS FOR EVIL above). In olden

times the leaves of *Ficus asperifolia* Sandpaper tree and *F. exasperata* Sandpaper tree were used to scrape and scarify the lips as punishment for evil-speaking (18). In a village, *Ceiba pentandra* Silk cotton tree is always planted to the southward, the direction from which the 'beneficial forces' come (48). The Temple of Zeus on Mt. Olympus was purified with water from *Verbena officinalis* Vervain, a shrub in the Verbenaceae family, since the plant was regarded as a panacea, able to deliver from all evil (47). According to Chinese mythology, *Magnolia* species suggest the fragrance of virtues; *Gardenia* species, ornamental shrubs with sweet-scented flowers, symbolize graceful charm; and trees often in glazed containers are meant to symbolize those qualities which enrich life (31). Honesty is one of the virtues of life, and there is a popular superstition that wherever *Lunaria annua* Honesty flourishes, those who grow it are exceptionally honest (49).

References.

1. Abaye, (pers. comm.); 2. Adibuer, (pers. comm.); 3. Adjanohuon & Aké Assi, 1972; 4. Agyeman, (per. comm.); 5. Akpabla, (pers. comm.); 6. Alhaji Asiedu Biney, (pers. comm.); 7. Asumadu-Sakyi, (pers. comm.); 8. Ayivor, (pers. comm.); 9. Badu, (pers. comm.); 10. Blumenthal & al., 2000; 11. Bouquet, 1969; 12. Bouquet & Debray, 1974; 13. Bowdich, 1819; 14a. Burkill, 1985; b. idem, 1994; c. idem, 1997; 15. Cofie, (pers. comm.); 16. Commeh-Sowah, (pers. comm.); 17. Cudjoe, (pers. comm.); 18. Dalziel, 1937; 19. De Feo, 1992; 20. Doe-Lawson, (pers. comm.); 21. Dokosi, 1969; 22a. Ellis, 1883; b. 1894; 23. Eshun, (pers. comm.); 24. Evans-Pritchard, 1937; 25. Everard & Morley, 1970; 26. Fabianu, (pers. comm.); 27. Ferry & al., 1974; 28. Galt & Galt, 1978; 29. Gamadi, (pers. comm.); 30. Gledhill, 1972; 31. Graf, 1978; 32. Gunn & Dennis, 1974; 33. Hepper, 1976; 34a. Irvine, 1930; b. idem, 1961; 35a. Kerharo & Adam, 1964; b. idem, 1974; 36. Kerharo & Bouquet, 1950; 37. Kingsley, 1897; 38. Lamptey, (pers. comm.); 39. Lanzara & Pizzetti, 1977; 40. Meek, 1931; 41. Mensah, (pers. comm.); 42. Monod, 1950; 43. Moro Ibrahim, (pers. comm.); 44. Nkwantabisa, (pers. comm.); 45. Noamesi, (pers. comm.); 46. Ntim, (pers. comm.); 47. Pamplona-Roger, 2001; 48. Pasque, 1953; 49. Perry & Greenwood, 1973; 50. Pitkanen & Prevost, 1970; 51. Rattray, 1923; 52. Richards & Kaneko, 1988; 53. N.W. Thomas, 1913-14; 54. Thompson, (pers. comm.); 55. Tremearne, 1913; 56a. Verger, 1967; b. idem, 1972; 57. Vergiat, 1970; 58. Walker & Sillans, 1961; 59. Warren, Buckley & Ayandokun, 1973; 60. Watt & Breyer-Brandwijk, 1962; 61. Willis, 1973.

15 FARMING

He that tilleth his land shall be satisfied with bread: but he that followeth vain persons is void of understanding.
Proverbs 12:11

Just as modern farming techniques rely on plant manure and chemical fertilizers for increased yield, traditional farming rites are primarily aimed at increased soil fertility with its accompanying high crop production. In addition, some of these farming rituals are believed to protect farm products from pilfering, or to protect the farm from trespassers, and from adverse remarks unfavourable to good harvest.

Rituals are performed during the preparation of the land. For instance, in Liberia, bundles of leaves of *Ixora aggregata,* a forest shrub or small tree with white flowers, are beaten on the ground during felling of the bush for farms as a juju to bring strong winds (Cooper fide 7). The seed-sowing could also be with rites. In N Nigeria, it is customary in Bauchi State to offer a sacrifice at the foot of *Milicia (Chlorophora) excelsa* or *M. regia* Iroko, high forest timber trees, at sowing time, and to tie a strip of cloth around the trees (21a). The Igala do likewise at harvest-time and discharge arrows at the cloth (21a). Sometimes the ritual is given a royal touch. 'In 2800 B.C., nearly 5,000 years ago, the Emperor Chin-nung of China ordained that the season of sowing would be opened by a special ceremony in which he himself would plant the first and best seeds. Seeds of four other kinds would then be sown by the princes of his family' (22). Rituals were also performed during harvesting. In N Nigeria, the chief cuts the first corn, smearing the reaping knife with the juice of the root-tubers of *Gloriosa simplex* or *G. superba* Climbing lily, in tribal agricultural rites among the Mumbake (21b). This is yet another royal ritual.

HIGH YIELD

In SE Senegal, the Tenda use the bulbils of *Dioscorea bulbifera,* in a secret preparation made by men only who eat it before dawn to ensure

peace and good crops (12). In a magical application the tribe lay twigs bearing fruits of *Crossopteryx febrifuga,* a savanna shrub, on *Solanum incanum* Garden egg or Egg plant to ensure abundant fruiting (12). In this country, Mistletoe taken from *Vitex madiensis,* a shrub or small savanna tree, or from *Ficus ingens,* a fig tree, is a fetish to the Tenda to place in a field of *Vigna (Voandzeia) subterranea* Bambara groundnut to ensure an abundant crop (12). In this country, *Ficus cordata,* a fig tree, is said to have magical properties, so the people place the fruits in a broken crock in the middle of a groundnut or bambara-nut field to encourage rainwater to drain towards it to produce a good crop (12). The tribe also place Mistletoe growing on *Spondias mombin* Hog plum or on *Stereospermum kunthianum,* a savanna tree, in groundnut fields to promote crop increase (12).

In parts of the Democratic Republic of the Congo (Zaire), *Celosia argentea,* an erect annual amaranth - often cultivated as a vegetable, enters into a form of witchcraft to ensure good crops where it is growing (16). An interesting parallel to this is recorded from New Britain in Polynesia, where the amaranth is planted in taro plots: the taro 'smells it' and produces large tubers (Blackwood fide 4a). In Ghana, some farmers of *Musa sapientum* var. *paradisiaca* Plantain carefully tend *Ficus exasperata* Sandpaper tree amid the crop to improve the harvest. The tree is highly prized when grown with plantain, but is a nuisance where maize is farmed (11). On the contrary, the presence of *Hilleria latifolia,* a perennial forest herb, in plantain farms is traditionally believed to reduce crop yield (19). In S Africa, the Southern Sotho burn *Cenia hispida,* a composite, in the fields in the summer to increase the crops; or burn *Rhus discolor,* a bush or small tree, in their cultivated lands to ensure a good crop (33).

In its cultivation area throughout the Region, some traditional farmers of *Ananas comosus* Pineapple make sure to dig only ones with the implement when planting the heads so that the crop fruits after only one year. The traditional belief is that if the planting hole is dug with two strokes of the implement, the crop too will bear fruit after two years - with three strokes, after three years, and so on (2). In S Nigeria, the occurrence of *Acalypha ciliata,* an annual herb, under planted cotton may not be altogether vicarious, for perhaps magical ends it is 'used to adopt good results' (4b). In N Nigeria, the leaves of *Echinops longifolius,* a herbaceous perennial composite with spinose-margined leaves, cut up and sown with seed-corn by the Fula ensures an increase yield (7). In SE Senegal, the leaves of a species of *Combretum,* a shrub or small tree, are hung on the culms of a fully grown maize to hasten ripening - two or

three plants in the corner of the field will suffice (12). In Sudan, medicine prepared from *Dyschoriste perrottetii*, an acanth herb or undershrub, is tied by the Azande tribe round the stalks of their maize so that it will be more fruitful (10).

In Liberia, the fruit of *Antidesma membranaceum*, a shrub or small tree in savanna region, tied on a stick put in the ground when a plot is cleared for farming by the Mano people is a charm to ensure that the rice will grow well (7). It is possible that the practices refer to *A. oblongum* rather than this species (4b). In this country, *Iodes liberica*, a forest climbing shrub or liane, has a place in juju: it is buried in rice-fields to make them yield more plentifully (4b). In SE Senegal, Mistletoe, an epiphytic parasite, growing on either *Terminalia avicennioides* or on *T. macroptera*, savanna trees, is put in rice-padis by Tenda farmers to improve yield. The tribe also burn *Leucas martinicensis*, an aromatic labiate, in millet fields to ensure good crop (12). In this country, *Bauhinia rufescens*, a scandent shrub or small savanna tree, is credited with magical properties to ensure the millet harvest (17b). In Niger, the seeds from dark purple variety of *Pennisetum glaucum (americanum)* Pearl, Bulrush or Spiked millet are purposely mixed with grain for sowing for magical purposes to ensure a good crop (33). In Japan, gardeners used to chop the leaves of *Viscum album* Mistletoe, an epiphytic parasite, with Pearl millet and other seeds (after prayers had been said over them) to ensure good harvest (25).

The Yedina of Lake Chad make offerings of fish and cereal to *Faidherbia (Acacia) albida*, a large savanna tree legume, as a propitiating sacrifice for an abundance of fish, cattle, *Bos* species; and grains (27). In the Hoggar, along the northern limits of the Region, small bags filled with dried whole plant of *Grangea maderaspatana*, an annual or perennial herb or shrubby composite, or *Centipeda minima*, a weedy herbaceous composite, are hung by the Tuareg to fig plantations to ensure development of the fruit (20). In Sierra Leone, the Mende call *Strophanthus gratus*, a liane with milky latex, 'sawa', in part meaning medicine to ensure good crops (4a). (See SECRETS: ASSOCIATED PLANTS). The Malays believe that *Myristica fragrans* Nutmeg trees will not bear unless they can hear the sea, and the trees must be fed with animal food (26). The beliefs are corroborated by the fact that trees grown near the sea and fed with animal food actually do produce the finest fruits. In Hausaland, *'Uwargona'* or *'Uwardawa'*, a female, is the goddess of agriculture (28).

Sympathetic Magic

Synsepalum dulciferum Miraculous berry

Dioscoreophyllum cumminsii West African serenpidity berry

Thaumatococcus daniellii Katemfe 1/2 natural size (symbol of female fertility).

Some of the rituals for high yield appears to be based on sympathetic magic. In The Gambia, to obtain a bumper harvest, farmers will pound up 100 fruits of *Ficus sur (capensis)*, a fig tree with abundance of fruits, for mixing with groundnuts or beans at sowing-time (15). This is surely an instance of sympathetic magic. In Senegal, certain agriculturists growing groundnuts stretch the runners of *Ipomoea asarifolia*, a perennial trailing plant, over the top ends of stakes in the fields as a 'good example' to the groundnuts to encourage the plants to emulate the long stems and many pods of the *Ipomoea* (29). Around Obuase, a gold-mining town in Ghana, the farmers of *Citrus sinensis* Sweet orange first wash the seeds with a macerate of *Synsepalum dulciferum* Miraculous berry or *Dioscoreophyllum cumminsii* West African serenpidity berry or *Thaumatococcus daniellii* Katemfe, all three plants with extreme sweetening properties, before the seeds are sown. The practice is a charm to transfer the sweetening property to the orange - a sure example of sympathetic magic. The mining town is famous for sweet oranges which is attributed by others to the gold in the land (24).

PROTECTION FROM THEFT

Some traditional farmers use charms, juju, fetish or a local deity as precaution to protect farm produce from stealing, or to prevent trespassers and other outside interference likely to affect the yield of the farm. (See also BURGLARY: PROTECTION OF FARMS AND FARM PRODUCE for more instances). Among many traditional groups, it is considered normal for any thirsty or hungry person to help himself or herself on the spot with some farm produce as need be, without the owner's permission. It is not for such people that 'prescriptions' against stealing farm produce are targeted. The real intention of these charms is rather to identify those who steal or intend to steal farm produce in commercial quantities, either to sell or for personal use later, or both; and to prevent such persons and accordingly, punish them severely.

Writing on charms against theft in farms, it has been observed, 'most of the coloured inhabitants of Sierra Leone, whether Christians or otherwise, believe in the efficacy of *gris-gris;* and in the open country about Freetown, in the cultivated plots of ground, one sees these charms raised on poles and sticks to prevent birds from devouring the crops, to keep off thieves, arrest high winds, and, in fact, protect them from every mischance' (9). Similarly, in Ghana it is on record that a protective deity placed in a field to stop thieves among the people of Akropong Akwapim is effective because the Akwapims fear the consequences if they do not take heed (13). 'Charms are not all worn upon the body, some go to the plantations, and are hung there, ensuring an unhappy and swift end for the thief who comes stealing' (18).

In S Sudan, 'the Azande often tie magical creepers round their garden. I saw a magician doing this. He said simply to the medicine that if any one comes to spoil his produce might the medicine break him. He then twisted a length of creeper, saying now and again, 'Break, break, break'. He then tied it to sticks thrust into the ground to support it. As he tied it, he said once again, 'Break, break, break'. Then he took another length of creeper and repeated the performance, and so on' (10). It is traditionally believed that any medicine made against theft in farms has to be cancelled by cleansing before the owner commences to harvest the crop and eat the produce of his cultivation. 'A man must not leave medicine at work in those cultivations in which one makes magic, and proceed to eat of their produce lest the medicine slay him. It is for this reason that they cancel the medicine made to protect cultivation' (10).

In Senegal, Basari medicine-men endow *Pteleopsis suberosa*, a small savanna tree, with fetish power of protection of millet: a few fragments

being added to the grains; consequently the tree is protected in the neighbourhood of millet fields (17a, b). This indeed is an effective traditional method of conserving biodiversity to prevent extinction of species. In Ghana, *Dioscorea dumetorum* Bitter yam is planted on farm roadsides as a protection of the farm from remarks of passers-by: if the farm is praised too much it might become weedy or the crops might die; the influence being that the presence of inferior yam will not tempt providence (4a) - the Twi name 'nkanfo', meaning 'remembrance' or 'praise'. Bitter yam is a poisonous wild yam with prickly bulbils, that twines clockwise, as opposed to anti-clockwise a characteristic of the edible species of *Dioscorea*. In the Yam Zone of W Africa, it is deliberately planted at the edge of farms to prevent the pilfering of the better crop within, and to punish would-be thieves and monkeys, species of *Cercocebus, Cercopithecus, Colobus, Erythrocebus, Papio, Piliocolobus* and *Procolobus.* 'In the fields it is sometimes planted at the edges, perhaps to divert a would-be thief or monkey from a better crop' (7). In Benin Republic, *D. sansibarensis,* a massive perennial tuber producing bulbils along the stem, is sometimes planted around sorghum granaries in, apparently, an act of superstition - probably with similar intentions (4a).

In the Region, *Sida linifolia,* an erect little malvaceous weed popularly believed to be fetish, is either tended or planted with rites in a farm as a preventive of theft and any evil intention (19); and for the same reasons, *Bryophyllum pinnatum* Resurrection plant is often planted near to the entrance of farms (1). Similarly, a band of red cloth, signifying 'danger', attached to any crop is an indication that the whole farm is spiritually protected against pilfering, and a warning to persons with such intentions (in their own interest) to keep off. In Sierra Leone, the fruit-pod of *Didelotia afzelii,* a small legume forest tree, is used in a superstitious practice to avert trespassers on field crops (7). In Ghana, the Nzimas use the curved fruit pod of *Pentaclethra macrophylla* Oil-bean tree to signify ownership of a plot, and to prevent others from encroaching on the farm or property (7). In S Nigeria, unauthorized entry into parcels of land for whatever purpose is barred when the tips of the young fronds of *Elaeis guineensis* Oil palm are displayed there (23). In the Plateau state of N Nigeria, *Tacca leontopetaloides* African arrowroot enters in the rite of *Kum gwaar* - the putting up of a scarecrow to protect a farm (Walu fide 4d).

In south-western Cameroon, *Kigelia africana* Sausage tree protects fields from trespassing and serves as a symbolic deterrent among the

Banen (8). In Ghana, the seeds of *Hygrophila auriculata,* a thorny acanth in marshy places, collected with rites, is ground with poultry feed by the Ga-speaking tribe as spiritual protection against any evil remarks that might cause the birds to die, and for the general good health of the birds (5). In Central African Republic, a piece of Mistletoe from *Tamarindus indica* Indian tamarind placed in the field in Ubangi is held to confer protection against bird-damage; and a small piece of the leaf of *Erythrophleum suaveolens* Ordeal tree stuck on the top of a stick placed in a plantation is deemed to be protection against thieves (31). In Nigeria, the fruit of *Sarcocephalus (Nauclea) latifolius* African peach is impaled on a stick out in a field in Ingalaland as a charm to prevent wild animals damaging crops (4c). In Gabon, *Canavalia ensiformis* Sword bean has a number of magical attributes, and it is grown as a protector of standing crops against marauders (32). A similar belief is found in the West Indies, perhaps taken thither by slaves. The plant is said to 'cut the eye' to prevent petty theft (34).

In Botswana, the Southern Sotho use the fruit of *Citrullus lanatus (Colocynthis citrullus)* Watermelon or *Equisetum ramosissimum* Horse-tail as a charm to drive away worms from crops such as maize (33). The tribe also use *Phygelius capensis,* locally called 'mafifi-matso', as a charm against hail damage to crops, and *Linum thunbergii* Wax flax as a charm to prevent accidents to cultivated lands. In this country, the Ngwaketse use a species of *Amaryllis,* a lily, as a medicine against the bewitchment of crops (33). In Lesotho, a root-decoction of *Asparagus africanus,* a climbing annual with numerous wiry, spiny branches, smeared on pegs placed at the corners of a field is considered to provide protection (14)

BELIEFS AND LORE

In parts of W Africa, the bitter taste of some varieties of *Manihot esculenta* Cassava is explained thus - 'the farmer flatulated during planting'. The slight bitterness noticed even after cooking *Phaseolus lunatus* Lima bean is sometimes superstitiously attributed by the people to a snake having climbed the plant (7). In Ghana, some farmers in the Awutu area, near Winneba, claim that 'weird murmurings' are sometimes heard in yam farms - particularly in farms of a purple variety of *Dioscorea alata* Water yam, locally called 'awuku'. Could this be attributed to the wind? In this country and Ivory Coast, the Akan name of *Maytenus senegalensis,* a thorny savanna shrub or small tree, 'kumakuafo' means 'it kills farmers'. Some species of *Combretum* with similar sharp, pointed thorns go by the same local name - 'kumakuafo'.

In S Nigeria, the numerous seeds fixed on the central axis of *Zea mays* Maize has symbolic significance. Its productive power is reflected in an Odu incantation in Yorubaland which runs that *agbado* (maize) has good luck: it goes into a field naked and returns home with 200 children and 200 sets of clothes (30). In this country, the appearance of *Termitomyces clypeatus,* an agaric mushroom, is deemed by the Yoruba farmer to indicate that the new yam crop should be available in the next two months (Oso fide 4d). Mambila farmers note that when the flowers of *Stereospermum acuminatissimum,* a forest tree, begin to fade, it is time to plant guinea-corn (Chapman fide 4d). In SE Senegal, the Tenda aver that if a field intended for cultivation is possessed by a certain spirit, *bi-yil,* it is necessary to plant *Canavalia ensiformis* Sword bean first (12). In this country, the Bedik decide that a field is cultivatable if seed of this plant germinates in a trial planting, but if its leaves are damaged 'lions', *Panthera leo,* will come and despoil the area (12). In Sierra Leone, the Sherbro say that if an ill-wisher throws a piece of *Ischaemum rugosum,* an annual grass, on a farm of his enemy, it will cover the farm (4c).

Low Yield

The presence of some plants on a farm is traditionally believed to signify low yield. For instance, in cocoa farming areas there is the general belief that the decorative epiphytic rain forest fern, *Platycerium elephantotis* or *P. stemaria* Stag-horn fern, is a bad omen when found on cocoa farms and an indication that the trees have long outlived their usefulness, and must be replanted for a better yield (3). The association of some crops is also believed to adversely influence the fruiting of one of them. For instance, farmers would not plant *Abelmoschus (Hibiscus) esculentus* Okro or Okra and *Lycopersicon lycopersicum* Tomato together. Also when *Elaeis guineensis* Oil palm and *Persea americana* Avocado pear are near each other, the latter does not normally fruit - especially when the palm fronds brush against it; and by coincidence, if palm nuts are stored with pear fruits, the nuts cause the fruits to rot (5). Crops grown in the vicinity of *Piptadeniastrum africanum* African greenheart, a forest tree legume, or under its shadow, fail for some reason. Could it be soil depletion or lack of sunlight? (See also FERTILITY).

References

1. Akoto, (pers. comm.); 2. Atwie, (pers. comm.); 3. Badu, (pers. comm.); 4a. Burkill, 1985; b. idem, 1994; c. idem, 1997; d. idem, 2000; 5. Cofie, (pers. comm.); 6. Cudjoe, (pers. comm.); 7. Dalziel, 1937; 8. Dongmo, 1985; 9. Ellis, 1881; 10. Evans-Pritchard, 1937; 11. Falconer, 1992; 12. Ferry & al., 1974; 13. Gilbert, 1989; 14. Guillarmod, 1971; 15. Hallam, 1979; 16. Hauman, 1951; 17a. Kerharo & Adam, 1964; b. idem, 1967; c. idem, 1974; 18. Kingsley, 1897; 19. Klah, (pers. comm.); 20. Maire. 1933; 21a. Meek, 1925; b. idem, 1931; 22. Moore, 1960; 23. Okafor, 1979; 24. Osei, (pers. comm.); 25. Perry & Greenwood, 1973; 26. Pitkanen & Prevost, 1970; 27. Sikes, 1972; 28. Tremearne, 1913; 29. Trochain, 1940; 30. Verger, 1967; 31. Vergiat, 1970; 32. Walker & Sillans, 1961; 33. Watt & Breyer-Brandwijk, 1962; 34. Williams, 1949.

16 FERTILITY

...that in blessing I will bless thee,
and in multiplying I will multiply thy
seed as the stars of the heaven,
and as the sand which is upon the sea shore.
Genesis 22:17a.

One of the greatest assets of a woman in traditional Africa is fertility; therefore the performance of puberty rites for girls as a symbol of feminine development and fertility is an important aspect of the culture among many ethnic groups. The ability to beget children is the true mark of womanhood in African societies; and infertility is a serious and devastating problem to women in many of our African cultures where children continue to be regarded as an assurance of personal immortality and old age insurance (28). Sterility is believed to be a curse - more often blamed on witchcraft. A very important cause of barrenness that cannot be ignored in Yoruba folk-medicine is the action of witches who are believed to be able to make a fertile woman sterile (37). The author adds that this accounts for about 40% of cases of infertility in Yoruba land; and that it can be treated with occult practices and atonement. Infertility and impotency are regarded as the greatest misfortune that can happen to anyone. Even in some tribal areas, a barren woman is denied a normal burial.

DEITIES OF FERTILITY

There are a number of fetishes or deities of fertility. For instance in Ghana, rivers contain part of the power and spirit of God, therefore they are worshipped as the spring of life in the Dormaa religious beliefs; water is regarded as the symbol of fertility and femininity; and '*amodini*' (meaning 'forbidden to be called by name') is the goddess of life and fertility (13). In its distribution area, *Adansonia digitata* Baobab is traditionally believed to be a god and symbol of fertility, and accordingly invoked, among prayers and libation, on festivals and ceremonial occasions for the fertility of the women folk (26). In Ivory Coast, *Dioscorea praehensilis* Wild yam is considered fetish and is eaten by women wishing to have children (20); and in S Nigeria, among the many incantations of *D. rotundata* Guinea yam or White yam is one to get a child (39a) (see also TWINS: CEREMONIES: Birth) - this species of yam sometimes has one large tuber with several small ones, so the practice is probably an instance of sympathetic magic. Contrary to the traditional or natural uses, however, modern science relies on hormone extracts of the genus for birth control. The steroidal sapogenine or diosgenin contained in *Dioscorea* can be effectively utilized as a means of contraception.

The human-form roots of *Mandragora officinarum* Mandrake, is associated, in Songs of Solomon, with an Ancient cult (17); and valued by the Ancients for various medicinal purposes - such as a cure for sterility as mentioned in Genesis (32). In Benin Republic, propitious offerings to the Legba or the Tegba fetish are supposed to remove barrenness (10b); and in S Nigeria, 'Orisha Oko' is the god of fertility among the Yoruba (10c). The Ancient Greeks dedicated *Artemisia absinthium* Wormwood, a composite, to Artemis (the Roman Diana), the goddess of fertility because of its notable effects on the uterus (31). In Ancient Babylon, Tamus was the god of fertility.

SYMBOLS OF FERTILITY

Some plants are also believed to be symbols of fertility among some cultures in the Region and elsewhere. In W Africa, *Thaumatococcus daniellii* Katemfe, a perennial forest herb with an extremely sweet aril, is symbol of female fertility among the pagan tribes as *Marantochloa* species is of male (8). Consequently, some tribal groups in its distribution area apply a paste of the crushed whole fruit of Katemfe to the abdomen during pregnancy. In Ghana, corn dough is the symbol of fecundity among the Ga-speaking people - to ensure the continued fruitfulness of the mother, it is rubbed on her abdomen during pregnancy (12). In S Cameroon, *Copaifera religiosa,* the 'oven tree', is considered the chief of all trees controlling fecundity, wealth, power and fame (23). Sometimes *Guibourtia tessmannii,* a tall forest tree with buttress roots, is also called 'oven tree'. In Kenya and Tanzania, the fruits of *Gomphocarpus physocarpus,* an asclepiad shrub, are put on the back of the bride by the Maasai to signify fertility (25). In Japan, the delicate beauty of the flowers of *Dianthus superbus,* a widespread hardy perennial, coupled with the strength of the plant has given it a long tradition in literature and a place as the symbol of the ideal woman - a combination of strength and grace .

Sympathetic Magic

Some of the symbols of fertility appear to be based on sympathetic magic. For instance, in Ivory Coast, the abundance of fruits on *Blighia sapida* Akee apple is considered symbolic of fecundity among the Ando, suggestive that the village women should bear many children (40). Usually plants that produce profusely are used to prepare fertility charms. For instance, the abundant clustered figs in some species of *Ficus*, especially *F. sur (capensis)* - locally called 'uwar yara' in Hausa, meaning mother of children (which is also applicable to *Euphorbia balsamifera* Balsam spurge), suggests the notion of, and symbolizes fertility. The plants are thus used in various ways as a charm to promote conception or yield of crops (8). Other species of *Ficus* with abundant fruits are listed in Table 1 below:

Species	Plant Family	Common Name
Ficus elegans	Moraceae	-
F. iteophylla	Moraceae	-
F. kamerunensis	Moraceae	-
F. sansibarica ssp. macrosperma	Moraceae	-
F. platyphylla	Moraceae	Gutta-Percha tree
F. ottonifolia	Moraceae	-
F. umbellata	Moraceae	-
F. vogeliana	Moraceae	-

Table 1. Other Fig Trees with Abundant Fruits.

In Ivory Coast, a decoction of the roots of *Ficus glumosa,* a fig tree, are used to treat sterility in women in the Koulango region of Bouna. However, an important condition is that only roots that cross a path are used to prepare the medicine; and it is also necessary to dig and cut the root with only one stroke of the cutlass (without cutting a second time) from either side of the path (1). The decoction is drunk warm or cold - twice from the right palm and twice from the left palm (1). In Senegal, the root and the fruit of this fig tree, collected necessarily with incantations and magical ceremony, also enter into preparations used by the Fula for female sterility (3, 19a, b). The cleansed root is heated in hot embers, then pounded with a newly spiked corncob. The abundance of the corn grain represents phallic imagery. The powder is taken with water or milk over a period of 2-3 months (3, 19a, b). In this country, *Ficus ingens,* is endowed with magical powers, as such various parts are put into medico-magical prescriptions for fecundity (19b). In S Africa, it is a prevalent belief in the Transvaal that eating the dried fruit of *Ficus carica* Common fig facilitates conception (42).

In S Nigeria, *Solanum incanum*, Egg plant or Garden egg is symbolic of fertility and used by barren women (8). In Botswana, the root of *Asparagus crispus*, a climbing lily, has tubercles on it and so is regarded by the Southern Sotho medicine-man as symbolic of fertility and prosperity in domestic animals (42) - an instance of sympathetic magic. Since *Cocos nucifera* Coconut palm always has abundance of fruits, the roots and bark, collected with rituals are used by many tribal groups in the Region with the prayer, 'as you always produce fruits, so let such-and-such a person bring forth many children' to induce conception and to have many children (26) - an instance of sympathetic magic. In the Region, *Salvadora persica* Salt bush, a small tree of the Saharo-sahel, is regarded by women as favouring conception (8). This is perhaps a sympathetic fantasy based on the abundance of the fruits (6d). In Ivory Coast and Burkina Faso (20), and in Congo (Brazzaville) (5) the bark-sap of *Ceiba pentandra* Silk cotton tree is given to sterile women to promote conception by reason of the fecundity of the seed. In W Africa and in Sudan, *Adansonia digitata* Baobab is worshipped as a 'fertility tree' - probably because of the abundant fruits, or the shape of the fruits (42). In fact, rock art in the Limpopo Valley depict women's breasts as Baobab pods (42). In Cameroon, the shape of the fruit of *Omphalocarpum procerum*, a forest tree, evokes language names meaning 'maiden's breast tree' (8); as

Cocos nucifera Coconut palm (National Plant of Malaysia and Polynesia)

Adansonia digitata Baobab (legendary and mythical plant; associated with traditional religion; symbol of fertility; forbidden to be cut; National Plant of Congo (Braz.)

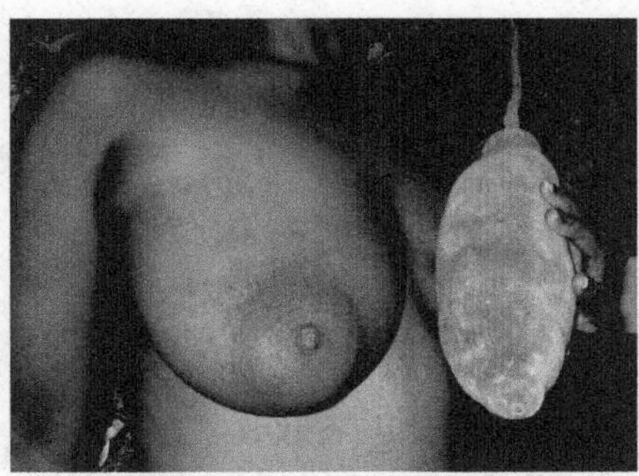

Breast-like fruit of Baobab

such the pulped bark is massaged on the breasts as a galactogene - an homeopathic concept (5).

In Greece and among the Arab peoples, the fruit of *Punica granatum* Pomegranate, serves as a symbol of fertility (22) and love (31) because of the large number of seeds in each fruit. According to Chinese mythology, Pomegranate represents fecundity and longevity (13). The well-filled pods of *Nymphaea lotus* Water lily suggests the cornucopia, emblem of fertility - an old symbol (32); and *Linaria cymbalaria (Cymbalaria muralis)* Kenilworth ivy, a creeping ground cover introduction from S Europe, has pale and purple flowers with yellow spots - often so abundant that the plant is sometimes known as Mother of Thousands (32). In Ghana and Ivory Coast, the Akan name of the many-fingered variety of *Musa sapientum* var. *paradisiaca* Plantain is 'apem', meaning a thousand. The word 'millet' is derived from the Latin mille, a thousand, indicating the fertility of these plants (27). A list of millets is given in Table 2 below. Some species of *Paspalum* and *Eleusine* are also called millet.

Nymphaea lotus, Water lily or Sacred lotus (symbol of fertility, chastity, truth and purity; symbol of immortality and resurrection; symbol of peaceful intent, regeneration and purification; biblical plant but also associated with Islam and Oriental religion; National Plant of Egypt)

Species	Plant Family	Common Name
Echinochloa crusgalli var. frumentacea	Gramineae	Japanese or Barnyard millet
Panicum miliaceum	Gramineae	Common millet
Pennisetum glaucum (americanum)	Gramineae	Pearl, Bulrush or Spiked millet
Setaria italica	Gramineae	Foxtail millet

Table 2. List of Millets

Fertility rites with fruit of *Kigelia africana* Sausage tree (symbol of fertility; associated with traditional religion)

Many of the ethnic groups in the Region regard *Kigelia africana* Sausage tree as a symbol of fertility and use it as such. As part of the puberty and fertility rites among some tribes in Ghana and Ivory Coast, girls touch their breasts against a young fruit on the tree, locally called 'nufuten' in Akan, meaning 'hanging breasts' - with the belief that the breasts would increase in proportion to the growth and development of the fruit (see also INFANTS: PHYSICAL DEVELOPMENT: Plant Prescriptions). When the breasts develop to the required size, the fruit is cut down. A parallel practice occurs with the male organ (see MANHOOD: POTENCY). In Ivory Coast, the Senoufo sometimes hung the fruit of Sausage tree in their huts as a superstitious use for fertility (20); and in Senegal, the decoction of the fruit pulp is similarly used to promote ample development of the breast - a practice said to be proven by recorded measurements (19a). The authors report that the roots also enter into medico-magical treatment for sterility. In Tanzania, however, the Sambaa use the bark to reduce swellings of the breasts (6a); and in Sierra Leone, the heated bark is applied to women's breast to hasten their return to normal after a suckling child has been weaned (8). In S Nigeria, a plant locally called 'isupeyinkota' in Yoruba, meaning 'yam sprouts but does not yield', is invoked to cause a girl's breasts to shrink (39a).

In Tanzania, the white latex of *Euphorbia tirucalli,* an unarmed shrub or small tree, on the Doctrine of Signatures is considered to be remedy for sexual impotence (42); and in Malawi, young girls smear their breasts with it to cause greater development, boys their penis as an added stimulant in masturbation (16). (See APPENDIX - Table 1 for a list of symbolic plants. See also TABOO: SPECIFIC FOR TARGET GROUPS).

PUBERTY RITES

In Sierra Leone, the seeds of *Abrus precatorius* Prayer beads are used in practices concerned with puberty ceremonies for girls (8). Among the

Dipo rituals among the Krobo tribe of Ghana

Krobo tribe of Ghana, rituals associated with puberty for girls, is called 'dipo'. In Nungua, parts of Tema and small sections of Teshie and Labadi (all Ga-speaking people in Ghana), girls have a paste of the powdered bark of *Baphia nitida* Camwood plastered on their heads during puberty rites (12). In S Africa, Mpondo girls undergoing initiation use the sedge *Eleocharis limosa* as bedding (6a); and during the initiation ceremony the Venda girl wears a tassel called 'thabu' which is woven from the bark of *Ximenia caffra* Sour plum to which the juice of *Annona chrysophylla* Wild custard apple is applied (6a). In Botswana, a lotion made from *Aloe kraussii*, or the root of *Asparagus declinatus* (along with *A. stellatus*), or *Kniphofia sarmentosa,* all lillies, enter into initiation rites of young Sotho girls (6a). In SE Senegal, the Tenda put the seeds of *Coix lacryma-jobi* Job's tears, a perennial grass by rivers and wet places, into petticoats made for girls to wear at the excision ceremony (10a). In Lesotho, *Asparagus africanus,* a climbing lily, is part of a protective medicine in the initiation ceremony of girls (15)

FERTILITY IN HUMANS

In Sierra Leone, a chicken, *Gallus domestica,* killed by a male child is cooked with the pounded bark of *Alstonia boonei* Pagoda tree, a tree with whorled branching system sometimes planted for religious purposes (6a).

The stomach (of the chicken) becomes exceedingly bitter and is taken by those, especially women, suffering from intestinal disorders (6a). The treatment is also followed for any barrenness in women over thirty years of age, and by women with umbilical suppuration (6a). In S Nigeria, *Chrysophyllum albidum* White star apple forms the focal point or venue for fertility in which young girls and childless wives celebrate a festivity in parts of Anambra and Imo State - eating, singing and dancing - for the sole purpose of praying to the gods of birth (29). In this country, the leaves of *Tetracarpidium conophorum,* a climbing shrub cultivated for its oil-rich fruits, is used to 'wash children to cause other babies to come' (36); and a plant known locally as 'ogodo' in Yoruba, meaning 'huge', is invoked by the tribe to obtain children (39a).

Young girls dancing

In Senegal, the Fula prepare a macerate or a decoction of *Trichilia emetica* ssp. *suberosa (roka),* a savanna tree, given to women 'to empty the stomach in preparation for treatment to make them fecund' (19a). In SE Senegal, *Afraegle paniculata* Nigerian powder-flask fruit, a tree in secondary thickets, has magical connotation to the Tenda people for they place beneath it sacrifices on behalf of women (11). In S Nigeria, the leaves of *Plukenetia conophora,* a woody liane, have magical use to wash children to cause their mothers to conceive - the Igbo name, 'okumu', meaning *babies call babies* (6b). In Gabon, consumption of the seeds of this woody liane by husbands of wives already pregnant is believed to mitigate the risk of miscarriage (41). In E Nigeria, a piece of the pulverized dead wood of *Pterocarpus soyauxii* Redwood is commonly used in fertility cults (8). Among many cultural groups in the Region, *Milicia (Chlorophora) excelsa* and *M. regia* Iroko, forest trees, are especially associated with fertility and birth (35). In Ivory Coast, the Maninka take the sap of *Uapaca togoensis,* a savanna euphorb with stilt roots, in the belief that this can result in having fine children (Adjanohuon & Aké Assi fide 6d).

In Botswana, *Aloe aristata,* a lily, is uprooted and placed on a shelf in the hut of a barren woman – if it flowers under these conditions she will become pregnant, but if it withers she remains barren (42). In E Africa, the Lobedu and other tribal groups use the leaf of *Melhania forbesii,* a tree, in two ways for barrenness in women: firstly by administering a cold infusion and secondly by fixing leaves above the door of the hut (42). In Zimbabwe, the Manyika use *Protea roupelliae* Sugarbush with other plants to facilitate conception in women (42). In Lesotho, *Commelina africana,* a prostrate herb, cooked with *Haplocarpha scaposa,* a composite, and another unspecified edible root is given as a medicine to a young woman supposedly barren, while an infusion of the plant is drunk and its ash rubbed over the loins as a fertility charm (42). In Kenya and Tanzania, ceremonies such as fertility rituals by the Maasai often take place under *Ficus cordata* ssp. *salicifolia,* a fig tree in rocky wet areas (25). In the Democratic Republic of the Congo (Zaire), *Vitex doniana* Black plum is ascribed with magical power, and it is said to be used to induce conception in women (2). In Zanzibar, women are said to eat the fruits of *Securinega virosa,* a shrub or small tree, in order to promote fertility (43).

Securinega virosa

In S Nigeria, *Kylinga erecta,* a sedge, is invoked in an Odu incantation to make a man's sperms stay in the body of a woman (39a, c). In this country, Yoruba put *Ouratea flava,* a shrub or straggling tree, to magical use to induce pregnancy by conversion of blood into a baby (39b). In N Nigeria, the expression *alyara ka ba ni nono (*Hausa: *nono,* breast) used by Gobir girls at puberty, and in the term *taya-ni-goyo* (Hausa: help me carry the child) which is used for *Euphorbia balsamifera* Balsam spurge, a tree with abundant milky latex, has superstitious connotations. It is also the name of a skin-affection usually on the back of a woman due as a result of carrying a child (8). In Congo (Brazzaville), the young leaves of *Glyphaea brevis,* a shrub of secondary jungle, are prepared with the terminal buds of *Musanga cecropioides* Umbrella tree, one of the early colonizers when the forest is cleared, as a vegetable to be eaten with a freshly killed chicken, *Gallus domestica,* by a married woman who has not conceived after several years, or having conceived is in threat of miscarriage (5). The woman must not drink any water on the morning of the treatment (5).

In Ivory Coast, the Guéré believe that when a woman wants children she has to abstain from sex for four days while eating the leaves of an *Aframomum* species, herbaceous perennials in the ginger family - probably A. *baumannii*, a fetish plant - and by doing so she is certain to be pregnant (20). In this country, the Bété and other tribal groups on the other hand, believe that salt and pepper added to the powdered whole plant of *Secamone afzelii,* a twiner with milky latex in savanna and secondary forest, can induce pregnancy (20). In this country, *Cyrtococcum chaetophoron,* a perennial grass, enters magical application by traditional herbal practitioners. In case of a difficult pregnancy due to demoniacal influences, it is necessary to draw four parallel lines from the chin to the navel of the patient with the ash of this grass (Bouquet & Debray fide 6b). In Liberia, an infusion of *Ricinodendron heudelotii* African wood-oil-nut tree, a soft-wooded forest tree, is either used to relieve the pains of women in labour or prevent miscarriage, and in the superstitious belief that it somehow prevents sterility (8). In the Plateau state of Nigeria, *Tacca leontopetaloides* African arrowroot enters into the rite of *Ja'am* (meaning twins) to bring wealth in the form of children and farm produce, and also protection (Walu fide 6d).

In Mexico, votive arrows are presented to *Solandra* species, trees on rocky cliffs believed by Huichol Indians to be the god-plant of mythology, to assist a barren woman in becoming fecund (21). In Peru, healers use the hallucinogenic effects of *Trichocereus pachanoi,* a cactus, in treating barrenness in women (9). In S Africa, *Datura metel* Hairy thorn-apple or Metel, a plant with hallucinogenic properties, is used by the Shangana-Tsonga in the Northern Transvaal for purposes of ensuring fertility among girls in puberty initiation schools (9). In Papua, New Guinea, *Biophytum petersianum* African sensitive plant, an erect annual, is believed to increase fecundity; women take it, and it is fed to pigs, *Sus* species (38).

Sympathetic Magic

Some fertility prescriptions are based on sympathetic magic. In upper Ivory Coast and Burkina Faso, the Senoufo and the Kirma women eat a macerate of the fruit of *Saba senegalensis* var. *senegalensis,* a savanna climber with milky latex, along with other drug plants as a sterility treatment (20). The practice may be a belief in the Doctrine of Signatures hinging on the ovoid 'pregnant' shape of the fruit (20). In Ivory Coast, a pulp of the still curled up croziers of *Pteridium aquilinum,* a terrestrial fern, is made in warm water given in an enema to combat sterility thought to be due to vaginal obstruction (20). This is based

on the Doctrine of Signatures, the fronds bearing some resemblance to the pudenda (20). In N Nigeria, a macerate of the crushed fruit of *Physalis angulata* Wildcape gooseberry, an erect herbaceous weedy annual, taken with milk is remedy for sterility, 'maganin haifua' in Hausa, because the plant retains its fruits during the barren season (8) - an instance of sympathetic magic. In the Region, traditional practitioners rub a macerate of *Phyllanthus ninuri* var. *amarus (amarus)*, an annual euphorb herb, collected with rites, on the breasts of sterile women to induce conception (26). The practice is linked directly to the many fruits at the back of the leaves.

In S Africa, a Mpondo female herbalist uses the root of *Senecio oxyriaefolius,* a composite, as a remedy for sterility in a married woman (42). The powdered root is formed into a cake with water and the patient is given a crumb of it (in crossed hands, in the way that sacrificial meat is received) morning and evening (42). In case of an infertile marriage the Zulu use the flower of *Pupalia* species. This is bruised and made into small balls, one being eaten by the man and the second being inserted into the vagina of the woman (42) - likely an instance of sympathetic magic based on the hooked burs of the plant. The glume of *Pseudechinolaena polystachya,* an annual grass, at maturity is armed with bristly hook. In Sierra Leone, the Mende commonly apply the name *nane* to plants with prickly fruit. The sophistry of the use in Congo (Brazzaville) to promote fertility may perhaps be by sympathetic magic (6b).

In Uganda, parts of *Pseudarthria hookeri,* an erect shrub, are worn by women as charm to promote child-bearing because the hairy leaves catch on hands and clothing (6c). In Yoruba mythology, *Termitomyces microcarpus,* a mushroom, is used in combination with other ingredients in a propitiation to the gods for increased population in towns and villages by reducing mortality rate (30). The power of the mushroom to grow in large numbers is believed to be the effective force here - another instance of sympathetic magic. In India, *Asparagus racemosus,* a lily locally called 'shatavari', meaning 'she who has a hundred husbands', is used by traditional Ayurvedic practitioners to strengthen the female reproductive system (4).

Predicting the Sex of Child

It is traditionally believed that medicine-men are capable of performing rituals to forecast the sex of the child. (See CHILDBIRTH: PRE-NATAL for other instances). In Ghana, before a girl is given into marriage among the Ga-speaking people, she picks a handful of ripe nuts of *Elaeis guineensis* Oil palm, and with other ingredients (but not fresh fish for some reason), prepares a meal and eats in a ceremony to induce fertility. If the nuts are picked with the right hand, it shows a preference for girls, if with the left - boys, if with both hands - both boys and girls (7). In Ivory Coast, *Hybanthus enneaspermus,* a small perennial in the Violet family, is believed to be fetish by the Guéré, the Kru and the Shien and it is used by traditional midwives to determine the sex of the child (20). In Central African Republic, after cooking and pounding the root of *Eriosema pulcherrimum,* an erect savanna legume herb, it is eaten by women desiring children, and according to the shape of the root eaten, the medicine-man will guess the sex of the child to come (6c).

Fertility Dolls

The use of *Akuaba,* fertility doll, usually decorated with strings of beads and ear-rings (symbol of fertility), is claimed by believers to have magical effects, and to induce conception in apparently sterile women. For instance, in Botswana, Sotho women carry a doll made from the root of *Massonia bowkeri* Abraham's book, a lily, as a remedy for sterility (42). This belief is not confined to Africans only. The fertility doll is reported to have helped foreigners who confided in it. On fertility dolls it has been observed that a pregnant woman should not look upon a monkey, species of *Cercocebus, Cercopithecus, Colobus, Erythrocebus, Papio, Piliocolobus* and *Procolobus;* or upon any deformity, even a badly carved wooden figure, 'lest she gives birth to a child like it' (33). She may, however, carry an *Akuaba,* the black Ashanti doll, 'because its long-shaped neck and beautiful head will help her to bear a child like it (33). (See also TWINS: CEREMONIES: Death for other uses of the fertility doll).

However, in N Nigeria, women who wish to have no more children wear *Adenium obesum* Desert rose, an erect succulent shrub in savanna often cultivated as an ornamental, as a girdle (8). In S Africa, Zulu youths wishing to harm an unmarried maiden, use the root of *Ansellia humilis,* an orchid, as a charm to prevent her ever giving birth to children (42).

FERTILITY IN CATTLE

In addition to human beings, plants enter into fertility prescriptions for cattle, *Bos* species. In N Nigeria, *Trochomaria macrocarpa (macroura),*

a trailing cucurbit, is prized by the Fula tribe as a superstitious medicine for cattle to promote fertility (8). In this country, the latex of *Euphorbia balsamifera* Balsam spurge, a succulent shrub commonly grown as a hedge in dry regions, is given by the Fula to cattle to promote fertility, a use probably mainly superstitious (18). In Niger, Fula cattle-men add a portion of a fig tree known as *ibbi* or *Polygala erioptera,* an annual herb, or *Bauhinia rufescens,* a scandent shrub or small savanna tree, or *Piliostigma reticulatum,* a dry savanna tree, along with other plants in a magical formulation called *fud'ngo* for promoting the fertility of their herds (23).

In S Africa, the Kgatla use a decoction of *Ximenia caffra* var. *natalensis* Natal plum in cattle fertility rites, or the root of *Peltophorum africanum* African blackwood is one of the ingredients in the Kgatla 'doctor's' mixture to promote the fertility and well-being of cattle - the prepared medicine being known as *leswalo* (42). Among the Ndebele and the Tswana in this country, *Vangueria venosa* Wild medlar is used in the cattle fertility rites - the Kgatla use it as one of the ingredients of a medicine to make their animals prolific (42). In Botswana, the Southern Sotho use *Asparagus asparagoides,* a climbing lily, as a charm to increase fertility in cattle; and the forked stick of *Celtis africana* White stinkwood is sometimes used by the tribe to stir meat during cooking - this is said to encourage the rapid increase in number of domestic stock (42).

In Mexico, *Solandra* species, trees on rocky cliffs, are believed to be the god-plant of mythology among the Huichol Indians. To increase fertility of cattle, drawings of the animals, which are done in brightly coloured wool pressed in wax, are brought to the plant (21). (See also CATTLE: INCREASE IN HERDS).

FERTILITY OF THE SOIL

Fertility rites for the improvement of the soil and the high yield of crops are also practiced by traditional farmers (see FARMING: HIGH YIELD for other examples). The numerous fruits of *Ficus sur (capensis)*, a fig tree, confers a suggestion of fertility, thus the tree has fetish attributes in the promotion of stock and crop increases. In E Africa, herdsmen plant the tree in sacred places and sites where sacrifices are made to ancestral spirits to promote rain, and to prevent famine and epidemics. Such offerings to the tree are deemed to be to *Earth* and *Forest,* the two great divinities of productivity (8, 42). In Gabon, the ability of *Bryophyllum pinnatum* Resurrection plant to regenerate from a single leaf, has acquired fetish properties to render the soil fertile (41). In Liberia, the fruits of

Deinbollia pinnata, a shrub or small savanna tree, are hung by the Mano in rice fields to ensure fertility (8). In N Nigeria, the Fula of Adamawa call *Vernonia nigritiana,* an erect herbaceous savanna composite, 'gorgo baigore', which means 'husband of pumpkin', and are said to plant or retain it in fields to increase a cucurbitaceous crop (8). In E Sudan and E Africa, the fruit of *Kigelia africana* Sausage tree or the ash mixed with seed-maize, is hung as a charm outside huts as a magic medicine to increase the crop (8).

By cutting off the apex of male *Carica papaya* Pawpaw, or by injuring it in some way, the tree has sometimes been caused to fruit (18). The practice is, however, not without the usual traditional superstitious touch, such as adorning the trunk and the leaf petioles with strings of beads, ear-rings and necklaces - symbolizing female fertility. (See also TALISMAN).

References:

1. Aké Assi, 1980; 2. Aubréville, 1959; 3. Berhaut, 1979; 4. Blumenthal & al., 2000; 5. Bouquet, 1969; 6a. Burkill, 1985; b. 1994; c. 1995; d. 2000; 7. Cofie, (pers. comm.); 8. Dalziel, 1937; 9. Dobkin de Rios, 1977; 10a. Ellis, 1881; b. 1883; c. 1894; 11. Ferry & al., 1974; 12. Field, 1937; 13. Fink, 1989; 14. Graf, 1978; 15. Guillarmod, 1971; 16. Hargreaves, 1978; 17. Hepper, 1987; 18. Irvine, 1961; 19a. Kerharo & Adam, 1964; b. idem, 1974; 20. Kerharo & Bouquet, 1950; 21. Knab, 1977; 22. Lotschert & Besse, 1983; 23. Maliki, 1981; 24. Mallart Guimera, 1969; 25. Maundu & al., 2001; 26. Mensah, (pers. comm.); 27. Moore, 1960; 28. Nyamwaya & Sober (Ed.), 1983; 29. Okafor, 1979; 30. Oso, 1977; 31. Pamplona-Roger, 2001; 32. Perry & Greenwood, 1973; 33. Rattray, 1927; 34. Richards & Kaneko, 1988; 35. Schnell, 1946; 36. N.W.Thomas, 1913-14; 37. Thomas, 1988; 38. Veldkamp, 1971; 39a. Verger, 1967; b. 1972; c. 1986; 40. Visser, 1975; 41. Walker & Sillans, 1961; 42. Watt & Breyer-Brandwijk, 1962; 43. Williams, 1949.

17 FETISH

O'er heathen lands afar
Thick darkness broodeth yet.
Arise, O morning Star,
Arise, and never set!
Lewis Hensley. 1824-1905

A fetish is looked upon as a spiritual guardian and worshipped in the belief that a powerful supernatural being dwells in it. The abode of the deity could take various forms: a clay model; a wooden carving; a calico-covered, blood-smeared structure; a river or the confluence of two rivers; a solid piece of rock or a group of rocks; a big tree; a relic of forest set aside for the purpose or a grove; an aggry bead (for the origin of the bead see INFANTS: PHYSICAL DEVELOPMENT: Other Products); and so on. It is, however, the indwelling spirit that is really deified and worshipped, but not the physical object *per se*.

DEFINITION

It has been admitted, 'I do not think anybody knows what fetishism is exactly, and I never met anyone who could tell me what the word 'fetish' means. It is so universally applied by the Fantis to everything partaking of the unknown and supernatural, that it can scarcely be defined. They worship a fetish, and certain objects are called fetish, meaning, I imagine, sacred objects; and all charms and amulets are generally termed fetishes' (9a). The author concludes, 'fetish means some great, overshadowing, and unseen power, which the natives worship with or without the attendant forms of small representative images. This power is also supposed to act through various objects, which are sacred to it and called also fetishes' (9a).

'A fetish *(suman)* is an object which is the potential dwelling place of a spirit of an inferior status, generally belonging to the vegetable kingdom; this object is also closely associated with the control of the powers of evil and black magic for personal needs' (34). Fetish has been described as the religion of the natives of the Western Coast of Africa, where they have not been influenced either by Christianity or Islam

(25b); or manmade objects in which mystic power resides (13); or as conglomerations or conjunctions of materials believed to be powerful, and initially activated by offerings made over them (28). A fetish is defined as denoting 'a general theory of primitive religion, in which external objects are regarded as animated by life analogous to man's' (Sir E.B. Tylor fide 34). The author quotes a modern definition as 'a limited class of magical objects in West Africa' (34).

ORIGIN AND MEANING

The word 'fetish' is of Portuguese origin, and is a corruption of *feitico,* an amulet or charm (9c). In support of this statement, it has been stated that the word fetish, which is derived from the Portuguese *feitico*, meaning a charm or made thing, has become current in present-day West African English (35). The reasons for the foreign name is explained thus: 'these worthy voyagers, noticing the veneration paid by Africans to certain objects, trees, fish, idols, and so on, very fairly compared these objects with the amulets, talisman, charms and little images of saints they themselves used, and called those things similarly used by the Africans *Feitico,* a word derived from the Latin *factitius,* in sense magically artful' (25b).

In fact, there is a lot in common, and very little difference between the images of the saints and fetish idols (9b). 'There is a strange resemblance between the grosser forms of Roman Catholicism, as seen in the Canary Islands, and the fetish worship of the Negro tribe in the Gulf of Guinea' (9b). It appears, however, that fetish images were not found further inland or north of the Guinea Coast, for it has been recorded, 'the Hausas themselves had no fetishes except for posts set up in the fields - they worshipped the spirits themselves which lived in wells or trees' (37).

POPULAR IDOLS.

Among the popular fetish idols in and around the Region include Akonodi, Gyabom, Kumkuma, Oten, Sabe and Tigare of Ghana; the Azoon, Bo, Ho-ho, Legba or Tegba and So of Benin Republic; and Arie-ogbo, Boofima, Choma, Gbangbani and Wonde of Sierra Leone. Others are Dioh, Ganfa and Gleyoh of Liberia; Sango (Thunder), Esu (Devil), and Sapona (Smallpox) of Nigeria; Ebyesso and Sakpata of Togo; Ichak and Socé of Senegal; Kwanalré of Ivory Coast; Sagara of Mali; Ydtaba of Burkina Faso; *Ngoye* and *Nzobi* of Congo (Brazzaville).

OWNERSHIP

A fetish could be for an individual person, a family or household, a whole village or clan, or for several villages collectively. It may be inherited from father to son or from generation to generation. It has however been observed, 'they are personal means of protection, in contrast to the *abosom* (deities) who watch over the welfare of families, village communities and state' (12). Nevertheless, the author admits, 'some *asuman* (fetish) carry names and come close in importance to *abosom* (deities)'.

CUSTODIAN

The custodian of the fetish is normally called a fetish priest or fetish priestess. However, he or she could also be a medicine-man or medicine-woman, a juju-man or a juju-woman, or a witch-doctor. Though these persons may have different titles or offices, they, in fact, do perform in practice, more or less similar or overlapping functions - see the preamble to FORTUNE-TELLING AND DIVINATION; and JUJU OR MAGIC.

Functions

The functions of the custodian includes consulting the spirits for counseling (after prayers, offering libation and performing the necessary customary rituals), on all sorts of problems or mishaps - such as finding the causes of diseases, healing diseases, invoking rainfall, ensuring a good harvest or a good catch, finding out a thief or a murderer, discerning witches, predicting victory or defeat in war, or some other forms of divination.

JUSTIFICATION FOR ACQUISITION

A fetish may be acquired primarily for protection. It could also be acquired for vengeance, and therefore, for spiritual power to eliminate an enemy. '*Asuman* (fetish) and *aduro* (protective medicine) always have a dual function. They are capable of averting misfortune, illness and influences of witches from their owners and to bring them good luck in all kinds of endeavours and success in their professional and private life. But to the extent that they protect their owner, they are *aduto* (evil medicine) against enemies, competitors and rivals' (12). Explaining the rational behind the belief in, and the use of protective deities

Fetish tree

among the Akan-speaking tribe of Ghana and Ivory Coast, it has been observed, 'Finding himself constantly threatened by dangers, exposed to miseries, and thwarted or obstructed in his designs or intentions by causes over which he had no control, he conceived all these misfortunes to be the acts of inimical or malignant beings, and so mentally peopled the earth with superhuman agents' (9c).

INITIATION

Among a series of complex training for fetish initiates in Ashanti includes two sets of washing for seven consecutive days each. The first wash is with a concoction of the leaves of *Spondias mombin* Hog plum or Ashanti plum, the leaves of *Triumfetta rhomboidea* Bur weed and the leaves of a species of *Tapinanthus* or *Phragmanthera* Mistletoe, epiphytic parasites, in a pot on a grave at night. The ritual is meant to strengthen the ankles for dancing, and to cause his god to stay with him (34). In addition, the eyes are either rubbed with the leaves of *Tragia* species, twiners with irritating leaves; or the leaves of *Costus afer* Ginger lily, or the leaves of *Momordica balsamina* Balsam apple or *M. charantia* African cucumber if the spirit of possession will not manifest itself to the novice (34). For the initiate who cannot 'hear his god's voice', the leaves of *Justicia flava*, a weedy acanth, locally called 'afema' in Akan, are put under his pillow (34). Plant extracts - usually in liquid form - are in fact applied to the eyes and ears of the novice during apprenticeship. Thus prepared and sensitized, the novice becomes able to see the *abosom* (gods) who are normally

Fetish woman

invisible to human beings, and to understand their language (12).

For the second wash - also for seven consecutive days - the medicine in the pot upon the grave is replaced with an epiphytic decorative fern, *Platycerium elephantotis* or *P. stemaria* Stag-horn fern, locally called 'sasabonsam-kyew' in Akan, which means 'Devil's hat' (34). After this bath, the body is rubbed with the mashed roots of *Paullinia pinnata,* a woody climber with tendrils, and the seeds of *Aframomum melegueta* Guinea grains or Melegueta (34). The novice then bathes three times a day and three times a night for several days with an infusion of leaves plucked at random from the left and right sides of a narrow forest path, together with cut pieces of the roots which run across a path (34). The bath is followed with a rub with a medicine made from the bark of *Milicia (Chlorophora) excelsa* or *M. regia* Iroko, locally called 'odum' in Akan, to bring the *nkomoa* (spirit of possession) upon the pupil (34).

In Ghana, the apprenticeship period among the Dormaa people is three years. 'The second half of the first year is filled with ceremonial ablutions with *aduro* (protective medicine), made from herbal and plant essences growing near the graveyard, which the novice must gather at night by himself (12). At the end of the first year, the novice has to stay on the graveyard for seven nights in a row all by himself. Such a close contact with the *nsamanfo* (souls of the dead) is meant to strengthen him - that is, his ego' (12).

In Gabon (31), and Congo (Brazzaville) (4), *Tabernanthe iboga* Iboga, an apocynaceous shrub with hallucinogenic properties is used in fetish initiation. 'The most interesting use of Iboga is as an hallucinogen. Probably the first report of such utilization is that of Guien, who describes an initiate in a fetishist cult: 'Soon all his sinews stretch out in an extraordinary fashion. An epileptic madness seizes him, during which, unconscious, he mouths words, which, when heard by the initiated ones, have a prophetic meaning and proves that the fetish has entered him'' (31). In Gabon, *Alchornia floribunda,* a semi-scandent shrub or small tree,

Fetish abode

is used as an aphrodisiac and stimulant before fetish rites, sometimes admixed with Iboga; and *Dracaena fragrans,* a shrub or small tree, is planted as a village fetish tree (40).

In Ghana, the black seeds of *Operculina macrocarpa (Merremia alata),* a stout climber with hollow stems, called 'ebia' or 'ayibiribi' in Fanti, are worn with other coloured beads as *banka* by fetish initiates among the coastal tribes in the Region (16) - (see TWINS: CEREMONIES: Anniversary). During initiation rituals into womanhood among the people of Avatime in Ghana, this black seed is worn on the neck with other coloured beads by the candidates for two weeks. The towns within this traditional area are Amedzofe, Biakpa, Dzogbefeme, Dzokpe (New & Old), Fume, Gbadzeme and Vane.

APPEASEMENT SACRIFICES

Sacrifices such as yam, eggs, alcoholic drinks, fowl, *Gallus domestica;* sheep, *Ovis* species; cow, *Bos* species; and sometimes even human beings are offered to the fetish on occasions - apparently as a thank offering or as 'payment' for its services, but the underlying objective is to revive and strengthen the spiritual power. Human sacrifice to a fetish is one of the causes of ritual murder. This ritual is now illegal in law. Human sacrifice did take various forms. (See also CEREMONIAL OCCASIONS: FESTIVALS: Human Sacrifice Aspect; and WEALTH: WITCHCRAFT).

Human Sacrifice

In Ghana for example, the traditional Black Stool of the people of Akropong Akwapim was blackened with the blood of a human being who had to be an important member of the matrilineage concerned - the blood being mixed with gunpowder and the web of a spider, *Nephilengys* species, as ingredients (13). The author adds that the young leaves of *Elaeis guineensis* Oil palm were formerly put round the neck of a person to be sacrificed for the ancestral stool. On other occasions, the human sacrifice had another objective. 'On the death of a king or the queen-mother - in former times - there were also human sacrifices in addition to libation, food sacrifices and other objects that were deposited with the corpse. It was believed that the ruler should not embark on the long and arduous journey to the kingdom beyond without his proper retinue' (12). The author concludes, 'usually,

human sacrifices had the purpose of providing the spirits of deceased human beings in life after death or on the journey to the realms of the dead with the spiritual powers thus released. Humans were never sacrificed for the sake of cruelty'.

In Nigeria, a form of ritual murder among the natives of the lower Niger is described: 'I have stood over a grave of a warrior Chieftain that had not been filled in many hours. In this grave had been buried sixteen living human beings and sixteen others had been slain at the same time and buried to escort their chief and master to the other world. And those that were slain were supposed to be sent direct, so that they might be there to prepare for his arrival' (3). The number of victims sacrificed could reach astronomic proportions during the customary butchery in Ashanti when a royal member died. 'The most celebrated custom remembered in Ashanti was that for the mother of the king immediately preceding Kofi Karikari (that must be Kwaku Dua 1), at which more than 3,000 slaves were sacrificed, at least 2,000 of them being Fanti prisoners' (9a).

Ritual murder did occur in other cultures outside the Region. 'Hundreds were slaughtered when a Pharaoh died. Likewise, at the death of Guayanacapa, the last Inca, one thousand persons of all ages were killed' (38). The author adds, 'it was the custom of the Peruvian Indians to bury their chiefs in the way the Egyptians buried their Pharaohs. Their wives, attendants, pets, treasures, clothing, food and wine were placed in the grave so as to be close at hand for use in the afterlife'.

LIMITATIONS

Fetish practice has defined powers and limitations - even for the fetish doctor who is universally admired and trusted. The liability of a fetish to fail is recounted: 'One day he himself fell sick, and he made juju against the sickness; but it held on, and he grew worse. He made more juju of greater power, but again in vain, and then he made the greatest juju man can make, and it availed nought, and he knew he was dying; and so with the remaining strength, he broke up, dishonoured and destroyed all the Fetishes in which the spirits lived, and cast them out into the surf and died like a man' (25a).

FETISH PLANTS

In traditional Africa, many plants are associated with fetish practice. The plants are here classified under trees, shrubs and lianes, herbs and food crops.

Trees

Throughout its distribution area in W Africa, *Ceiba pentandra* Silk cotton tree is believed to be an important fetish and sacred tree - probably because of its size and height (some trees measure 200 ft or 67 m high in the forest zone). In S Nigeria, many initiation ceremonies, offerings and meetings of the sect occur under the foot of the Cotton tree (18). Similarly, in the ancient Indian cultures of C America, the tree had a great mythological significance (27). In Burkina Faso, *Adansonia digitata* Baobab is left standing when clearing the bush as a fetish tree (24); and it is even believed in certain parts of Africa that old Baobab trees ignite spontaneously and consume themselves (10). It is also believed that the Baobab destroys all life around it, and it is surrounded by superstition and legend dating back to prehistoric times (15). Similarly, Mungo Park (1769-1806), the Scottish surgeon, plant collector and pioneer explorer of the Niger River, called *Pandanus* species Screw pine, a river bank tree with stilt roots, *'Fang jani* or self-burning tree', and maintained that it was burnt spontaneously by some internal process (22). The species in the Region include *P. abbiwii* and *P. aggregate (candelabrum)*.

In Ivory Coast, *Okoubaka aubrevillei*, a rare forest tree, is considered such a powerful fetish that no other tree can grow near it (24). This is the general belief in Ghana as well. In support of this observation and belief, it has been recorded that this tree has the reputation of killing the surrounding trees (20, 22). However, a single, heavily debarked tree growing on top of the Atewa Range Forest Reserve near Kibi in the Eastern Region of Ghana has a pure association of *Costus afer* Ginger lily growing under its shade, and on the branches a heavy colony of epiphytic orchids. Among

Okoubaka aubrevillei 1/4 natural size (associated with traditional religion; legendary and mythical plant; forbidden to be cut)

Sacrificial offerings

the several magical attributes of *Okoubaka* - which literally means 'tree of death' among the Anyi people of Ivory Coast - is the traditional belief in its power to 'resurrect' when cut down - and the consequent taboos forbidden felling. Sometimes a white calico is wound round the trunk of the tree. All parts of the tree - the leaves, the bark, and especially the fruits - enter into various charms, juju and medico-magical prescriptions. In addition, it is believed that when the shadow of the tree is on one side, one has to walk on the other. No bird flies over it alive, except the eagle, *Polemaetus bellicosus,* nor animal walks under alive, except the brush-tailed porcupine, *Artherurus africanus,* locally called 'apese' in Akan; and the rat, *Cricetomys gambianus.* In Ashanti, the 'apese' is credited with very great courage and fierceness. It will run, they say, even through fire.

Similarly, in Madagascar, 'according to legend, birds which fly over *Erythrophleum couminga,* a toxic legume tree used as ordeal poison, were reported to die immediately (33).

In Burkina Faso, *Afzelia africana* African teak, a savanna tree legume, is a fetish to the Bobo (Baldry fide 6c). In Ivory Coast, the tree is also considered a fetish along with other trees of similar status such as *Ceiba, Daniellia, Milicia* and *Khaya*. Medicine made from it must be made according to certain strict practices, and patients similarly conforming to rote such as taking the medicine by the right hand or the left hand as prescribed (5, 24). In Ghana, *Cordia millenii* Drum tree may be grown as a village fetish tree (21b); and *Oxyanthus unilocularis,* a shrub or small tree in forest or fringing forest in savanna, is used as a fetish tree in Akwapim and in Ashanti (21b). In the savanna region, *Blighia sapida* Akee apple is planted or tended as a fetish tree and used in important magic medicine (24). In French Sudan, *Elaephorbia drupifera,* a tree with fleshy branches and containing abundant very caustic white latex, is spontaneous and sometimes planted as a fetish tree (7). In Senegal, the tree is scarce where it is considered a fetish, and where complainants come and ask the tree the judgement of God (24). (The high forest equivalent of this tree is *E. grandifolia).* Throughout the Region, *Ficus lyrata* is sometimes planted in villages as a fetish tree (8); and in Gabon, *F. sansibarica* ssp. *macrosperma* has fetish property (40).

In Ghana, *Cola nitida* Bitter cola (and also *C. acuminata* Commercial cola nut tree) is associated with the Tigare fetish, and a concoction is prepared which is used in 'drinking fetish' (36). In Ivory Coast, *Lannea welwitschii,* a big forest tree with characteristic pitted bark like gun-shot wounds, is a fetish to some tribes (24). In Ghana, *Millettia thonningii,* a small tree legume with very decorative purple flowers when deciduous - between January and March - has been planted within a fetish compound in Labadi, a suburb of Accra, the capital. The seeds of *Garcinia kola* Bitter cola, known locally as 'minchingoro' in Hausa - meaning 'male kola nut' - the product of a popular chewing stick tree called 'tweapea' in Akan, have unspecified superstitious uses (most likely a fetish to acquire money - see WEALTH: CHARM PRESCRIPTIONS). An object that stuck in the food pipe of a patient, and was removed in a major operation at the Korle-Bu Teaching Hospital was found to be this seed tied to a bundle of the red tail feathers of the grey parrot, *Psittacus erithacus,* a common paraphernalia of juju-men.

Many of the vernacular names of *Alstonia boonei* Pagoda tree, a tree with whorled branching sometimes planted for religious rites, do relate to the use of a stick, with its whorl of twigs, as a stand for holding a bowl before a fetish shrine (14). In Ghana, *Sterculia tragacantha,* African tragacanth, a gum-producing tree, is considered fetish and sometimes planted before a chief's house instead of the more usual fig (36). In Senegal, the Pulaar consider *Acacia sieberiana* African laburnum, a savanna tree with decorative yellow flowers, a fetish plant (23b); and the Non consider African laburnum so sacred that any use is restricted to prescriptions by senior Muslim priests (23b). In this country, *Entada africana,* a savanna tree, is deemed to have fetish powers (23b, c); and Basari medicine-men prepare a fetish called *achak* which is considered very powerful, but the purpose of its use is not made clear (23b). Also in this country, the Socé consider *Faidherbia (Acacia) albida,* a large savanna tree legume, a fetish for 'soul' magic (23b). In the Sédhiou area of Casamance, *Pterocarpus erinaceus* African kino, a savanna tree legume, is considered fetish so that the bark and roots are a general panacea (23a, b, c).

In Liberia, *Loesenera kalantha,* a tall shrub or small tree, is reputed

fetish reviver, and in the olden days the most potent and effective charms were made from it (Cooper fide 8). In the Region, *Kigelia africana* Sausage tree is often planted as a village 'palaver tree' and roadside shade-tree, and for the flowers and fruits which are in places regarded as fetish (6a). The fruits (especially) are prized as fetish emblems in many parts, and are commonly sold in markets for the fetish and medicinal values (6a). In Central African Republic, the wood of Sausage tree is regarded as holy and fetishes are cut from it (39). In Senegal, the Basari rotaries of the Ichak fetish consider *Spondias mombin* Hog plum or Ashanti plum as a fetish-protector of millet (23c). In S Nigeria, the Ekoi believe *Croton zambesicus,* a small tree often planted in villages, is a fetish (8) (see MEDICINE: OTHER SUPERSTITIOUS USES).

Products The fibre bast of *Raphia* species Wine palm is used extensively throughout W Africa as a fetish garment or kilt, and for threading beads in fetish practices. The palms in the Region are listed in Table 1 below:

Species	Plant Family	Distribution
Raphia farinifera	Palmae	N Nigeria, W Cameroon
R. hookeri	Palmae	Guinea, Sierra Leone, Ivory Coast, Ghana, Benin Rep., S Nigeria W Cameroon, Bioko (Fernando Po), Rio Muni, Gabon.
R. humilis	Palmae	Benin Rep.
R. palma-pinus	Palmae	Senegal, Gambia, Portuguese Guinea, Guinea, Sierra Leone, Liberia, Ivory Coast, Ghana.
R. regalis	Palmae	S Nigeria
R. sudanica	Palmae	Senegal, Gambia, Mali, Guinea, Sierra Leone, Ivory Coast, Ghana.
R. vinifera	Palmae	Benin Rep., S Nigeria, Bioko (Fernando Po), E Cameroon, Congo (Braz).

Table 1. Wine Palms in the Region and Their Distribution

Shrubs and Lianes

In Ghana, the wood of *Ochna afzelii* and *O. membranacea,* shrubs or small trees in savanna or forest margins, called 'koliawatso' in Adangbe, are used for making walking-sticks by fetish priests; and the vernacular names are stated to mean 'the fetish stick or sacred tree of the Ada people' (21b). (See also EVIL: SPIRITUAL PROTECTION; and TABOO: GENERAL FOR ALL AND SUNDRY: Sweeping Broom). In Central African Republic, *Drimiopsis bulbifera,* a savanna plant in the Lily family, is cultivated in villages in Middle Shari region as fetish (7).

In Congo (Brazzaville), *Schumanniophyton magnificum,* a forest shrub or treelet, has magical significance as part of the fetish *Ngoye;* and the *Nzobi* sect never move their fetish without offering it a mastication of the leaves of *Carpolobia lutea,* a forest shrub, with some kola-nut (4). In Mali, *Euphorbia paganorum,* a low fleshy shrub of rocky savanna, is planted as a fetish plant (2). In Ivory Coast, fetish practice by the Kyama requires that one does not touch anything on which a piece of *Adenia rumicifolia* var. *miegei (lobata),* a large forest climber, has been placed (32). In E Cameroon, *Olax gambecola,* a shrub of closed forest, has use as a fetish plant (6d); and in Gabon, *Lasianthera africana,* a shrub, in secondary forest, enters into certain fetish rites (40).

In Ghana, the natives chew the wood of *Ehretia cymosa,* a savanna shrub with red fruits, along with the seeds of *Cola verticillata,* a forest tree, from which is produced a red colour which is used for fetish-ornaments, amulets, and similar objects; and *Baphia nitida* Camwood as a dye is used by the natives for fetish ceremonies and amulets (Thonning fide 17). In Ivory Coast, Camwood is a well known magic plant, which is looked on as a guardian of the fetish house in the Agniblekrou region where it decorates the fetish house of sickness - called Kwanalré (24). In this country, *Paullinia pinnata,* a woody climber with tendrils, is tended as a fetish plant in addition to its medicinal uses (24). In S Nigeria, *Entada rheedei (pursaetha)* Sea bean or Sea heart, a tropical drift dissemule, is grown as a fetish in villages (8); and in Angola the plant is used as a fetish (19). In Liberia, *Sabicea ferruginea,* a climbing shrub with very large leaves, is a fetish plant of the Kru tribe - the dried leaves placed on top of houses are believed to stop rain from falling (24). In this country, *Vismia guineensis,* a small tree in secondary forest, known as 'ge-an' in Basa, is regarded as a fetish (8). (See ORDEAL: PLANT TESTS: Adultery and Lesser Crimes).

Herbs

In many parts of the Region, *Momordica balsamina* Balsam apple and *M. charantia* African cucumber are important fetish plants - most fetish ceremonies generally finish with washing the body with water in which consecrated leaves are laid (Thonning fide 17). African cucumber is such a powerful fetish plant that a bowl of water in which the plant has been immersed (together with other assorted plants as ingredients) may be invoked in incantations to possess initiates who

carry this bowl (29). (See TWINS: CEREMONIES: Spirit Possession). The peculiar vegetative character of *Bryophyllum pinnatum* Resurrection plant in which detached leaves produce regenerating shoots, suggests superstitious uses as a fetish plant, and is used as such. The leaves are also put in water and the liquid used in sprinkling fetish objects. (See FERTILITY:FERTILITY OF THE SOIL; and LOVE: HOSPITALITY AND FRIENDLINESS). The leaves of *Solenostemon monostachyus,* a labiate, is placed along with Resurrection plant in fetish pots (8). Among some coastal tribes *Oldenlandia corymbosa,* a prostrate weedy herb, placed under a fetish is believed to attract followers to the cult and make it popular (29).

In the Region, *Boerhaavia diffusa* Hogweed is one of the commonest fetish plants which is used by the natives for their cleansing baths in sickness and other cases (17). In Ghana, *Gomphrena globosa* Bachelor's buttons is cultivated in some villages, mostly in coastal districts as a curiosity or protective fetish (8). In Ivory Coast, *Lepianthes peltata* Cow-foot leaf, a forest shrub, is considered by the Guéré tribe as a fetish, and the leaves are put in water to sprinkle on masks and fetish (24). In Gabon, some fetish rites require the use of *Englerina gabonensis* Mistletoe, together with sweet bananas and the leaves of Cow-foot leaf (40b). In this country, *Cymbopogon densiflorus,* a tufted perennial grass, has fetish attributes - its inflorescence is burnt in fumigations required in certain rituals, such as, to rejuvenate and restore the efficacy of a fetish, an amulet or a talisman when the owner has violated a taboo (40b). In this country, leaders of the Bwiti sect grow *Ansellia africana,* a beautiful epiphytic orchid, as a fetish (6d).

In Senegal, the Socé ascribe to *Cymbopogon giganteus,* a tufted perennial grass, fetish value in the case of illness when a piece is suspended inside the hut (23b). The tribe also consider *Eragrostis ciliaris,* a tufted annual grass, to be fetish for longevity (23b). In the Region, several tribal groups plant *Sporobolus pyramidalis* Cat's tail grass, a very common weed, as a fetish, or take their oath upon it (8). Having reached a certain age (? menopause), they deem it good to hang a piece below the bed (23b). In Ghana, *Bergia capensis,* an aquatic herb, has been found in a fetish pot in Shai, and appears to be used as a fetish plant (6b). The people also plant *Scadoxus (Haemanthus) cinnabarinus* and *S. multiflorus* Blood flower or Fire-ball lily in villages, probably for fetish purposes on account of the bright red flowers (8).

In W Africa, *Mirabilis jalapa* Marvel of Peru is scarcely used except as an ornament or a fetish plant; and *Caralluma russelliana*

(retrospiciens), a succulent shrubby asclepiad in savanna, is generally avoided by animals and regarded as a fetish by some tribes (8). In Ghana, the leaves of *Gardenia ternifolia* ssp. *jovis-tonantis,* a savanna shrub popularly believed to conduct lightning, and *Tragia* species - especially *T. benthamii* Climbing nettle, *T. preussii* and *T. volubilis,* trailing plants with irritating hairs - are used in fetish practice (21a). Fetish doctors mix the leaves with eggs and wash themselves with the mixture; and they also spread the plant on the ground and lay their objects on top of it (21a). (Since *Tragia* species are well known for the irritating stinging hairs, it is most surprising that fetish-men wash themselves with it without any apparent signs of irritation. It could probably be a manifestation of one of their numerous tricks). Throughout the Region, *Costus afer* Ginger lily and *C. lucanusianus,* perennial forest herbs in the ginger family, feature in various fetish practices among many of the tribal groups. Other species are *C. deistelii, C. dubius,* and *C. schlechteri.*

In the Region, some wild species of *Aframomum,* perennial ginger-like herbs; and *Culcasia,* climbing or epiphytic or sometimes erect perennial aroids, are looked on as fetish plants (8). For instance in Ivory Coast and Burkina Faso, the seeds of *A. melegueta* Guinea grains or Melegueta are put into amulets, fetish charms and the like, usually in 7s or multiples of seven (24); and in the former country, an unidentified species of *Culcasia* is considered to be fetish and a tonic (6a). In Ghana, *Indigofera tinctoria,* a softly woody leguminous plant formerly cultivated as a dye plant, is reported to be used by the coastal tribes as *bosom* (god or fetish); and *Cardiospermum grandiflorum* and *C. halicacabum* Heart seed or Balloon vine are both used as fetish - especially the former (Thonning fide 17). In Gabon, *Ipomoea quamoclit,* a very decorative twiner with small tubular bright red flowers, is a fetish for eloquence, and *Elytraria marginata,* a small acanth with rosette leaves, a fetish of prostitution (40b). In this country, *Drynaria laurentii* and *Platycerium stemaria* Stag-horn fern, epiphytic ferns, are hung on a piece of timber in the centre of a village as a fetish-protector for the whole community (40b). In W Africa, the leaves of *Ancylobotrys amoena,* a climber with white latex, have unspecified fetish uses (30). In Ghana, *Lagenaria siceraria* Calabash or Bottle gourd is called 'apebentutu' in Fante, meaning a small tubercled gourd used only by men for holding their fetish (21a). In Mexico, the root cuttings of *Solandra* species, the god-plant of mythology, are regarded by the Huichol Indians as being powerful fetish objects; but more powerful by far, is said, are the fruits of the plant which are very rare (26).

Food Crops

Some pagan tribes in the Region regard *Cajanus cajan* Pigeon pea with superstition, planting it near fetish home, but refusing it as food except as votive offerings (8). This probably explains why the beans are not an important food crop in the Region. Similarly, the Hausa do not use *Canavalia ensiformis* Sword bean much as food, but plant it probably with superstitious idea. In Ghana, *Talinum triangulare* Water leaf, a wild vegetable, is used in fetish ceremonies, placed in a pot with eggs, among other objects (8). In this country, *Pennisetum glaucum (americanum)* Bulrush, Pearl or Spiked millet is a crop never grown in the Ga-speaking area for other than ceremonial use (11). In Central African Republic, *Dioscorea bulbifera* Potato yam is cultivated both in edible and known poisonous forms, the latter as fetish in Ubangi-Shari region (7).

Cajanus cajan Pigeon pea

FETISH GROVES AND PLANT CONSERVATION

Traditional medicine-men tend to protect fetish plants by either hiding them in their compounds or in groves. For instance in the Ivory Coast, *Plumbago zeylanica* Ceylon leadwort is often planted by the fetish on a small scale and hidden from those who have no respect for it (24). Throughout the Region, *Newbouldia laevis* is a common tree around fetish groves and an important one. Among other properties, its presence is believed to prevent the entry of evil (1). The tree is also used in the fetish cult (Thonning fide17). *Antiaris toxicaria* Bark cloth tree is also often found in fetish groves. These groves are important in the religious life of the villages, as they are believed to be the abode of ancestral spirits and other deities (16). Indirectly however, fetish groves are an effective traditional means of conserving biodiversity and protecting trees from over-exploitation and ultimate extinction.

COUNTER-FETISH AND PROTECTIVE-FETISH PLANTS

Some plants are known to be used in rituals to counteract malicious fetish practices, or to neutralize the power of these forces. For instance, among the Ewe-speaking people of Ghana, Togo and Benin Republic, an infusion of *Croton lobatus,* a common weedy annual herb, collected with the necessary rituals, macerated in water and used to wash the whole body amid prayers for guidance, for assistance and for protection, is believed to thwart the spiritual machinations of any enemy who seeks the services of a destructive fetish (29). Among the tribe, *Abrus precatorius* Prayer beads, similarly collected and used, has the power of exposing all evil conspirators and of punishing them severely (29). This is probably an instance of sympathetic magic with reference to the Common Name of the plant.

Throughout the Region, *Ficus sur (capensis),* a common shade tree with abundant cauliferous fruits, is considered as a fetish tree, so a decoction of the bark and branches is drunk and used as bath water for protection from danger. For a similar reason, in Ivory Coast, the Guéré use *Sterculia tragacantha* African tragacanth to protect children from danger and sickness (24). In this country, *Rothmannia longiflora* and *R. whitfieldii,* small forest trees, have protective attributes for fetish or from malign influences (5). In Gabon, *Ficus ovata,* is often planted in villages to serve as a fetish protective tree; and *Entada gigas* Sea bean or Sea heart, a stout forest liane, is planted near villages as it is endowed with fetish protection against misfortune, danger and accidents. (This account applies to *E. rheedei (pursaetha)* Sea bean or Sea heart (40b).

Other plants feared by the dark or evil forces include *Sida linifolia,* a little malvaceous weedy annual of cultivated land, believed to be fetish itself; and *Momordica balsamina* Balsam apple or *M. charantia* African cucumber, trailing herbaceous cucurbits, and popular fetish plants among many cultural groups. The list includes *Phyllanthus niruri* var. *amarus (amarus),* an annual weedy euphorb of cultivated land; *Newbouldia laevis,* a tree used extensively by fetish-men for various rituals, but traditionally believed to be inimical to the forces of darkness and of evil; and *Elaeis*

Leptadenia hastata

Pergularia daemia

guineensis Oil palm. In Senegal, roots of *Annona glauca,* a shrub, are put as girdles onto young children as a fetish protection against illness (23a). However, by far the most potent of these plants are probably *Leptadenia hastata* and *Pergularia daemia,* both trailing asclepiads often tended in and around villages and settlements as magic plants. (See also CARVINGS AND IMAGES; CHARM; JUJU OR MAGIC; and RELIGION).

References:

1. Agbordzi, (pers. comm.); 2. Aubréville, 1950; 3. Brown, 1878; 4. Bouquet, 1969; 5. Bouquet & Debray, 1974; 6a. Burkill, 1985; b. idem, 1994; c. idem, 1995; d. idem, 1997; 7. Chevalier, 1932; 8. Dalziel, 1937; 9a. Ellis, 1881; b. 1885; c. 1887; 10. Everard & Morley, 1970; 11. Field, 1937; 12. Fink, 1987; 13. Gilbert, 1989; 14. Gledhill, 1972; 15. Graf, 1978; 16. Hall, 1978; 17. Hepper, 1976; 18. Hives, 1930; 19. Holland. 1922; 20. Hutchinson & Dalziel, 1927-36; 21a. Irvine, 1930; b. 1961; 22. Keay; 1961; 23a. Kerharo & Adam, 1962; b. 1964; c. 1974; 24. Kerharo & Bouquet, 1950; 25a. Kingsley, 1897; b. 1901; 26. Knab, 1977; 27. Lotschert & Besse, 1983; 28. McLeod, 1981; 29. Mensah, (pers. comm.); 30. Pichon, 1953; 31. Pope, 1990; 32. Portères, 1935; 33. Rasoanaivo, Petitjean & Conan, 1993; 34. Rattray, 1927; 35. Swithenbank, 1969; 36. Taylor, 1960; 37. Tremearne, 1913; 38. Van Sertima, 1976; 39. Watt & Breyer-Brandwijk, 1962; 40a. Walker & Sillans, 1953; b. idem, 1961.

FISHING

Behold, I shall send for many fishers, said the Lord, and they shall fish them...
Jeremiah 16:16a

A number of plants which have the property of either paralyzing fish or killing them are employed as fish-poison by tribal fishermen throughout the Region and elsewhere. The practice is now illegal in many communities because many fingerlings are killed or paralyzed during the operation. There is also the high probability of contaminating the water, with its consequent hazards to both man and beast. Besides the poisonous principle in the plants, superstitious charms may sometimes be involved in the use of some fish-poisons. There are also charms that protect fishing traps from theft. In order to induce a bumper catch, fishermen in the Region also do regularly present sacrificial offerings to the gods.

FISH-POISONS

The use of some fish-poisons may be accompanied by traditional rituals. For instance, in Burkina Faso, the bark of *Balanites aegyptiaca* Desert date or Soap berry, a popular fish-poison plant, taken for fish-poisoning may only be removed from the branches by men who must avoid meeting pregnant women (8). The bark is knocked off with a cadence of blows accompanied by songs and dancing imitating swimming fish, and is then placed in the stream with cries of *ibo! ibo!* (die, die quickly) (8). Similarly, *Tephrosia vogelii* Fish poison, another popular plant for killing fish, enter into fishing rites in W Africa - the seeds are planted by an old man while reciting incantations lying on the ground, then getting up repeating *kap? kap? kap? kap?* (8). Many other plants are employed by fishermen to poison fish in the Region. Table 1 below lists some of these.

Species	Plant Family	Common Name	Part Used
Anacardium occidentale	Anacardiaceae	Cashew nut	Bark
Annona muricata	Annonaceae	Sour so	Powdered seed
A. squamosa	Annonaceae	Sweet sop	Powdered seed
Cassia sieberiana	Caesalpiniaceae	African laburnum	Fruit and root
Croton tiglium	Euphorbiaceae	Croton oil plant	Seed
Balanites aegyptiaca	Balanitaceae	Desert date, Soap berry	Bark
Diospyros mespiliformis	Ebenaceae	West African ebony	Bark
Dodonaea viscosa	Sapindaceae	Switch sorrel	Leafy twigs
Hura crepitans	Euphorbiaceae	Sandbox tree	Latex
Jatropha curcas	Euphorbiaceae	Physic nut	Latex
Lepidium sativum	Cruciferae	Common cress	Whole plant
Luffa cylindrica	Cucurbitaceae	Vegetable sponge	Fruits
Mammea africana	Guttiferae	African mammee- apple	Bark
Mundulea sericea	Papilionaceae	Fish-poison bush	Whole plant
Pentaclethra macrophylla	Mimosaceae	Oil bean tree	Bark and seed
Pseudocedrela kotschyi	Meliaceae	Dry-zone cedar	Bark with gum
Scadoxus cinnabarinus	Amaryllidaceae	Blood flower or Fire-ball lily	Bulb
Scadoxus multiflorus ssp. katerinae	Amaryllidaceae	Blood flower or Fire-ball lily	Bulb
Strophanthus hispidus	Apocynaceae	Arrow poison	Seed
Swartzia madagascariensis	Caesalpiniaceae	Snake bean	Root
Tephrosia vogelii	Papilionaceae	Fish poison	Whole plant
Turraeanthus africanus	Meliaceae	Avodire	Bark with leaf
Zanthoxylum senegalense	Rutaceae	Candle wood	Root bark
Ziziphus mauritiana	Rhamnaceae	Indian jujube	Fruit

Table 1. Some Fish-poison Plants

CHARM PRESCRIPTIONS

Plants which are non-poisonous to fish and, therefore, not normally applied as fish-poison are, however, sometimes also used as a charm to catch fish by tribal fishermen. For instance, in Gabon, the vines of *Ipomoea cairica,* a slender twiner, are believed to be a lucky charm for those who wish to catch big fish (15). In this country, *Asplenium africanum,* an epiphytic fern, is a fetish plant to bring good luck in fishing (15). In N Nigeria, *Crotalaria arenaria,* a woody prostrate legume in savanna, was a component of a fish-poison in Katagun (4). The preparation is mainly made of the fruits of Desert date and that of *Parkia biglobosa (clappertoniana)* African locust bean; and the woody legume is added

doubtless for superstitious reasons (4). Among traditional fishermen in the Region, the leaves of either *Desmodium adscendens,* a prostrate herbaceous legume on forest floor, or of *Vernonia cinerea,* Little ironweed, an annual composite weed, or of *Triumfetta cordifolia* Bur weed collected with rituals, are attached with incantations to the fishing net as a charm to attract fish (10). Alternatively, the leaves may be ground and sprinkled on the fishing gear with prayers, accompanied by the necessary ceremonial rituals, as a fish-lure (10).

In N Nigeria, fishermen are reported to smear their bodies with the milky juice or to rub themselves with the leaves of *Taccazea apiculata,* a forest climber with latex, in the belief that fish would be attracted (4). In S Nigeria, the bulb of a species of *Crinum,* a water lily, is a popular fish-poison in the Benue River (13); and it is probably used with others as a charm by fishermen (4). In N Nigeria, the leaves of this water lily with a piece of white calico enter into fishing rituals to attract big fish to one's net (12). The leaves are collected early on a Friday morning before talking to anyone, then dried and sprinkled on the fishing gear (12). In Sierra Leone, fishermen and hunters employ *Gouania longipetala,* a scandent forest shrub or liane, to bring forth good luck and abundant catches (Boboh fide 2b). In Senegal, *Pupalia lappacea,* a weedy amaranth with fruits that stick to the legs and clothing, is used in a fish-lure in the Senegal River, not as a fish-poison but in a manner not clearly defined (7a, b). Traditionally, the plant is reported to have many superstitious uses (4). It could also be a case of sympathetic magic hinged on the curved hooks on the fruits.

Tropical beach

In Central African Republic, there is a belief in Ubangi, that the leaves of *Tragia preussii,* a twining sub-shrub with irritating leaves, bundled into a package and dropped into a river attract fish (14). In this country, before one goes out fishing, one soaks the nets in water in which *Eleusine indica,* a tufted annual grass, has been crushed for a bumper catch (14). The same people rub the stems of *Indigofera conjugata,* an erect savanna herb, on fish nets to bring luck in the catch (2a). In Ghana, the whole plant of *Solenostemon monostachyus,* an erect annual weedy labiate, collected with the necessary ceremony; with either *Momordica balsamina* Balsam apple or *M. charantia* African cucumber, twining cucurbits; the charred head of Sea bream, *Pagrus pagrus,* a large reddish fish locally called 'tsire' in Ga; and kaolin all ground together is sprinkled with rites and incantations by the Ga fisher-folks on the fishing lines and hooks for a good catch of this fish (3). Kpone, Ningo and Prampram fishermen are said to be good at catching 'tsire'. In this country, African cucumber was used in ancient times by the prophet *Lomo* of La, a suburb of Accra, in fishing rituals after which abundant fish were caught (6).

PROTECTION FROM THEFT

Where the stealing of fish or of fishing traps occurs - the practice is more common in inland fishing - charms may be displayed to warn would be thieves of the consequences of stealing someone else's catch; and to prevent any such intentions. These 'prescriptions' generally work, because the local tribesmen believe in them, fear them, and respect them. A description of one such 'prescription' and the instant punishment to any would-be thief is given: 'a fetish charm is then secured to it to say that any one who interferes with the trap, save the rightful owner, will 'swell up and burst', and the trap is left for the night, the catch being collected in the morning' (9).

RITUAL OFFERINGS

As a prelude to the fishing season among the Ewe thunder-worshipping fishing communities of Ghana, Togo and Benin Republic, a large bowl lined with the leaves of *Newbouldia laevis,* a popular fetish tree, and containing a ceremonial corn meal mixed with palm oil is rowed out to sea annually, together with a sacrificial cow, *Bos* species. Amid prayers, incantations and libation the bowl with its contents and the cow are thrown overboard as a sacrificial offering to the gods for a bumper catch (1). Barely a week after the ceremony, dolphins, *Delphinium*

species, are seen jumping out of the sea to push the fish stock nearer to the shore. This ritual is likely a modification of an outmoded one in which a man was sacrificed. In addition to offerings of corn, cowry shells, palm oil or a cow, *Bos* species, human sacrifices to the gods of the ocean was formerly practiced in the Region. 'Sometimes, too, the King of Dahomey (now Benin Republic) sends an ambassador, arrayed in the proper insignia, with gorgeous umbrella and a rich dress, to his good friend the ocean. The ambassador is taken far out to sea in a canoe, thrown overboard and left to drown or to be devoured by sharks, *Carcharhinus amboinensis* or *C. leucas*' (5).

On the ritual uses of products obtained from *Zea mays* Maize, it has been observed, 'in a fiesta for the gods of fishermen, popcorn was scattered like flowers over the water by the Aztecs' (11). (See also BURGLARY; and TABOO).

References.

1. Avumatsodo, (pers. comm.); 2a. Burkill, 1995; b. idem, 1997; 3. Cofie, (pers, comm.); 4. Dalziel, 1937; 6. Ellis, 1885; 6. Field, 1937; 7a. Kerharo & Adam, 1964; b. idem, 1967; 8. Kerharo & Bouquet, 1950; 9. Kingsley, 1901; 10. Mensah, (pers. comm.); 11. Moore, 1960; 12. Moro Ibrahim, (pers. comm.); 13. Stauch, 1966; 14. Vergiat, 1970; 15. Walker & Sillans, 1961.

19 FORTUNE-TELLING AND DIVINATION

There shall not be found among you...
a consulter with familiar spirits.
Deuteronomy 18:10a, 11a

The act of divination, a craft in which the future is foretold by means of magic, astrology or other supernatural means, is an ancient practice and extremely widespread. In addition to magic and astrology, possession by an evil spirit or a demon could also account for the practice - as observed in Acts 16:16 of the *Holy Bible*. Today, diviners, like magicians, still form an important part of the council in a king's court in traditional Africa and elsewhere - as it was in biblical times. However, the craft is not generally practiced as a separate profession in Africa, but rather forms a part or additional function of juju-men or medicine-men; fetish priests or priestesses; witch-doctors; 'possessed subjects'; nomadic Fula-tribesmen *(aduro-aduro);* and sometimes ordinary people (say, after a dream); and so on. The diversity of these various groups and the absence of statistics on them, makes it impossible to estimate the number of fortune-tellers in the Region, much more the number of their clients.

According to the spokesman for the congress of fortune-tellers in Paris, however, 'four million French (people) go to psychics every six months'. It is also reported that in the United States there are an estimated 175,000 part-time astrologers and 10,000 full-time. They are also numerous in Great Britain, where they have their own schools. Fortune-tellers and diviners are not only patronized by the lower class people or by private individuals, but surprisingly, the élite and officials as well. For instance, Madame Soleil, a famous French astrologer reveals, 'They all come to me, whether rightist or leftist, politicians of all points of view, and foreign chiefs of state. I even have priests and communists'. When the magician Frédéric Diendonné died, *Le Figaro,* a French newspaper, disclosed that he attracted a very large clientele of personalities, ministers, high officials, writers and actors.

The story is much the same in W Africa. For various reasons - insecurity, prospects in life, promotion at work, success in business,

Divining with cowry shells

or in political aspirations, or in marketing, or in academic studies; for wealth and luck at games of chance; for excellence in athletics, boxing, wrestling or in football; or to unravel some social mishap or mystery - high officials leave their modern premises to consult these fortune-tellers and diviners in their ramshackle huts, in the hope of securing solutions to these problems.

Fortune-tellers and diviners such as gypsies and occultists may profess through a crystal ball, a magic ring, a glass of holy water or the palm of the subject. In traditional Africa, by far one of the most popular medium of divination is the cowry shell. Sometimes a mirror is used (26a, b). However, plants may also be used in forecasting and fortune-telling, or as ingredients in the craft or to cleanse the divining dice. Many methods of divination with plants are practiced - such as throwing shells or fruits or seeds as with cowry shells or with the dice, the poison oracle, and the rubbing-board oracle - to unravel past events such as detecting a murderer, or diagnosing an illness, or finding out a missing person or a stolen property. There is also general divination to foretell the future - such as an impending danger or to predict the outcome of a war.

UNRAVELING PAST EVENTS

The secrets behind specific events, past occurrences or incidents, witchcraft and sorcery, the causes of disease, among others; may be unraveled or revealed by the supernatural through the medium of plants. For instance, in its distribution area the stems of *Thalia welwitschii,* an erect herb usually in wet places, were at one time used as divining rods in detecting a murderer (15a). The observation that *Trachyphrynium braunianum,* a straggling forest herb, has many superstitious uses (27), could include divination, since the two plants are very similar. In Sierra Leone, *Palisota hirsuta,* a robust perennial forest herb, is carried in the hands in a superstitious practice among the Temne in witch-divination

(4). However, by far the most widely used method of witch-divination in the Region is by ordeal. (See ORDEAL: PLANT TESTS: Witch Trial).

Causes of Disease

Among the traditional healers' job is to divine the causes of disease - either a malicious person, a witch, a local deity, an evil or ancestral spirit, or from natural causes. In the case of a spirit-influenced ailment, this information enables him or her to mediate on behalf of the patient. In addition, the healer is able to determine the most appropriate herbal prescription to be administered, the method of collection, the preparation and the administration.

In Ghana, the fruit of *Kigelia africana* Sausage tree is said to be used, painted in various colours, by fetish-men in divining the cause of disease (4); and in Angola, the seeds of *Cassia sieberiana* African laburnum are met with in all markets of the coast, used by native medical men to divine (the cause of) disease, but not to cure it (14). In Botswana, the Sotho diviners use *Ilex mitis* Cape holly in conjunction with the divinatory dice to prevent a sick person from being bewitched; and the Sotho medicine-man uses *Haplocarpha scaposa,* a composite, when consulting the divining dice (32). In S Africa, the Southern Sotho medicine-man uses *Crabbea hirsuta,* an acanth, in conjunction with the divining dice (32). In S Sudan, *Gardenia ternifolia* ssp. *jovis-tonantis,* a shrub traditionally believed to redirect lightning, together with *G. vogelii* and *Rothmannia whitfieldii,* shrubs or small savanna trees, are used in detecting whether a sick person will recover (7). 'When you go to divine on behalf of a sick man, burn it and add its ashes to your magical paste. If, while dancing, tears begin to swell up in your eyes, you will know that he will not recover from his sickness' (7).

In N Peru, *Trichocereus pachanoi* and *T. peruvianus* San Pedro, cacti, are used in divining the cause of an ailment and its remedy (5). The author observes, 'these two species, known as 'huacuma', 'aquacolla' or 'cimarron', are the most important magical plants used in shamanism. They are picked according to special rituals by a 'curandero' who has followed a strictly vegetarian diet in the three days preceding the harvest. The San Pedro is used to induce medianistic trance and is itself the reincarnation of the most potent spirit (the 'Inca') who leads the 'curandero' to see the origin of the sickness and the remedy for it' (5).

GENERAL CASES

In Mozambique, the divining dice of the Shangana diviners include the nut of *Sclerocarya caffra* Cider tree which represents the vegetable kingdom or 'medicine'; and the Venda of S Africa make divining bowls and drums from the tree (32). In S Africa, the Zulu have a charm use of *Crabbea hirsuta,* an acanth, where the froth overflowing from a cold infusion of the root, beaten up, indicates whether a disputant will win or lose (32). In this country, the Southern Sotho 'doctor' seldom uses any medicine without consulting the divining bones which are 'doctored' with *Hermannia depressa* Rooi opslag (32). The tribe use *Polygala gymnocladia,* locally called 'ntsebele', as a charm when using the divining dice (32). Also decoctions of the root of either *Galium dregeanum,* locally called 'scharane', or of *G. rotundifolium,* locally called 'moriri-wa-leshala', enter into ritual of the tribe's witch-doctor to wash the divining dice, if he loses a patient by death (32).

In E Africa, the Swahili use *Microsorium scolopendria,* an epiphytic forest fern, in magic medicine. For instance, a witch-doctor carrying out a divination will use it during a seance to ensure full attention of the congregation (3c). In Botswana, the Sotho witch-doctor chews *Heliophila suavissima* Bloubekkie before using the divining dice; and also diviners wash their dice with the decoction of *Brunsvigia radulosa,* a lily, in the belief that this imparts greater accuracy to them (32). In S Nigeria, an infusion of *Lentinus tuber-regium,* a large rigid toadstool with a cup-like cap, is used by Yoruba fortune-tellers to wash their face (21).

<u>*Hallucinogenic Induced Divination*</u>

As in traditional religion (see RELIGION: RELIGIO-HALLUCINOGENIC PLANTS), hallucinogenic plants are sometimes used by fortune-tellers before consulting the oracle or throwing the dice, to induce a state of divination. For instance, in Gabon, it is emphatically recounted that to eat *Tabernanthe iboga* Iboga, an apocynaceous shrub with hallucinogenic properties, is the only way to see the vision of Bwiti, according to the lore of the cult (25). Similarly, taken in sufficient quantities, the root of *Monadenium lugardae,* a succulent euphorb, is believed to produce hallucinations and delirium (32). In S Africa, the *sangoma* or diviner sometimes swallows a bit of the root before a big 'indaba', when it is supposed to make them see visions and to prophesy under its influence (32). The stimulating effects of *Datura stramonium* Jamestown weed are said to have been used to inspire the pythoness of the Delphian oracle (32). In Mexico,

Panaeolus campanulatus var. *sphinctrinus,* a fungus, was used by the Mazatec Indians of Oaxaco under the name of teonanacati as a narcotic for divination (32).

PREDICTING THE FUTURE

In S Nigeria, the dry dead wood of *Pterocarpus soyauxii* Redwood darkens in colour, and when pulverized is used as a fetish medicine called 'iye' in Yoruba, meaning sawdust or dust formed by boring beetles, *Apate monachus* and *Minthea obsita;* and under the name 'iyi-ifa', meaning good luck or the god of divination, the powder is spread on divining tables used by fortune-tellers (4). In Burkina Faso, *Merremia hederacea,* a climbing twiner with small yellow flowers, is known as 'sapu pora', meaning herb of soothsayers, and it is used there, along with *Sclerocarya birrea,* a savanna tree with edible fruits, in fortune-telling (13). In Senegal, the Tenda use the bark of *Spondias mombin* Hog plum or Ashanti plum for divination (10); and in the Democratic Republic of the Congo (Zaire), the wood of *Tessmannia africana,* a forest tree legume, contains some resin-canals which yield a little copal used by the Turumba for divination (17a, b). In Gabon, *Lapportea aestuans* and *L. ovarifolia,* herbaceous stinging plants of waste places, produce divinatory powers to priestesses who can foretell the future, uncover malefactors and resolve knotty problems (30).

In Nigeria and Cameroon, the split shells of the fruit of *Irvingia gabonensis* Wild mango are used in divination by semi-Bantu tribes - giving a favourable omen if one falls flat and the other convex side upward (4). The fruit shells are thus used in the same manner as cowry shells. In S Nigeria, the fruit shells of *Detarium microcarpum* and *D. senegalense* Tallow tree, tree legumes in savanna and forest outliers in the moist savanna region, respectively; and the kernels of *Dacryodes edulis* Native pear are equally used as items for the oracle (soothsaying) by tribesmen (20). In The Gambia, the fruit of *Abelmoschus esculentus* Okro or Okra is held to have magical properties; however, persons with second sight risk losing this skill if they eat it (28). In Ghana, the seeds of *Afzelia africana*

Afzelia africana African teak 1/3 natural size (tree forbidden to be used as fuelwood or cut).

African teak, a tree legume in savanna and fringing forest, are used in divination (15b) - probably because of their curious appearance.

In Ivory Coast, the masticated or the chewed leaves of *Hoslundia opposita,* an erect labiate traditionally believed to be a fetish plant, enables one to foresee the future - particularly an impending danger and an intended poisoning (16). One of the secrets used by fetish priests for clairvoyance is to instil the leaf juice of *Ocimum gratissimum* Fever plant crushed with seven fireflies, species of *Lampyris* or *Photinus,* into both eyes (18) - the sympathetic magic appears to be the luminous fire-flies. The informant did not, however, reveal the method of reversing this 'prescription'.

The Yoruba name of *Elaeis guineensis* var. *idolatrica* King oil palm, 'ope-ifa', has reference to the god of divination, and the nuts are used by priests (4). The nuts are said 'to talk' and thus used for the purpose of divination (3b). The Yoruba name *'Orunmila'* is that of the God of Divination; and it is applied to any branched palm which thereby is considered sacred (3b). It has earlier been observed (8b) that, in Yoruba mythology, ' 'Ifa', the God of Divination (who is usually termed the God of Palm nuts, because sixteen palm-nuts are used in the process of divination), comes after 'Shango', God of thunder, in order of eminence' - (see EVIL: PLANTS FOR EVIL for the full story about this latter deity). In W Africa, seven beans are used for divination (8b). In Yoruba folk-lore, *Lentinus tuber-regium,* an agaric, large rigid toadstool with a cup-like cap, was the first crop plant on earth by 'Eli-Ogbe', the premier disciple of 'Ifa', the God of Divination, a messenger of *'Orunmila',* the supreme deity representing God on earth (21).

In S Sudan, a poisonous creeper called *benge* is used in divination. 'Azande catches some chicken, *Gallus domestica,* today and takes them to *benge.* He mixes *benge* with a little water and seizes a chicken and pours *benge* into its beak and addresses *benge* thus: *'Benge, benge,* you are in the throat of the chicken. I will die this year, *benge* hear it, twist the fowl round and round and lay down the corpse. It is untrue, I will eat my eleusine this year and year after let the fowl survive'. If he will not die the fowl survives. If he will die the fowl dies in accordance with the speech of *benge*' (7). 'The substance is a strychnine poison, so that the chicken will either be killed or will vomit it; and the operation cannot be controlled by the amount which is given to the chicken' (12). From this observation, it is most likely, therefore, that the creeper benge is a species of *Strychnos.*

Another Azande oracle is called *mapingo.* Three pieces of wood are required for each question asked. Two pieces of sticks are placed side

by side on the ground, and a third piece placed on top and parallel to them. The little pieces are generally arranged just before nightfall. The oracle gives its answers by the three sticks remaining in position all night or by the structure falling (7). Another one is the *dakpa*, or the termite, *Macrotermes* species, oracle. A man has only to find a termite mound, and insert two branches of different trees into one of their runs and return next day to see which of the two the termites have eaten (7).

Yet another one - the most used of all Azande oracles - is the *iwa*, the rubbing-board. It consists of miniature table-like construction, carved out of wood of various trees, the more usual being *Cola cordifolia*, a savanna tree. They have two parts, female or flat surface of the table supported by two legs and its tail, and the male, or the piece which fits the surface of the table like a lid. 'When the operator jerks the lid over the table it generally either moves smoothly backwards and forwards, or it sticks to the board so firmly that no jerking will further move it, and it has to be pulled upwards with considerable force to detach it from the table. The two actions - smooth sliding and firm sticking - are the two ways in which the oracle answers questions' (7).

To try out the art of divination oneself, a lover takes the stem of *Holarrhena floribunda* False rubber tree and splits it. If the parts are equal he takes courage, if unequal he awaits a more favourable time - the plant serving as a fortune-teller, 'bakin mayi', in Hausa (4). The Hausa name of False rubber tree, 'gaman sauwa', implies a happy conjunction in the rites of forecasting. Similarly, *Cassytha filiformis* and *Cuscuta australis* Dodder, both parasitic epiphytic twiners, feature in forecasting (15a) - (see LOVE: ASSOCIATED PLANTS). The initials of a future partner is also believed to be predicted by throwing an orange peel over the shoulder (see MARRIAGE: SELECTION). In Turkey, a newly-wed woman may predict the number of her offsprings, it is believed, by throwing the fruit of *Punica granatum* Pomegranate on the ground and counting the scattered seeds (9).

Among the Akan-speaking tribe of Ghana and Ivory Coast, a very common mode of divination is to throw palm-wine on the ground, future events being foreshadowed by the figures it forms (8a). Yet another is to take at random a handful of the seeds of *Dioclea reflexa* Marble vine, and let them fall - the future course of events being foreshadowed by their number being either odd or even (8a). The fruit of *Citrus aurantiifolia* Lime or the seed of *Cola acuminata* Commercial cola nut tree or *C. nitida Bitter* cola cut in two halves may also be used to divine the course of events. (For the details, see DEATH AND DYING: FUNERAL OBSEQUIES: Separating the Dead from the Living). In Mali, the fruit of

Bitter cola is part of a prescription known as *saraka* issued by soothsayers (6). The leaves of two adjacent plants of *Kalanchoe integra* var. *crenata* Never die, a fleshy herb, may also be employed in a ritual to foretell events by squeezing the juice of *Hilleria latifolia,* a perennial herb near villages in the forest region, onto each leaf while a question is asked. A shake of the *Kalanchoe* leaves signifies a positive answer (1).

In S Africa, powdered *Begonia sutherlandii,* a herb, locally called 'uqamamawene' in Zulu, is used as a charm to find out whether a kraal is threatened by an enemy (32). The procedure is to sprinkle a half calabash with the powder and place a grinding stone on it. This process is repeated until there are three stones on the calabash (32). If it withstands the weight, no enemy threatens; but if the calabash breaks, it is certain that ill will befall the kraal (32). During the Middle Ages, *Verbena officinalis* Vervain was used by sorcerers and fortune-tellers as a magic herb (22). In the temperate regions, the use of *Hamamelis virginiana* Witch hazel, a cultivated ornamental shrub and native of N America, in divination used to be very popular (24). 'By the help of the hazel's divining-rod the location of hidden springs of water, precious ore, treasure, and thieves may be revealed, according to old superstition. Cornish miners, who live in a land so plentifully stored with tin and copper lodes, say they can have had little difficulty in locating seams of ore with or without a hazel rod, scarcely ever sink a shaft except by its direction' (24).

Animals in Divination

Sometimes animals may also be used in divination. For example, there is record of two instances when a fowl, *Gallus domestica,* was killed in divining the coarse of events. 'Then followed the ritual killing of the fowl and a divination. Suddenly, I heard sighs of relief from the people around me, which meant that I had passed the test and that the deity had responded in my favour. Both the fowl's kidneys were white, which proved my sincerity and honesty in dealing with the healers. Black would have been symbol of betrayal, whereas a white and black kidney would have predicted an accident for me in the near future' (11). In the second instance - which established the loyalty of a newly-ordained priest or priestess during the three-year apprenticeship - the author writes, 'the fowls are slaughtered and thrown into the air. If the novice has observed all the rules without violating any taboos such as sexual abstinence, the fowls will die with their chest against the sky. If the chest points to the ground, this proves that the novice has violated one or several taboos' (11).

The Founding of Kumasi

Perhaps one of the most historic uses of plants in divination in W Africa is that surrounding the founding of Kumasi as the capital of the Ashantis. Okomfo Anokye (a priest) made Osei Tutu (the King of Ashanti) plant two young *kuma* or *kumaniwa* trees, requesting him to make his new capital town at the place where either one grew and flourished (31). One of them, sure enough, died; and its place has ever since been called Kumawu, meaning 'the dead *kuma* or *kumaniwa* tree'. The other lived and grew, and Osei Tutu built his new town of Kumasi, meaning 'under the *kuma* or *kumaniwa* tree', under its shade (31). The botanical identity of the *kuma* or *kumaniwa* tree is not absolutely certain. However, Okomfo Anokye is reported to have obtained the seedlings from Akyem Okumenini, which is named after the tree, and means 'the male *okum*' or 'the male *okumaniwa*'. The most probable tree species is *Lannea welwitschii,* a forest tree with characteristic pitted bark similar to gun-shot wounds, locally called in Akan 'kumenini', meaning 'python killer'.

However, according to Nana Kwame Kwamin, Chief of Obuasi, Okomfo Anokye caused to be planted two *kumaniwa* trees. While the tree at Kwaman (now Kumasi) drooped, that at Agyaase (now Kumawu) bloomed. But the elders at Kwaman visited Agyaase and stealthily uprooted the survived tree and replaced the withered one. The former was replanted at Kwaman which as a result later derived the name Kumasi. According to Dormaa oral records however, the name Kumasi has a different origin and a different meaning altogether, and existed long before the days of Osei Tutu. 'By the end of the 17th century when the Dormaa settled down near contemporary Kumasi in Asuonya, all the Akan chiefs and kingdoms in this region were under Denkyirahene as *akuma-ase-hene* and had the task to cut firewood for the Denkyirahene and the queen mother with a special axe during certain festivities, and to present this at court. *Akuma* meant axe, and the *akuma-ase-hene* was the royal who served with an axe. The Dormaa therefore derive Kumasi from *akuma-ase,* he who serves with the axe, dating back to the time when the Asantehene was obliged to pay tribute to the Denkyirahene' (11).

DREAMS AND VISIONS

Another means of divination is through prophetic or wilful dreams. Should one desire to dream about a specific subject or incident, *Evolvulus alsinoides*, a low spreading hairy herb, collected with rites, is macerated in a container of water, placed in the open with seven grains of maize around it, and observed until a fowl, *Gallus domestica,* comes along to peck at the maize and to sip the water. After the seventh sip, the fowl is driven away. The subject of the dream is now recited, and the remaining water drunk for wilful dreams (18). Alternatively, the juice of *Solenostemon monostachyus,* an erect herbaceous labiate, collected with rites, is squeezed into the eye before retiring to bed; or *Alternanthera sessilis* Sessile joyweed, a herbaceous semi-aquatic amaranth, collected with rites is ground and mixed with Florida Water, a type of lavender, then sprinkled in the room at bed-time - after which a candle is lighted and kept burning before retiring to bed (18).

In Congo (Brazzaville), the leaves of *Renealmia africana,* a perennial forest herb in the ginger family, placed under a bed can make one see hidden things in dreams (2). On the contrary, a bark preparation of *Corynanthe pachyceras,* an under-storey forest tree, is taken as a stimulant to prevent dreams (2). In this country, 'to obtain pleasant dreams' one burns on the hearth in the evening some flowers of *Symphonia globulifera,* a forest tree, with some other sweet-smelling plants (3a). To banish bad visions however, one needs to sprinkle the eyelashes with powdered charcoal made from the fruit of *Pachyelasma tessmannii,* a forest tree legume believed to be full of strong juju (2). In this country, *Smilax anceps (kraussiana)* West African sarsparilla, a prickly forest climber, is believed to have magic power, so when placed under the ear of a sleeping person it will make him talk in his sleep (2). On the other hand, a tisane of young leaves of *Pteridium aquilinum,* a terrestrial fern believed to have magical uses, taken before going to sleep is said to prevent one talking in one's sleep (2). In S Nigeria, an unidentified plant locally called 'inaeyinfun' in Yoruba, meaning 'fire of egg is white', is used to induce troubled sleep in someone (29).

In the Central African Republic, *Greenwayodendron (Polyalthia) suaveolens* var. *gabonica,* a forest tree, is used by Pygmy witch-doctors to induce dreams and clairvoyance, by instilling a cold water macerate of the root-bark to the eyes (19). In Tanzania, witch-doctors smoke the flower panicles of *Cymbopogon densiflorus,* a tufted perennial grass, either alone or with tobacco to induce dreams to foretell the future (3a). In S Africa, the Kgatla dice thrower (bone-thrower) uses the powdered root of *Mundulea sericea* Fish-poison bush as a charm to ensure that he becomes a good diviner. His dice-throwing apparatus is put into a dish with the powder, water is added and stirred vigorously. The froth thus formed is drunk and the dice are kept near the owner when he sleeps.

If he dreams of them, the charm has been effective (32). In Botswana, the Sotho uses a root infusion of *Pittosporum viridiflorum* Stinkbas to give accuracy to their divining bones (32).

In W Africa, the scraped root of *Boerhaavia* species Hogweed or *Abutilon mauritianum* Bush mallow, a semi-woody perennial malvaceous bush, ground with water and instilled into the eyes induces the second sight; or *Bryophyllum pinnatum* Resurrection plant is 'talked to' and offered two cowry shells as compensation. Two of the leaves are picked, some fragrant powder and lavender sprinkled on, and placed under the pillow with prayers for visions. One condition is that no one is talked to just before retiring to bed (18). (For wilful visions, see WITCHCRAFT: PLANTS AGAINST WITCHCRAFT).

In Europe, the thick roots of *Mandragora officinarum* Mandrake often shaped in human form, were valued by the Ancients to foretell the future and render other notable services; and magical powers ascribed to *Viscum album* Mistletoe, an epiphytic parasite, which has long association with British folklore, include prophetic dreams (23). *Aipyanthus (Arnebia) echiodes* Prophet flower, a native of Asia Minor, Iran and Caucasus, is so called because the brownish spots which mark the petals of young flowers are supposed to represent the fingers of Mohammed, but they fade and curiously disappear as the flower ages (9). (See also DEATH AND DYING; KINGSHIP; and ORDEAL).

References.

1. Abban, (pers. comm.); 2. Bouquet, 1969; 3a. Burkill, 1994; b. idem, 1997; c. idem, 2000; 4. Dalziel, 1937; 5. De Feo, 1992; 6. Diarra, 1977; 7. Evans-Pritchard, 1937; 8a. Ellis, 1887; b. idem, 1894; 9. Everard & Morley, 1970; 10. Ferry & al., 1974; 11. Fink, 1989; 12. Gluckman, 1963; 13. Guilhem & Herbert, 1965; 14. Holland, 1922; 15a. Irvine, 1930; b. idem, 1961; 16. Kerharo & Bouquet, 1950; 17a. Leonard, 1950; b. idem, 1952; 18. Mensah, (pers. comm.); 19. Motte, 1980; 20. Okafor, 1979; 21. Oso, 1977; 22. Pamplona-Roger, 2001; 23. Perry & Greenwood, 1973; 24. Pitkanen & Prevost, 1970; 25. Pope, 1990; 26a. Rattray, 1923; b. idem, 1927; 27. Saunders, 1958; 28. Tattersall, 1978; 29. Verger, 1967; 30. Walker & Sillans, 1961; 31. Ward, 1956; 32. Watt & Breyer-Brandwijk, 1962.

20 GHOST AND SPIRIT WORLD

There shall not be found among you...
a necromancer.
Deuteronomy 18:10a, 11b

Ghost stories, ranging from haunted houses, allegations of dead relatives and friends seen in the flesh, to messages and gifts from the dead and so on, are common in the folklore of many countries. The stories are not only among simple, primitive pagans, but also among the urbanized, developed, educated and religious people. Britain is reported to be famous for its ghost stories. It has been observed that in the British Isles there are more ghosts seen, reported and believed in than anywhere else in the world (39). The spirit world, in folklore, is believed to be capable of influencing events in this world. In Africa alone, for instance, the lives of countless people revolve around the belief that dead ancestors have the power to protect their descendants or to chastise them with minor illness or misfortune. Consequently, the 'good' spirits are honoured and praised; but the 'bad' ones are grouped among the possible causes of any disaster, loss of lives, general low morale and poor living standards within the family.

As tradition demands, at the beginning of occasions or to herald in anniversaries, sacrifices are offered and libation poured to the beneficial ghosts or ancestral spirits for their continuous support, blessing, concern, guidance and protection of the living in the community. With the vengeful spirits however, the necessary ceremonies are performed to exorcize them accordingly (see below under EXORCISM). Naturally, plants feature in the various rituals associated with sacrificial offerings to the departed, and with the exorcism of haunted ghosts. Some plants are traditionally believed to be haunted or to haunt. Other plants are named after the underworld or the spirit world.

EVIDENCE OF THE BELIEF

Some practices and customs in traditional Africa indicate the general belief in ghosts and the spirit world. For instance, in parts of W Africa, drivers automatically blow the horn of the vehicle before crossing a bridge or a fatal accident spot. The traditional belief is that the practice signals an approaching vehicle and warns the spirits of the stream or the ghosts of the accident victims, respectively, to disperse. There are many who sincerely believe that deliberate failure to give this signal might result in the vehicle running over these spirits and off the road into a ditch. Again, an elder pours libation to the departed fathers on occasions as custom demands, and as a sign of gratitude; or presents an offering to the ancestral spirits. The fetish priest administers a herbal preparation believed to be inimical to evil spirits before applying a prescription to drive away these forces. Among Christians, both the authodox and spiritual churches burn incense to exorcize haunting ghosts.

OFFERING OF SACRIFICE

It is customary to offer sacrifice to the departed - as a routine, or on ceremonial occasions. In Ghana and Ivory Coast, the practice of pouring libation and offering mashed unsalted yam, 'eto', with eggs to ancestral spirits, 'nananom', forms part of the traditional custom among the Akans and many other tribes throughout the Region. The nuts of *Cola acuminata* Commercial cola nut tree and *C. nitida* Bitter cola have many religious and magical uses, and as such they are commonly used as an offering to the spirits and genies (24, 6). In Gabon, the roots of *Quassia (Simaba) africana,* a forest shrub, enter into an offering to the ancestral spirits (11); and the seeds of *Aframomum alboviolaceum (latifolium),* a perennial in the ginger family, are added to incense offered to the spirits and shades of departed ancestors (41). In this country, the strongly-scented roots

Mashed yam

of *Rinorea subintegrifolia,* a forest shrub, is included in aromatic offerings to the souls of departed ancestors (41); or people place offerings to the Shades of the Departed on leaves of *Lepianthes peltata* Cow-foot leaf, a forest shrub (42).

Sometimes special abodes are constructed for such offerings. For instance, in both Kenya (Magogo & Glover fide 8a) and Tanzania (Tanner fide 8a), the culms of *Cyperus latifolius,* a sedge, are used to make spirit houses. In Ghana, *Pennisetum purpureum* Elephant grass is used instead. In Kenya, *Ficus natalensis,* a fig tree, is considered to be the abode of ancestral spirits, and people pray and make offerings to their gods under its shade (8c).

ANCESTOR

It is the belief in traditional Africa that not all the dead qualify as ancestors. 'But not everyone is granted entry into the *asamandu* and is accepted into the group of ancestors. This privilege is reserved to those who previously had lived an exemplary life, reached old age, had children and died, a 'good' (natural) death' (18). In listing those disqualified to be ancestors, the author continues, 'a person suspected of *bayie* (witchcraft) who dies as a result of its own mischief or is killed by the gods is buried on the same day without any funeral ceremony. This also happens to a person who has committed *hyeakomfo* (suicide), died childless, or drowned. All of these types of death are considered as 'bad', and the persons concerned are denied entry into the *asamandu* (realm of the dead) as well as the transformation into an ancestral spirit. Their souls are considered evil and are feared by the living. It is believed that they roam about as unpacified spirits who at night lie in wait for travellers who are plagued by them and led astray' (18).

EXORCISM

Haunting ghosts are normally exorcized - usually by the clergy, but sometimes also by traditional medicine-men, fetish priests or priestesses and by witch-doctors - after the performance of the necessary customary rituals. In Senegal, the Wolof name of *Stereospermum kunthianum,* a tree of savanna woodlands, 'yetudomo', meaning the wand of the sorcerer, thus they bestow on the tree power to exorcize ghosts (23). In this country, the tribe endow the roots of *Neocarya (Parinari) macrophylla* Ginger plum with powers to relieve one possessed or tormented by Shades of the Departed (23). In Congo (Brazzaville), *Anonidium mannii,* a medium-

sized tree with wide-spreading crown, and *Afrostyrax lepidophyllus,* a forest tree, are ascribed with magical properties to ward off evil spirits and ghosts, quell nightmares, protect a house and its occupants, and to treat illnesses caused by evil spirits. All that is necessary is to sprinkle a bark decoction about the house, and put some leaves in the roof or a small piece of the trunk under the threshold (6).

Throughout W Africa, *Ocimum canum* American basil and *O. gratissimum* Fever plant are used to exorcize evil spirits and juju (24). For instance, the natives have a superstition that the spirits of their deceased relatives can appear to them and bring them all kinds of illnesses. When a native attributes his illness to this cause, American basil and other strongly scented plants are boiled in water with which he washes himself, and one besprinkled the place where he is staying with the powerful odour to expel the spirits (Thonning fide 21). *Lippia multiflora,* Gambian tea bush, a savanna shrub, is used similarly as *Ocimum.* In S Nigeria, the Ijo of the Niger Delta have a superstitious use of *Phyllanthus odontadenius,* a sub-woody herb, to drive away bad spirits (8b). A person who suffers fever every evening which is attributed to the spirit of a dead person should urinate on the plant, pick the leaves for addition to a bath with local soap, and be relieved of the spirits (8b).

In Ivory Coast and Burkina Faso, *Elaephorbia drupifera,* a savanna tree with fleshy branches containing abundant very caustic white latex, is used to drive away ghosts and bad spirits from a village (the forest equivalent is *E. grandifolia*); and in Ivory Coast, the leaves of *Hyptis pectinata,* an aromatic herbaceous to woody labiate, in water is sprinkled in the room by the Shien, the Anyi and the Kru to exorcize bad spirits (24). The authors add that in this country, *Ageratum conyzoides* Billy goat weed along with other plants confer protection against ghosts and evil spirits. In Ghana, a similar use of this weed is reported (15). In Congo (Brazzaville), the leaf sap of *Caloncoba glauca,* a small tree, diluted in water is sprinkled over tombs to keep the spirit of the dead away; and *Eragrostis ciliaris,* a tufted annual grass, is considered able to chase away spirits (6). In Gabon, the wood of *Barteria nigritana* ssp. *nigritana,* a slender forest tree, is burnt before occupied houses in order to exorcize the spirit of a departed member of the family (41).

Among the Akan-speaking people of Ghana and Ivory Coast and other tribes, an assembly of carefully arranged, assorted articles are burnt in a ritual ceremony by a fetish priest to exorcize a vindictive ghost. 'Medicine' made of powdered poultice of the leaves of American basil, Fever plant, *Momordica balsamina* Balsam apple or *M. charantia* African cucumber, *Amaranthus hybridus* ssp. *incurvatus* Love-lies-bleeding;

Hyptis species, aromatic labiates; and the stem of *Gouania longipetala,* a forest climber locally called 'homa-biri' in Akan, is placed on the leaf of *Baphia nitida* Camwood and an egg is stood on this (15). The ingredients include three fruits of *Capsicum annuum* Pepper, a bulb of *Allium cepa* var. *aggregatum (ascalonicum)* Shallot, three needles, three hooks and a large black ant with an offensive odour, *Palthothyreus* species, locally called 'adam' in Akan. These are arranged around the egg, sprinkled with salt, covered with two more leaves of Camwood and burnt among incantations and libation (15).

Medicine-men in the Region, consider the whole plant of *Emilia sonchifolia,* a weedy composite, and African cucumber inimical to ghosts; and *Gomphrena celosioides,* a common weedy amaranth, burnt with milled corn in the middle of the compound of a house, is a ritual employed by traditional healers to drive away ghosts (28). A special ritual is performed by a hunter after killing someone accidentally in the forest. He washes himself with a maceration of the leaves of *Newbouldia laevis* in water. The tree is popularly believed to be fetish in many parts of the Region. The ritual is aimed at preventing the spirit of the dead person from following the hunter to the house to wreak vengeance (2). In S Africa, *Boophone disticha* Candelabra flower, a lily, is thought to keep the 'mudzimu' away after a death. The bulb has also been administered to 'drive out spirits'; and in Egypt, *Epilobium hirsutum* Willow herb is used as a charm against ghosts (42).

Emilia sonchifolia

The trapped air between the nodes of bamboo expand and explode when a clump is set on fire. 'Early Chinese found that the exploding bamboo were good for chasing away wild animals and for routing ghosts and goblins that haunt primitive people. It was their custom to build night fires near bamboo groves so that the heated stems might explode and frighten off such malevolent spirits' (37). In Mexico, *Gnaphalium* species Cudweed, a composite herb, whose name in Algonquin means 'to smoke out spirits' is kept in every house by the Kickapoo Indians to drive away ghosts (13). The authors add, 'it is burned after the death of a person to restrain his spirit from entering the house. Clothes, as

well as all articles belonging to the deceased received by those who bathe and bury the body are smoked with the herb. When a person is ill and begins to dream of deceased relatives, it is believed the spirits of the relatives have arrived to take the sick person away. Cudweed is burned to drive out any spirits with this intent' (13).

In the ancient Roman festival of the *Lemuria,* black beans were cast upon graves, and beans were also burnt (10). It was thought that this, and other associated practices, such as beating of drums and uttering of magical words, would prevent the ghosts of the dead from troubling the living (10). On the other hand, legumes seemed to have been consumed in a reverend manner at funerals, from which it appears that their occult properties were such as link the living and the dead (10). Similarly, in its area of cultivation in W Africa, a meal prepared with *Vigna (Voandzeia) subterranea* Bambara groundnut (a bean) is served at funerals. Families therefore cultivate this crop as a necessity because the funereal meal has to be obtained from the household, but not bought outside or from the open market (3).

In S Africa, a Zulu marrying his deceased brother's widow protects himself from contamination of death by applying a lighted brand of *Ptaeroxylon utile (obliquum)* Sneezewood, a tree, to various parts of the body (42). In this country, the Southern Sotho use a decoction of *Cucumis myriocarpus* Bitter apple, a cucurbit, to purify a man who is about to marry a widow, in the event of her not having undergone the necessary rites - the widow must also bathe in the lotion and anoint herself with an ointment of the powdered plant (42). *Scabiosa atropurpurea* Sweet scabious or Mournful widow is a decorative plant with fragrant deep crimson flowers; and in the language of flowers Scabious signifies 'I have lost all', so it was considered 'an appropriate bouquet for those who mourn for their deceased husbands' (34)

Widowhood Rites

In traditional Africa and elsewhere there is a general belief that after the death of the husband or the wife, their spirit sometimes haunts the living partner unless the necessary customary rituals are performed to separate the marriage spiritually. For instance in S Nigeria, if a Yoruba widow dreams of her dead husband, a medicine is prepared with *Clerodendrum splendens,* a climbing shrub in forest, and placed on the pillow to prevent his appearing (11). In Ghana, wreaths of *Cardiospermum grandiflorum* and *C. halicacabum* Heart seed or Balloon vine, locally called 'asuani' in Akan, meaning tears, passed over the shoulder and crossed, passing under the arm, and also as a head band is used among the Ashanti to protect widows from the ghost of their late spouse (35b).

Other antidotes against the ghost of ones husband are the seeds of *Tetrapleura tetraptera*, a forest tree; the leaves of *Ocimum canum* American basil and that of *O. gratissimum* Fever plant - all being discarded in the widow's pot on the day the body is buried (35b). In the Region, *Clausena anisata* Mosquito plant is hung in houses or used as a fumigant, also to dispel or warn off revengeful spirits such as the *sasa* (revengeful spirit) of a dead husband or that of a dangerous animal (11). In its distribution area, *Eryngium foetidum,* a spiny-leaved perennial herb with unpleasant smell and fleshy rootstock exuding white latex, sometimes cultivated or tended in villages and the compound of spiritual churches - especially among the Nzima-speaking people - is used by fetish priests in fumigation against the *sasa* of a dead husband. In Gabon, the inflorescence of *Cymbopogon densiflorus,* a tufted perennial grass, is burnt in fumigations required in certain rituals such as in making incantations to chase away malign spirits, or to cleanse those who have lost their spouse (41).

The 40th Day It is the belief among many ethnic groups in traditional Africa that the spirit of dead relations depart from the family on the 40th day after death. The belief and practice could probably be linked to the Ascension of Our Lord, Jesus Christ, 40 days after the Resurrection. Consequently, on the 40th day of the death of a husband, a solemn ritual ceremony of purification is held at midnight to separate the widow from the ghost of the dead husband. American basil, Fever plant with the dried fruit husk of *Elaeis guineensis* Oil palm, and assorted spices including the dried fruits of *Capsicum annuum* Pepper with those of *Piper guineense* West African black pepper or *Aframomum melegueta* Guinea grains or Melegueta (all 'hot' spices deliberately added to render the fumes unpleasantly offensive), are set ablaze in an earthenware pot, carried by the widow - in the company of family members, friends and sympathizers (all elderly women or widows) - and thrown into the sea - or a river in the case of inland dwellers. The ceremony is accompanied with a chorus of shouts and yells warning people to keep off. It is sacrilegious for anyone, especially a married woman, to walk the streets during the ceremony or to meet with this procession. The belief is that any woman who meets this blazing pot would lose her husband within the year. (This traditional ceremony has, in recent times, come under severe criticism

as cruelty against women, though those who enforce the custom are themselves all elderly women. In Ghana, widowhood rites have been banned under Criminal Offences Act, 1960, Act 29; Section 88 (A)).

Other plants that enter into purification rituals for a widow after the 40th day include *Senecio (Crassocephalum) biafrae,* a climbing herbaceous forest composite; either *Momordica balsamina* Balsam apple or *M. charantia* African cucumber, climbing cucurbits popularly believed to be fetish; *Costus afer* Ginger lily, a robust forest herb traditionally used for purification rituals; *Ricinus communis* Castor oil plant, a soft-wooded euphorb; and *Jatropha gossypiifolia* Red physic nut, another soft-wooded euphorb (19). In Ghana, the leaves of *Hyphaene thebaica* Dum palm are woven and worn by the widow as symbolic beads during the widowhood period among the Ada people, a coastal tribe, and discarded into the sea after the rituals ending the 40th day mourning (5).

Hyphaene thebaica Dum palm or Gingerbread palm

Death of Successive Partners

By some mysterious means, some wives lose one husband after another till the fourth dies before there is a stable partnership; while some husbands similarly, lose three wives. The traditional belief and explanation among others, is that such persons are already spiritually married to a river god or goddess. Among the Hausa of N Nigeria, the Zarma of Mali and Niger and the Fula, when couples lose their partners successively as a result of a curse or some evil or spiritual influence, the dried nuts of *Cyperus articulatus, C. maculatus, Kylinga erecta, K. pumila, K. squamulata* or *K. tenuifolia,* all sedges (collectively locally called 'kaajiijii' or 'turare-wuta' in Hausa) with the dried leaves of *Calotropis procera* Sodom apple are used for fumigations to dispel the forces responsible for the deaths (30). Similarly in Ghana, to prevent the misfortune of the death of one husband after another among the Ga-speaking tribe, a preparation of either *Endostemon tereticaulis* or

Orthosiphon pallidus (incisus) or *O. rubicundus* by misidentification - since the three labiates are very similar - with *Kalanchoe integra* var. *crenata* Never die, a fleshy herb, and cowry shells - all collected with the necessary ceremony - enter into prescriptions for cleansing rituals to expel the vindictive deity blamed for the deaths (9).

However, in some cultures the misfortune of the death of successive partners is directly blamed on witchcraft. The accused (more usually the woman) was summarily sentenced to death after the death of the fourth husband. She was first paraded through the village with hooting and insults, then tied to the stake and burnt. The practice is now illegal, but the belief nevertheless persists.

Accident Victims

The spirits of those who die through unnatural causes - that is, death from causes other than old age and sickness, are traditionally believed to be vindictive spirits and to take delight in haunting the living. In such situations, the necessary customary rituals are accordingly performed in anticipation to counteract any spirits with such intentions. For instance, during the Ashanti wars, several of the hearts of the enemy were cut out by the fetish-men who followed the army. The blood, with small pieces of the heart muscle were mixed up (with much ceremony and incantations) with various consecrated herbs (unidentified); and all those who had never killed an enemy before ate portions. For it was believed that if they did not, their vigour and courage would be secretly wasted by the haunting spirits of the deceased (7). In Ghana, the *Gyabom* fetish of Ashanti, a powerful charm for driving away evilly disposed, disembodied human spirits, consisted of the leaves of *Ocimum canum* American basil with a variety of ingredients dyed a deep red tree bark colour with eggs (35a, b). The fetish was set upon the knee of a condemned man while his head was cut off, to prevent the *sasa* (revengeful spirit) of the victim from returning to wreak vengeance on his executioners or upon the king who had ordered the execution (35a, b).

Alternatively, the condemned men were gagged for the same reasons. If they are thought desperate characters, a knife is trust through their mouth to keep them from swearing the death of any other (7). There is also record that prior to these executions, the condemned men were deliberately entertained to a heavy dose of alcoholic liquor, and while in this state of intoxication and stupor,

were led happily chanting to the gallows. Similarly, when Aztec war prisoners were sent to their death, they were generally given hallucinogenic mushrooms (such as *Amanita muscaria* and *Psilocybe mexicana)* to make them gay and happy before the sacrifice (12). In Mexico, the powder of T*agetes lucida* Marigold, a decorative strongly scented composite often cultivated as an ornamental, was thrown in the face of captives, among the Huichol Indians, to 'dull their senses' before being sacrificed to Heuheuotl (37). (For the burial of those who die through unnatural causes, and its spiritual significance to bad juju, see GRAVES AND CEMETERIES: SUPERSTITIOUS VALUES).

HAUNTING TREES

In folklore some trees are believed to haunt or to be haunted. For instance, *Milicia (Chlorophora) excelsa* and *M. regia* Iroko; and also *Morus mesozygia,* all forest trees, are believed to possess spirits which are able to haunt one at noon or at midnight alike (20). Thus broken pottery from appeasement sacrifices is found around the foot of these trees (20). The odour of *Costus afer* Ginger lily is regarded as inimical to ghosts and evil influences, and it is offered to a spirit occupying a tree (11, 22a). It has been confirmed (35a) that the smell of Ginger lily is sometimes said to drive away ghosts . In SE Senegal, Tenda folk-lore has it that 'santiyu' a spirit who lives in the shade of a species of *Parinari* with edible fruits, terrifies children, thus one does not climb the tree to collect the fruits, but waits till the spirit throws them down (17). The belief in the rebirth of human souls into trees could be an explanation for the belief in this haunt by some trees. (See CHILDBIRTH: PARTURITION, and POST-NATAL; DEATH AND DYING: THE LAST MOMENTS; KINGSHIP: CEREMONIAL RITUALS).

INFLUENCE OF BATS

In W African folklore and mythology, there are superstitious beliefs about *Adansonia digitata* Baobab being the haunt of spirits which, like *Ceiba pentandra* Silk cotton tree, it owes to the activities of bats, such as *Eidolon helvum* and *Epomophorus gambianus,* and their nightly screeching during the flowering period (20). (The Silk cotton tree flowers from October to December). Bat-pollinated flowers characteristically have a very strong fermenting or fruit-like odour (36). The bats are obviously attracted by the strong scent during the opening of the flowers just as night falls; and that by the time it is completely dark the tree is full of

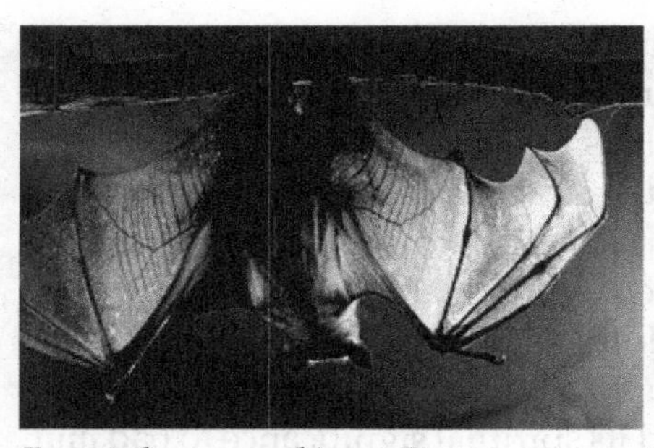
Epomophorus gambianus Bat

squeaking, scuffling bats, crawling over the flowers and flying from branch to branch (16).

Bats are considered sinister, and their unexpected nightly visitations are believed to be signs of ill omen in some societies (4). In many primitive societies bats have been associated with the dead, who are believed to possess souls that have the ability to fly freely at night when the living are asleep (4). For instance in Burkina Faso, the bat is an important figure in local philosophy as the bearer of illnesses and the souls of the departed (8b). As part of a rogation ceremony held in the dry season among the Dagaari of this country, a small bat is stuffed full with pap (locally called *saab*) prepared from grains of *Pennisetum glaucum (americanum)* Pearl, Bulrush or Spiked millet till the bat is so full it cannot fly, and thus cannot spread disease nor remove the spirits of the dead (8b).

The trees visited by bats are listed in Table 1 below:

Species	Plant Family	Common Name	Remarks
Adansonia digitata	Bombacaceae	Baobab	Tree, nectar attractive
Anacardium occidentale	Anacardiaceae	Cashew nut	Fruit tree
Azadirachta indica	Meliaceae	Nim or Neem	Medicinal
Carica papaya	Caricaceae	Pawpaw	Fruits edible
Ceiba pentandra	Bombacaceae	Silk cotton tree	Tree, nectar attractive
Ficus umbellata	Moraceae	Fig tree	Shade tree
Khaya senegalensis	Meliaceae	Dry-zone mahogany	Bark medicinal
Kigelia africana	Bignoniaceae	Sausage tree	Bat-pollinated
Mangifera indica	Anacardiaceae	Mango	Fruit tree
Maranthes (Parinari) polyandra	Rosaceae	-	Savanna tree
Parkia biglobosa (clappertoniana)	Mimosaceae	West African locust bean	Bat-pollinated
Psidium guajava	Caricaceae	Pawpaw	Fruit tree
Spathodea campanulata	Bignoniaceae	African tulip	Flowers decorative
Terminalia catappa	Combretaceae	Indian almond	Fruit tree

Table 1. Trees Visited by Bats.

The first suggestion indicating that *Maranthes (Parinari) polyandra*, a savanna tree, may be bat-pollinated (27) was made on the observation of the co-incidence of white flowers, evening opening, strong unpleasant

smell and copious nectar. Further research, also in the Mole Game Reserve in Ghana established this observation (25). Bats are also agents of seed dispersal. In the Amazon forest of Brazil, fruit bats are the most important of all agents of seed dispersal. Bats are the most important seed dispersers of any animals - they are voracious and can fly long distances (32). Plants visited and dispersed by bats are listed in Table 2 below:

Species	Plant Family	Common Name	Remarks
Anacardium occidentale	Anacardiaceae	Cashew nut	Fruits edible
Ficus species	Moraceae	Fig tree	Shade tree
Mangifera indica	Anacardiaceae	Mango	Fruits edible
Melia azedarach	Meliaceae	Persian lilac	Medicinal
Morinda lucida	Rubiaceae	Brimstone tree	Medicinal
Musa sapientum	Musaceae	Banana	Fruits edible
Psidium guajava	Myrtaceae	Guava	Fruits edible
Spondias mombin	Anacardiaceae	Hog plum	Fruits edible
Terminalia catappa	Combretaceae	Indian almond	Nuts eaten

Table 2. Plants Visited and Dispersed by Bats

Tyto species Owl

Like bats *(Eidolon helvum, Epomophorus gambianus,* and species of *Cynopterus, Glossophaga* and *Pteropus)*, owls (species of *Alba* and *Tyto*), are also associated with the dead in the folklore of many countries. 'The cry of an owl heard over the house is believed to be prognostic of death of one of the inmates; this superstition which appears to be almost universal, is doubtless due to the nocturnal habits of the bird, and its strange appearance and cry, as seen and heard in the gloom of the night' (14).

INFLUENCE OF MUSHROOMS

In W Africa and many other countries, mushrooms, toadstools, stinkhorns, puffballs and shell fungi are generally objects of misconception and superstition. This is especially so with the latter four, and with inedible or toxic mushrooms. These are traditionally associated with the dead and spirit world. Besides the poisonous nature of some of them, they are often of unpleasant colour. They are also normally associated with putrescence and decay, and may be attended by a foetid smell. Puffballs were once believed to be sown by spirits (Rolfe & Rolfe fide 33); and in Mexico, the Kickapoo Indians used the dried powdery substance of *Calvatia* species, a mushroom, for painting the face and mocassin of the deceased (13). In Yoruba mythology, *Termitomyces microcarpus*, an edible mushroom, is likened to a chameleon, *Chamaeleon vulgaris.* There is a belief that this mushroom is capable of changing its form to assume the stature of the first person to see it when just emerging from the soil (23). In Ghana and Ivory Coast, the Akan name of all inedible mushrooms is 'saman-mbire', meaning 'ghost mushroom'.

PLANT NAMES AND THE SPIRIT WORLD

The local plant names and English names alike link some plants with ghosts and the spirit world - for various reasons. These may either be the colour, or the structure and habit, or the taste and poisonous properties, or some other superstitious beliefs. For example, *Encephalartos barteri* Ghost's palm (the only cycad indigenous to W Africa) is commonly known as 'palm of the spirits', and used as Shade for the Departed, as opposed to *Elaeis guineensis* Oil palm being the one given to the living (Schönfeld fide 11). In Europe, *Epipogium aphyllum* Ghost orchid, a northern temperate orchid with yellowish flowers, is so called because the plant has translucent stem (31); and in the Far East, *Davidia involucrata* Ghost tree, a native of western China, has long pendulous creamy white bracts (26).

Encephalartos barteri Ghost's palm

In W Africa, some people call *Thonningia sanguinea* Crown of the earth, an obligate root parasite, Ghost pineapple (1); and in S Nigeria, the Igbo tribe call *Smilax anceps (kraussiana)* West African sarsparilla, a

scrambling prickly forest shrub with tuberous root, 'ji abana mmwo', meaning *Dioscorea alata* Water yam of ancestral spirits (8d). In this country, the Efik name of *Caladium bicolor* Heart of Jesus, a decorative perennial aroid, is 'mkpon ekpo', meaning cocoyam of ghosts (8a). In Ghana, the Asante name of *Megaphrynium macrostachyum*, a perennial semi-woody herb, 'saman awuram', alluding to 'spirits' is given in the belief that spirits hide under the leaves (8c); and the Awuna name of *Scaevola plumieri*, a low shrub on sandy shores near high-water mark, 'ngoli foyi', means an edible blue fruit of the spirits (11). The Twi name of *Lagenaria (Adenopus) breviflora*, a perennial climbing cucurbit, 'asaman-akyeakyea', means spirits' Water melon, but as it is not eaten by man, it is said to be consumed by spirits (22a). *Coccinea (Physedra) barteri* and *Ruthalicia (Physedra) longipes*, herbaceous climbing cucurbits, are known by the same Twi name, 'asaman-akyeakyea' (8a). In Mexico, *Colocynthis lanatus (vulgaris)* Water melon must be served at the summer feast for the dead among the Kickapoo Indians (13).

In Ghana and Ivory Coast, the Akan name of *Mucuna sloanei* Horse-eye bean or True sea-bean and *Physostigma venenosum* Calabar bean or Ordeal bean, 'saman-ntew', and the Aowin name of *Cathormion altissimum*, a tree legume by stream in forest, 'enwomei-ate', means spirits' marble as opposed to *Dioclea reflexa* Marble vine, 'ntew', which is for the living (8c). *Leptoderris brachyptera*, a climbing shrubby legume, is also known as 'nwomele-ate' in Nzima, meaning spirits' marble plant (8c). The Akan names of *Anthonotha macrophylla*, *Berlinia* species, *Bussea occidentalis*, *Calpocalyx brevibracteatus*, *Chidlowia sanguinea* and *Xylia evansii*, all tree legumes, 'samanta' or 'samantawa', mean the oil bean of the spirits (22b). *Pentaclethra macrophylla* Oil-bean tree or Atta bean, another tree legume, locally called 'ataa' or 'atawa', the edible oil-bean being for the living (11).

The Akan name of *Rothmannia longiflora*, a shrub or small tree, 'saman-kube', means ghosts' coconut; and that of *Citropsis articulata* African cherry orange, 'saman-anka', means ghosts' orange (22b). The Fanti name of *Bersama abyssinica* ssp. *paullinioides*, a tree in lowland and montane forest, 'saman-gya', means ghosts' fire, because of the poisonous nature of the plant (22b). *Clausena anisata* Mosquito plant is called 'samandua' in Akan, and 'samanyobli' in Ga, both names meaning ghosts' tree, because the leaves are burnt in Ghana and Ivory Coast to drive away ghosts from haunted houses, in addition to repelling mosquitoes (22b). *Sophora occidentalis*, a maritime legume, goes by the same name 'samandua', probably in reference to the toxic beans (8c).

In Usumbura, in Burundi, *Sphenostylis stenocarpa* Yam-bean, a legume

cultivated in Mukenge both for the seed and the tuber, is known as 'bean for the departed' (11); and in S Nigeria, *Piper guineense* West African black pepper is symbolic and sacred to ghosts among the Ekoi (38). In this country, the Yoruba say that a ghost does not go out without a whip of *Glyphaea brevis*, a shrub of secondary jungle, in his hand to ensure calm in the house (40). In Ghana, 'a log of *Celtis* species, locally called 'issa', was always kept alight in the palace of the Ashanti king in the olden days, and this was for the *samanfo* (spirits or ghosts)' (35a). In Congo (Brazzaville) *Cissus aralioides*, a forest climber, can be used to assemble spirits at a predestined spot, and even to serve as a reconciliation between spirits and humans (6). (See also DEATH AND DYING; and GRAVES AND CEMETERIES).

References.

1. Adams, 1957; 2. Agbordzi, (pers. comm.); 3. Alando, (pers. comm.); 4. Ayensu, 1974; 5. Ayim, (pers. comm.); 6. Bouquet, 1969; 7. Bowdich, 1819; 8a. Burkill, 1985; b. idem, 1994; c. idem, 1997; d. idem, 2000; 9. Cofie, (pers. comm.); 10. Crow, 1969; 11. Dalziel, 1937; 12. Dobkin de Rios, 1977; 13. Dolores & Latorre, 1977; 14. Ellis, 1887; 15. Eshun, (pers. comm.); 16. Ewer & Hall, 1972; 17. Ferry & al., 1974; 18. Fink, 1987; 19. Gilbert, (pers. comm.); 20. Gledhill, 1972; 21. Hepper, 1979; 22a. Irvine, 1930; b. idem, 1961; 23. Kerharo & Adam, 1974; 24. Kerharo & Bouquet, 1950; 25. Lack, 1978; 26. Lanzara & Pizzetti, 1977; 27. Lock & Marshall, 1976; 28. Mensah, (pers. comm.); 29. Moore, 1960; 30. Moro Ibrahim, (pers. comm.); 31. Novak, 1965; 32. Osburn, 1865; 33. Oso, 1977; 34. Perry & Greenwood, 1973; 35a. Rattray, 1923; b. idem, 1927; 36. Raven, Evert & Curtis, 1976; 37. Siegle, Collins & Diaz, 1977; 38. Talbot, 1926; 39. Underwood, 1971; 40. Verger, 1967; 41. Walker & Sillans, 1961; 42. Watt & Breyer-Brandwijk, 1962.

21 GRAVES AND CEMETERIES

Only the action of the just
Smell sweet and blossom in their dust.
James Shirley, 1596-1616

Cemeteries and graveyards throughout the world are usually planted with decorative trees, shrubs and annuals. In African tradition, the presence of some of these plants is not only for their decorative and aesthetic values, but also for their spiritual significance. The traditional belief is that these plants have influence on the departed ones - either to keep the spirits of the dead away or to immortalize them. Plants growing on, or around graves, therefore, enter into magical and superstitious prescriptions such as bad juju to kill an enemy, or to induce the ghost to haunt and drive men mad. Even a sample of grave-soil, collected with the necessary ceremony, could be invoked for similar purposes. In India, *Papaver somniferum* Opium poppy is specially cultivated on graves for the spiritual potency.

Plants may also serve as a living pedestal of unwritten epitaph or a commemorative plaque of lost loved ones. Some trees may serve as graves.

GRAVE MARKER

In S Nigeria, *Euphorbia deightonii,* a shrub with milky latex forming large clumps, was formerly used to mark Christian graves at Oyo; and *Aloe buettneri, A. macrocarpa* var. *major* and *A. schweinfurthii,* perennial lilies, are sometimes planted on graves (7). In Africa, *Jatropha curcas* Physic nut, a soft-wooded shrub popularly used for live fence in villages, is often planted in graveyards (17, 32); and in Sierra Leone, an unidentified *Nopalea (Opuntia),* a shrub-like xerophytic plant, has been found planted, perhaps to afford protection, as a memorial over a child's grave at Gbinti (Deighton fide 4). In Liberia, *Pterocarpus santalinoides,* a tree legume on river banks, is sometimes planted over a grave (7); and in Tanzania, *Adenium obesum* Desert rose, an ornamental shrub with

Adenium obesum Desert rose

milky latex, is planted to mark the position of graves (13). In this country, *A. coetanum* is used by the Bena to mark a grave; and in the east side of S Pare Mountains, *Calotropis procera* Sodom apple is planted to mark the site of graves (34).

In N Nigeria, *Erythrina senegalensis* Coral flower or Bead tree is common in scrubby lands of Borgu where it is planted over graves by the natives; and *Maerua angolensis*, a shrub or small tree with white flowers, is often planted on graves in Nupe area (7). In this country, *Moringa oleifera* Horse-radish tree or Oil of Ben tree is planted on graves in the belief that it keeps away hyenas, *Hyaena hyaena;* and other wild animals (7). In parts of the Region, *Pedilanthus tithymaloides* Redbird-cactus or Jew-bush or Slipper-flower, a cactus-like shrub introduced from tropical America, is often found planted on or around graves - particularly by the Ewe-speaking tribe of Ghana, Togo and Benin Republic; the Fula of Sierra Leone and Niger and some other tribal groups. In Ghana, *Amorphophallus johnsonii* Johnson's arum, a tuberous aroid, has been found growing on graves at Pokuase (15).

In S Africa, the thorny branches of *Ziziphus mucronata* Buffalo thorn are used in Swati burial rites, being probably placed over the grave as a protection; and in Ethiopia, twigs of *Catha edulis* Abyssinian tea or Chirinda redwood are placed on a grave in Harer for seven days, and a visitor to the grave chews a twig (34). In Mexico, *Lasianthus (Eustoma) nigrescens* 'Flor de Muerto' or Funeral flower of Mexico, a shrubby plant with drooping blue-black flowers, is so called because it is used as a decoration for graves in the southern part of the country (9). In Europe, *Malva moschata* Musk mallow, an European plant with purplish-pink flowers and a musky fragrance when bruised, was formerly much planted on graves (27). *Spiraea prunifolia* 'Plena' is a beautiful shrub from China where it is frequently planted on graves; and *Acaena inermis* and *A. microphylla*, both from New Zealand, and *A. myriophylla* from Chile, all shrubs in the Rose family, are grown on graves (27). In the Middle

East, *Iris florentina* Orris root, a vivacious plant of the Iridaceae family, is planted on Arab/Moslem tombs (16b) .

According to legend, before leaving for St. Helena, Napoleon Bonaparte asked for permission to visit Empress Josephine's grave where he picked *Viola odorata* Violets, which after his death was found in a locket around his neck (27); and over the grave of the Emperor on the island of St. Helena was planted *Salix babylonica* Weeping willow (replaced after it had been torn to pieces by enthusiastic relic hunters), *Cupressus* species Cypress, and other funereal trees (8). Carvings from this Willow were the centre of curiosity and commerce. 'Souvenirs like walking-sticks, boot-jacks, wooden paper-knives, snuff-boxes, card-cases, pen-holders, and so on, allegedly carved from an inexhaustible willow-tree planted over the Emperor's grave were sold at fancy prices by coloured natives of the island, to passengers of mail-steamers and troops-ships which called, to earn a living' (8).

Upon the grave of Mary Slessor (1848-1915), the Pioneer Missionary of Calabar, in S Nigeria, were placed crosses of purple Bougainvillea and white and pink Frangipani, and the earth was planted a slip from the Rose bush, that it might grow and be symbolic of the fragrance and purity and beauty of her life (22). Among Christians, death was symbolized by *Taxus baccata* Yew, which was grown in burial grounds (6); and in Europe, *Vinca minor* Bowles variety or 'La grave' has blossoms several centimetres wide and is so called because it is reported to have been collected by Mr. E.A. Bowles (one of Britain's best-known horticulturists) at La Grave in France growing on a grave (27).

In W Africa, three brothers called Ana, Afona and Eku, who all died in a battle, according to folklore, were buried in a fetish grove at Jankama, a village near Aburi, in Ghana. When their parents visited the graves later they saw *Hilleria latifolia,* a perennial forest herb near villages, locally called 'anafoannaeku' growing on them (14). The local name is after the three brothers. Burial still takes place in the fetish grove at Jankama, but *Musa sapientum* var. *paradisiaca* Plantain or other exotics like *Codiaeum variegatum* Garden croton may be planted to mark the position of the grave. A similar story is told from Polynesia, in the South Pacific, about the origin of *Piper methysticum* Kava kava, a herb. 'A very long time ago, orphan twins, a brother and sister, lived happily on Maewo. One night, the boy, who loved her sister very much, had to protect her from a stranger who asked to marry her but whom she had refused. In the struggle, the frustrated suitor loosed an arrow which struck the boy's sister and killed her. In despair, the brother brought his sister's body home, dug her a grave and buried her. After a week,

Grave in N Ghana,

before any weeds had grown over her tomb, there appeared a plant of an unusual appearance which he had never seen. It had risen alone on the grave' (24).

Some tribal groups do not mark their graves, however. For example, in Cameroon the Ntumu tribe do not permanently mark their graves. Instead, after the first heap of soil over the coffin, it is covered with broken bottles before it is filled up to the top with soil and dirt. Anyone who sees broken bottles while digging knows that someone has been buried there before (31).

AESTHETIC AND REMEMBRANCE VALUES

In W Africa, *Plumeria rubra* var. *acutifolia* Frangipani or Temple flower, sometimes also called Forget-me-not, is usually planted in cemeteries, not really for superstitious purposes, but more of an affection for, and remembrance of, the dead. The plant is similarly used in Asia (4); and in Ceylon, India as well as SE Asia (23); and associated with burial grounds in the Old World (9). Its singular ability to burst into leaf and flower suggests immortality so it is frequently planted near graves, including those of Buddhists and Muslims (27). 'In Java, if the root of Frangipani is dug out of a cemetery it is thought to be toxic, but is commonly regarded as non-poisonous' (34). In W Africa, *Spondias mombin* Hog plum or Ashanti plum is often grown in graveyards (17). The living stems of *Dracaena arborea*, a palm-like tree popularly believed to be fetish and often planted around shrines, are used as posts in fences in cemeteries (17); and *Lonchocarpus sericeus* Senegal lilac, a tree legume with pale purple or lilac fragrant flowers, is sometimes planted as shade tree in cemeteries (7). In Japan, *Lycoris*

Lonchocarpus sericeus Senegal lilac

radiata Equinox flower, locally called 'Higan-bana', is often planted in graveyards because it is an ancient belief that it is one of the flowers found in Paradise (29).

SUPERSTITIOUS VALUES

Besides the commemorative, decorative or aesthetic values of plants, their association with burial, graves or cemeteries generally do also have superstitious intentions. For example, in N Nigeria, the leaves of *Momordica balsamina* Balsam apple are put in water used for ceremonious washing after digging a grave among the Ngizim in Bornu (25). Among the Ewe-speaking tribe of Ghana, Togo and Benin Republic, *Newbouldia laevis,* a medium-sized tree associated with fetish practices, is used instead (20). Similarly, in S Nigeria, *Newbouldia* is found in Efik and Ibibio graveyards and sacred places - and also in Gabon (33), and Ivory Coast (30) - a tree being planted near the tombs and in villages as a protective talisman. Many tribal groups rather prefer to use *Momordica charantia* African cucumber in purification rituals associated with both grave digging and burial ceremonies. Among the Bulsa tribe of NW Ghana, grave diggers are believed to be subjected to certain dangers since they touch dead bodies. Thus they are forced during their ordination to eat six different types of herbs (unspecified) that protect them spiritually and give them power to bury 'difficult bodies' - such as the bodies of witches, pregnant women, lepers and decayed bodies (21).

Normally, a section of the cemetery is set aside for the burial of accident victims and 'difficult bodies' (in line with traditional beliefs, and at the request of these spirits). This is the section that juju-men and other spiritualists are normally attracted for invocation rituals. (For the consequences of non-compliance with this directive see DEATH AND DYING: FUNERAL OBSEQUIES: Cremation). In The Gambia, it is said that when people of the Mandinka race are buried, some leaves of *Guiera senegalensis* Moshi medicine, a small shrub in sandy wastes and semi-desert areas, are placed in the grave with them (Fox fide 14). In ancient Egypt, the wood of *Cordia myxa* Assyrian plum has been used in tombs (7); and archeologists believe that the timber used in Egyptian tombs is *Ficus sycomorus (gnaphalocarpa)* (2).

In W Africa, *Commiphora africana* African bdellium, a shrub or small tree, is venerated by the Tuaregs, who regard it as a symbol of immortality and place it on graves (Rodd fide 4). Similarly, the Fula, in a superstitious token lay sticks of *C. kerstingii,* a tree - often planted, across graves (18) - perhaps in the sense of conferring protection. In Congo

(Brazzaville), *Dioscorea dumetorum* Bitter yam is planted on graves to drive away spirits and ghosts (3); and throughout the Region, *Sanseviera liberica, S. senegambica* and *S. trifasciata* African bowstring hemp may be found in native compounds or planted as a fetish on graves (7). In front of the present-day European cemetery in Kumasi, Ghana, still stands the historical trees known all over Ashanti as 'akuakuaanisuo' and 'wama' trees; the cemetery occupies the site of the mausoleum - the last resting place of the Ashanti kings and was destroyed in 1895 (28). (The trees are *Spathodea campanulata* African tulip tree and *Ricinodendron heudelotii* African wood-oil-nut tree, respectively).

TREES AS GRAVES

A tree may serve as a fetish medium or a grave in itself. For instance, in N Ghana, the trunk of *Adansonia digitata* Baobab is used as a grave for important chiefs (1); and in Sudan as a burial place for the dead (17). It has been observed that in Ghana, the inhabitants in Ashanti, Akim and Akwapim always bury their principal dead secretly and often in this tree, especially in times of war, when they fear the enemy will discover the body and keep the bones on his drum as a sign of victory and of his enemy's disgrace (Thonning fide 16a). Secret burials are still practiced in traditional Africa and elsewhere - not only for chiefs but also for political figures and dictators. Usually there is an open burial - but in reality the coffin contains nothing but stones or a piece of log. The corpse is secretly buried somewhere else. Commenting on secret burials, it has been observed that in case of a great chief the head is cut off and buried with great secrecy somewhere else by the Calabar people of S Nigeria (19a). In sharp contrast to this, a picture in *Out of Africa* shows a funereal custom of the Quissama tribe of Angola in which the heads of the deceased are cut off and placed on their graves (10). Among the Bubi tribe of Bioko (Fernando Po), burial customs are exceedingly quaint in the southern and eastern districts, where the bodies are buried in the forest with their heads just sticking out of the ground (19a). This implies that the dead were buried vertically upright.

Mary Livingstone, wife of David Livingstone, was buried beneath a Baobab in April 1862 on the banks of R Zambezi in present-day Mozambique. 'Her remains still lie beneath the great Baobab tree where they buried her' (26). When the British Missionary and explorer himself died in May, 1873 near Lake Bangwenlu in present-day Zambia, it has been observed that 'Livingstone's body was cut open, and the heart and innards were removed and buried under a tree (unspecified), on which

Jacob Wainwright carved Livingstone's name. The body itself was conveyed to Westminster Abbey and given a hero's burial the following year ' (26). The Baobab trunk also serves as a sort of dehydrating chamber for the dry-preservation of human corpse (11); or a place where a body denied burial may be suspended between earth and sky for mummification (7). The old hollow trunks of baobab are also used as houses, prisons or water reservoirs (12).

Writing about minstrels who frequent market towns and for a fee sing stories, it has been recorded that 'they are not buried as other people are; they are put into trees when they are dead - may be because they are 'all same for one' with those singers the birds' (19b). In parts of Brong Ahafo region of Ghana, when one of the fraternity (the priest of a well-known god called *Dame*) dies, he is dressed in white and crucified on *Ceiba pentandra* Silk cotton tree, being fastened to it with staples which are placed round the legs and arms. When the flesh has disappeared from the bones, the skeleton is then buried (28). The Silk cotton tree is also planted at tombs where offerings may be made to the Shades of Ancestors or to protective genie (4). (See also DEATH AND DYING; and GHOST AND SPIRIT WORLD).

References.

1. Akpabla, (pers. comm.); 2. Berhaut, 1979; 3. Bouquet, 1969; 4. Burkill, 1985; 5. Cooper & Record, 1931; 6. Crow, 1969; 7. Dalziel, 1937; 8. Ellis, 1885; 9. Everard & Morley, 1970; 10. Freyberg, 1935; 11. Gledhill, 1972; 12. Graf, 1978; 13. Greenway, 1941; 14. Hall, 1978; 15. Hall & Lock Herb. GC; 16a. Hepper, 1976; b. idem, (pers. comm.); 17. Irvine, 1961; 18. Jackson, 1973; 19a. Kingsley, 1897; b. idem, 1901; 20. Klah, (pers. comm.); 21. Kroger, (pers. comm.); 22. Livingstone, 1914; 23. Lotschert & Besse, 1983; 24. Martin, 1995; 25. Meek, 1931; 26. Mountfield, 1976; 27. Perry & Greenwood, 1973; 28. Rattray, 1927; 29. Richards & Kaneko, 1988; 30. Schnell, 1950; 31. Sheppherd, 1988; 32. Visser, 1975; 33. Walker & Sillans, 1961; 34. Watt & Breyer-Brandwijk, 1962.

22 HUNTING

...and after will I send for hunters, and they shall hunt from every mountain, and from every hill, and out of the holes of the rocks.
Jeremiah 16:16b

Some hunters attribute their success at night to the phase of the moon - claiming that when there is no moonlight, the animals leave their hideouts to feed. They are thus easily spotted and shot at. However, on moonlight nights, animals are cautious and reluctant to go out and feed. Another factor is the rains. When new leaves start to flush during the rainy season, animals come out into the open to feed; whereas during the dry season they tend to hide away. Furthermore, when the forest floor is dry, the slightest noise from dry, broken twigs as the hunter walks stealthily through the forest scares away game. Others attribute their success to sheer luck; while yet others attribute theirs to rites of abstinence. Traditional hunters may, however, use other charms for general success and luck in hunting. It could be charms to improve the shooting skill, or to hypnotize and spot the game more easily, or to render the hunter invisible in the presence of fierce animals like the elephant, *Loxodonta africana;* the lion, *Panthera leo;* the tiger, *Felis tigris;* the bongo, *Tragelaphus euryceros;* and others.

Some hunting charms or magic are plants or plant material. These charms are either applied to the hunting implement, or worn as talisman or girdle, or taken in food, or used as a wash, and so on. There are charms that aid the sense of smell in hunting dogs, *Canis* species. On the contrary, some plants are believed to protect hunted animals from hunters. There are also said to be instances when an animal is just bullet-resistant. Finally there are deities in hunting. In S Nigeria, *Erythrina senegalensis* Coral flower or Bead tree is supposed to be dedicated to the god of hunting (11); and in Hausaland, *'Uwardowa'*, a female, is the goddess of hunting (33). There is a fetish of the hunt as well.

SHOOTING CHARMS

On the Liberia and Ivory Coast border, elephant hunters prepare charcoal from the bark of *Ricinodendron heudelotii* African wood-oil-nut tree, powder it and apply to their guns and foreheads before going hunting - in the belief that by doing so they will kill all the elephants, *Loxodontha africana*, they meet (21). In Ghana, the hunter believes that a few leaves of *Dracaena aubryana (humilis)*, a low forest shrub, put into his gun will make him a good shot (18a); and the dried leaves of *Cocos nucifera* Coconut palm pushed into a flint-lock gun - before and after the pellets - is equally reported to aid hunters to shoot straight (3). In N Nigeria, if *Crotalaria pallida (mucronata)*, a legume under-shrub, is gathered early in the morning and mixed with other suitable medicines, it brings luck in shooting whether the finder swallows it or not (11).

In the northern parts of Ivory Coast, Banda (Ligbi) hunters tie a piece of the root of *Parinari curatellifolia*, a savanna tree, to the stocks of their guns for success in the chase (5); and in SE Nigeria, the Ijo tie *Eleusine indica*, a tufted annual grass, to a gun to prevent it misfiring (10b). In SE Senegal, Tandanké hunters carry out a ritual thought helpful to meet a bush buck, *Tragelaphus scriptus*, and to kill it. To achieve this the hunter will macerate some leaves of *Mitragyna inermis* False abura, a shrub to small tree of damp perennially flooded sites, for a few days and with it will wipe down and clean the barrel and stock of his gun (20). In this country, hunters also put the leaves of *Vernonia colorata*, a savanna composite shrub, in water as a charm to help the hunt for bush buck (13). In N Nigeria, the Pulaar seek good luck in hunting by making a paste of the bark of either *Diospyros mespiliformis* West African ebony tree or *D. chevalieri*, a forest shrub, with the belly-fat of a ram, *Ovis* species, which is rubbed on their hunting bows (19). In The Gambia, hunters and warriors believe that to eat the bean of *Parkia biglobosa (clappertoniana)* West African locust bean renders them more susceptible to spear-wounds (32).

In Senegal, the Bedik consider *Cissus populnea*, a liane of woody savanna, to be a hunter's charm (13); and *C. producta*, a large herbaceous forest climber, is credited with magic powers, as such the Tenda prepare a decoction to wash a hunter's gun as a charm (13). In this country, Tenda hunters consider both *Baissea multiflora*, a forest climber with milky latex, and *Guiera senegalensis* Moshi medicine, a savanna shrub in sandy wastes and semi-desert areas, as charms for the hunt - to attract game (13). In N Nigeria, the Fula wash in water in which the roots of this shrub, have been soaked to ensure good hunting (19). In Ivory Coast, hunters in the

Gagnoa area crush the leaves of *Oplismenus burmannii,* an annual grass, in water found in the crotch of a tree, and rub the liquid over the face to ensure they will meet game and be the better able to see it (8). Some races in the Region attribute magical properties to the bark of *Lecaniodiscus cupanioides*, a small tree of savanna and forest outliers, bringing good luck in hunting (7). In Congo (Brazzaville), hunters will wash in a bark macerate of *Enantia chlorantha* African yellow wood before setting forth on a hunt to ensure meeting game and shooting straight (7).

In S Africa, a stick of *Terminalia sericea* Assegei wood, a savanna tree, is stuck, with apparently magical significance, in the floor of a shrine in which homage is paid to the ancestral spirits when a hunting party sets out; and a Xhosa hunting party is sprinkled with an infusion of the roots of *Rhamnus prinoides* Dogwood before setting out (35). In this country, the tribe burn *Popowia cafra,* locally called 'dwaba', before setting out on a hunting expedition, and when the first animal is killed 'dwaba' ash is strewn around it to charm more game to the hunters (35). In the Democratic Republic of Congo (Zaire), hunters place a branch of *Thomandersia hensii (laurifolia),* a shrubby acanth with white or red flowers, in their huts to ensure success in the hunt (17). In Gabon, Fang hunters are reported to eat the leaf of *Gloriosa superba* Climbing lily, chopped up with the seeds of a small pumpkin and the meat of a pig, *Sus* species, as an exercise in magic to bring luck in hunting wart-hogs, *Potamochoerus porcus* (34); and *Mussaenda erythrophylla,* a forest shrub or climber, is considered to be a good-luck charm for hunters (34).

Cissus quadrangularis Edible stemmed vine

In N Nigeria, Hausa hunters tie the roots of *Vetiveria nigritana,* a robust perennial grass, up in leather armlets, locally called *kamba* in Hausa, as a charm against injury; and the leaf and roots of *Tephrosia bracteolata,* a herbaceous annual savanna legume, are ingredients of charms used against injury by hunters and warriors (11). In Ivory Coast, *Cissus quadrangularis* Edible stemmed vine, a climbing plant with square stems, serves as a protective fetish for the hunters

of panther, *Panthera* species from 'gnama', a spirit released from any panther they may kill. Should any hunter become tainted by 'gnama', they must purify themselves lest they become mad and run amok in their village (21). In Senegal, the Tenda use the leaves of *Ozoroa (Heeria) insignis*, a savanna shrub or small tree, or the whole plant of *Hymenocardia acida*, a savanna tree, to make a charm used by hunters; or hunters wash with water in which the roots of *Stereospermum kunthianum,* a savanna tree, have been soaked to ensure seeing plenty game (13). Protective properties are attributed to *Psorospermum senegalense,* a savanna bush, by the Tenda. It is made into an amulet to protect hunters from meeting a lion, *Panthera leo* (13).

In N Nigeria, hunters prepare a magical prescription of the berries of *Phragmanthera* or *Tapinanthus* species Mistletoe, epiphytic parasites, ground with shea-butter and a kind of rock salt called *gallo* in Hausa; if some of this is eaten every morning, the hunted game will be drowsy and easy to kill (11). In Nigeria and Senegal, Mistletoe growing on *Acacia senegal* Gum-arabic tree, a low-branching shrub or small savanna tree, has magical powers to assist hunters. An infusion of the Mistletoe is used by the Fula in Nigeria as a body-wash by hunters especially when hunting the roan antelope, *Hippotragus equinus,* locally called 'kooba' (19). In this country, huntsmen soak the root of *Waltheria indica,* a savanna under-shrub, in water used as a wash to ensure good hunting (19). In Senegal, the Tenda huntsmen use the crushed bark of the Mistletoe to have a quick clear vision (13); and Mistletoe taken from *Quassia undulata,* a medium-sized soft-wooded tree in savanna and forest, makes a hunters charm - the water-macerate with millet serving as a wash for themselves (13). The superstitious uses of Mistletoe among many of the ethnic groups in the Region do vary according to the nature of the host plant. For example, in S Nigeria, Mistletoe growing on *Calotropis procera* Sodom apple is reported to be especially valuable in the Benue region (11). (For examples, see INVISIBILITY AND VANISHING CHARMS below; and WEALTH: CHARM PRESCRIPTIONS). (For a list of the epiphytic parasites in the Region see DISEASE: LEPROSY: Table 1).

In Senegal, Tenda hunters wash in water in which the leaves of *Sarcocephalus (Nauclea) latifolius* African peach have been crushed to help them approach their quarry; and *Ziziphus mucronata* Buffalo thorn or *Mitragyna inermis* False abura, a shrub to small tree of damp perennially flooded sites, serves as a hunter's charm for catching the antelope, *Adenota kob kob* (13). In Ghana, one hunting charm or juju consists of the hair of black duiker, *Cephalophus niger,* locally called 'iwi'; the tail of the

brush-tailed porcupine, *Artherurus africanus,* locally called 'apese'; ten nail clippings of the hunter; the leaves of *Ocimum canum* American basil and that of *O. gratissimum* Fever plant; and a dead chameleon, *Chamaeleon vulgaris.* All these are charred and mixed with shea-butter and seven pellets in an empty shell among incantations (12). The informant adds that for success in the hunt, libation is poured on the charm to Nana Twema (a local deity) before setting out; and that the hair of the animal to be killed will appear on the juju. The conditions attached to the juju, however, is that the first game killed should not be eaten by the hunter himself, but sold. The consequence of non-compliance is that though there may be further successes in hunting, the hunter will also successively lose his children (12).

In S Nigeria, Yoruba hunters prepare a magical prescription of *Termitomyces globulus,* an edible mushroom, chewed with seven seeds of *Aframomum melegueta* Guinea grains or Melegueta and the leaf of *Phyllanthus muellerianus,* a shrub or woody climber (28). When this is spat on the palm and rubbed on the bow and arrow with some incantations, the hunted game becomes drowsy and easy to kill (28). In N Nigeria, the root of *Cyanotis caespitosa,* a grassland herb often flowering after fires, is a magical potion by hunters of the canerat, *Thryonomys swinderianus,* locally called 'gyazbi' in Hausa - to enable them to come upon it unawares (11). In this country, a preparation of the flowers of *Vernonia galamensis (pauciflora)* Ironweed, a herbaceous composite in woodland savanna, crushed along with scent is used by hunters, either smeared on the body or worn as a charm to attract 'beef' (gazelle, *Gazella* species; locally called 'barewa'); or other antelope, *Adenota kob kob* (11).

In N Nigeria, *Ficus platyphylla* Gutta-Perch tree and *F. sycomorus (gnaphalocarpa)* are held to have magic powers. Fulfulde hunters when out after the roan antelope, *Hippotragus equinus,* locally called 'kooba', give themselves a body-wash with a bark decoction to bring good luck (19). In this country, Fula hunters wash in a bark infusion of *Terminalia avicennioides,* a savanna tree, especially when going to hunt the roan antelope (19); or use as a body-wash water in which the bark of *Ficus polita* has been steeped in a belief that this will bring luck especially when hunting the reed buck, *Redunca redunca,* locally called 'padula' (19). In this country, *Vitex doniana* Black plum, a savanna tree with edible fruits, is credited with magical powers. Water in which the bark has been soaked is used by Fula hunters as a body-wash (19). In Gabon, the seeds of *Monodora myristica* Calabash nutmeg are regarded as a lure for hunting manatees, *Trichechus senegalensis* (34). In this country, *Adenia rumicifolia*

var. *miegei (lobata)*, a large forest climber, is ascribed with magical powers to bring good luck in hunting (34).

In Botswana, the ashes of the whole plant of *Dipcadi polyphyllum*, a lily, or a decoction of species of *Dipcadi* is rubbed into incisions on the thumb and index finger of a Sotho to give him accurate aim in hunting and fighting (35). In S Africa, the root of *Schrebera saundersiae*, locally called 'sehlulamanye', is used by the Swati as a hunting charm; and in olden times the Xhosa bound seeds of *Calodendron capense*, a tree, around the wrist when hunting for the purpose of charming game to fall easy prey to his weapons (35). In Gabon, magical good luck in hunting can be achieved by eating a mixture of the leaves of *Trema orientalis*, a small tree of forest clearing, with cane sugar juice (34). In Congo (Brazzaville), *Barteria fistula*, a slender forest tree, is used to improve a hunter's luck (7); and in Central African Republic, witch-doctors of Ubangi make a whistle from the wood of *Brevia sericea*, a forest tree, to promote the hunt (Motte fide 10c). In this country, the seeds of *Canavalia regalis*, a perennial scandent legume, are strung together into a bracelet to bring luck when shooting (29).

Burning the vegetation to trap animals is very often practiced in the savanna zones - especially during the dry harmattan season from November to March. Sometimes the practice may be given a superstitious touch. In N Nigeria for instance, the Chawai of Zaria province set the bush on fire before a hunt by the ritual method of drilling two pieces of *Ficus dekdekena (thonningii)*, a fig tree locally called 'che'diya' (24). The manufacture of hunting equipment may also have a superstitious touch. For instance, in N Nigeria, the hard roots of *Sida cordifolia*, a weedy malvaceous perennial in waste places, is sometimes an ingredient in arrow-poison for hunting (probably a superstitious usage) (11). For a similar reason, in Ivory Coast, the Guéré tribe use *Lepianthes peltata (Piper umbillatum)* Cow-foot leaf, a forest shrub in moist places, as an ingredient in the preparation of arrow-poison (21). In its distribution area in the Region, the juice of *Pupalia lappacea*, a common weedy amaranth with sticking fruits, collected with rituals and applied to the bow and arrow, or the gun, enters into personal charms for hunting (25). This is an instance of sympathetic magic - hinging surely on the clinging burs of the fruit.

HUNTERS' FETISH

Fetish of the hunt is practiced in the Region and elsewhere. For instance, *Alchornia floribunda,* a semi-scandent shrub or small tree in forest undergrowth, is a hunter's fetish; and *A. cordifolia* Christmas bush enters into many superstitious practices (11). In western Senegal, the Basari prepare a leaf-macerate of *Baissea multiflora,* a woody climber with milky latex, in a bath as a stimulant and fetish for the hunt (20). Basari hunters also consider *Morinda geminata,* a tree of secondary jungle and scrub, a fetish of the hunt, and wash in water in which the roots have been steeped (20). In Ivory Coast, the Tagonana use *Abrus* species - *A. canescens, A. precatorius* Prayer beads and *A, pulchellus,* leguminous twiners - as a fetish for hunting and against arrow poison (21). In Congo (Brazzaville), hunters deem *Chytranthus atroviolaceus,* a shrub or small tree, has fetish powers - a hunter must wash his hands in bark macerate to ensure catching much fish or game (7); and in the Democratic Republic of the Congo (Zaire), *Hilleria latifolia,* a perennial forest herb, has fetish attributes for hunting (6). In Gabon, huntsmen use *Cymbopogon densiflorus,* a tufted perennial grass, as a fetish lure for taking game (34); and in S Nigeria, a plant locally called 'olorofo' in Yoruba - the name of a bird - is invoked by the tribe to protect a hunter from evil at midnight (Verger fide 10c).

INVISIBILITY AND VANISHING CHARMS

By far one of the most astounding phenomenon in nature is the invocation of plants or other substances, with or without ingredients, as charms reportedly to effect invisibility or to 'teleport' oneself, that is vanish from one spot and appear at another - sometimes many miles apart. Invisibility charms are distinct from vanishing charms. While the subject is present but cannot be seen in the former, he or she disappears from the scene altogether in the latter - appearing somewhere else. Besides hunters, this juju is reported to be practiced by prisoners and others to escape from their adversaries, or by witches on their nocturnal assignments. Fetish-men are also known to use invisibility and vanishing charms. For instance, in Ivory Coast and Burkina Faso, the leaves of *Dissotis grandiflora,* a small woody plant with purple flowers, serve some fetish-men to vanish (21); and in Gabon, the broad leaf of *Lepianthes peltata* Cow-foot leaf, a forest shrub in moist places, when rubbed on a fetish-man, is said to render him invisible (9). The practice is said to be used by robbers.

In Ghana, hunters are believed to use the seeds of *Okoubaka aubrevillei,* a rare forest tree popularly believed to be fetish, as a vanishing protection against wild animals like the tiger, *Felis tigris;* and the leopard, *Panthera pardus* (2). In this country, *Achyranthes aspera* Devil's horsewhip, a common weedy amaranth, collected with rituals, enters into invisibility prescriptions among some of the tribal groups - the informant adds. In S Nigeria, *Phallus aurantiacus,* a gasteromycete stink-horn toadstool, is commonly used by Yoruba hunters to make a charm called *egbe* which makes one invisible. It is useful in the chase and in time of danger (28). In W Africa, the dried fallen leaves of *Musanga cecropioides* Umbrella tree kept in ones pocket may be invoked to effect invisibility; and *Uraria picta,* a perennial legume in dry savanna, and generally all plants traditionally associated with lightning and thunder serve as juju for vanishing (25). (See LIGHTNING AND THUNDER for a list of these plants). In Zambia, the Nyanja use *Psorospermum febrifugum* Rhodesian holly as a charm to prevent a person being seen by wild animals, for which purpose it is worn in the hair; and in Botswana, the Sotho use *Urginea capitata,* a lily, with the belief that a person can glide among his enemies unnoticed and unharmed (35).

The leaves of *Imperata cylindrica* Lalang grass, a notorious perennial savanna weed, collected with the necessary rites, with assorted ingredients, including the scrapings of a mortar, a pestle, a wooden ladle, a stick of broom and a broken fragment of a hearth (all traditionally believed to possess power over life - see TABOO: GENERAL FOR ALL AND SUNDRY: Sweeping Brooms) are ground in lime-juice and dried as a poultice, locally called 'dufa' in Akan. This is worn as a talisman and a charm for instant vanishing - as and when the owner desires (12). Some vanishing/invisibility charms are said to be triggered by sudden fear. In Liberia, the bark of *Tieghemella heckelii* Makore, a forest tree commercially used for high-class furniture, is used by the Krus as an amulet in attracting and shooting elephants - probably by rendering the hunters invisible (18b). In Gabon, *Pollia condensata,* a stout herb in forest, is a fetish of elephant hunters (33); and in N Nigeria, Bornu hunters use *Monocymbium ceresiiforme,* a tufted perennial grass, to conceal themselves (11) - perhaps by vanishing. In N Nigeria and Niger, a decoction of the aerial roots of *Ficus* species taken regularly by draught is believed by the Hausa, the Zarma and the Fula to confer vanishing powers (26). *Ficus* species with aerial root system are listed in Table 1 below:

Species	Plant Family	Remarks
Ficus anomala	Moraceae	Aerial roots system normally present
F. barteri	Moraceae	Aerial roots system normally present
F. congensis	Moraceae	Aerial roots system normally present
F. dekdekena (thonningii)	Moraceae	Aerial roots system normally present
F. elegans	Moraceae	Aerial roots system normally present
F. ingens	Moraceae	Aerial roots system normally present
F. leprieurii	Moraceae	Aerial roots system normally present
F. lyrata	Moraceae	Aerial roots system normally present
F. natalensis	Moraceae	Aerial roots system normally present
F. polita	Moraceae	Aerial roots system normally present
F. populifolia	Moraceae	Aerial roots system normally present
F. tessalata	Moraceae	Aerial roots system normally present
F. umbellata	Moraceae	Aerial roots system normally present
F. vogelii	Moraceae	Aerial roots system normally present

Table 1. Indigenous Ficus species Normally with Aerial Root System

Ficus dekdekena (aerial roots)

In addition to the above, it has been observed that some other *Ficus* species like *F. ovata* and *F. platyphylla* Gutta-percha tree which normally do not have aerial roots, do produce same when injured.

Some introduced *Ficus* species to the Region have aerial root system. These are listed in Table 2 below:

Species	Plant Family	Common Name	Country of Origin
Ficus altissima	Moraceae	-	India to China and the Philippines
F. bengalensis	Moraceae	Banyan	India
F. benjamina	Moraceae	-	India
F. craterostoma	Moraceae	-	S Africa
F. elastica	Moraceae	India-rubber fig	India
F. retusa	Moraceae	-	India
F. rubiginosa	Moraceae	-	Australia

Table 2. Exotic Ficus species with Aerial Root System

Sympathetic Magic

Some of the prescriptions appear to be based on sympathetic magic. For instance, the leaves of *Phragmanthera* and *Tapinanthus* species Mistletoe, epiphytic parasites, with ingredients enter into prescriptions by medicine-men for invisibility because of their unique position. This is an instance of sympathetic magic. However, an important ritual is that during collection the epiphyte must not touch the ground (16). Among the Hausa of N Nigeria, the Zarma of Mali and Niger and the Fula mistletoe growing on *Diospyros mespiliformis* West African ebony tree - a very rare occurrence - is an extremely lucky find indeed (see WEALTH: CHARM PRESCRIPTIONS). The dried, ground plant taken in milk serves as a charm for invisibility (26). Similarly, the epiphyte growing on *Manihot esculenta* Cassava finds use in prescriptions for vanishing among these tribes (26). The black spots on the seeds of *Abrus precatorius* Prayer beads suggests the pupil of the eye and they are therefore an ingredient in charms to ensure invisibility (11). This is an instance of sympathetic magic. In the Region, the whole plant of *Chamaecrista (Cassia) rotundifolia,* a scrambling legume, collected with the necessary ceremony, and previously sprinkled with a few drops of the blood of a black hen, *Gallus domestica,* then burnt to a black powder, is blown from the palm of the hand in an incantation by traditional medicine-men to effect vanishing (25). The 'black' in both instances - black hen and black powder - suggests sympathetic magic.

Among the Ewe-speaking tribe of Ghana, Togo and Benin Republic, the whole plant of *Triclisia subcordata,* a trailing forest menisperm, woven with white cloth into an arm bangle and studded with cowry shells is besmeared with an egg and worn as a vanishing charm (25). It is, however, important that the charm be prepared in hiding (25). This tribal group also chars *Commelina lagoscensis,* a herbaceous weed of cultivation, with a black cat, *Felis domestica,* and a red-coloured mangrove crab, species of *Cardiosoma, Goniopsis* or *Sersama,* amid incantations as vanishing prescriptions (25). The black cat does feature in other invisibility charms. This is most likely an instance of sympathetic magic hinged on the colour. For instance, *Egyptian Secrets* mentions another invisibility charm prepared with the ear of a black cat that has been boiled in the milk of a black cow, *Bos* species. The charm is worn as a thumb cover. Similarly, one of the lower vertebra of a black cat carried on ones person, is a charm for invisibility. Only one of the twenty or twenty-one thoraco-lumbar vertebrae has this unique property. To identify this, the aspirant holds each vertebrae in turn while looking at his or her image in a mirror. Immediately the correct vertebra is chosen, the image in the mirror vanishes instantly (25). (See also BURGLARY: COUNTER CHARMS).

HUNTING DOGS

The characteristics that dogs need for hunting success are scent-accuracy, speed, enthusiasm and stamina (22). In Sierra Leone, *Crotalaria glauca,* an erect leguminous herb, is squeezed on the nose of a dog, *Canis* species, in Sherbro to aid its sense of smell in hunting (11); and in Ivory Coast, the pounded, moistened leaves of *Bonamia thunbergiana (cymosa),* a woody twiner, are inserted into the nostrils of hunting dogs to improve their ability to scent (Portères fide 10a). In this country, *Annona senegalensis* var. *onlotricha,* a small savanna tree, and *Lepianthes peltata (Piper umbillatum)* Cow-foot leaf, a forest shrub in moist places, enter into a fetish among the Guéré tribe to ensure catching plenty of game; and also make a sort of dope to improve the hunting ability of dogs (21); and the Tagwana are reported as putting *Hyptis suaveolens* Bush tea-bush, a strongly aromatic labiate, into medications used to prepare dogs for hunting (21).

In Trinidad and Tobago, plants used for success in hunting dogs have been divided into four categories (22). The first is called steaming. A wasp, *Sphex* species and its prey the spider, *Nephilengys* species, are put in rum with *Aframomum melegueta* Guinea grains or Melegueta on a Friday, and the solution is either given to the dog or included in the bath water for the stimulant effect (22). In the second category, assorted plants are placed in the dog's nose to act as a nasal and chest decongestant and the dog will subsequently have a better sense of smell and improve its ability to follow the scent (22). The third category is based partly on the Doctrine of Signatures in which a plant characteristic is considered to have a desirable quality or to have a physical property that resembles the desired game. This desirable quality is claimed to be transferred to the dog after the plant is used in a bath (22). In the forth category called 'cross', the dog goes in the opposite direction from the game. The dog is first faced upstream the river and bathed and rubbed with the crushed leaves of seven different plants; and then turned to face downstream to start the hunt (22).

Some hunters are so fond of their dogs that they are given a fitting burial. For example in Gabon, the Mitsogo people make domes of the

stems of *Trachyphrynium braunianum,* a rhizomatous forest herb with bamboo-like sub-woody stems, to place over the graves of their hunting dogs (34).

THE HUNTED PROTECTED

Hunting of wild animals is among the practices that threaten the survival of species (15). A full list of these threats (15) is given in Table 3 below:

Threat categories
1. Afforestation (establishing timber plantations)
2. Agriculture
3. Alien plant infestation
4. Collection (removal of medicinal plants and poaching)
5. Damming
6. Deforestation (land clearing of woody cover)
7. Desiccation (drying of wetlands)
8. Fire
9. Forestry exploitation (removal of woody species)
10. Grazing (by cattle; or browsing by elephants)
11. Habitat degradation (applied in general or specific terms)
12. Harvesting (removal of certain plants, e.g. for medicine)
13. Mining
14. Pests/pathogens
15. Road network
16. Salinisation
17. Siltation
18. Soil erosion
19. Urban expansion (expanding human settlements)

Table 3. Activities That Threaten the Survival of Species

As an additional means of protecting endangered animals from indiscriminate hunting, African tradition, among others, surrounds such animals with myths, taboos and legends. The practice is, in effect, a means of conserving biodiversity for sustained yield. (For another instance, see ANIMALS: SPIRITUAL SIGNIFICANCE). Hunted animals do also enjoy the protective benefits of plants. For instance, in S Nigeria, the Yoruba name of *Mutinus bambusinus* and *Phallus aurantiacus,* gasteromycete fungi, is 'akufodewa' which reflects the bad smell the plants have. It is said that the stink will halt hunters in their search for game to find the dead animal causing it (28).

According to folklore, *Milicia (Chlorophora) excelsa* and *M. regia* Iroko, and *Morus mesozygia,* all forest trees, are supposed to possess a guardian spirit of wild creatures against hunters. It is also strongly believed that when a game is shot at near *Securidaca longepedunculata* Rhode's violet, a savanna shrub, the animal will definitely be missed, should the gun fire at all. Firearm placed against the shrub is immobilized

Securidaca longepedunculata Rhode's violet (legendary and mythical plant).

(25). In Ghana and Ivory Coast, the local name of this shrub in Akan is 'kyirituo', which literally means 'against firearms'. The bark of the shrub collected with rites (see PROTECTION: FIRE ARMS), enters into prescriptions for immunity against gun shot (27). It is traditionally believed that game are, by instinct, aware of the inherent protective charm in this shrub, and accordingly take advantage of it by quickly running to stand near 'kyirituo' when they spot a hunter (4). *Salacia debilis,* a lofty undulating forest liane traditionally believed to be fetish, is similarly credited with protecting game from hunters; with the bark entering into bullet-proof charms (31). The local name in Akan is 'homa-kyereben', meaning 'mamba-like vine'. Hunters also claim that when the bush cat, *Viverra civetta,* locally called 'obrebi' in Akan is on *Musanga cecropioides* Umbrella tree, it is impossible to shoot it (12).

There are said to be other animals that are simply just against gun shot. Among veteran hunters, it is the belief that when the seeds of *Picralima nitida,* a forest tree with an extremely bitter seeds - source of the alkaloid akuammine - are chewed by some animals, the seeds develop into 'dufa' or poultice in the system of the animal. Among others, the poultice renders the animal fearless and resistant to gun shot (12). The buffalo, *Synceros caffer,* is among the animals traditionally believed to be immune to bullets. In Ghana, one such animal caught in a trap in Akropong Akwapim could not be shot with three different guns - each gun failing to fire even a single round. When the animal was eventually run down by hunting dogs, *Canis* species, and hacked to death by hunters, a muscle-like tissue near the heart was observed to be still beating rhythmically several hours after the animal had been killed and butchered into pieces (1).

Animals with Vindictive Spirits

Another group of animals may only be hunted after the necessary rituals before and after the chase. Legends often describe people who suffer misfortune because they have killed an animal without first asking permission from its soul (23). In Ghana, every Ashanti hunter who goes after the bongo, *Tragelaphus euryceros,* has in his wallet some of the root of the tree called 'atwere nantem' which is a *sasa* antidote (30). (The botanical name of the tree is *Diospyros monbuttensis,* a tree of the drier semi-deciduous rain-forest). If a hunter kills a *sasa* animal (those animals with vindictive spirits) he immediately performs a series of rituals for purification. First, the meat is passed and repassed through a split liane which is joined and spliced; then the hunter will wash himself with 'medicine'; and he may not himself eat the meat (30).

Animals with vindictive spirits are the bongo, *Tragelaphus euryceros;* the elephant, *Loxodonta africana;* the roan antelope, *Hippotragus equinus;* the waterbuck, *Kobus ellipsiprymnus;* the duiker, *Sylvicapra grimmia;* the black duiker, *Cephalophus niger;* the yellow-backed duiker, *C. sylvicultor,* locally called 'okwaduo'; and surprisingly a very small antelope, the royal antelope, *Neotragus pygmaeus,* locally called 'adowa' in Akan (30). 'Of all these *sasa* animals, the bongo is the most dangerous and most feared' (30). The aardvark, *Orycteropus afer,* a Guinea-savanna mammal and the bare crow or the bare-headed rock fowl, *Picathartes gymnocephalus,* a very rare forest bird that lives among rocks, are similarly surrounded by hunter's lore and superstition among ethnic groups in their distribution areas.

Included among the animals with vindictive spirits are wild cats, *Profelis aurata,* locally called 'okra'; and squirrels, *Protocerus* species, locally called 'opuro' (14) in Akan. 'If any of these animals are killed during the hunt, a ritual sacrifice is required before they can be brought to the village and used as food or for the fabrication of *asuman* (amulets) and *aduro* (protective medicine). A hunter who fails to comply with this rule will inevitably be punished by the *sasa* of his prey with disease of the category of *sasa yedee*' (14). In Ivory Coast and Burkina Faso, *Gardenia erubescens,* a savanna shrub, is used by the fetish priest to protect hunters from sickness they could contact from killing some animals (21).

ASSOCIATED PLANT NAMES

The names of some plants have reference to hunting. For example, in Ghana, *Lantana camara* Lantana or Wild sage, a notorious weed of cultivated land, is called 'odelamanyi' in Awuna, meaning a hunter does not eat (11). In N Nigeria, *L. rhodesiensis,* a woody savanna labiate, and *Rourea (Byrsocarpus) coccinea,* a climbing shrub, are known in Hausa as 'kimbar mahalba', meaning hunter's spice, and by the Fula tribe as 'urdi loho' be', meaning hunter's scent (also applicable to *Lantana*) (11). In S Nigeria, the Yoruba tribe call *Stereospermum kunthianum,* a savanna tree with beautiful pink or purple flowers, 'ayada' or 'ajade' - being an honourable title among hunters (10a). (See also CHARMS; JUJU OR MAGIC; PROTECTION; and TALISMAN).

References.

1. Akoto, (pers. comm.); 2. Akpabla, (pers. comm.); 3. Annan, (pers. comm.); 4. Ashieboye-Mensah, (pers. comm.); 5. Aubréville, 1959; 6. Balle, 1951; 7. Bouquet, 1969; 8. Bouquet & Debray, 1974; 9. Bowdich, 1819; 10a. Burkill, 1985; b. idem, 1994; c. idem, 2000; 11. Dalziel, 1937; 12. Eshun, (pers. comm.); 13. Ferry & al, 1974; 14. Fink,1989; 15. Golding, 2002; 16. Hammond, (pers. comm.); 17. Heine, 1966; 18a. Irvine, 1930; b. idem, 1961; 19. Jackson, 1973; 20. Kerharo & Adam, 1964; 21. Kerharo & Bouquet, 1950; 22. Lans, 2001; 23. Martin, 1995; 24. Meek, 1931; 25. Mensah, (pers. comm.); 26. Moro Ibrahim, (pers. comm.); 27. Noamesi, (pers. comm.); 28. Oso 1977; 29. Piper & Dunn, 1922; 30. Rattray, 1927; 31. Sereboe, (pers. comm.); 32. Tattersall, 1978; 33. Tremearne, 1913; 34. Walker & Sillans, 1961; 35. Watt & Breyer-Brandwijk, 1962.

23 INFANTS

When I was a child I spake like a child,
I understood as a child, I thought as a child...
1 Corinthians 13:11a

In traditional Africa and elsewhere, rituals are performed by many of the tribal groups to strengthen children physically or protect them spiritually. Plants feature in these ceremonies. These plants may also be medicinal or intended to render the child fearless, brave and intelligent. A nursing mother would normally have among her baby's bathing toiletry at least prescriptions to strengthen the child physically, if not also spiritually. Children are bathed in *aduro* (herbal medicine) from time to time to ensure strength and good health and to protect them from illness and bad influence (10). African domestic life, in anthropological summaries, seems to be full of special observances, and ceremonies to mark the advancement of individuals from birth, through weaning and puberty, to maturity and old age (11).

PHYSICAL DEVELOPMENT

It is the natural duty of nursing mothers to assist in the growth and physical development of the child - either with plant prescriptions and charms or with other products. Some of these prescriptions appear to be based on sympathetic magic.

Plant Prescriptions

In Liberia, midwives prepare a decoction from the leaf of *Polcephalium (Chlamydocarya) capitatum,* a forest liane, for putting into baths and make a draught for weakling babies to make them strong (7); and in N Nigeria, the root of *Bergia suffruticosa,* a shrub, is used by the Hausa as an ingredient in *dauri* prescriptions at weaning time to strengthen infants and prevent sickness (7). In Ivory Coast and Burkina Faso, a bark and leaf prescription of *Hymenocardia acida,* a savanna tree, is used in baths and lotion to strengthen debilitated children (16); and in Ghana, the leaves of *Jatropha curcas* Physic nut are an ingredient in enema preparations, and are prepared along with palm-oil fruit, for an injection administered to weakly children (4b). In Senegal, the Socé give a root-macerate of *Gossypium herbaceum* Cotton to new-born babies, and to sickly or rachitic children so that they will grow big and strong (15); and in Sierra Leone, the seeds of *Ongokea gore*, a forest tree with edible fruits when ripe, are worn by children as a necklace charm to keep their bowels open (4d).

In S Africa, a piece of the fruit of *Kigelia africana* Sausage tree is rubbed over the baby by the Luvale to make it fatter, but avoiding the head for fear of producing hydrocephalus (24). (See also FERTILITY: SYMBOLS OF FERTILITY: Sympathetic Magic). In W Africa, the Ewes of Ghana, Togo and Benin Republic, the Gas of Ghana, and other coastal tribes wash their babies with a decoction of the whole plant of *Paullinia pinnata*, a climber, in the belief of achieving the same effects, again avoiding the head for the same reasons (6). Among the Hausa of N Nigeria, the Zarma of Mali and Niger and the Fula a whole plant decoction of *Anchomanes difformis* or *A. welwitschii*, herbaceous annual aroids with tuberous rhizomes, is similarly used - also avoiding the head (19). In the Region, the unopened leaves of *Costus afer* Ginger lily, a perennial forest herb in the ginger family, ground with kaolin is applied by nursing mothers to children after bathing for bravery, but avoiding the head, breast and joints (6).

In Sierra Leone, the Kono believe that if the stem of *Pandanus aggregate (candelabrum)* Screw-pine is put in water in which a small baby is to be washed, the baby's head will become very big (13). In Ghana (22) and Ivory Coast (1), the ointment extracted from the powdered gum of *Balanites wilsoniana*, a copal-producing forest tree, is applied to newborn or suckling babies to make them grow big. In the Upper East Region of Ghana, a root decoction of *Grewia cissoides*, a savanna shrub, is used by the Mamprusi tribe in washing babies for strength (5). In Tanzania, *Trichilia emetica* ssp. *suberosa (roka)*, a savanna tree, has use as a powder to rub on children to make them grow big and strong (4c). In this country, *Sporobolus pyramidalis* Rat's tail grass, a densely tufted perennial grass, is made in a cryptic prescription given with other medicines and applied as charms, amulets and magic 'to make babies strong' (Culwick fide 4b). In S Africa, a lotion made from the crushed root of *Agapanthus africanus* var. *minor*, a lily, is applied by the Southern Sotho to a newborn child to ensure strength (24).

In Sierra Leone, *Ficus sur (capensis)*, a common shade tree with abundant cluster of fruits on the stem, is used to facilitate closure

Alternanthera pungens Khaki bur or weed

of the frontal suture in babies among the Mende; however, the drug should be applied with a small stick and never with the hand (17). In its distribution area, poultice prepared from the leaves of *Alternanthera pungens (repens)* Khaki bur or weed, a common weedy amaranth, and those of *Baphia nitida* Camwood, both collected with the necessary ceremonial rituals, is applied to the fissures in the skull of newborn babies by many nursing mothers (18). In N Togo, *Asystasia gangetica,* an erect or straggling herbaceous acanth, is a magic medicine to make young children fearless and brave (7). The main tribes in this region include the Kabre, the Kotokori, the Losso, the Moba and the Tsokosi. The plant is similarly used for babies in NE Ghana (Williams fide 4a). In Congo (Brazzaville), the Koyo bathe their children in a leaf decoction of *Tylophora glauca,* a forest climber with milky latex, to make them handsome, big and strong (2); and in S Nigeria, the leaf of *Combretum hispidum,* a scandent shrub or liane, is placed upon a child's head by the Igbo tribe in a superstitious way so that the child may grow (N.W. Thomas fide 4a).

Sympathetic Magic. Some of the prescriptions are based on sympathetic magic. For instance, in S Nigeria, goats, *Capra* species, eat *Rhaphiostylis beninensis,* a scandent or twining forest shrub, evoking in Igbo the epithet *ike,* meaning 'strong'. The idea of imparting strength is also manifest in the root and bark which are sold in Lagos medicine market for the purpose of preparing an infusion taken by pregnant Yoruba women to strengthen the baby, such preparation also being given as tonic to infants up to the age of 2-3 years (7). In W Africa, thunder stone or the bark of a tree that has been struck by lightning and collected with rituals, is an important ingredient among many of the tribal groups in charms worn and used in bath water for bravery in children (6). In N Nigeria on the contrary, if *Crotalaria macrocarpa,* a leguminous herb with woody base, locally called 'sa furfura' should be unwisely mixed with other medicines and given to an infant, the child's hair will turn grey before puberty (7). The name in Hausa means 'causing to turn white' - likely an instance of sympathetic magic. In traditional medicine, it is the belief that mixing plants during collection may have adverse effects (see MEDICINE: RITUALS).

In N Nigeria, *Crotalaria arenaria,* a woody leguminous plant in dry savanna, locally called 'manta uwa', has been found as an occasional ingredient in superstitious practice to wean a child, hence the name, meaning 'forget mother'. This term is properly used for epiphytic orchids such as species of *Angraecum, Ansellia, Bulbophyllum, Listrostachys, Polystachya,* among others (7). In SE Senegal, *Phragmanthera* and *Tapinanthus* species Mistletoe, epiphytic parasites, on *Combretum glutinosum*, a savanna tree, is added to the food of a Tenda child whose mother has abandoned it, and will prevent the child from joining her (9). In S Africa, the Shangana think that the root of *Annona chrysophylla* Wild custard apple has the property of making people forgetful and that, if put in the kaffir corn while it is being cooked, it assists a child who is being weaned to forget its mother's breast (24). Similarly, in Mozambique, the Tonga use the root of this plant to wean a child from its mother's breast as it is supposed to cause forgetfulness (24). In Central African Republic, the Suma name of *Sopubia simplex,* a savanna annual, means 'wash illness away', and they pound the leaf in water which is used to wash a child suffering a chronic illness (4d).

In E Africa, the Lobedu wash the newborn with a cold infusion of the powdered root of *Cussonia spicata* Cabbage palm daily until the infant leaves the hut for the first time. This is usually after five to seven days, but the washing may extend over one and half months (24). The idea underlying this application is to prevent skin irritation and pimples and to make the infant strong and fat. The latter is an example of sympathetic magic because the root is soft and fat (24). Similarly, in Zambia and Zimbabwe, the Manyika have a magical use of the bulb of *Raphionacme hirsuta,* an asclepiad, which is large and fat, to make the baby large and fat (24). In Ghana, weak children may be strengthened by washing them daily with water in which the fruits of either *Dioclea reflexa* Marble vine or a piece of the bone of an elephant, *Loxodonta africana,* has been immersed (18). The latter is an instance of sympathetic magic because of the size and strength of the elephant. In S Africa, the Luvale tribe also use the hard wood of the sacred *Vangueriopsis lanciflora* Wild medlar for making a spoon and a drinking cup for an infant to make it grow strong (24). This is likely an instance of sympathetic magic hinged on the strength of the wood.

Other Products

In addition to plants, other materials may be used to enhance the physical development of the child. For example, in Ghana, if a child is rubbed over with a powder made of crushed aggry beads, it is believed that his maturing will be hastened (8a). Children who are successors to a stool, or any inheritance in Ashanti are rubbed with this powder daily after washing (3). Aggry beads are products of a lost art, and said to have originated in biblical times from Phoenicia or Egypt. One bead was said to be worth the price of seven slaves or twice its weight in gold dust (8a). It has been observed that besides their intrinsic values, they are greatly esteemed by the natives on account of the cabalistic virtues which are attributed to them (8a). Similarly, in the then Gold Coast (now Ghana), the earth collected from the spot where the Portuguese governor and his countrymen were executed by the Gas for exercising the greatest cruelty and enormities among the natives, was used to rub a newborn child, in order that he or she might grow up to be strong and fearless - and also to commemorate the event (3).

Aggry beads

Teething Problems

In S Nigeria, the branches of *Diospyros monbuttensis,* a forest tree with spines, to which perhaps the Yoruba name 'erikesi', meaning 'pigs teeth' refers, are made into an infusion which is given, probably on the premise of sympathetic magic, to alleviate teething pains in children (25). In this country, the Yoruba invoke *Alternanthera pungens (repens)* Khaki bur or weed, a weedy amaranth, in an incantation to aid children in teething, and to assuage gum tenderness (23). In S Africa, a necklace of the glumes of *Coix lacryma-jobi* Job's tears, a perennial grass usually by streams and wet places, is placed on an infant with the idea of warding off teething troubles (4b, 24). This custom is not unique to Africa but found elsewhere. For instance, in Hawaii it is used as a curative charm (24). In England, the seed of *Entada gigas* or *E. rheedei (pursaetha)* Sea bean or Sea heart, tropical drift dissemules (for the meaning see RELIGION :SYMBOLIC PLANTS: Sea-beans), was used for a teething ring (12); and bead necklace made from the dried root of *Paeonia officinalis* Common paeony, an introduction from southern Europe, strung on leather were worn by infants to aid teething (21).

SPIRITUAL PROTECTION

Some plants also serve as spiritual protection for the child. In Nigeria, *Eleusine indica,* a tufted annual grass, is commonly held to have magical powers (N.W. Thomas fide 4b). In S Nigeria, the Igbo tribe at Okpanam believe so; and in SE Nigeria, the Ijo will tie a piece of the plant around the waist of a child as protection should that child have to visit a place where an old person has died (4b). In Sierra Leone, Mende herbalists give a decoction of *Sarcocephalus (Nauclea) latifolius* African peach to a new-born infant to make it thrive and to exorcize all evil influences (14). In Ivory Coast, the Guéré consider *Sterculia tragacantha* African tragacanth protective of children, and a paste of eight leaves, intact without holes, and clay is applied to their faces in various ciphers each morning (16).

Sarcocephalus (Nauclea) latifolius African peach

In Senegal, the seeds of *Desmodium gangeticum,* an erect undershrub legume in secondary forest, are considered by the Tenda to have magical properties (9). They are tied to the arm of an infant as a protector charm (9). In Senegal, *Phragmanthera* and *Tapinanthus* species Mistletoe, epiphytic parasites, growing on *Acridocarpus spectabilis,* an under-shrub in savanna, is steeped in water which is used to wash infants attacked by sorcerers (9). In this country, Mistletoe growing on *Vitex doniana* Black plum, a savanna tree with edible fruits, is used by the Bedik to prepare a bath for a new-born baby, and for one which cries too much if one is seeking to find out whether the baby is a reincarnation (9).

In S Nigeria, an unidentified plant locally called 'sekiseki' in Yoruba, meaning 'short, short', is invoked by the people to make a dying baby stay on earth (Verger fide 4d). In this country, the genus *Terminalia* - mostly savanna trees, but there are two species in the forest zone - is invoked by the tribal group in an incantation under the general name of *idi,* 'the closing' of the paths of evil and death, for the safe conduct of an 'abiku' child during its life on this earth (23). The Yoruba also invoke *Lantana camara* Wild sage, a notorious weed of cultivated land, or *Crotalaria lachnophora*, a legume shrub of wooded savanna, in an incantation to keep an 'abiku' child on earth (23). The tribe also imprecate *Microdesmis puberula,* a small forest tree or bush, in an incantation to give life to an aborted foetus (23). 'Abiku' - predestined to death - is a word used to mean the spirits of children who die before reaching puberty, and also a class of evil spirits who cause children to die (8b). A child who dies before twelve years of age is called 'abiku', and the spirit or spirits, who cause the death being also called 'abiku' (8b).

In parts of traditional Africa, plants may symbolize the zodiac. For instance, in Burkina Faso, children of the Ela born under the sign of *Adansonia digitata* Baobab, locally called 'kukulu' in Lyela, are given the patronymic *kukulu,* boys; or *ekulu,* girls (20). In this country, the Lyela tribe believe that a *Ficus* species named *epiku* contains vital forces capable of initiating life. Women passing a tree will always make an offering of a little food to the spirits living in it, and if pregnant, a child born, of either sex, would be given the patronym *epiku.* (20). (See also CHILDBIRTH. For intelligence or mental development of the child, see INTELLIGENCE: PRESCRIPTIVE CHARMS).

References.

1. Adjanohoun & Aké Assi, 1972; 2. Bouquet, 1969; 3. Bowdich, 1819; 4a. Burkill, 1985; b. idem, 1994; c. idem, 1997; d. idem, 2000; 5. Casey, (pers. comm.); 6. Cofie, (pers. comm.); 7. Dalziel, 1937; 8a. Ellis, 1881; b. idem, 1894; 9. Ferry & al., 1974; 10. Fink, 1989; 11. Gluckman, 1963; 12. Gunn & Dennis, 1979; 13. Huynh, 1988; 14. Joru, 1973; 15. Kerharo & Adam, 1964; 16. Kerharo & Bouquet, 1950; 17. Macfoy & Sama, 1983; 18. Mensah, (pers. comm.); 19. Moro Ibrahim, (pers. comm.); 20. Nicholas, 1953; 21. Perry & Greenwood, 1973; 22. Taylor, 1960; 23. Verger, 1976; 24. Watt & Breyer-Brandwijk, 1962; 25. White, 1957.

24 INTELLIGENCE

The heart of the prudent getteth knowledge,
and the ear of the wise seeketh knowledge.
Proverbs 18:15

Intelligence is the ability to reason and understand, and to adapt to new situations. It is acquired with age and experience. Environmental and genetic factors equally influence ones intelligence. It is also traditionally believed that some plant prescriptions prevent forgetfulness, improve the memory and therefore, intelligence. For instance, *Centella asiatica* Gotu kola or Indian navelwort, a prostrate herbaceous umbel, has been mentioned in ancient Indian literature for its intelligence promoting property (11). Some plants symbolize intelligence, while other plants enter into charms to counteract intelligence. Some of these prescriptions could be described as general superstition and others as instances of sympathetic magic - the uses that appear to be based on the observable qualities of the plant.

Centella asiatica Gotu kola or Indian navelwort

PRESCRIPTIVE CHARMS

In Nigeria, *Ludwigia octovalvis* ssp. *brevisepala* Primrose-willow, a hairy herb with yellow flowers usually growing in damp places, enters into prescriptions among both the Hausa and the Yoruba to prevent forgetfulness (6); and at Lagos, a leaf infusion of *Sabicea calycina*, a slender climbing forest shrub, is said to be good for the memory (Punch fide 3). In N Nigeria, species of *Phragmanthera* and *Tapinanthus* Mistletoe, epiphytic parasites, growing on *Balanites aegyptiaca* Desert date, locally called 'kauchin aduwa' in Hausa, forms part of a magical prescription, of which the hoopoe bird, *Upupa epops,* and washings of the Koran text are necessary ingredients, which imparts zeal or intelligence to youthful scholars (6). In the Guinea-savanna zone of the Region, *Biophytum petersianum* African sensitive plant, a small annual plant with sensitive leaves, collected with the necessary rites, then boiled with assorted ingredients, and taken internally and also used regularly as bath water, enters into prescriptions among some of the tribal groups for intelligence at school (9).

Among the Hausa of N Nigeria, the Zarma of Mali and Niger and the Fula the leaves or seeds of *Senna (Cassia) occidentalis* Negro coffee, collected with customary rites, and brewed as beverage is drunk for intelligence (10) - probably superstitiously. The nomadic northern Fula tribe also inhale a snuff prepared from the bark of *Erythrophleum suaveolens* Ordeal tree with a charred frog, *Bufo* species, and the seeds of *Aframomum melegueta* Guinea grains or Melegueta for intelligence (9). In S Nigeria, the Yoruba invoke *Chloris pilosa,* an annual grass, in an incantation to overcome forgetfulness (16a); and Yoruba witch-doctors use *Spondias mombin* Hog plum or Ashanti plum in magical procedures to develop intelligence (16b). In Ghana, the *popo* ceremony is performed at first pregnancy among the Ga tribe. Among others, it is to prevent the child from being born stupid (7).

Aframomum melegueta Guinea grains (fruits)

In Botswana, decoctions of the root of *Galium dregeanum,* locally called 'scharane', and that of *G. rotundiflorum,* called 'moriri-wa-lenala', enters into the ritual of the Southern Sotho witch-doctor to ensure intelligence and judgement in the aspirant witch-doctor (17). In E Africa, a piece of the foliage of *Gardenia neubria* Kaffir-cherry is picked by the Lobedu in order that previous thoughts may be forgotten (17). In S Africa, *Chamaecrista (Cassia) mimosoides* Tea senna, a herb or low shrub legume, is used as a charm to induce stupidity in an enemy (17).

In S Nigeria, the Yoruba invoke *Piliostigma thonningii,* a savanna tree legume, in an incantation to disturb the mind of an enemy and make him unable to calculate (16a). In this country, the tribe endow *Cocos nucifera* Coconut palm with 'intelligence', and invoke it in an incantation against people who do not keep a promise (16a). However, in Japan, legend says that if you eat *Zingiber mioga,* a herbaceous perennial locally called 'myoga', you will become forgetful (15).

Sympathetic Magic

Some of the prescriptions appear to be based on sympathetic magic. For instance, the weaver bird, *Ploceus* species, is supposed to be one of the most intelligent birds judging from how it weaves its nest, so black powder prepared from the charred nests is taken with honey by people in the bird's distribution area in the Region for intelligence (8). (The nest of the weaver bird is usually woven with strips of *Cocos nucifera* Coconut palm or *Elaeis guineensis* Oil palm fronds). The prescription, with the addition of honey both suggests sympathetic magic. Similarly, the powdered root-bark of *Abrus precatorius* Prayer beads, a twining legume with decorative red and black seeds, collected and ground with rites, and taken in honey - twice teaspoonful daily is a prescription to improve the memory (4). In Ghana, the ground leaves of *Tridax procumbens* Coat buttons, a very common weedy composite, with the brain of the cat, *Felis domestica;* apple-cider, vinegar and honey, taken twice a spoonful daily, enters into rituals among the Ga-speaking tribe to improve intelligence (1). The addition of a cat's brain is likely an instance of sympathetic magic.

Tridax procumbens Coat buttons

In S Nigeria, *Uapaca heudelotii,* a stilt-rooted tree, is used by Yoruba medicine-men in medication to promote intelligence: *oye,* from which the Yoruba name of the tree is derived (16b). In this country, Yoruba witch-doctors use *Hydrolea palustris,* a glabrous herb with spongy stem and root of wet places, and *Maranthes robusta,* a forest tree, in a treatment to develop intelligence, hence *oye* Yoruba for 'intelligence', is compounded in the Yoruba name of the plants 'oniyeniye' and 'aiye', respectively (16b). *Clerodendrum capitatum,* an erect or scrambling shrub of savanna and forests, enters into magical applications. Medicine-men use the plant to develop intelligence, or *oye,* a derivative

from 'iyu', the Yoruba name of the plant (16b).

Among the Akan-speaking people of Ghana and Ivory Coast and the Nzima people of Ghana, *Baphia nitida* Camwood and *B. pubescens,* legume shrubs or small trees, locally called 'odwen', meaning 'he thinks' or 'she thinks' or 'thoughtful', is symbolic of intelligence and wise counseling. The heated leaves of 'odwen' are therefore applied to the head of newborn infants in the belief that this will make them thoughtful (8). This is surely an instant of sympathetic magic. Among many ethnic groups in the Region, a sea water decoction of *Momordica balsamina* Balsam apple or *M. charantia* African cucumber, trailing cucurbits that feature regularly in traditional rituals, mixed with honey, is taken to sharpen the memory (9). Also the whole plant of *Imperata cylindrica* Lalang grass, a notorious perennial weed popularly used as thatch material, charred into black powder, then mixed with honey and taken - invoking the plant in an incantation, 'a man who is roofing never forgets about Lalang grass, why should I forget?' - is a prescription used by traditional medicine-men for retentive memory (9). Again, the sympathetic magic appears to be the honey in both instances - and in the latter, the roofing grass as well.

The leaves of *Croton zambesicus,* a small tree planted in villages and believed to be fetish, and that of *Sida linifolia,* a common malvaceous weed popularly believed to have magical properties, enter into juju prescriptions for intelligence by traditional medicine-men in the Region (9). The leaves, collected with the necessary ceremony, are put in an enclosed container with some native black soap (soft kotokoli soap) and sponge as ingredients, and placed on the grave of a highly educated person in the evening with a specific invocation and an offering of a few coins (9) . The container is collected at dawn the next day, and the sponge with the soap used for a ritual bath every morning for seven consecutive days. The whole ritual is to impart the intelligence of the deceased to the aspirant (9).

The association of the words 'wise' and 'sage' arose from the belief that *Salvia officinalis* Sage, a decorative introduced annual labiate, strengthens the memory (14); and bananas were formerly named by Linnaeus, a Swedish botanist, *Musa sapientum,* 'fruit of the wise men', because Pliny said that wise men of India lived on them (14). It has also been observed that the scientific name of Sage comes from the Latin word *salvare,* which means to save, since Sage was believed to heal almost all diseases with the exception of death (12). The ancient Greeks used *Rosmarinus officinalis* Rosemary to strengthen memory function; and scholars wore garlands of the plant during examinations

in order to improve memory and concentration (2). In the Middle Ages, honoured poets - and later men of letters - were crowned with a wreath of berried *Laurus nobilis* Sweet bay or Laurel, hence the term 'Poet Laureate' (13). Students studying for degrees at Universities were called Bachelors, from the Latin *baccalaureus* (laurel berry); and were forbidden matrimony lest this enticed them from their studies (13). A Neo-platonic philosopher, Iamblicus, explained the round leaves and spherical fruit of *Nymphaea lotus* Water lily as symbolic of intellect (5). (See also KINGSHIP: CEREMONIAL RITUALS).

References.

I. Blankson, (pers. comm.); 2. Blumenthal & al., 2000; 3. Burkill, 1985; 4. Cofie, (pers. comm.); 5. Crow, 1969; 6. Dalziel, 1937; 7. Field, 1937; 8. Irvine, 1961; 9. Mensah, (pers. comm.); 10. Moro Ibrahim, (pers. comm.); 11. Nalini, Arboor, Karanth & Reo, 1972; 12. Pamplona-Roger, 2001; 13. Perry & Greenwood, 1973; 14. Pitkanen & Prevost, 1970; 15. Richards & Kaneko, 1988; 16a. Verger, 1967; b. idem, 1972; 17. Watt & Breyer-Brandwijk, 1962.

25 JOURNEY

And the children of Israel journeyed from Rameses to Succoth, about six hundred thousand on foot that were men, besides children...Now the sojourning of the children of Israel, who dwelt in Egypt, was four hundred and thirty years.

Exodus 12:37, 40

Before the advent of motor transport, long journeys had to be made on foot with the luggage either on the back or on the head. The journey could sometimes be through difficult terrain and hostile tribesmen. To overcome this arduous zeal and possible danger, the people sometimes resorted to charms and deities with the belief of obtaining some relief and protection - both physically and spiritually. In Ghana, two deities are mentioned (4) among the people of Dormaa: 'Boam' (help me) protects its owner on journeys and takes care that there is no misfortune at home during his absence; and 'Atenka', which is tied to the left knee on journeys, helps the person wearing it to recognize dangers on time and prevents him or her from falling down (4). There are also plant charms believed to induce the compulsory return of a traveller home. Some plant names are also associated with a traveller.

PROTECTION AND GUIDANCE

In N Nigeria, the fruit of *Gardenia erubescens*, a savanna shrub with edible fruits, was used together with *Cordia africana (abyssinica)*, a small tree; *Detarium senegalense* Tallow tree, a savanna tree legume; *Hibiscus cannabinus* Hemp-leaved hibiscus, a cultivated fibre plant; *Hymenocardia acida*, a savanna shrub or small tree; *Pericopsis (Afrormosia) laxiflora* Satinwood; and the root of *Sanseviera liberica* or *S. senegambica* or *S. trifasciata* African bowstring hemp with

Momordica charantia African cucumber

other plants in a complex prescription of many ingredients in the preparation of a tonic and stimulant taken to impart strength or moral stimulus when undertaking a journey or other enterprise (3). In Ghana, when a native on his journey finds *Momordica balsamina* Balsam apple or *M. charantia* African cucumber, trailing cucurbits associated with traditional rituals, he gladly hangs a piece round his neck with the idea that it will assuredly preserve him from unfortunate happenings (Thonning fide 6).

In Congo (Brazzaville), the leaf-sap of *Rhektophyllum mirabile,* a stout climbing aroid in forest, applied to the soles of the feet and to the legs of one setting out on a journey is considered to offer protection; and to avoid storms and rain when on a journey, *Tylophora sylvatica,* a herbaceous twiner with milky latex, is tied round the waist (1). In this country, the leaves of *Alchornea floribunda,* a semi-scandent shrub or small tree, are crushed to ensure favourable circumstances on setting out on a journey or entering a strange house (1). In Senegal, *Scoparia dulcis* Sweet broomweed, a shrubby herb in waste places, is believed to confer protection to those going on a journey; and the 'Socé' leaving a village soak a small piece in water with which they wash (7a, b).

In Gabon, the natives manage to double both the length of their day's march and the weight of what they are carrying without noticing the extra effort required by eating the roots of *Tabernanthe iboga* Iboga, an apocynaceous forest shrub with hallucinogenic properties (Brzezicki fide 15). In S Africa, the leaf of *Tarchonanthus camphoratus* Camphor wood is chewed by the African to ward off evil influences, especially when a person is journeying through a strange country; and in Botswana, a traveller encountering *Anthospermum pumilum,* locally called 'masopolohane', regards it as a good omen, and if he stops and repeats a few words, good food and welcome awaits him at the end of his journey (17). In Kenya, the Digo say that if a traveller eats the ash of *Biophytum petersianum* African sensitive plant, an annual herb with sensitive leaves, children will play around him wherever he goes and ask their mother to bring him food. Thus anywhere he will be welcome (2b).

In Ivory Coast, the Dyiminis believe that massaging the body with the leaves of *Leptadenia hastata,* a savanna twiner with clear light green latex sometimes tended around villages, protects travellers from evil spirits (8). In this country, the root of *Leea guineensis,* an erect or sub-erect shrub, is believed to confer a beneficial disguise, and also one taking a long journey is recommended to take lengths of the liane to suck during the travel to prevent accidents and intervention of possible evil spirits (8). Among some of the coastal tribes in the Region - especially the Ewe-

speaking people of Ghana, Togo and Benin Republic - it is believed that a leaf of *Hyssopus officinalis* Hyssop, a Mediterranean labiate; or *Allium cepa* var. *aggregatum (ascalonicum)* Shallot; or the fruit of *Abelmoschus (Hibiscus) esculentus* Okro or Okra collected with rites, invoked to that effect, and then kept in one's pocket, is a protecting charm from injury while on a journey (12).

In Nigeria, *Polycarpaea corymbosa, P. eriantha* and *P. linearifolia,* erect herbs in savanna woodland, are used to make an infusion which is drunk to combat fatigue while on the march, probably by psychological effect (3). In Sierra Leone, the Mende use a cryptic name for the fruit of *Landolphia dulcis,* a stout climber with milky latex and edible fruits, that is, meaning 'a handsome youth' (Boboh fide 2a). It is the belief that if a hungry traveller refers to the plant by its proper name, the fruit will turn rotten and maggoty; but if the cryptic name is used, the fruit can be picked sound and palatable (Boboh fide 2a). In Ivory Coast, leafy stems of an unidentified species of *Culcasia,* an aroid believed to be fetish, are macerated for six days, and the filtered liquid is taken (one or two glasses daily) to give strength and to preserve one's well-being, and while on a journey as an envigorant (8).

Traditional medicine-men in the Region prescribe a poultice made from the ground leaves of *Solenostemon monostachyus,* a herbaceous labiate, mixed with a few drops of the blood of a white fowl, *Gallus domestica,* then dried in the sun, to be kept on one's person as a protecting charm while travelling (12). Alternatively, the chewed seeds of *Cola verticillata,* a forest tree with slimy kola-nuts, is spat on the labiate before it is collected and prepared into a poultice, then kept in one's pocket as a charm for the safe return from a journey (12). A black powder prepared from *Gongronema latifolium,* a twining asclepiad, enters juju prescriptions to enable the traveller to walk safely over water when in peril (12). In Mexico, peyote pilgrims among the Huichol Indians seeking aid on their journey, or those seeking aid on long journeys, offer tobacco gourd *(yekwe)* or miniature huaraches *(kakai)* respectively to *Solandra* species, trees on rocky cliffs believed by the people to be the mythological god-plant (9).

COMPULSORY RETURN HOME

When a ward or a dependant has been abroad for a long time without remitting the parents or guardians with whose sweat and toil it was made possible to travel, and he or she does not intend to return home, the necessary juju or charms may be set in motion to compel this traveller to return home where he or she belongs. In one such prescription by traditional medicine-men in the Region, *Tridax procumbens* Coat buttons, a common composite weed, is offered in a ritual with a young chick, *Gallus domestica,* (that is sexually indeterminable) killed by pulling the head off between the bigger toe and the next one, and the plant besprinkled with the blood (12). Some coins are offered to symbolize the fare, and libation is poured with a spell. The traveller is said to buy a ticket and fly home a few days after the ritual (12). This ritual offering could also be performed, with slight modifications, to induce the return home of a runaway wife. (For charms against runaway wives, see MARRIAGE: STABILITY IN MARRIAGE).

Among the Hausa of N Nigeria, the Zarma of Mali and Niger and the Fula *Combretum micranthum,* a shrubby plant of Guinea-savanna forest, locally called 'gaiza' in Hausa, is collected with the necessary ceremony by medicine-men and used to induce the compulsory return of a traveller back home - through physical contact. The roots are powdered in lavender which is applied to the palms prior to a hand-shake with the victim (13). In these countries, *Gardenia ternifolia* ssp. *jovis-tonantis,* a savanna shrub popularly believed to conduct lightning also enters in rituals by medicine-men to induce the return of a traveller. Cut pegs of the shrub, with the name or the initials of the victim inscribed on the top surface, are driven into the ground with a wooden mallet at crossroads in the evening, invoking the plant in an incantation and a spell in which the name of the victim is mentioned (13). If the victim is a male, three pegs are driven into the ground - if a female, four pegs. This ritual serves also to prevent a wife (with such intention) from running away from the husband, or to induce the return of a runaway wife (13).

In Botswana, the Southern Sotho use *Silene undulata,* locally called 'letomokoane', as a charm when he is afraid to return home after a long absence (17). In addition to these charms above, there are also sinister prescriptions or bad juju or an evil spell believed to prevent someone from travelling - by changing his or her mind - even at the very last moment of boarding a plane.

Sympathetic Magic

Some of the prescriptions appear to be based on sympathetic magic. Alternatively, to demand the immediate return of a relative from his or her travels, *Bryophyllum pinnatum* Resurrection plant, a fleshy herbaceous perennial, is approached by medicine-men with

invocations and a spell in which the name of the relative is repetitively mentioned (12). After pouring libation nearby, seven leaves of the plant are picked amid incantations, and placed in a fire to scorch, but quenched with water soon afterwards. The victim is said to feel uncomfortable wherever he or she might be, due to the scorching fire, and immediately packs bag and baggage to seek comfort back home (12). The sympathetic magic appears to be the plant's natural resistance to fire.

To recall home a relative from abroad or a murderer from the forest among the Akan-speaking tribe of Ghana and Ivory Coast, scrapings of two trees that rub together with assorted ingredients, including the hair of a fairy (said to be obtainable from medicine markets - see DWARFS AND FAIRIES: FAIRIES : Nature and Characteristics for the 'trickery' employed to acquire it), the chipped bit of a dead wood from the trunk of a tree that has fallen by itself across a river, and a bottle of Schnapps are buried at crossroads with incantations, libation and a spell, mentioning the person by name. The buried items are then stamped with the right foot (16). It is apparent that the sympathetic magic lies with both the two trees that rub together and the log across the river.

ASSOCIATED PLANT NAMES

Some plants are named after travellers. Perhaps the most famous is *Ravenala madagascariensis* Traveller's palm or Traveller's tree, a native of Madagascar, and the National Plant of this country. (For a list of National, State or Emblematic Plants, see APPENDIX, Table IV). This island country is very rich in flora with about one hundred indigenous palms (11). The distichous leaves of this palm tree are borne on long sheathing stalks, closely overlapping, so that water collects in their hollow base. This is the feature that gives the tree its popular name, and travellers in need could drink from it (10). It has been confirmed that the cup-shaped leaf bases hold healthy drinking water for

Ravenala madagascariensis Traveller's palm (National Plant of Madagascar).

thirsty travellers (5). When an old specimen of the palm (there were then two planted) growing in front of the Balme Library, University of Ghana, Legon, was blown down in a rainstorm, water was observed flowing out of the leaf base. The name of the palm does not refer to the direction of the fan-like arrangement of the leaves as others erroneously claim.

In N Nigeria, the Igala confer the name 'ode doona' on *Striga asiatica,* a root parasite of crops, implying it has a protection value for travellers on the road (2c). *Echinocactus simpsonii* Barrel cactus has saved the life of many a prospector and miner travelling through the desert areas. It has a tremendous reservoir of liquid which quenches the thirst (14). A wild species of temperate *Clematis* is known as Traveller's joy.

References.

1. Bouquet, 1969; 2a. Burkill, 1985; b. idem, 1997; c. idem, 2000; 3. Dalziel, 1937; 4. Fink, 1989; 5. Graf 1978; 6. Hepper. 1976; 7a. Kerharo & Adam, 1964; b. idem, 1974; 8. Kerharo & Bouquet, 1950; 9. Knab, 1977; 10. Lanzara & Pizzetti, 1977; 11. Lowry, P. (pers. comm.); 12. Mensah, (pers. comm.); 13. Moro Ibrahim, (pers. comm.); 14. Pitkanen & Prevost, 1970; 15. Pope, 1990; 16. Sereboe, (pers. comm.); 17. Watt & Breyer-Brandwijk, 1962.

26 JUJU OR MAGIC

And in all matters of wisdom and understanding, that the king enquired of them, he found them ten times better than all the magicians and astrologers that were in all the realm.
Daniel 1:20

Juju is the concept of African magic. It is the act of influencing events by supernatural powers, or by the occult. In traditional Africa, some plants are believed to be juju and accordingly, feature in juju practices. The practice could be a double-edged sword - with advantages and disadvantages. While bad juju may be used for evil purposes to destroy and to kill, good juju could be used for the advancement of mankind, such as the healing of diseases. Furthermore, the traditional preservation of plants feared to be juju is an effective conservation of biodiversity and protection of otherwise endangered species. The practitioner - who is the custodian of the spiritual force or deity - is technically a juju-man, but he could also be called a medicine-man or a fetish priest or a witch-doctor.

ORIGIN AND MEANING

There are two schools of thought on the origin and meaning of juju. 'The term juju means the same as fetish or obeah. It is not an African word, but derived from the French *jeu*, a play; though of course it is more than a play, it is a religion (24). However, it has also been observed that juju is from the French word *joujou*, meaning 'toy' (1). This latter observation is confirmed thus - 'juju, is, for all the fine wild sound of it, only a modification of the French word for toy or doll, *joujou*' (29b).

JUJU PLANTS

In folklore and in practice some trees, shrubs, climbers and herbs are believed to be juju and are used generally as such. Other plants have specific juju uses, such as for tapping palm-wine. Yet some other plants enter as ingredients or accessories in the practice.

Trees and Shrubs

Throughout the Region and beyond, *Ceiba pentandra* Silk cotton tree, which occurs in both the Old and New World, has become an important sacred and fetish tree because of its enormous size, and the tree is accordingly held to have magical properties by many tribal groups (8a). Similarly, the odd appearance of *Adansonia digitata* Baobab, a savanna tree, has also resulted in magical and supernatural uses (8a). In Africa, *Spathodea campanulata* African tulip tree is still connected with the practice of magic (33) - probably due to the strong scent of garlic when freshly debarked. In Ivory Coast, *Cola acuminata* Commercial cola nut tree and *C. nitida* Bitter cola are used in making juju, and in the preparations of powders or magic drink (28). In Gabon, the Fang and other tribes use pieces of the bark of Bitter cola along with eggs in acts of magic (45). In the soudanian zone, *Parinari curatellifolia,* a savanna tree, enters into magic (N.W. Thomas fide 8a); and in W Africa, the wood of *Prosopis africana,* a legume, is credited with soporific properties and used in magic (13). In SE Nigeria, the seeds of *Canarium schweinfurthii* Incense tree are strung for attaching to calabashes as musical instruments - called *abago juju* in Igbo, and *jujunam* by the Isibori - the implication though uncertain, appears magical (13).

Spathodea campanulata African tulip tree

In the Region, *Baphia nitida* Camwood is held to have magical properties (22); and in S Nigeria, the Igbo use *Helictonema (Hippocratea) velutina,* a climbing forest shrub, as a magic to keep white ants, *Macrotermes* species, away from yams (N.W. Thomas fide 13). In Liberia, *Eugenia whytei,* a shrub or small tree, is held to have magical powers for those who are friendly to it (8d). In S Nigeria, *Voacanga africana,* a shrub with milky latex, has been recorded with magical attributes to 'cleanse' the house (N.W. Thomas fide 8a); and in N Nigeria, *Crotalaria naragutensis,* a savanna shrub, is used for magical purposes (8c). In Ivory Coast, the Guéré put the leaves of *Macaranga barteri,* a small tree, into magical powders (28). In this country, both *Cnestis ferruginea,* a forest shrub with bright red fruits, and *Blighia sapida* Akee apple have many magical uses (25). In Sierra Leone, *Distemonanthus benthamianus* African

Blighia sapida Akee apple (symbol of fecundity among the Ando; associated with traditional religion)

satinwood, a tree legume with characteristic red bark, is a much-feared juju-tree to the Temne who aver that it makes noises like stormy night, and anyone walking round it will die (8c). In S Africa, *Celtis africana* White stinkwood is much used in the magical background of the Bantu; and in Zimbabwe, *Piliostigma thonningii,* a leguminous shrub or small tree in savanna, is used in local magic (46).

In N Nigeria, the Fula claim that a leaf-infusion of *Cochlospermum planchonii,* a low floss-producing shrub, (and probably also *C. tinctorium*) bestows magical protection (26); and in Liberia, *Musanga cecropioides* Umbrella tree enters into magic (11) - (see LIGHTNING AND THUNDER: PREVENTIVE PLANTS). In its distribution area, the wood of *Ochna afzelii* and *O. membranacea,* shrubs in savanna, are traditionally used as a magic wand by some coastal tribes (35). In Liberia, *Cynometra ananta,* a forest tree, is regarded in native superstition as a powerful juju (Cooper fide 13); as such, if a snuff made from the dried leaves of the tree is blown from the hand in the direction of a far-off friend, calling his name or pronouncing a message, he will come or will hear the message (Cooper fide 44). In this country, *Vismia guineensis,* a forest tree, is important in juju to which the Basa name, 'ge-ahn', (grave or spirit world) from which no living person returns, owes reference; and is one to which the believer will turn to obtain advice (11). In Senegal, *Combretum molle,* a small savanna tree, is ascribed with magical properties (27b, 28) - (see WAR : COURAGE AND SUCCESS CHARMS). In Congo (Brazzaville), *Combretum platypterum,* a forest liane, has magical application to afford protection to a man against afflictions that can arise through sleeping with a woman in broad daylight (Sandberg fide 8a).

In Europe, both *Cinnamomum camphora* Camphor and *Liquidambar orientalis* Storax, a native of Turkey, were used in magic for ceremonies connected with the moon; and clove oil, obtained from the flower buds of *Eugenia caryophyllus (caryophyllata)* Clove was used in magical

ceremonies connected with the planet Mercury (12). All the three trees are commercially important as sources of fragrant balsam or essential oils.

Climbers and Herbs

In W Africa, magical properties are ascribed to *Leptadenia hastata,* a dry savanna twiner with clear light green latex (8a); in Ivory Coast to *Secamone afzelii,* a climbing or scrambling shrub in secondary forest or savanna with milky latex (8a); and to *Amaranthus spinosus* Prickly amaranth by some Gabonese races (8a). In Ghana, *Boerhaavia diffusa* and *B. repens* Hogweed may be hung as juju in the house to keep away lice, *Pediculus humanus;* and in N Nigeria, the Hausa use the roots in an unspecified superstitious practice (13). In S Nigeria, *Crassocephalum crepidioides,* a composite, has magical attributes as a 'medicine for wrestling' at Ibuza (N.W. Thomas fide 8a). In the Region, *Crotalaria microcarpa,* a variable diffusely branched legume herb in savanna, has apparently some magical associations (13) - (see INFANTS: PHYSICAL DEVELOPMENT: Plant Prescriptions : Sympathetic Magic). In the Cap Vert area of Senegal, the grain of *Cenchrus biflorus,* an annual grass, has undefined magical use (27b); and in Ivory Coast, *Phyllanthus niruri* var. *amarus (amarus),* a herb, has a number of magical uses (7). In Senegal, the Fula of Dianguel ascribe magical properties to the fruit of *Mukia (Melothria) maderaspatana,* a trailing cucurbit, and in this sense use them as poison-antidotes (27a, b).

In the Region, the cryptic habit of *Thonningia sanguinea* Crown of the earth, a root parasite, has led it being endowed with magical attributes and fanciful names (8a). In S Sudan, *Phragmanthera* and *Tapinanthus* species Mistletoe, epiphytic parasites, are ascribed with magical properties among the Azande. Writing about plants with such properties, such as bulbs, *ranga,* characteristic of members of the Amaryllidaceae family, it has been observed that 'another category is *ngbimi.* These are arboreal parasites and are material from which the most potent whistles and charms are manufactured' (18). 'A third category are creepers, *gire,* which figure frequently in magical rites, particularly to enclose gardens, and for winding round the wrist of a man as a charm' (18). In E Africa, the Lobedu with magical intention plant *Hypoxis villosa,* a lily, on mounds dedicated to the gods (46); and in Tanzania, the roots of *Dalechampia scandens* var. *cordofana,* a slender twiner of grazing savanna, enters into magic (8b). In Europe, one of the nine magic herbs is given as Waybread, the old country name

Clerodendrum thomsonae

for *Plantago major* Plantain (5). A dwarf rhizomatous *Begonia* species is commonly called Black magic (23) - the reason is uncertain.

In W Africa, magic power is attributed to *Clerodendrum thomsonae,* a climbing shrub introduced into cultivation as a decorative plant (36); and in S Nigeria, *C. volubile,* a climbing shrub, is held to have magic protection against cuts (8e). In Ghana, Togo and Benin Republic, *Eclipta alba,* a herbaceous composite in wet places, is an important juju plant among the Ewe-speaking and some of the other coastal tribes. In the Region, the leaves of *Bryophyllum pinnatum* Resurrection plant are chewed by the fetish priest, medicine-men or the witch-doctor to give magical effect to any spoken word, especially to prayers addressed to the gods and idols (35); and the seeds of *Coix lacryma-jobi* Job's tears, a perennial grass by river sides and wet places, have a place in magic (9). In China, the hard, white to lead-coloured false fruits of Job's tears which hang down like tears when ripe, are credited with miraculous powers and furnished into rosaries (37). In the Hoggar area, *Panicum turgidum,* a perennial tussock-grass, is held to have mystical properties (47). In Congo (Brazzaville), *Tylophora sylvatica,* a slender herbaceous twiner with thick sap, is used in magic (6) - (see JOURNEY: PROTECTION AND GUIDANCE; and MEDICINE: MEDICO-MAGICAL TREATMENT).

Palm-wine Tapping

There is a belief among palm-wine tappers that successes in palm-wine tapping could sometimes be attributed to juju. For instance, with some professional wine tappers in the Region, *Platycerium elephantotis* and *P. stemaria* Stag-horn fern, decorative epiphytic forest ferns, collected with the necessary ceremony, enter into juju prescriptions to induce a copious flow of the sap not only from the felled palms, but also from all the surrounding unfelled palms. The fern, with seven pieces of the unopened fronds of *Elaeis guineensis* var. *communis* or var. *repanda* Crazy oil palm - the variety with reddish orange fruits and greenish yellow tips when ripe - is placed in the water used to sprinkle

on the stone for sharpening the cutlass used to slice the surface of the felled palms. As part of the ritual the cutlass has to be placed back in the water. The practice is said to result in abundant flow of wine. As proof of the real source of this wine, none of the surrounding palm trees exude any sap when tapped until after three years (32).

Similarly, *Ageratum conyzoides* Billy goat weed, an annual weedy composite, enters into juju for tapping palm-wine among the Akan-speaking tribe of Ghana and Ivory Coast. The whole plant collected with rituals is charred with the roots of a variety of *Musa sapientum* var. *paradisiaca* Plantain, known locally as 'apem' meaning 'a thousand', and that of a male *Carica papaya* Pawpaw, in water which is used to sprinkle the stone to sharpen the knife or cutlass used for slicing the cut surface of the palm trees (17). The sympathetic magic appears to be the Plantain - for this particular variety could have as many as a thousand or more fingers on a bunch. This black powder could also be sprinkled with incantations in the fire for distilling local gin - *akpeteshi* or *ogogoro* for a copious flow of the distillate. However, once the ritual is performed it cannot be reversed, so in instances where the distilling apparatus is not one's own, but hired or rented, no advantage is gained by performing the rites. In Senegal, the Tenda powder the debarked and dried root of *Ficus sur (capensis),* a fig tree usually with abundant fruits, and put this powder in the collecting vessels of palms when tapping to encourage increased flow of wine (20). This is likely an instance of sympathetic magic, since the tree bears abundance of fruits.

'AFRICAN ELECTRONICS'.

In traditional African folklore, and that of some other countries, there is a strong belief and fear that juju-men are supernaturally powerful and capable of invoking spiritual forces or powers or local deities to kill others. This evil way of eliminating people is commonly dubbed 'African Electronics' or 'remote control', since the practice is more often effected from a distance with apparently no visible connection between the practitioner and the poor victim. Revengeful individuals, therefore, openly threaten or secretly resort to bad juju, black magic, fetish, sorcery or some other local god or deity to eliminate an enemy or a rival. It is always a quick death, such as from an accident or snake-bite, or an illness that lasts one or two days only (21).

Thus stories of the sudden death of otherwise healthy persons under inexplicable and mysterious circumstances are sometimes told, and

allegedly attributed to juju. While death by juju is sudden, death through witchcraft is slow (18). 'If a man becomes suddenly and acutely ill he may be sure that he is a victim of sorcery not witchcraft. The effects of witchcraft lead to death by slow stages, for it is only when a witch has eaten all the soul of a vital organ that death ensues. This takes time, because he makes frequent visits over a long period and consumes only a little of the soul of the organ on each visit, or, if he removes a large portion, he hides it in the thatch of his hut or in a hole in the tree and eats it bit by bit. A slow wasting disease is the type of sickness caused by witchcraft' (18). The reader is reminded of one of the reasons (with possible repercussions) why the 'prescriptions' below should not be blindly copied to settle old scores (see INTRODUCTION: THE TEXT). Furthermore, instances of bad juju bouncing back to kill the originator are fairly open secrets. Though the end result is death, various methods are employed by traditional medicine-men to achieve this end. These methods may be classified broadly into two main groups. The spiritual gun or *Tukpui* or adaptations of this, and sinister rituals with or without parts of the victim's body as vital ingredients.

Spiritual Gun or Tukpui.

This 'gun' is believed to spiritually fire assorted objects at a victim from a distance. It is the traditional belief that unless these objects are successfully removed from the body, the victim is supposed to die. One such juju to kill an enemy is the Fangara charm used by the Mende of Sierra Leone. 'Different (unspecified) herbs soaked over a fire and powdered with a dead man's bone, anthill earth and charcoal as ingredients is parcelled. A small animal is tied onto it and placed on a frame of bamboo sticks close to a path. As the victim passes, a dry stick is broken which causes him to turn and look on the charm; at which moment the murderer hits the animal calling his victim's name, and the charm is on him' (29b). A miniature adaptation of this juju (locally called *Tukpui*, in Ewe, meaning spiritual gun) is practiced by the Ewe-speaking tribe of Ghana, Togo and Benin Republic, and some other tribal

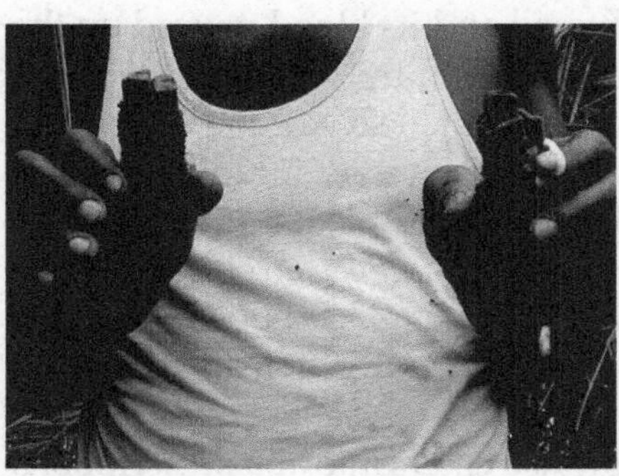

Tukpui or Spiritual gun

groups in the Region and elsewhere. The 'gun' is either pointed at the victim by the assailant, or it is hidden in the pocket while the index finger is used as a pointer (3). It is feared that the showy black and red seeds of *Abrus precatorius* Prayer beads are used as bullets in spiritual gun (2). The tribe uses *Kalanchoe integra* var. *crenata* Never die, a thick fleshy perennial, as one of the important ingredients in prescriptions by medicine-men to neutralize the effects of *Tupkui*, and to extract any objects said to have been spiritually shot into the body (35). In S Africa, the Luvale sorcerer, when plotting to kill a person, makes an effigy of the intended victim and inserts seeds of Prayer beads in place of each ear (46).

The extraction of these objects by the Azande of S Sudan is described. 'In treating a leech generally makes a poultice from the bast of the *kpoyo* tree *(Grewia mollis)* or from twisted *bingha* grass (*Imperata cylindrica* Lalang grass) and places this on the affected part, where he has previously made one or two incisions with his knife. He then massages the part with his hands and eventually withdraws the poultice and searches in it - with invariable success - for a bone or a piece of charcoal or some such object, which he shows to the sick man's family. The object is called *hu mangu,* thing of witchcraft, and it is believed that a witch has shot it into the sick man and that it is the cause of his sickness' (18). In W Africa, the *sabe* or *oten* fetish of the Ashantis of Ghana in which needles are stuck into the fruit of *Citrus aurantiifolia* Lime in an incantation performed either at midday or at sunset to kill someone is reported as sympathetic magic (39).

Sinister Rituals

Among some of the ethnic groups in the Region, the leaves of *Manihot esculenta* Cassava macerated in water to wash the head at midnight is part of a ritual ceremony to kill an enemy or a rival. A goat, *Capra* species, is then slaughtered and its blood poured on the leaf residue. The meat is cooked and eaten by all present without breaking a single bone. The bones with the leaf residue are buried with affirmations in a ceremony accompanied by incantations and libation. The victim is supposed to swell up, bloat and die shortly after the ceremony (35). It appears that *Sida linifolia,* an erect common malvaceous weed popularly believed to be fetish among the Ewe-speaking tribe, and called 'woduowogbugbo', meaning 'they planned but retreated' is a double-edged sword. As opposed to its normal use against evil and witchcraft, the plant also enters into juju prescriptions

to kill someone among the tribal group (40). A previously selected plant is visited at night with the same ingredients required for a love charm (see LOVE: CHARM PRESCRIPTIONS). However, instead of lavender, a knife is driven through two of the leaves, simultaneously pronouncing the name of the victim in a sinister incantation (40). The victim is said to collapse and die a few hours after the ceremony.

Throughout its distribution area, it is the belief that should *Spiropetalum heterophyllum,* a lofty undulating forest climber with blood-red exudate, be slashed through with a cutlass just at the moment a victim's name is mentioned in an incantation with libation and a spell, the one will surely die (34). Traditionally, it is a taboo to cut this liane - locally called 'homakyem' in Akan, meaning mamba-like vine (see TABOO: FELLING TREES). Similarly, it is the general belief among many of the tribal groups in the Guinea-savanna parts of the Region that anyone whose name is mentioned to *Piliostigma thonningii,* a savanna tree legume or shrub, in a sinister incantation at midnight is supposed to die (35). In addition, the gum of the tree is reported to have magical uses (25). Sadly, contestants in *akpeteshi* or *ogogoro* drinking competition are believed to use juju not only to enhance their performance but also to kill the opponent. *Euphorbia hirta* Australian asthma herb, collected and chewed with rites, is invoked to transfer any quantity of alcohol consumed direct into the opponent's stomach. The resultant overdose kills the victim.

In Mexico, the Huichol Indians invoke *Solandra* species to kill someone (30). 'A sorcerer wishing to do someone harm, goes alone to one of the god-plants with his sorcerer's paraphernalia. He offers the special sorcerer's arrows and sings all night in front of the plant. In the morning he leaves his arrows and, depending upon the power of the sorcerer's magic, his victim is supposed to take from just a few days to several weeks to die' (30). In S Africa, a species of *Stapelia,* either *S. gigantea* Carrion flower or *S. nobilis,* perennials with swollen stems, enter into rites of the Zulu sorcerer, being used in conjunction with earth from a grave to prepare a medicine capable of causing another death in the same kraal (46); and in Botswana, the Sotho use *Urginea capitata,* a lily, to send illness and death to one's enemies (46).

In S Africa, the root of *Terminalia sericea* Assegai wood is said to be poisonous and so it is used in Zululand by sorcerers who symbolically administer the poison by boiling the root in a potsherd, dipping the fingers in the mixture and smearing the fluid on the assegais (spear) which are then thrown in the direction of a person it is intended to bewitch. When the assegais (spear) are thrown, the operator does not look towards the intended victim, who, though he may be miles away, dies of wounds and coughing. *Combretum salicifolium* Bush willow and *Adenia gummifera*, a climbing shrub, are also Zulu charms for harming an enemy (46). In Zambia and Zimbabwe, the root of an *Asclepias* species resembling *A. decipiens* or *A. fruticosa* Milkweed has a charm use among the Manyika - the powder is thrown on the door of the hut of a person to whom harm is intended (46).

In W Africa, *Senna (Cassia) hirsuta* Hairy cassia, a herbaceous legume, collected with rites and ground between two stones (one big, the other small) in a ceremony at crossroads calling the name of the victim, is one of the prescriptions employed by juju-men either to kill someone or to inflict the person with disease. If the plant is thoroughly ground, the victim is supposed to die, but if it is only coarsely ground, then the victim is attacked with protracted sickness (35). Similarly, *Portulaca quadrifida* Ten o'clock plant enters into juju to kill someone. The whole plant, collected with ceremony, is ground, mentioning the victim's name, and taken in an alcoholic drink with the invocation, 'I want to swallow him, but he should not choke me' (35). *Datura innoxia,* a herb with hallucinogenic properties (and probably also *Brugmansia (Datura) suaveolens* Angel's trumpet or Moon flower) is used in juju prescriptions to kill an enemy. The leaves, picked with the necessary rites, are ground and rolled into seven balls which the spirit of the victim is invited, in an incantation, to eat. He is supposed to die a few days afterwards (35).

Among many cultures in Africa, parts of a person's body such as nail clippings or a few strands of the hair or even the clothes one uses are believed to be vital parts required in juju prescriptions against the individual. As a general rule, therefore, such parts are properly disposed of as a precautional measure, and to thwart the evil intentions of enemies. This belief and fear is reported in other cultures. The use of a person's private castoffs - such as hair, nails, a rag of his garment, or drops of his blood - for magic directed against him was also an American practice (42). 'The magician kneaded these relics into a lump of wax, which he molded and dressed in the likeness of the intended victim, who was then at the mercy of his tormentor. If the image was exposed to fire, the person whom it represented would immediately fall into a burning fever. If it was stabbed with a knife, the victim would feel the pain of the wound' (42).

In S Nigeria, the Yoruba invoke *Cymbopogon citratus* Lemon grass in a magical incantation to kill an enemy (42). Another method of eliminating an enemy in N Nigeria is recorded. 'If in a mixture

of (malam's) ink and water there be soaked - with the appropriate words, of course - a piece of wood taken from a tree which has been struck by lightning, a very powerful potion is produced. If a person washes his own body with this, his enemy (not he himself) will die' (41). This is likely an instance of sympathetic magic hinged on the tree that is struck by lightning. It has been observed from N Nigeria: 'the Hausa is not the only one who kills with a written charm. Only last year I heard of an English society lady who had hidden a paper in a drawer for sometime with a wish written upon it, in order to cause an injury to someone who had offended her, and she quite believed that it would act' (41).

In the northern Peruvian Andes, *Coleus blumei* Cimorilla, a labiate; *Iresine herbstii* Cimora senorita, an amaranth; *Acalypha macrostachya* Cimora leon, an euphorb; and *Sanchezia* species Cimora, an acanth are among plants that promote 'suchaduras'. They suck the vital spirit from the enemy, causing psychological and physical disorders till death occurs (14). Another species of *Coleus* Cimorilla dominatora is used to kill enemies. It is mixed with *Ceiba pentandra* Silk cotton tree, *Capsicum annuum* Pepper, salt, alcohol and holy water; and the solution is sprayed from the mouth naming the designated victim (14).

The majority of medicine-men, juju-men, fetish priests and fetish priestesses do practice openly, and are normally known to use their spiritual powers for the benefit of humanity. These benefits include protection from evil influences, witchcraft, bad juju and black magic. Others are casting away spells, and diagnosing the causes of disease as a prelude to healing these ailments. However, there are also known to be a few of these practitioners who do employ these powers - from their hideouts - for evil and destructive purposes like causing illness or even death. Spiritual power ones used for evil and destructive purposes - such as killing a fellow human being - cannot be used for any good purposes like healing again.

MEDICO-MAGICAL TREATMENT

Some of the traditional medicinal practices are based on magic or juju. The manner in which the plants are collected, or the form in which the prescription is prepared and administered suggests this method of treatment. The collection of the medicinal plants is preceded by certain rites and strict taboos. Similarly, the preparation is accompanied by definite incantation and a spell - words of direction uttered to medicine

linking them with desired ends. When a juju-man cures an ailment with a plant medicine, the common question often asked is whether it is the juju at work or the plant is responsible for the cure or both of them (18). It must be appreciated however, that the optimum conditions for the most effective influence of the plant medicine on the disease are achieved through the traditional rituals. The efficacy of magic lies in the medicine and the rite, and not any power outside these (18).

In Liberia, *Bersama abyssinica* ssp. *paullinioides*, a tree in lowland and montane forest, is locally called 'je-ra-kpar', or bone that is broken; and this refers to the magical use of the bark and wood as a dressing applied to the leg of a fowl, *Gallus domestica*, that is deliberately broken as a vicarious treatment for a man's broken limb (Cooper fide 13) (see MEDICINE: SYMPATHETIC MAGIC). In Senegal, the nomadic Fula consider *Nothosaerva brachiata*, an annual herb, a magical plant and use it in various medico-magical formulae for both stock and man (27a, b). In this country, the root and leaf of *Ximenia americana* Wild lime, a thorny savanna shrub, are ascribed by Wolof medicine-men to have medico-magical properties. Plants whose roots in the mysteries of magic become passive during winter will pass their potentials to the shrub which remain active during the season. As such Wolof medicine-men refer to the shrub as 'mother of roots' (27b). The fruits and flowers of *Balanites aegyptiaca* Desert date or Soap berry are used by the tribal groups in certain magical prescriptions. For example, the flowers and young leaves are boiled, and next morning at day-break they are mixed with *Parkia biglobosa (clappertoniana)* West African locust bean, locally called 'daudawa' in Hausa; salt and pepper are added and eaten ceremoniously (13). The ailment for which the prescription is prepared is not mentioned.

Ximenia americana Wild lime.

In SE Nigeria, *Pachyelasma tessmannii*, a forest tree legume, is full of strong juju; so in the Oban area a fruit-pod sticking upright in the ground is thought to provide very strong medicine (8c). In Tanzania and in parts of Ivory Coast, the root of *Pergularia daemia*, a twining herbaceous asclepiad with milky latex, has unspecified magical use; - thus in the latter country, it is often left to grow freely round fetish shelters by Guéré villagers, where it has more magical than medicinal application

in treatment of various illnesses (Koritschoner fide 8a). In Liberia, the plants used for magical or medico-magical purposes include the bark of *Isolona cooperi*, a small tree (11). The bark and leaves of *Scottellia coriacea*, a tall forest tree, are also reported to have superstitious uses in this country in ceremonies to revive a fainting juju, charm or medicine (Cooper fide 13). (For other instances of medico-magical treatment of diseases, see MEDICINE: MEDICO-MAGICAL TREATMENT).

BIOLUMINESCENCE

In the folklore of many countries, bioluminescence is either associated with magic, or with evil or ill-fate or with witchcraft. This is the ability of an organism to produce visible light in the dark. The phenomenon is widespread among bacteria, plants and animals such as the firefly, species of *Lampyris* or *Photinus*. Many fungi in the Basidiomycetes are known to luminesce - the fruiting body, the non-fruiting mycelium or both may be luminous, depending on the species. In the tropics, the species with luminous fruiting-bodies are more common than in the temperate regions. The bracket fungi, *Fomes lignosus* and *F. noxius;* and the cream-coloured mushroom, *Armillaria mellea* belong to this group of fungi - the latter possessing luminous mycelium. The glow of these fungi from rotten roots seems to have given rise to belief in magical power of trees through the equation of light and power, and the effect of these inexplicable lights on superstitious minds must have been profound (10).

It has been recorded that the phenomenon was witnessed while descending the great peak of Mt. Cameroon. 'There is a very peculiar look on the rotten wood on the ground round here. Tonight it has patches and flecks of iridescence like one sees on herrings or mackerel that have kept too long. The appearance of this strange eerie light in among the bush is very weird and charming. I have seen it before in dark forests at night, but never so much of it' (29a). Another instance of a tree with such property has been recorded. 'The *Apa*, frequently called the African mahogany, is inhabited by an evil spirit, and is commonly seen encircled with palm-leaves, and with earthen pot at its foot to receive the offerings of woodcutters. It is believed to emit a phosphorescent light by night' (16). In addition to the roots, some flowers are reported to glow at night. For instance, it is said that hunters of wild gingseng do their searching at night, because the flower of the plant is claimed to emit a phosphorescent glow (38).

CONSERVATION

The fear and respect of the people for trees professed to have supernatural powers and the subsequent associated taboos, eventually tends to preserve these species. It has been observed that 'it is not surprising that some degree of conservation has been exercised as a result of the thorough exploitation of *Balanites aegyptiaca* Desert date or Soap berry through attributing to it magical and ceremonial properties' (22). It has also been observed that 'in some cases these most primitive peoples have developed some concept of protecting wild plants by means of a magical or religious sanctions' (4) . For instance, *Okoubaka aubrevillei*, the rare forest tree traditionally said to be fetish; *Spiropetalum heterophyllum*, the forest liane with blood-red exudate also said to be fetish; and all sacred trees or plants either forbidden to be used as fuel-wood or even cut are, in effect, all traditional methods of conserving biodiversity for sustained yield. (See SACRED PLANTS. For trees forbidden to be cut or used for fuel-wood, see TABOO: FELLING TREES, and FUEL-WOOD. & APPENDIX - Table V).

For similar reasons, due to the great fear among Huichol Indians of Mexico that non-natives might seek and destroy the god-plant, *Solandra* species, or steal its secrets, the knowledge about these plants is usually a closely guarded secret, and many individuals will go so far as to deny any knowledge of their very existence. This fear is well founded, as, in the past centuries, over-zealous Catholic priests, from whom today the existence of these plants is especially well concealed, probably destroyed many of the plants in their unsuccessful effort to stamp out idolatry in the region (30). The roots of *Solandra* species must be tricked by singing into the open, then cut (30) - probably for protection and conservation purposes. In Europe, *Mandragora officinarum* Mandrake, a northern temperate plant, provided the most important anaesthetic drug known to the Ancients; as such the herb-gatherers probably perpetuated the story that it could only be uprooted at midnight and at great risk, in order to protect the stock from excessive exploitation by the uninitiated (19). A similar precaution is observed by diggers of the roots of *Paeonia officinalis* Common paeony, a native of southern Europe (37).

These observations are, in fact, quite true, important and of scientific interest in respect of fetish or sacred groves as well. They are most often perfectly intact relics of virtually virgin forests, in sharp contrast to the endless stretches of over-exploited secondary vegetation or abandoned

and depleted farmlands next door. Even in the savanna woodland, where the destructive effects of bush fire is ever imminent, especially during the harmattan season - from November to February - fetish groves are protected from fire. In Cameroon sacred groves are the only forested areas that remain (31).

The domestication of food plants is also initially attributed to medico-religious protection. This concept of virtually protecting a wild food plant has been... the first step towards the establishment of agricultural systems based on vegetatively propagated food plants such as yams and aroids (4). (See also FETISH; OCCULT; and WITCHCRAFT).

References.

1. Adegbola, 1983; 2. Amato, (pers. comm.); 3. Avumatsodo, (pers. comm.); 4. Ayensu & Coursey, 1972; 5. Back, 1987; 6. Bouquet, 1969; 7. Bouquet & Debray, 1974; 8a. Burkill, 1985; b. idem, 1994; c. idem, 1995; d. idem, 1997; e. idem, 2000; 9. Busson, 1965; 10. Cooke, 1977; 11. Cooper & Record, 1931; 12. Crow, 1969; 13. Dalziel, 1937; 14. De Feo, 1992; 15. Dobkin de Rios, 1997; 16. Ellis, 1894; 17. Eshun, (pers. comm.); 18. Evans-Pritchard, 1937; 19. Everard & Morley, 1970; 20. Ferry & al, 1974; 21. Gilbert, 1989; 22. Gledhill, 1972; 23. Graf, 1978; 24. Hives, 1930; 25. Irvine, 1961; 26. Jackson, 1973; 27a. Kerharo & Adam, 1964; b. idem, 1974; 28. Kerharo & Bouquet, 1950; 29a. Kingsley, 1897; b. idem, 1901; 30. Knab, 1977; 31. Koagne, 1986; 32. Kulefianu, (pers. comm.); 33. Lanzara & Pizzetti, 1977; 34. Lordzisode, (pers. comm.); 35. Mensah, (pers. comm.); 36. Moldenke & Moldenke, 1983; 37. Perry & Greenwood, 1973; 38. Pitkanen & Prevost, 1970; 39. Rattray, 1927; 40. Sofar, (pers. comm.); 41. Tremearne, 1913; 42. Van Sertima, 1976; 43. Verger, 1967; 44. Voorhoeve, 1965; 45. Walker & Sillans, 1961; 46. Watt & Breyer- Brandwijk, 1962; 47. Williams & Farias, 1972

27 KINGSHIP

And he hath on his vesture and his thigh a name written,
KING OF KINGS, AND LORD OF LORDS.
Revelations 19:16

Chieftaincy, as an institution, is the highest hierarchy of power in traditional Africa, and some other cultures of the world. This traditional figure symbolizes unity, culture, strength, history, wealth and all that contribute to the identity of a people. A chief is a historic figure, and is therefore of special importance. The throne as a secular symbol of power, legitimizes the king's role as political leader (7). The king is not only the political leader but also the supreme servant of the ancestral stool (7). Since ancestral stools are at the same time insights of sacral kingdom, the king also becomes the spiritual and religious leader of his people (7). 'He was responsible for carrying out certain ceremonies and for employing magicians in order that the nation might have adequate rains, good crop, freedom from epidemics, and victory in war. For these purposes he also - and he alone - could approach his ancestral spirits who are believed to be partly responsible for the peace and the prosperity of the nation' (9). 'If there is drought, or an epidemic among men and cattle, crop-failure or locust swarm, the monarch is held responsible for failing to exhibit the virtues appropriate to his office' (9).

Some plants are used in chieftaincy rituals, or are named after the institution, or are connected with the selection of a successor to the throne. Some other plants are believed to have identical virtues or qualities of a chief.

CEREMONIAL RITUALS

Traditionally, some plants are associated with the ceremonial rituals in connection with the institution. In both Yorubaland and Hausaland for instance, the leaf of *Newbouldia laevis,* a medium-sized tree commonly associated with fetish, is held in regard, and called chieftaincy leaf. A leaf is placed on the head of a new chief as a symbol of his authority. And

therefore cutting the tree with an axe or burning it as fuel is avoided (4). (For other trees that may not be cut or burnt as fuel-wood, see TABOO: FUEL-WOOD, and FELLING TREES; & APPENDIX - Table V).

In Ghana, *Momordica balsamina* Balsam apple or *M. charantia* African cucumber is worn by chiefs round the neck for spiritual protection (see EVIL: SPIRITUAL PROTECTION). In this country, *Sterculia tragacantha* African tragacanth is sometimes planted in front of a chief's palace as an alternative to the more usual *Ficus* species (20) - for spiritual protection or other superstitious reasons, or for the pseudo-whorled branches - like those of *Alstonia boonei* Pagoda tree. Similarly, *Baphia nitida* Camwood, a small tree legume, is often planted in front of a chief 's palace or an elder's house among the Nzima of Ghana and the Akan of Ghana and Ivory Coast for its traditional values. Among others, Camwood is traditionally associated with intelligence, and therefore wise counseling (see INTELLIGENCE: SYMPATHETIC MAGIC). In Ashanti in Ghana, shady trees are associated with the concept of kingship and with the spiritual 'coolness' or peace of the town (8).

In S Africa, the Tlhaping regard *Acacia giraffae* Giraffe thorn as a mighty tree not to be used by common people. Only chiefs and other prominent men are permitted to cut the timber for cattle folds; and *Rhoicissus caneifolia* Wild grape is used by the Southern Sotho chiefs as a charm when establishing a new village; and in Natal, only chiefs are allowed to carry a knobkerrie made of the wood of *Rhamnus zeyheri* Red ebony (22). In Botswana, *R. prinoides* Dogwood is widely used as a protective charm to safeguard the courts of Sotho chiefs (22). In this country, the Sotho crush *Indigofera fastigiatus,* a legume, and mix it with other plants for use as a charm to give prestige to a chief, who himself uses *Psoralea polysticta*, another legume, to bathe his body in order to keep the prestige of his subjects. For the same reason, chiefs are vaccinated with *Urginea capitata*, a lily (22).

In W Africa, the flowers of *Lonchocarpus laxiflorus,* a savanna woodland tree legume with purple or lilac flowers, and flowers of other trees allied to it are called 'furen yan sarki', meaning princess flowers and enter into certain superstitious practices. One who hopes to be king has the flowers and leaves prepared with meat and butter, and partakes of the dish in solitude (4, 5). In the Region, superstitious ideas in connection with *Stereospermum kunthianum,* a savanna woodland tree; *S. acuminatissimum,* a tall forest tree often planted as an ornamental; and *Parkinsonia aculeata* Jerusalem thorn, a tree, are common. In N Nigeria, the title *'dan sarkin itatuwa'* or *'jiri dan sarkin itatuwa'*, meaning 'prince of trees' or 'son of the chief of trees' - a superstitious epithet, is sometimes given to

Stereospermum kunthianum (forbidden to be used as fuel-wood or cut)

them in Sokoto and Katsina (4). In Europe, *Iris florentina* Orris root, a temperate plant is always considered to be a noble herb, symbolizing power and majesty (1).

Throughout traditional Africa, tree euphorbias are planted to commemorate events (see TWINS: CEREMONIES: Birth). Often these trees persist long after the dwellings have fallen into decay. In S Africa, the Zulu chief, Dingaan, is said to have had his throne built beneath a particularly fine *Euphorbia ingens* specimen in Zululand, near the site where Piet Retief and his Boer followers were slaughtered on 6th February, 1838 (6). Other tree euphorbias are *E. candelabra,* E. d*awei, E. excelsa* and *E. quadrialata*. In this country, the African in the neighbourhood thought that the spirit of the long-deceased African Chief Magoeba dwelt in the swelling of the stem of *Borassus flabellifer* var. *aethiopum* Brab tree, a palm (22). In S Nigeria, there appears to be a similar belief among the Yoruba-speaking tribe. 'The souls of the dead are sometimes reborn in animals, and occasionally, though but rarely, in plants' (5).

In W Africa, other plants associated with the institution of chieftaincy include *Ficus platyphylla* Gutta percha tree for settlement of disputes (see COURT CASES AND FAVOUR: CHARM PRESCRIPTIONS); *Costus* species, ginger-like herbs in forest, among the Nzima of Ghana and the Akan of Ghana and Ivory Coast - for protecting the stool regalia; and the fruit of *Citrus aurantiifolia* Lime for spiritual protection when a chief is carried in a palanquin (see EVIL: SPIRITUAL PROTECTION). In a cosy sort of demonstration of kingship in N Nigeria, the Minta people of the Pabir call *Afzelia africana* African teak, a savanna tree, *minta* after themselves, or vice versa. Some other associated plants are *Senecio (Crassocephalum) biafrae*, a climbing forest composite, (see CEREMONIAL OCCASIONS: PURIFICATION AND SACRIFICE: Purification); and *Portulaca oleracea* Purslane or Pigweed (see MEDICINE: PURIFICATION). The two latter plants are both for purification rituals as well.

ASSOCIATED PLANT NAMES

In African folklore and that of other countries, some plants are believed to be royal, and are accordingly named. There are many possible factors that could influence this belief. These include the size or posture of the plant, or the colour or fragrance of the flowers, or the economic or social importance of the plant, or some other qualities or values of the plant that may be likened to a king. In N Nigeria for example, *Adenia venenata,* a sub-arborescent climber, is known among the Chamba as 'gantimi', meaning 'prince's tree' (4); and the Hausa name of *Leonotis nepetifolia* var. *nepetifolia,* and var. *africana,* robust herbaceous labiates often cultivated, 'tutar 'yan sarki', means 'prince's flag' or 'state umbrella' (4) - likely in reference to the whorled or verticillate arrangement of the inflorescence. In S Nigeria, *Rungia grandis,* a herbaceous acanth with purple-veined flowers in conspicuous spikes, is known as 'adade oko' in Yoruba, meaning 'prince of the field' (4).

In Ghana, *Elaeis guineensis* var. *idolatrica* King oil palm is called 'abehene' and 'de-fia' in Twi and Ewe or Krepi respectively, meaning 'chief palm'; and *Bombax buonopozense* Red-flowered silk cotton tree is 'ahenegeen tso' in Ga, meaning 'white dried sticks for chiefs' (4, 13). In N Nigeria, the Hausa name for *Saccharum officinarum* Sugar cane is 'karan sarki', and means 'cane of chief' (4). In S Nigeria, the Yoruba name of *Cymbopogon citratus* Lemon grass and *C. giganteus,* tall perennial grasses, 'koko oba', means 'king of grasses' (4); and *Ipomoea asarifolia,* a plant with large flowers, 'gboro ayaba', meaning 'long and stately queen' (4). In this country, the Yoruba name of *Luffa cylindrica* Loofah or Vegetable sponge, 'kanrinkan-ayaba', means 'queen's sponge' or 'king's sponge'; and that of *Cocos nucifera* Coconut palm, 'unoba', means 'nut of the chief' (23). In Ghana and Ivory Coast, the sponge from *Momordica angustisepala,* a cucurbit, is known as 'ahensaw' in Akan - also meaning 'queen's sponge' or 'king's sponge'.

Lagerstroemia speciosa (flos-reginae) Queen crape-myrtle

The species name of *Delonix regia* Flamboyante or Flame tree, *Lagerstroemia speciosa (flos-reginae)* Queen crape-myrtle with beautiful blooms, and *Strelitzia reginae* 'Bird of paradise', a remarkably beautiful plant from Transkei

Republic in S Africa, all have references to kingship. The generic name of *Victoria amazonica,* the largest and most remarkable water-lily known, commemorates Queen Victoria (6). *Protea cynaroides* has the largest flower heads and goes under the appellation of Giant or King protea; and *Selenicereus grandiflorus* Queen of the night, a straggling Mexican climbing cactus has great trumpet-shaped blooms which expand at dusk with a rich vanilla fragrance (16). Two other night-blooming cactus plants - *Epiphyllum oxypetalum* from Brazil, and *Nyctocereus serpentinus* from Mexico and Nicaragua, are known as Queen of the night.

The list includes *Fritillaria imperialis* Crown imperial a decorative lily of the Middle East; and *Cypripedium reginae* a beautiful N American orchid with large white-petalled flowers (18). Of the many descriptions of *Disa uniflora,* an orchid from S Africa, 'Queen of terrestrial orchids' seems an appropriate accolade (18). *Meconopsis regia* is a Poppy from Napal; *Sinningia reginae* from Brazil with trumpet-shaped nodding violet flowers; and *Juglans regia* Walnut, a native of W Asia is useful for its nuts (16). *Sauromatum venosum (guttatum)* Monarch of the East or Voodoo lily is an aroid from India and parts of Africa (16). (See DEATH AND DYING: COMMUNICATING WITH AND OFFERINGS TO THE

Roystonia regia Royal palm (National Plant of Cuba and Haiti).

DEAD: Communicating; and also GRAVES AND CEMETERIES: GRAVE MARKER). *Paeonia officinalis* Common paeony, a native of southern Europe is often designated King of flowers alongside a Phoenix - King of birds (16). The poet Sappho called *Rosa damascena* Damask rose the Queen of flowers. *Roystonia (Ereodoxa) regia* Royal palm, an introduction from Cuba and the National Plant of this country, is widely cultivated in the Region. (For a list of other National, State & Emblematic Plants, see APPENDIX - Table IV). Palms are generally known as 'Princess of the vegetable kingdom' because of the diversity of ways in which the plants and their products are useful to man. Indeed there is hardly any part of *Elaeis guineensis* Oil palm that man does not find some use for in a way or the other. Even the dead tree trunk serves as a medium in which edible mushrooms grow. *Triticum aestivum* Wheat, the world's most important food, is the Queen of cereals; and *Zea mays* Indian corn or Maize is the King of crops (14). The shape of the persistent sepals of *Punica granatum* Pomegranate, an introduction to the Region and intimately involved with the life and myths of the Mediterranean world, is believed to have inspired that of King Solomon's Crown (6). In Japan, *Lespedeza bicolor* Bush clover, a tree legume, the only plant in the courtyard of the Seiryo-den of the Kyoto Gosho Palace is a supreme example of harmony and reverence for nature within the exuberance of court life (17)

THE QUESTION OF SUCCESSION

The choice of a successor to the crown, in the event of the death of a king, is made by the king-makers - based on certain traditional guidelines and regulations. These vary from region to region. A candidate must, however, necessarily first be in the line of succession, and secondly be both physically and mentally sound - that is not deformed in anyway. Both parents must be natives of the traditional area; and left-handed candidates are automatically disqualified. Some traditional areas even disqualify circumcised candidates. With civilization, literacy is an added advantage. The selection could be simple and automatic - especially where there is only one line of succession; or it could be complicated where there is more than one line of succession - as is often the case. The selection of the wrong person, for one reason or another, or the deliberate manipulation of the selection committee to impose a person (who is not in the line of succession) just because of his social influence, position, power or wealth, is among the causes of some of the many chieftaincy disputes.

An unusual selection in Karagwe (partly in present Ruanda Burundi and partly in N Tanzania) is recorded: 'When Dagara died, and Rumanika, Nuanaji and Rogero were the only three sons left in the line of succession, a small mystic drum of diminutive size was placed before them by the officers of the state. It was only a feather weight in reality but, being loaded with charms, became so heavy to those who were not entitled to the crown, that no one could lift it but the one person whom the spirits were inclined towards as the rightful successor. Now, of all the three brothers, Rumanika alone could raise it from the ground; and while his brothers laboured hard, in vain attempting to move it, he with his little finger held it up' (15).

It is on record that: 'Among certain people subject to Argungu (to the north-west of Zungeru) the new chief was chosen as follows: The bull, *Bos* species, was killed as soon as the old chief was dead, and the corpse was wrapped in it, and then placed on a bed, and carried out into the open. The dead chief's relatives were then made to stand in a circle round the body, and the elders of the town spoke thus: 'O Corpse, show us who is to be chief, that we may live in peace, and that the crops may do well'. The bearers then took the body round the ring, and it would cause them to bump against the man it wished to succeed. It was then buried seven days afterwards, and the new chief was installed amidst rejoicing' (21). On other occasions, circumstances surrounding the birth of a chief could equally be influenced by the deity. It is recorded from the then Gold Coast (now Ghana) (12) that the most famous occasion on which Otutu, the god of Berekuso was successfully consulted was when Manu, the only sister of Obiri Yeboah, the Asantehene, came to beg Otutu for a child. The son she later bore (in 1697) was called Osei Tutu after the god of the shrine, Otutu, and he became founder of the powerful Ashanti state in present Ghana.

QUALITIES

In many African myths and stories, the tree is portrayed as an ancestral symbol of wisdom, authority and custom, providing a bond between the dead and the living (19). In other stories the tree often symbolizes a mediator and a judge (10). Similarly, according to Chinese mythology, the Bamboo symbolizes virtues as purity of mind, fidelity, humility, wisdom and gentleness; *Pinus* species Pine denote integrity and dignity; and the orchid stands for culture, refinement and nobility of character (11). In Japan, there are rows of Pines standing firm on the grass on the south side of the Imperial Palace, Tokyo. An appropriate place for the

king of trees (17). Among Christians, *Commiphora molmol* or *C. myrrha* Myrrh represents continence and *Nymphaea lotus* Water lily chastity (3). Both virtues are surely qualities required of a king or a queen. (See also FORTUNE TELLING AND DIVINATION).

References.

1. Back, 1987; 2. Baumer, 1975; 3. Crow, 1969; 4. Dalziel. 1937; 5. Ellis, 1894; 6. Everard & Morley, 1970; 7. Fink, 1989; 8. Gilbert, 1989; 9. Gluckman, 1963; 10. Gorog-Karady, 1970; 11. Graf, 1978; 12. Hall, 1978; 13. Irvine, 1961; 14. Moore, 1960; 15. Mountfield, 1976; 16. Perry & Greenwood, 1973; 17. Richard & Kaneko, 1988; 18. Stewart & Hennessy, 1981; 19. Studstill, 1970; 20. Taylor, 1960; 21. Tremearne, 1913; 22. Watt & Breyer-Brandwijk, 1962; 23. Zeven, 1964.

28 LACTATION

Thou shalt not seethe a kid in his mother's milk.
Exodus 23:19b

There are instances in traditional medicinal practice where some plants are used (sometimes with rites) to induce or increase milk production in both nursing mothers and in cattle. In most cases, the plants are either so named or do bear semblance to the female breast. Other plants - especially those in the milky latex families like Apocynaceae, Asclepiadaceae, Caricaceae, Euphorbiaceae, Moraceae, Periplocaceae and Sapotaceae; and a few others outside these plant families enter into similar prescriptions. These uses are often based on sympathetic magic and therefore, probably superstitious.

A few plant exudates are used as actual milk-substitutes.

SYMPATHETIC MAGIC

Some tribal groups in Africa use *Sarcostemma viminale,* a leafless twiner with milky latex, to increase lactation, both in women and in cows (15). This is probably an instance of sympathetic magic. The application of the latex of *Euphorbia convolvuloides,* a herbaceous plant in savanna and waste places, to breasts in the belief that it increases lactation (3b); and a rub of the leaves of *Adenia dinklagei,* a forest climber in the Passion flower family, on the breasts among the people of Akropong Akwapim in Ghana after childbirth to stimulate flow of milk (Herb GC fide 8a) - are similarly, both probably cases of sympathetic magic.

The latex of *Euphorbia balsamifera* Balsam spurge is given by Fula herdsmen to cattle to increase their milk-supply - a use probably mainly superstitious (8b); similarly, the use of the copious latex by women to promote the flow of milk is probably sympathetic magic. In Cameroon, the application of the latex of *Alstonia boonei* Pagoda tree, a tree with whorled leaves and branches and abundant white latex, by the Bakwiri women to increase lactation (4); and in Ivory Coast, the use of the latex

Carica papaya Pawpaw

of *Landolphia dulcis*, a climber with edible fruits and white latex, as lactogenic - are probably both cases of sympathetic magic. The latex from the crushed fruits of *Carica papaya* Pawpaw is massaged on the breasts, or the breasts exposed to boiling fruits to induce the flow of milk (8b) - probably an instance of sympathetic magic. Sometimes the fruit decoction is taken in cold doses during the day. Similarly, the use of the young leaves of *Ficus sur (capensis)*, a small tree with white latex and abundant cluster of figs borne on the trunk, in palm soup by nursing mothers in Ghana to promote a copious flow of milk is probably an example of sympathetic magic (8b).

In S Africa, the Zulu give an infusion of the leaf and bark of this fig tree to a cow with insufficient yield of milk, as a matter of mimetic magic; however, in Botswana, *Ornithogalum ecklonii*, a lily, is used by the Sotho as a charm by means of which an enemy can cause a milch cow to go dry or one's cow to miscarry (15). In Tanzania, the expissation of the gum of *Harungana madagascariensis* Dragon's blood tree, a shrub or small tree, is likened to the flow of milk; as such the bark or root is used in a treatment to stimulate breast-development (15). In Congo (Brazzaville), the sap of *Tetracera alnifolia* and *T. potatoria* Sierra Leone water tree, a scandent shrub or liane in forest, is given to lactating mothers as a galactogogue (1) - probably an instance of sympathetic magic; and in Ivory Coast, 'Tiger-nut milk', a preparation from the tubers of *Cyperus esculentus* Tiger nut is considered in the Kong area, perhaps on the Doctrine of Signatures, to be lactogenic (2). In S Sudan, the ripe fruits of an unidentified plant locally called 'vuruma' in Azande, round and velvety, and full of milky sap, bears a resemblance to the breasts of a woman who has just given birth to a child. The root of this plant is given to a mother in an infusion if she is unable to supply her child with sufficient nourishment (6).

A legend says that the white stains on the leaves of *Silybum marianum* Blessed Mary thistle, a dark-leaved shrubby composite, are drops of milk fallen from the breast of the Virgin Mary when she hid her Son from Herod's persecution (12). On this basis, Middle Age physicians recommended this plant for increasing milk production on breast-feeding women (12).

RITUAL INDUCEMENT

Sometimes a 'formal request' has to be made to the plant and 'approved' before any part may be removed to induce milk production. Very often the collection of plants for other medicinal purposes may also be preceded by a similar ritual, prayer, request or libation.

In its distribution area, *Milicia (Chlorophora) excelsa* or *M. regia* Iroko, timber trees popularly believed to be fetish, may be the centre of rituals to induce breast-milk in a foster mother (on the death of a nursing mother). The tree is approached with an egg. As the reason for the call is being recited, the egg is thrown at the trunk. Should the two halves of the egg shell fall down showing the inside, and in addition latex oozes out of the bark when the tree is slashed, it is a sure sign of acceptance. In which case a bit of this bark may be cut off to prepare the necessary decoction to be taken by the foster mother. On the contrary, should the egg stick to the tree trunk, while at the same time the bark does not exude any latex when slashed, it is unfortunately a sign that the request has been rejected (14). Normally, nursing mothers in the Region drink the palm-wine tapped from *Raphia hookeri* Wine palm in the belief that it induces lactation (8b).

A similar ritual ceremony as above is performed before a sample of *Spiropetalum heterophyllum,* locally called 'homakyem' in Akan, is cut for medicinal purposes. The plant is a lofty undulating forest liane traditionally believed to be a fetish and therefore sacrilegious to cut. (See APPENDIX- Table V for full list of plants forbidden to be felled or used for fuel-wood). Other lianes covered by this ritual include *Salacia debilis, Dalbergiella welwitschii* West African blackwood and *Adenopodia scelerata,* all forest climbers with bright red exudate like 'homakyem' when slashed, and therefore believed to be fetish. This traditional practice of protecting some species from being cut or used for fuel-wood is in effect a means of conserving biodiversity for sustained yield. It is worthy of support and encouragement.

MILK-SUBSTITUTE

The raw or cooked stem of *Sarcostemma viminale,* a leafless shrubby latex-producing plant in dry savanna and Sahel areas, is eaten by the Somalis and much liked by pregnant women (Gillet fide 3a). It is reported to be given to both humans and cows, *Bos* species, as a galactogogue; and recorded being actually used as a milk-substitute in India (15). The

Singalese use the sap of *Gymnema lactiferum* Celanese cow plant as a substitute for milk; and *Oxystelma esculentum,* an asclepiad of the Old World tropics, also has drinkable milk (13).

ASSOCIATED PLANT NAMES

Some plant names have reference to breasts or milk production. For instance, in N Nigeria, *Glossonema boveanum* ssp. *nubicum,* a weedy herb of cultivation in dry areas, called 'tafo ka shamamarka' or 'mama' in Hausa, means a woman's breast (3a) - an epithet referring to the use of the plant as a galactogogue. In Ghana, Togo and Benin Republic, the local name of *Euphorbia hirta* Australian asthma herb in Ewe is 'notsigbe', meaning milk plant (4); and used by women to increase the flow of milk. This is likely an instance of sympathetic magic. To induce milk production faster, some of the herb is half cooked with maize for the nursing mother to eat. The concoction or the water used for the boiling is also drunk, and the remaining plant is then ground and applied to the breasts (11). Among the tribe, *Calotropis procera* Sodom apple, another latex-producing plant, enters into similar preparation and use (11).

In Liberia, the Basa name of *Massularia acuminata,* a shrub or small tree, 'neh-mle-chu', means 'maiden's breast tree' (4); and in Ghana, the Adangbe name of *Uvaria ovata,* a scandent shrub forming dense thickets in savanna, 'nyotfo', means 'breast tree', from the shape of the edible fruits (8b). In N Nigeria, *Heliotropium bacciferum,* a sub-erect or prostrate herb in dry places, is called 'ba-filatana' in Hausa, meaning a 'Fula woman' (4) - alluding to the galactogenic properties; and in Sierra Leone, *Carica papaya* Pawpaw is known as 'nyini-fikati' in Mende, meaning 'woman's breast' (3a), that is pear-shaped.

In Ghana and Ivory Coast, the local name of *Kigelia africana* Sausage tree in Akan is 'nufuten' meaning 'hanging breasts' (8b). In Ivory Coast, the fruit with pimento is made into a decoction which is taken in draughts as a galactogogue, while the breasts are anointed and massaged with ointment of the fruit pulp in *karité* butter (10). In S Africa, the Luvale rub a piece of the fruit over the breasts to increase milk-flow (15). In Senegal, the Diola name of *Secamone afzelii,* a climbing or

Secamone afzelii

scrambling shrub with milky latex in savanna and secondary forest, 'bubeben mil', meaning 'medicament in milk', indicates the use of the plant as a galactogogue (3a). For this purpose, the leaves are macerated and the macerate is drunk and rubbed on the breasts daily for a month (9a, b). A similar application is made in Ivory Coast by the Guéré and the Kru who drink a decoction of the whole plant and massage the lees onto the kidneys, loins and back, as well as onto the breasts (10). In Ghana, the whole plant is taken with roasted corn as a galactogogue by nursing mothers (5).

In the temperate zone, *Polygala serpyllifolia* Common milkwort is an European plant, and the common name dates from the time when the plant was reputed to induce the flow of milk in nursing mothers (7). *Polygala vulgaris* Milkwort suggests the virtues ascribed to it as an assistant to wet-nursing; and *Asclepias* species are known as Milkweeds since many contain a milky latex in the stem, roots and leaves (13). (See also CATTLE; and FERTILITY).

References.

1. Bouquet, 1969; 2. Bouquet & Debray, 1974; 3a. Burkill, 1985; b. idem, 1994; 4. Dalziel, 1937; 5. Eshun, (pers. comm.); 6. Evans-Pritchard, 1937; 7. Everard & Morley, 1970; 8a. Irvine, ex Herb. GC; b. idem, 1961; 9a. Kerharo & Adam, 1962; b. idem, 1974; 10. Kerharo & Bouquet, 1950; 11. Mensah, (pers. comm.); 12. Pamplona-Roger, 2001; 13. Perry & Greenwood, 1973; 14. Sereboe, (pers. comm.); 15. Watt & Breyer-Brandwijk, 1962.

29 LIGHTNING AND THUNDER

And the temple of God was opened in heaven...and there were lightnings and voices, and thunderings, and an earthquake, and a great hail.
Revelations 11:19

In traditional Africa and elsewhere, it is the belief that plants and other materials may be invoked by spiritually powerful individuals, medicine- or juju-men, fetish priests and the like either to discharge lightning to strike a predestined object at a predestined time, or to redirect lightning from its original course. Some of the people therefore view lightning more as a curse and a punishment for some evil or wrong doing, than a natural phenomenon.

Various preventive and protective measures are practiced. Some plants may be planted on, or around houses, at villages or at settlements, to serve as preventive of lightning or to protect the people from its effects; while other plants enter into prescriptions as a charm. These may be worn, burnt, placed in bath water or pointed towards the lightning. It is the belief that these plants serve as preventive of lightning and thunder. There are equally plants believed to ward off thunderstorms, or to invoke or prevent rainfall. Ritual ceremonies to 'neutralize' an object that has been struck by lightning to make it 'safe' again is also practiced.

PREVENTIVE PLANTS

As part of the tradition and culture of many tribal groups, certain specific plants, either by name, symbol, belief or some other attributes, are said to be preventive of lightning and accordingly planted to prevent lightning or to protect the people from its effects. For example among some W African peoples, *Ricinus communis* Castor oil plant is regarded as preventing injury by lightning (34). In Liberia, the Mano call it *thunder tree* (3b): and in Gabon some of its vernacular names mean *thunder tree* or *thunder trap,* the plant supposedly conducting the lightning flash safely to the ground (35). In the Region, *Jatropha gossypiifolia* Red physic nut (and probably also *J. multifida*) is often planted in villages and believed

to ward off lightning (8); or has protective values against lightning (33); or features in an incantation among the Yoruba to prevent someone from being beaten (33). In the N Province of Sierra Leone, *Crotalaria pallida (mucronata),* an erect legume under-shrub, is regarded as a protection against thunder (Glanville fide 8); and in Liberia, pieces of the bark of *Musanga cecropioides* Umbrella tree are planted over a doorway as a supposed preventive of, and protection against lightning (7).

The specific name of *Gardenia ternifolia* ssp. *jovis-tonantis,* a savanna shrub, (originally called *Decameria jovis-tonantis*) is due to the native superstition in Angola and neighbouring countries, that the plant is a protection against lightning. This is the general belief in W Africa as well. Like *Gardenia nitida,* a shrub or small tree in dry forest and sometimes planted in villages, branches are placed on the roofs of huts - a usage known also in the western Sudan (8) as well as Angola (14). For the same reason, leafy branches of the shrub may be worn as a talisman. In W Africa, *Lecaniodiscus cupanioides,* a shrub or small tree, is traditionally believed to have the same protective properties against lightning as *Gardenia* and is used as such. In Sierra Leone, the Temnes associate *Colocasia esculenta* Eddoes with another Araceae, *Caladium bicolor* Heart of Jesus, whose variegated and coloured leaves are regarded as a protective against thunder and lightning (8). In this country, an unidentified *Nopalea (Opuntia),* a shrub-like xerophytic plant, is planted by the Mendes and the Temnes to ward off thunder (Deighton fide 8). In the S Province of this country and Upper Cavally of Ivory Coast, a tuft of *Afrotrilepis (Catagyna) pilosa* Devil grass is planted on top of the huts to keep away lightning (Deighton fide 8). In northern Sierra Leone, a plant of this species is put on top of the main post of each house in the Mabonto-Bumbaa area - also presumably in a superstitious sense (3a).

In W Africa, *Dracaena arborea* and *D. fragrans,* palm-like trees often planted around dwellings and fetish places, locally called 'sunya' by the Ewe-speaking tribe, meaning 'against lightning', enters into juju to prevent lightning. Among the tribe, a cutting of *Dracaena* is usually planted with rituals to the east and the west of the compound over a live cat, *Felis domestica,* and a few grains of *Sorghum bicolor* Guinea corn in a ceremony to protect the household from lightning, and to redirect the evil to the originator (17). In N Togo, *Adenium obesum* Desert rose, an erect ornamental with milky latex, is held to have magical attributes of keeping away lightning - hence the Bimoba name 'saa tesaga', meaning 'tornado thunder' (3a). In W Africa, *Elaephorbia grandifolia,* a forest tree with caustic white latex, is widely planted as a fetish tree in Senegal (16), Sierra Leone (3b), Mali (8) and Ivory Coast (1, 16, 27) for protection

against thunderbolts. (The savanna species is *E. drupifera*).

In Botswana, the African thinks that *Ziziphus mucronata* Buffalo thorn has the property of warding off lightning and that a person taking refuge under one of these trees during a storm will not be harmed (36); and no one should cut down a plant after the first summer rain or drought will follow (3d).

In Mozambique, *Trichilia emetica* ssp. *suberosa (roka)*, a savanna tree, is used to ward off lightning by the Thonga; and in Zambia, the leaves of *Hymenocardia acida*, a savanna shrub or tree euphorb, are placed on the roof of a house by the Nyanja as a protection against lightning (36). In S Africa, *Plumbago capensis* Blue plumbago is used by the Xhosa and Zulu to ward off lightning (36); and in Lesotho, *Portulacca oleracea* Purslane or Pigweed is protection against illness and lightning (12). In Congo (Brazzaville), *Cleistopholis patens* Salt and oil tree is planted in compounds to ward off thunder-bolt (2); and in Tanzania, the Bahaye people have a superstitious use for *Peddiea fischeri*, a small tree of montane forest, as a lightning conductor. Poles 2m long and charred at one end are set outside huts (3e).

In Europe, *Sempervivum tectorum* Houseleek, a herbaceous decorative, is referred to by Greek and Roman writers and has long been grown as a charm to prevent lightning from striking the roof of dwellings. The Emperor Charlemagne, for example, directed it to be planted on the houses of his gardeners (11). In Japan, *Iris tectorum* Roof iris, a decorative garden plant with lilac flowers, was believed in older times to give protection from gales, and was planted on thatched roofs in country districts (25). In Mexico, *Targetes lucida* Marigold, a composite locally known as *yyahutli* or *hierba de nube* (cloud shrub) is described as stimulating venereal appetite and alleviating crazy people and those astonished and frightened by thunder (Hernandez fide 28).

CHARMS AGAINST LIGHTNING

In W Africa, the leaves of *Rourea (Byrsocarpus) coccinea*, a shrub or climber in secondary forest or savanna thickets, carried on one's body, are claimed by fetish priests, medicine-men and other practitioners as protection against lightning as one walks in the rain; and *Tradescantia* and *Zebrina* species Wandering Jew, perennial straggling decorative commelinaceous plants, placed in the mouth is believed by the people to be protective against lightning (21). *Uraria picta*, a small woody legume in savanna, is used as charms to resist the effects of lightning (21). Among the traditional uses of *Okoubaka aubrevillei*, a rare forest tree occurring mainly in Ivory Coast and Ghana, and forbidden to be felled as it is widely believed to be fetish, is a charm against lightning. The egg-like seed placed in bath water with the necessary rituals and incantations, protects one from the effects of lightning (6).

In Sierra Leone, the young leaves of *Albizia adianthifolia* West African albizia, a tree, have superstitious use in the making of a wash for children as protection against thunder and lightning (26). In this country, *Crotalaria pallida (mucronata)*, a savanna shrub, is full of strong juju. It is said to confer protection against thunder (3c); and in Senegal, the Socé think *Tamarix senegalensis*, a savanna shrub, is a fetish protection against lightning (15a). In S Nigeria, *Elaeis guineensis* var. *idolatrica* King oil-palm or Juju-palm is held to have power of protection. A portion of a leaf of Juju-palm placed over a child will protect that child from harm (8, 37a, b). Evil-doer may be required to swear innocence standing near a palm while holding a piece of its leaf in his hand. Should he commit perjury, it is expected that he may be struck down by a thunderbolt: hence the Igbo name 'nkwu-kamanu' meaning 'oil-palm of thunder' (8, 37a, b).

In S Africa, the Zulu in the Mfongosi district use *Athanasia acerosa*, a composite, to ward off lightning. For this purpose the plant is burnt on a piece of broken pot, either inside the hut or just above the kraal (36). In this country, the leaf and root of *Vernonia natalensis*, a composite, locally called 'ndhlamnhloshane' is a Zulu charm against lightning (36). It is added slowly to glowing charcoal in an earthenpot in such a way as to produce smoke without catching alight; and *Pentanisia prunelloides* Wild verbena is used by the Southern Sotho and the Xhosa as a lightning charm (36). The Zulu think that the strong-scented rhizome of *Kaempferia* species, a ginger-like plant locally called 'indungulu', is good for warding off lightning; or use *Crassula harveyi* or *C. rosalaris*, fleshy perennials, as a charm against lightning (36). Similarly, *Rhamnus prinoides* Dogwood is widely used as a protective charm, especially by the Xhosa and the Zulu of S Africa and the Sotho of Botswana as a protection against lightning (36).

In Botswana, the African smears a mixture of *Hypoxis multiceps*, a lily, and *Ipomoea oblonga*, a creeping convolvulus, on pegs placed in the ground around a kraal as a charm against lightning; or the dry bulb of *Nerine angustifolia*, a lily, mixed with parts of other plants, is used as a charm against lightning (36). In this country, the Southern Sotho burn *Striga elegans* Witchweed, a root parasite, during stormy weather in order to ward off lightning; or mix the leaf of *Turbina oblonga*, locally called 'mothokxo', with tobacco in making a snuff and use the plant as a charm to drive away lightning (36). The Southern Sotho also use

Scirpus burkei, a sedge, as an ingredient in a charm preparation against lightning (36).

In S Africa, the Ndebele insert a stick of *Salix woodii* Wild willow medicated with a mixture, in which the fat of a black goat, *Capra* species, is the main ingredient, to protect the hut from lightning, and a similar stick is planted in the direction of an oncoming storm (36). In this country, the Xhosa use medicated pointed sticks of *Acridocarpus natalitius,* a shrub, to ward off lightning, the sticks being stuck in the thatch with the points in the direction of the storm (36). Similarly, small sticks of *Capparis citrifolia* Cape caper are thrust into the thatch over the door of a hut to ward of lightning; and among the Bomvana, the leaf is placed on the fire in a hut to ward off lightning (36). Pegs of the wood of *Grewia flava* Raisin tree are used by the African as a protection against lightning (36). In this country, *Brachycorythis ovata* ssp. *schweinfurthii*, a terrestrial orchid, has use as a protective charm of general sort, and in particular against storms and lightning (19, 24).

In Botswana, the Southern Sotho use *Eriospermum* species, a lily, as a charm for protection against lightning; or burn, crush and boil *Aloe saponaria* Soap aloe, a lily, and sprinkle the mixture round the village as a charm against lightning (36). The tribe also use *Stephania umbellata,* locally called 'lesibo', as a magic medicine to prevent a person being struck by lightning which has been sent by an enemy (36). The Southern Sotho also use *Hibiscus malacospermus,* locally called 'boboyana', as a charm to prevent villages being struck by lightning; or use *Scilla lanceifolia* Wild squill, a lily, as a charm to drive away lightning (36). In this country, *Lebeckia sericea,* a legume, is an ingredient in the medicine-man's horn of medicine against thunder, lightning and storms (36). To the Southern Sotho, *Cyperus fastigiatus,* a sedge, is an ingredient in a charm preparation used against lightning and storms - the same preparation being also rubbed into scarifications on the first day of the initiation ceremony (36). In ancient times, Roman Emperors, athletes and victorious warriors wore wreaths of *Laurus nobilis* Laurel on their heads, which were supposed to protect them from lightning and evil forces (22). In its distribution area, *Sempervivum tectorum* Houseleek, forms a thick lawn around it, therefore it has been used to cover and give strength to earth roofs, as well as protect houses from thunderstorms (22).

INDUCIVE PLANTS

Other plants are traditionally believed to enter into rituals to induce lightning. This has been the secret of the lightning and thunder worshippers. For fear of possible misuse or abuse by the uninitiated, the identity of these plants is not normally revealed. The Vagla tribe of NW Ghana and the Lofa tribe of Liberia for instance, do traditionally swear and curse by lightning which is said to be sent like a 'Registered Mail' to strike the intended victim for sure with the precision of a Global Positioning System (GPS) if the warning is not heeded. The Ewe-speaking tribe of Ghana, Togo and Benin Republic, the Gas of Ghana and other tribal groups in the Region invoke *Vernonia cinerea* Little ironweed, a weedy annual composite, among its numerous attributes, in an incantation to induce or divert lightning (5). To induce lightning, a piece of broken earthen pot with the whole plant are tied together and whirled vigorously and continuously in a sinister incantation, mentioning the enemy's name repetitively - the sling being released suddenly to fling in the direction of the victim (5). In its distribution area, *Cassytha filiformis* and *Cuscuta australis* Dodder, parasitic epiphytic leafless twinning plants, are also believed to be similarly invoked, among other attributes, by medicine-men to induce lightning to destroy the house of an enemy (29). In S Africa, the Tembu believe that *Kiggelaria africana* Natal mahogany should not be touched as it attracts lightning (36). In this country, the Tlhaping believe that *Acacia mellifera* var. *detinens* Hookthorn is supposed to have the power of enticing and detaining the 'weather spirit' - the tribe also believe that *A. giraffae* Giraffe thorn is supposed to attract lightning, and charred portions of the tree which has been so struck, are mixed with the fat of a goat, *Capra* species, by the tribal 'doctor' for use as a protective charm (36). In Tanzania, the Sukuma remove any *Gloriosa superba* Climbing lily appearing near their houses as the plant is believed to attract lightning (3c).

Vernonia cinerea Little ironweed

RAINFALL

Among cultures worldwide, traditional ceremonies and rituals are performed by medicine-men either to prevent a threatening rain on a specific occasion like a football match; or more usually, to induce rainfall in arid areas when the lack of it is causing crops to fail and livestock to perish.

Rituals To Prevent Rainfall

It has been observed among the Azande of S Sudan: 'they pluck a leaf of a climbing yam ('mere') and twist *'bingha'* grass (*Imperata cylindrica* Lalang grass), to it and hang it from a stick with the *'bingha'* and say; 'You are leaf of yam, you drive away rain'' (10). An axe-head may also be used to prevent rain. 'It is knocked into the ground with its sharp side upwards...then uttered a spell, then strewed manioc leaves around the axe-head, and then made a ring of ashes around it' (10). A third method is described: 'A man takes green leaves of a tree and throws them onto a burning log on the ground. The leaves send up smoke so that the rain may remain in the sky. If later rain again threatens more leaves are thrown on the log. I witnessed this rite, but did not record the name of the tree which furnished the leaves. Probably any tree will serve, since the essential part of the rite is the action of the smoke going upwards' (10). In Ivory Coast, *Sabicea ferruginea,* a climbing shrub, is a fetish. Dried leaves placed on hut-roofs under certain conditions prevent rain falling (16). In Congo (Brazzaville), *Dorstenia elliptica,* a shrublet is ascribed with magical power: if hung from the fork of a tree, it will prevent rain falling (3c). In this country, a witch-doctor is on record as claiming that by cutting *Treculia obovoidea,* a small forest tree, he can impeach rainfall for a specified period (2). The decapitated head of an aardvark, *Orycteropus afer,* a Guinea-savanna mammal, hidden in a tree with the necessary rites prevents rain falling in the whole locality until it is removed (21).

Rituals To Induce Rainfall

In Senegal, *Capparis tomentosa,* a thorny climbing savanna shrub, is used in the invocation of rainfall or to ward off storms (15a, b); and in Ghana, the Mamprusi tribe use *Diospyros mespiliformis* West African ebony tree for rain juju - to induce rainfall (4); and the Dagomba use *Gardenia ternifolia* ssp. *jovis-tonantis,* a savanna shrub, to induce rainfall

in addition to its popular use as a lightning conductor. In N Nigéria, *Cissus quadrangularis* Edible stemmed vine, a climber with square stem, features in a Mumuye rain-making ceremony in Adamawa (20b). In Sierra Leone, *Dodonaea viscosa,* a shrub or small tree, has superstitious use. A leaf branch is put with due ceremony into a stream in order to induce rain should the weather be dry after planting rice (3e). In Mexico, the Huichol Indians identify *Solandra* species, trees on rocky cliffs, with 'Kieri Tewiari', god of wind and sorcery - believing that this god is an indirect bringer of rains through its control of the winds (18). The species include *S. brevicalyx, S. guerrerensis, S. guttata* and *S. nittida.* In southern N America, *Datura innoxia* Hairy thorn apple had importance in rites calling for rain (3e).

In Zimbabwe, the Manyika grow *Boophone disticha* Candelabra flower, a lily, outside their hut as a charm to bring rain; and in S Africa, when hailstorms threaten, the Southern Tswana draws a branch of *Ehretia rigida* Cape lilac along the ground in the neighbourhood of their gardens to protect them; and the Ngwaketse use the tree for rain-making in a very exclusive ceremony at which the chief is assisted by his rain-making expert and a few carefully chosen assistants, as the smoke from the fire of the wood is supposed to have the power of forming heavy rain clouds (36). Similarly, the ripe fruit of *Solanum americanum (nigrum)* Black nightshade, a common annual weed, is black which is the reason for its inclusion in rain-making medicine of the Southern Sotho and Tswana rain-doctors - the symbolism being that of dark clouds (36). In the E Usambaras of Tanzania, *Biophytum petersianum* African sensitive plant, a slender annual herb, and *Entada abyssinica,* a savanna tree, feature in a rain-making ceremony (3c).

Solanum americanum Black nightshade

Other plants used in connection with rain-making ceremonies (37) are given in Table 1 below:

Species	Plant Family	Common Name	Local Name	Remarks
Anacampseros rhodesica	Crassulaceae	-	-	Northern Sotho & Tswana tribe
Chenopodium batniyi	Chenopodiaceae	-	-	Sotho and Tswana tribe in Botswana
Datura stramonium	Solanaceae	Jamestown weed	-	Tswana tribe
Flagellaria guineensis	Flagellariaceae	Kanot grass	-	In Tanzania
Hirtella butayei var. greenwayi	Rosaceae	-	-	In Tanzania
Panicum maximum	Gramineae	Guinea grass	-	In Uganda
Rhus divaricata	Anacardiaceae	-	koditsana	Southern Sotho tribe
R. erosa	Anacardiaceae	-	tsilabelo	Southern Sotho tribe

Table 1. Some Rain-making Plants

Tabebuia rosea Pink poui

In Mexico and S America, the white clouds of tobacco smoke are also suggestive of, and associated with rain clouds, and play an important part in many ceremonies for securing rainfall (Mason fide 28). *Tabebuia rosea (pentaphylla)* Pink poui, an introduced ornamental tree from tropical and S America and especially the W Indies, burst out several times in a season so giving substance to a saying of the Tobago islanders that rain will not fall until the Poui blooms three times in succession (23). It has been observed that in C America: 'Quetzalcoatl' was the rainmaker god - the symbol of water or moisture produced by rain, which after a long drought, awakens vegetation to new life' (32). In medieval Mali, the equivalent of 'Quetzalcoatl' is the 'Desiri' of the Bambara...who is addressed with the words: 'Preserve us from evil doers...above all give us rain, without which the harvest is

Tacca leontopetaloides African arrowroot.

impossible (32). (The Bambara tribe is found in the savanna areas of Ivory Coast, Benin Republic, Niger and Senegal as well as in Mali).

In N Nigeria, an Aten rain-maker goes into the bush, in a rain-making ceremony, to address a prayer to the sun while holding in his hand a branch of *Ficus dekdekena,* the sacred *chediya* tree (20a); and in the Plateau State of this country, *Tacca leontopetaloides* African arrowroot is used in the rite of *Fa'an* (meaning rain) to bring rain (Walu fide 3e). There is a saying in The Gambia that if the fruits of *Parkia biglobosa (clappertoniana)* West African locust bean hang down a tree, the rain will be poor, but if the fruits are high up, the rains will be plentiful (13).

The Rain Dance

Among many of the tribal groups in the Region and elsewhere, the rain dance is an important part of traditional rituals performed in rain-making ceremonies. Rain-making rituals in both C America and W Africa included dancing with water-filled perforated pipes in the mouth and blowing the water through the perforation at the sky, throwing water from a vessel held high in the air so that the falling spray descended on the drought-stricken land on the suppliant crowd, squirting water in a fine stream from the mouth onto a field, and watering the graves of twins, who were thought in both cultures to have some mysterious connection with the weather (32). The rain-makers in both culture areas make use of black rainstorms, and in their sacrifices slaughtered black animals (32).

GOD OF THUNDER

In S Nigeria, *Distemonanthus benthamianus* African satinwood, a forest tree legume, seems to be the tree known in Yoruba folklore as the one on which the mythical deity 'Shango' (founder of Oyo and later deified as god of thunder) is said to have hanged himself, and from the wood of which 'Shango' clubs are made (8). (See EVIL: PLANTS FOR EVIL for the full story about 'Shango' and the *anyan-tree*). Evidence of the antiquity of the divine association of *Punica granatum* Pomegranate, a decorative introduction to the Region, is provided by the fact that it is thought to have symbolized the Babylonian and Syrian god of thunder and storms, 'Ramman' (11). In Hausaland, *'Gajjimare'* is the god of rain and thunder (31).

It has been observed that: 'Perhaps the fetish most dreaded in Benin

is So, the god of thunder and lightning, as what are considered to be the effects of his anger are frequently both seen and felt; so being supposed to strike with lightning those who disbelieve in his power or presume to scoff at him. It is unlawful for any person who has been killed by lightning to be buried, and it is commonly believed on the Slave Coast that the bodies of those who have met their death in this manner are cut up and eaten by the priests of So' (9) - but see DEATH AND DYING: FUNERAL OBSEQUIES: Burial Ceremony: Exceptions. In E Africa, the root of *Capparis* species, thorny scrambling bushes, is burnt by the Lobedu as the smoke is supposed to ward off storms (36). (For lightning and thunder worshipping, see RELIGION: NATURE AND DEITIES).

CHARMS AGAINST THUNDER

Some plants are traditionally believed to ward off thunder and enter into charms to that effect. For instance in Botswana, *Hypoxis haemerocallidea,* a lily, is one of the ingredients of a medicine used as a charm against thunder and storms; and in this country, the Southern Sotho also use *H. villosa* as a charm against thunder or the rootstock of *H. villosa* var. *scabra* is used by the African herbalist and diviner as a charm against thunder (36).

In Zimbabwe, the Manyika use the leaf of *Myrsine africana* Cape myrtle as a charm to ward off thunder by washing the body with the leaf, while standing in the river and allowing them to float away downstream (36); and in this country, the ash of the cone of *Becium obovatum* var. *hians,* a labiate, mixed with salt is regarded by the Manyika as a good protection against an evil wisher who tries to hit one with thunder (36). In Botswana, smoke from the fire of *Vernonia oligocephalus* or *V. pinifolia,* composites, is supposed to direct an approaching hailstorm (36). Among the Southern Sotho the plants, tied to a stick and waved towards an approaching hailstorm, has the charm to redirect the hail; and in Botswana, the smoke is thought by the tribe to drive away an approaching hailstorm - the smoke from the burning plant of *Lycium acutifolium,* locally called 'more-tlo', being similarly used (36). In S

Ceiba struck by lightning

Africa, the Zulu burn the leaf of *Oldenlandia caespitosa* var. *major*, a straggling weed, in their lands on the approach of a thunderstorm to ward off lightning; or a species of *Talinum*, probably *T. caffra*, a very thick fleshy herb, is a Zulu charm for warding off thunder and lightning (36). In E Africa, the Lobedu use *Asparagus virgatus*, a climbing lily, as a thunderstorm charm; and *A. plumosus* suitably 'treated' is planted upright or hung by the tribe to ward off a thunderstorm (36).

'NEUTRALIZING' THE CHARGES.

It is the traditional believe among many tribal groups in W Africa and elsewhere that objects such as trees that are struck by lightning should better be avoided. For instance, where such a tree stands near a footpath, this path is diverted to avoid the tree. More often such objects have to be traditionally 'discharged' in a ritual ceremony of purification to render them 'safe' again. It is considered dangerously sacrilegious to go near such an object, or touch it before the purification ceremony. The belief is that any deliberate defaulter would to be afflicted with skin rashes or rheumatism or paralysis or stroke.

With many of the culture groups in the Region, water from two bowls in which the leaves of *Dracaena arborea* or *D. fragrans*, palm-like trees commonly associated with fetish practices, and that of *Newbouldia laevis*, a medium-sized tree which is equally believed to be fetish, are immersed separately (representing the 'positive' and the 'negative' charges respectively) is sprinkled around the object in a ritual to 'neutralize' the charges of the lightning (30). The ceremony is usually preceded by prayers, the traditional libation and the sacrificial offering of a fowl, *Gallus domestica;* or a sheep, *Ovis* species; or a goat, *Capra* species; to the gods and the ancestral spirits (30). Alternatively, water in which *Phyllanthus niruri* var. *amarus (amarus)*, a weedy euphorb annual, has been immersed enters into a sprinkling ritual around the vicinity of the object to 'neutralize' or 'discharge' the forces. At the completion of the 'neutralization' ceremony, an emission of smoke is observed from the grounds around the object (23).

In S Africa, when a member of the kraal has been struck by lightning, the Zulu drink a cold drink infusion of the root of *Sansevieria thyrsiflora* Pile root as a protective charm; while in Botswana, an infusion of *Kohautia amatymbica*, a prostrate weed, is used by the Southern Sotho as an emetic in lightning stroke; and *Asclepias stellifera*, locally called 'mohlatsisa', is an African charm, used by the tribe when a person has been struck by lightning (36). (See also TABOO).

References.

I. Adjanohoun & Aké-Assi, 1972; 2. Bouquet, 1969; 3a. Burkill, 1985; b. idem, 1994; c. idem, 1995; d. idem, 1997; e. idem, 2000; 4. Casey, (pers. comm.); 5. Cofie, (pers. comm.); 6. Commeh-Sowah, (pers. comm.); 7. Cooper & Record, 1931; 8. Dalziel, 1937; 9. Ellis, 1885; 10. Evans-Pritchard, 1937; 11. Everard & Morley, 1970; 12. Guillarmod, 1971; 13. Hallam 1979; 14. Irvine, 1961; 15a. Kerharo & Adam, 1964; b. idem, 1974; 16. Kerharo & Bouquet, 1950; 17. Klugah, (pers, comm.); 18. Knab, 1977; 19. Lawler, 1984; 20a. Meek, 1925; b. idem, 1931; 21. Mensah, (pers. comm.); 22. Pamplona-Roger, 2001; 23. Perry & Greenwood, 1972; 24. Rayner, 1977; 25. Richard & Kaneko, 1988; 26. Savill & Fox, 1967; 27. Schnell, 1950; 28. Siegle, Collings & Diaz, 1977; 29. Sofar, (pers. comm.); 30. Sasu-Yawlule, (pers. comm.); 31. Tremearne, 1913; 32. Van Sertima, 1976; 33. Verger, 1967; 34. Walker, 1952; 35. Walker & Sillans, 1961; 36. Watt & Breyer-Brandwijk, 1962; 37a. Zeven, 1964; b. idem, 1967.

30 LOVE

Faith and hope and love we see,
Joining hand in hand agree;
But the greatest of the three,
And the best, is love.
Christopher Wordsworth, 1807-86.

Love is a feeling of warm affection for, or delight in oneself, other persons, pet animals, objects, subjects and concepts; and the desire to sustain this feeling. Both in and outside marriage, instances of men and women resorting to a fetish or similar institution for a love-charm or juju 'prescription' to win a partner or strengthen and secure an affection or to keep rivals out of a marriage partnership is quite often the gossip. The practice is strictly a secret affair, and often unknown to the other partner. Some people, therefore, tend to believe that a successful marriage has necessarily been influenced by this enchantment. Love-charms are also popular among boys and girls.

Some plants are traditionally associated with love in many cultures worldwide, while others are symbolic of hospitality and friendliness. Plants prescribed as love-charms may be worn, taken in food, added to bathing water, or applied during intercourse. Others are effected in normally inexplicable means.

CHARM PRESCRIPTIONS

Aspilia africana Haemorrhage plant

In N Nigeria, *Aspilia africana* Haemorrhage plant, a weedy composite, (and probably also *Melanthera scandens,* another composite) is used among the Hausa as a love-charm. The charm tied around the forehead attracts the 'glad eye' ('kalankuwa' the Hausa name, means 'a head band'), or a youth hides the plant in a maiden's house (13). It is

believed in Kordofan that if a youth secretly places a fresh piece of *M. scandens,* a weedy composite, in his girl-friend's home, his courting will be advanced (8c); and *Mitracarpus hirtus (scaber),* an annual herb, is held in Sokoto to have some magical power as a love-charm - the Hausa name 'magoori', meaning a vendor of love-philtres (8b, 13). In Gabon, *Aspilia kotschyi* has similar application (40). In this country, *Psilanthus mannii,* a shrub or small tree, is deemed to have magic properties. It is put in love-charm, and it is said to act as a charm to assuage anger (39, 40).

In S Africa, young African men rub *Gymnosporia buxifolius* Staff tree over the body, hands and face, after ceremonial emesis, as a charm to assist courting; and the Zulu either take an infusion of *Oldenlandia caespitosa* var. *major,* a prostrate weed, or take an infusion of *Pimpinella* species, an umbel, as an emetic before courting; or use the root of *Hippobromus pauciflorus* Horsewood as a love charm (41). In Botswana, *Cynoglossum lanceolatum* Hound's tongue, or *Linum africanum* Wild flax is used by the Southern Sotho as a love-charm; and in Zimbabwe, the Manyika use the root of *Ipomoea crassipes* Wild patata as a love-charm (41). In Tanzania, the Zigua drink a leaf-decoction of *Chamaecrista (Cassia) mimosoides* Tea senna, a herb legume or low shrub, to make a person like one (41). In Congo (Brazzaville), if the crushed roots of *Desmodium tortuosum,* an erect annual herb, together with those of *Boerhaavia diffusa* Hog weed, a common annual weed, are eaten with the appropriate incantations, one may be assured of the love of any woman one may desire (5).

In S Africa, the dry leaf of *Helichrysum cooperi,* a composite, made into an ointment which is applied all over the body after a bath, is a Zulu love-charm - as a result the desired lady finds the man irresistible (41). In E Africa, a young unmarried Lobedu woman when seeking a sweetheart, uses *Wedelia natalensis,* another composite, as a charm to counteract any medicine, which she suspects may have been given her to prevent this; and in Botswana, *Cymbopogon marginatus,* a coarse grass, is used by the Southern Sotho as a love-charm - while bathing a youth or maiden sprinkles the powdered root on the water calling the desired person by name, which is thought to attract the person to the spot (41). In other to attract the opposite sex in S Africa, a Zulu makes a cold infusion of the root of *Tephrosia lucida,* a legume, with *Dianthus crenatus* Wild pink, and washes the face with the froth that rises; or uses the tubers of *Bowiea volubilis,* a lily, in preparing a love lotion to attract women; or uses the bulb of *Tulbaghia violacea* Wild garlic, a lily, as an emetic love medicine (41). Among the tribe, the leaf ash of *Kalanchoe* species, probably *K. oblongifolia,* a thick-leaved herbaceous plant, is rubbed into two or three scarifications on the supraorbital ridges of young men who are courting, as this is thought to fascinate any young woman to whom attention is directed (41).

In W Africa, a grub which is sometimes found inside the stem of the pale-leaved form of *Crotalaria pallida (mucronata),* an erect legume shrub, is particularly fortunate as a love-charm. With proper ceremony and selected additions, it is made into a salve to rub the eyes, which will render the subject attractive to all men (13). In Congo (Brazzaville), *Biophytum petersianum* African sensitive plant, an erect annual, is put into love-charm to assure the exclusive holding of love (5); and in Ivory Coast, the plant is used by the Koulangos and the Abrons to prepare a love-charm placed in the food of the one to be attracted or to be charmed. The plant is approached slowly, uttering the words *'yabagdio toa'*, meaning 'old lady shut up', as the contracted leaves are plucked (23). In N Nigeria, *Ectadiopsis oblongifolia,* a savanna shrub, is worn round the waist by Fula youths as a charm to attract affection (13). In Ghana, the whole plant of *Gomphrena celosioides,* a prostrate amaranth weed, collected while calling the name of the loved one to be charmed, then pounded with black soap which is used in bathing while again calling the name of the desired one, serves as a charm among the Ga-speaking tribe to attract this lover (9).

Biophytum petersianum African sensitive plant (opened)

(closed)

In its distribution area, *Okoubaka aubrevillei,* a rare forest tree occurring mainly in Ivory Coast and Ghana and popularly believed to be fetish, enters into prescriptions for a love-charm. (The distribution extends from

Ivory Coast to Cameroon, with a distinct variety in the Democratic Republic of the Congo (Zaire) (21). *Okoubaka aubrevillei* var. *glabrescentifolia* has been recorded in Congo (Brazzaville) (18)). The seed shell is ground with a type of lavender known as 'Six Flowers' and used as a cosmetic while calling the name of the loved one (10). In S Nigeria, the Ijo use *Emilia coccinia* Yellow tasselflower (and probably also *E. praetermissa*) as a love-charm; a man urinates on the leaves and calls the woman's name four times, and he must then contrive to get the leaves into her bath-water (8a). In this country, *Picralima nitida,* a tree and source of the alkaloid akuammine, features in a Yoruba Odu incantation to attract a woman (38).

In N Nigeria, the Ebina people of Adamawa prepare a 'cheating cream' from the leaves of *Senna (Cassia) singueana,* a shrub or small tree legume, which must be picked on a Sunday between 7 am and 9 am. They are ground up and mixed with the meat of a donkey, *Equus* species, and the produce is rubbed on the body. It stops a man's wife from cheating him, and it has the additional capacity of enabling one to obtain favours of a whore without payment (6). In S Nigeria, *Senna (Cassia) obtusifolia,* an annual legume, has magical attributes. It is imprecated in a Yoruba Odu incantation to gain love (38). In SE Senegal, the branches of a Mistletoe found growing on *Combretum paniculatum* ssp. *paniculatum,* a savanna tree, are reduced by the Tenda to a powder which is mixed with *karité* butter and rubbed on the face, to ensure a boy's great success with the girls (16).

In N Nigeria, the fruit of *Maerua angolensis,* a small savanna tree with white flowers, is used by youths as a love-charm, mixed with galena or so-called 'antimony' and rubbed on the eyelids it is believed to render the youths irresistible to girls (13); and the leaves of *Polygala arenaria,* a widely distributed annual, are occasional ingredients in preparations to ensure popularity with women (13) - the practice is also recorded in S Nigeria (13). In N Nigeria, *Polycarpaea linearifolia* and *P. corymbosa,* erect herbaceous annuals or perennials, are used as a charm to ensure success in love; and in the northern parts of the country, the seeds of *Whitfieldia elongata,* an under-shrub acanth with conspicuous white flowers, or an undetermined Acanthaceae locally called 'bi ta swaiswai' or 'bi ta zaizai' in Hausa, are used by youths of this tribe and those of other tribes generally as a love-charm to secure affection, the underlying idea being urgent persistence in following the subject of desire (13).

In N Nigeria, the thickened roots of *Ipomoea nil (hederacea)* and *I. aitonii (pilosa),* annual and perennial twiners respectively, are used in Damagaram as a love-charm under the Hausa name of 'hantsar gada',

meaning the fruiting spadix of *Anchomanes difformis* or *A. welwitschii,* aroids with red berries (13). In this country, and also among the Zarma of Mali and Niger, and the Fula a twig of this tree or the root of *Pergularia daemia,* a climbing asclepiad often tended as a magic plant, collected with the necessary rituals and used as a chewing stick is a charm to secure the affection of a maiden (28). In the Region, traditional healers prescribe *Cissampelos mucronata* or *C. owariensis,* climbing shrubs in savanna and forest respectively, in rituals for a love-charm. The woman's name is called out at the time the plant is knotted - but not uprooted (9). In its distribution area, the dried leaves of *Lawsonia inermis* Henna with natron (native sesquicarbonate of soda) in water as a drink is a love-charm (9) (see WEALTH :CHARM PRESCRIPTIONS). In Senegal, a bath of the macerate leaves of *Guiera senegalensis* Moshi medicine, a small shrub in sandy wastes and semi-desert areas, is a charm to further affairs of the heart (22a, b).

In S Africa, the Malay and the Hottentot have a superstition about *Viscum* species Mistletoe in preparing a love-charm; and a decoction of *Acridocarpus natalitius,* a shrub, is widely used among the African in the Union as a love charm, applied by sprinkling (41). In this country, the Zulu youth sometimes fixes a piece of the leaf of *Ansellia humilis,* an orchid, under the arm bangle when out courting; and in Botswana, *Polygala rarifolia,* an erect annual, or *P. usambarensis* or *P. gymnocladia* is used by the Southern Sotho as a love-charm (41). In southern Africa, men take the root sap of *Eulophia angolensis,* a ground orchid, to ensure success in courting (25, 35). In Congo (Brazzaville), *Urginea altissima,* a bulbous lily, has use in magical preparations and love-charm (6). In this country, the seeds of *Garcinia kola* Bitter cola enter into many medico-magical remedies taken with palm-wine 'to cleanse the stomach and to give strength in love' (5). In Nigeria, mastication of the seeds of Bitter cola is held to be as effective (1).

In W Africa, many epiphytic orchids like species of *Angraecum, Ansellia, Bulbophyllum, Listrostachys,* and *Polystachya,* are used as a love-charm especially among the Hausa of N Nigeria (13). The charm is supposed to have special efficacy if the orchid is found on certain trees such as *Diospyros mespiliformis* West African ebony tree (13). In this country, *Evolvulus alsinoides,* a spreading hairy herb, and *Scoparia dulcis* Sweet broomweed are sold principally as a charm worn as a girdle or amulet to secure affection, fame or esteem; and the Hausa name of both plants 'ka-fi-malan', literally means better than a malan for such purposes (13). In Ghana, the Hausa similarly use Sweet broomweed in love-potions (19); and with many of the coastal tribes or tribal groups

in its distribution area generally, the plant is rubbed on the palm of the hands before a handshake with a lover to charm her (27).

In S Nigeria, a plant locally called 'asimawu' in Yoruba, meaning 'we are pleased', is used by medicine-men in an invocation to be loved (38). In N Nigeria, *Kohautia senegalensis,* an erect annual, is reported as a popular love-charm; or the buds (pondi) and flowers (chode) of *Leptadenia hastata,* a perennial climbing shrub in savanna, are used by Fula women to prepare a comestible (used apparently as a love-charm) (13). The plant appears to act as a love charm for both sexes. In S Africa, an infusion of the root of *Burchellia bubalina* Wild pomegranate is a Zulu love-charm, being used as an emetic, body and nasal wash, during which ceremony the swain calls out his sweetheart's name - as a result she is supposed to be unable to resist his advances (41) The tribe uses the powdered root of an *Oldenlandia* species, a straggling herbaceous weed, as a love-charm - a pinch is put into the mouth and blown through a reed while the name of a favourite girl is called out, and as a result she dreams pleasantly of him (41).

In S Nigeria, *Euphorbia prostrata,* a slender prostrate herb, under the Yoruba name 'ewe biyemi', meaning 'family fits me', enters an incantation of hope to obtain several women at home (38); and in N Nigeria, a Fula man will take an infusion of the pounded leaves of *Euphorbia hirta* Australian asthma herb, a decumbent herb with milky latex, before going to a woman to ensure she will love him (20). In W Africa, *E. convolvuloides,* a herb with milky latex in savanna and waste places, is used as a love-charm by youths and maidens (13); and women use the fruit of *Solanum incanum* Egg plant or Garden egg superstitiously in some love-charm (13). In N Nigeria, a form of special food, called in Hausa *fatefate* or *mayemaye* is prepared from the leaves of *Vernonia amygdalina* Bitter leaf and *V. colorata,* shrubs or small composite trees in derived savanna, along with butter and condiments, and taken by women in the belief that it renders them sexually more attractive; and an infusion of the leaves of *Abrus precatorius* Prayer beads as a wash is used by youths, and a paste of the pulverized seeds added to the cosmetic *galena,* 'antimony' or *kwolli,* is smeared on the eyelids by women to attract love, or made up to wear round the neck as a charm (13).

In Europe, the powdered whole plant of *Vinca* species Periwinkle with *Sempervivum tectorum* Houseleek wrapped with earthworms, *Lumbricus* species, and taken in meals induces love between man and wife; and the human-form roots of *Mandragora officinarum* Mandrake was valued by the Ancients as a love-charm among a multitude of other attributes (33). It has been confirmed that 'for making potions for love and fertility no plant was more highly regarded than the Mandrake' (11). 'Mandrake has been used since time immemorial as a constituent of love potions, due to its energizing action on the sexual glands' (34). The powdered root of *Iris florentina* Orris root is an ingredient in 'love potions' (34). In the northern Peruvian Andes, *Thevetia peruviana* Exile oil plant or Milk bush or Yellow oleander; *Cyperus articulatus* Piri piri, a sedge; and *Salmea scandens* Hayme-huayme, a composite; are used to promote love. Their magical activity works when they are held in the hands while speaking to the person desired (14). In Ancient times, *Viola tricolor* Heartsease, a herb, was often used in love-charms, for it was believed to be a most potent herb (4).

Sympathetic Magic

Some of the prescriptions for love appear to be based on sympathetic magic. Among many ethnic groups in the Region, scrapings from the bark where two branches of a non-poisonous tree rub against each other are used as a love-charm to win the heart of a girl, provided she can be made to eat these scrapings in her food unknowingly (36) - an instance of sympathetic magic. The charm is more effective when the bark of *Ficus platyphylla* Gutta-percha tree is added (see also COURT CASES AND FAVOUR: CHARM PRESCRIPTIONS). Among the Hausa of N Nigeria, the Zarma of Mali and Niger and the Fula the latex of this tree collected in drops from a knife wound, then mixed with *kwolli* (a black cosmetic) and applied as make-up around the eyes, is used by youths as a love-charm (28). Alternatively, among the coastal tribes in the Region, the scrapings of the tree are ground with the seeds of *Xylopia aethiopica* Ethiopian pepper. Some of this powdered mixture is placed in the mouth before talking to the woman to attract or to court her (27). Among the Ewe-speaking tribe of Ghana, Togo and Benin Republic, *Sida linifolia,* a small erect malvaceous weed believed to be fetish, enters into love-charms (37). A selected plant, visited earlier and cleared of other weeds, is approached at mid-night while completely naked with the back to the plant. Two cowry shells, powdered corn dough and lavender are offered to the plant, while a request for the affection of a loved one (mentioned by name) is made, with a promise to reward the plant accordingly if the request succeeds within three to seven days. Two leaves of the plant are then collected into the lavender bottle (without looking at the plant) and taken home, still with the back to the plant (37). The part played by the lavender is likely an instance of sympathetic magic.

Medicine-men in the Region use *Caladium bicolor* Heart of Jesus, an introduced decorative perennial aroid, as a love-charm. The underground stem or rhizome is collected with the necessary traditional rituals, then dried and ground with shea-butter and lavender in a white plate, and left in the sun to melt. The preparation is applied as a cosmetic - the name of the desired partner being mentioned in the process - to attract her (27). This is likely an instance of sympathetic magic with reference to the common name of the aroid, reflected in the heart-shaped leaf.

As a symbolism in human emotions, *Pupalia lappacea,* a common weedy amaranth, is used as a love-charm (8a). The juice of the ground whole plant squeezed by partners into each other's eyes is believed to strengthen the bond of friendship and love (27). The sympathetic magic is based on the clinging burs. To be most effective, the plant is first sprinkled with corn meal on a specific day and picked later when ants have infested it (27). The weed may also be ground into balls with the kidney of a sheep, *Ovis* species, a drop of one's own blood and lime juice, then given to the wife to forget about other men (27). It may also be charred into black powder with shea-butter and applied before intercourse to induce her to stay for good (27). The fruits of *Pupalia* may also be employed in a sinister incantation by an enemy or a rival to sow seeds of dissension among lovers (24). The fruits are clandestinely stuck on the clothes of the intended victim with a spell. Should this victim remove the clinging fruits and then throw away, then the action would trigger the charm. However, should the fruits be chewed or burnt, this would counteract the charm (24). Therefore when one has not been out in the fields, but still finds *Pupalia* fruits clinging on his clothing, evil intension is suspected. The fruits are either chewed or burnt accordingly - but never discarded (24).

Similarly, in Gabon, *Ataenidia conferta,* a rhizomatous herb in wet places, is a well known medium for sowing discord between families, and also to provoke quarrels and fights. It is but necessary to wrap a fish or other food for cooking in one of the leaves, or to burn the handle of a pestle and turn it in the direction of the house to be disrupted (Koechlin fide 8b, 41).

In Ghana, a preparation of the powdered leaf and seeds of *Uraria picta,* a woody legume herb in grassland, the whole plant of *Solenostemon monostachyus,* a labiate, and the leaves of *Mucuna pruriens* var. *pruriens* Cow itch all collected with the necessary rites, then mixed with honey and applied to the male member before intercourse is used by the Ga-speaking tribe and others in the Region as a charm to win the affection of the woman (9). The inclusion of the honey is most likely an instance

of sympathetic magic. The dried powdered leaf of the labiate with that of *Abrus precatorius* Prayer beads and *Scoparia dulcis* Sweet broomweed applied to the male member also enter into magic prescriptions for affection by the tribe (9). Alternatively, the request for a named partner is made over the powdered mixture, then burnt (9). If marriage is intended, then the leaves of *Lawsonia inermis* Henna, Sweet broomweed, and *Crotalaria pallida (falcata)*, a shrubby legume often in coastal sands, with the leaves of Prayer beads are used instead (9). In the Region, the whole plant of Prayer beads, with *Triumfetta cordifolia* Bur weed, *Uraria picta*, a legume in grassland and *Paullinia pinnata*, a woody or sub-woody climber with tendrils, all collected with the necessary rituals, are used in juju prescriptions to win a person's love (27).

Medicine-men in the Region use Prayer beads in another love-charm ritual. A coin is placed near the plant as a sacrificial offering while a request for the hand of a named maiden is made to the spirit of the plant. Two or three whole leaves are then picked and chewed, and before talking to anyone, the maiden is approached with the request which is said to be invariably successful. The sympathetic magic is hinged on the sweetening after effect experienced when one chews the leaves of the plant (4). Prayer beads is reported (2) in another love-charm especially among the Akwapim-speaking tribe. A preparation from the scrapings from the poultice of the ground leaves moistened with water and honey, is applied to the forehead and licked as primaries. In addition, some of the poultice is rubbed repeatedly and with force against the thighs during ejaculation as charms to secure the affection of the partner (2). The sympathetic magic lies with the honey.

For sympathy as well as for love, the nut of the white-seeded variety of *Cola acuminata* Commercial cola nut tree or that of *C. nitida* Bitter cola - (but never the nut of the red-seeded variety) is added as ingredient to the prescription. The red-seeded variety is believed to have the opposite effect (9). All the constituents and the ingredients are dried, powdered, mixed and sieved, then applied to the skin in either water or shea-butter as a vehicle for affection among the Gas and other coastal tribes (9). With some ethnic groups of the Guinea-savanna, the juice of *Leptadenia hastata*, a savanna twiner, with honey applied to the male member before intercourse is a charm to prevent the woman from ever deserting (28). Again, the addition of honey is likely an instance of sympathetic magic. In Ghana, *Mimosa pudica* and *Schrankia leptocarpa* Sensitive plant, collected and prepared with rituals, are used in love-charm by both sexes in the Nzima area to secure the affection of the other (12). This is likely an instance of sympathetic magic. In its distribution area, the ground

leaves of *Vernonia cinerea* Little ironweed, a herbaceous weedy composite, collected with the necessary rites, with assorted ingredients in lavender is sprinkled on a handkerchief as a charm for love by youths (27); and *Portulaca oleracea* Purslane or Pigweed used with white sponge and sweet-scented soap as bath for seven or seventeen consecutive days is a charm for love. The name of the sweet-heart is mentioned while bathing (27). The latex of *Milicia (Chlorophora) excelsa* or *M. regia* Iroko mixed with one's semen in a partner's food is a charm to induce her to stay (27). All three instances are most likely based on sympathetic magic.

In Congo (Brazzaville), *Helixanthera mannii* Mistletoe, an epiphytic parasite, enters in magic practice to improve one's status in courting based on generally widespread sophistry of the Doctrine of Signatures, not on any phytopharmacognosy (5). In this country, *Lomariopsis guineensis, L. hederacea* and *L. palustris,* epiphytic ferns in swamps and fringing forest, enter charms for love. In courting, a young man will burn the plant to a powder which he rubs over his body, saying to his girl 'just as the fern sticks to the tree, so I wish that you will stick to me' (5). In Gabon, the leaves of *Plagiostyles africana,* a shrub or small tree, are pounded for use as a fish-poison; and the bark made into talisman for trappers, or (by an interesting extension of sophistry) to attract suitors (40). In SE Senegal, Tenda girls who like their body to be as smooth and speckled like the fruit of *Lagenaria (Adenopus) breviflora,* a cucurbit, will carry them on their person in a play on sympathetic magic (16). In N Nigeria, *Desmodium velutinum,* a legume, is used in a superstitious practice, as a charm or potion, to secure love - the sympathetic magic referable to the clinging burs (13).

ASSOCIATED PLANTS

Cassytha filiformis Dodder (symbol of love)

By tradition, various plants are associated with love - either by name or function or symbol or legend - among different cultures. *Rosa* species Rose however, appears to be universally known and accepted as symbolizing love and charity.

The superstitious uses of *Cassytha filiformis* and *Cuscuta australis* Dodder, parasitic epiphytes, are similar. They are locally called 'dome atre' in Akan or 'soyayya' in Hausa - meaning 'if you love me spread' and 'mutual affection', respectively (13). However, the Nzima name, 'brezinle nyema', on the contrary means 'thread of a divorced woman' (13). In its distribution area, both plants are used by youths as well as adults in love-charms. If the plant survives and spreads when placed on a tree he is hopeful, if not he is regarded as rejected. Dodder is a general symbol of love in various tribes (19). The parasites enter into rituals for a love-charm among many other tribal groups in the Region. For this purpose, the plants are collected with both hands into a container of water which is left in the open over night. The water is then used for bathing either on a rubbish heap or at cross-roads. The name of the loved one is mentioned both during collection and bathing (27). *Amaranthus hybridus* ssp. *incurvatus* Love-lies bleeding, a robust annual herb, is so called because of the red leaves; and *Mukia (Melothria) maderaspatana,* an annual trailing cucurbit, known as 'maalamin maataa' in Hausa, indicates the superstitious use of the plant by youths as a love-charm (8a). With some of the people, *Amaranthus lividus* or *A. viridis,* herbaceous weeds, ground with the skin of a wolf, *Lupus* species, and soap for bathing are used as love-charms (27). Similarly, *Quisqualis indica* Rangoon creeper or 'Love and innocence', a woody climber or scandent shrub often cultivated for the attractive flowers, probably implies its use in affairs of the heart; and *Antigonon leptopus* Corallita or Chain of love or Love-vine, an introduced decorative climber, is known as Wedding flowers, and used as such. In Ghana, Corallita sometimes escapes cultivation in and around Accra. In the Mediterranean region, *Nigella damascena* Love-in-the-mist, a common annual garden plant, is believed to symbolize love; and *Paris quadrifolia* Herb Paris or True love, an European and Russo-Asian plant, has reference to love.

Antigonon leptopus Corallita or Chain of Love or Love-vine.

Some grasses are also similarly named for various reasons. For example, the introduced lawn grass, *Chrysopogon aciculatus* Love grass is so called because the brown seeds stick to clothing. In Malaya, this concept is extended to rubbing the seeds with some lime-juice on a decoy female elephant to entice solitary males (7). Other grasses so named

include *Eragrostis amabilis* Feather love grass, *E. curvula* Weeping love grass, and *E. tenella* Japanese love grass.

In Achimota School, Ghana, a secret Rose gift (the emblem of love) from a girl, means 'amorous affection'. On the contrary, the bright-red double-petalled variety of *Hibiscus rosa-sinensis* Shoe flower, locally dubbed 'ope', from a girl, is the school's traditional 'open regret' symbol. In its distribution area in the tropics, however, women use Hibiscus flower for adornment; and in Hawaii particularly, they are fashioned into the necklaces of blooms (leis) used to greet visitors (33). In S Africa, the root of *Hibiscus pusillus*, locally called 'uguqukile', is used by Zulu men as part of the preparation for courting (41). In Ivory Coast and Burkina Faso, women use *Datura metel* Metel to reject unwelcome amorous advances (23).

In Europe, *Catananche coerulea* Cupid's love, a Mediterranean composite, refers to its one-time use in love potions and charms; and *Lycopersicon lycopersicum (esculentum)* Tomato was introduced from its native Mexico under the name of Love apple, since it was thought to be an aphrodisiac (33). In Europe, *Elaphomyces granulatus*, a temperate fungus, was also formerly regarded as an aphrodisiac and used in the preparation of love potions (31). (An aphrodisiac is any food, drink or drug that arouses sexual desire). The fungus is named after the Greek goddess Aphrodite, called Venus by the Romans (11). *Myrtus communis* Myrtle was dedicated to Venus, the Goddess of Love in ancient times - an appreciation which has lingered in the use of Myrtle sprays in bridal bouquets; and *Cyclamen repandum* Common sowbread, another Mediterranean plant, was reported to make the shy more amorous (33).

In Europe, *Myosotis scorpioides (palustris)* Water forget-me-not, or *Veronica chamaedrys* Germander speedwell, both temperate decorative plants, are by tradition believed, in their growing areas, to keep lovers' memory fresh (15). From a German legend, a knight who, while gathering the flowers of a riverside plant for his lady love, fell in and was swept away by the current, crying as he vanished - Forget-me-not (33). *Plumeria rubra* var. *acutifolia* Frangipani or Temple flower, an introduction from the East and usually planted in cemeteries, is also called Forget-me-not (see GRAVES AND CEMETERIES: AESTHETIC AND REMEMBRANCE VALUES).

According to Greek legend, the flower of *Nymphaea* species Water lily arose from a nymph, who had died from unrequited love for Hercules (26). The indigenous species in the Region are *N. guineensis, N. lotus,*

N. maculata, N. micrantha and *N. rufescens. N. caerulea* is an introduction from NE Africa.

HOSPITALITY AND FRIENDLINESS

In N Nigeria, the Hausa pull apart the leaf of *Cyperus papyrus* Papyrus, a sedge, and the nature of the tear indicates either rupture *(sa ani)* or cementing of friendship. The leaves of *Tephrosia bracteolata* and that of *T. purpurea,* under-shrub legumes in dry waste places and open country, are used in a similar manner (13). In Ghana, a piece of the leafy shoots of *Securinega virosa,* a savanna shrub, or the whole plant of *Euphorbia hirta* Australian asthma herb, collected with the appropriate ritual invocation, and carried in one's pocket, it is believed, influences the acceptance of friendship (27).

In S Africa, the bark of *Ekebergia meyeri* Dog plum is used as an emetic and enema by any Zulu who is disliked by his fellows, in order to establish better relationship; or, instead, he or she uses a decoction of the bark of *Acacia gerrardi* Red thorn - sometimes, in addition, the vapour from the hot decoction is inhaled (41). However, in Botswana, the Southern Sotho believe that the smoke of *Lasiosiphon anthylloides,* a shrub, locally called 'moomang', makes people quarrelsome and that the inmates of a hut in which the plant has been burnt will eventually quarrel (41); and in SE Senegal, the Tenda believe that to burn a leaf or the wood of *Combretum molle,* a savanna tree, in a village will promote squabble (16). In Congo (Brazzaville), the presence of *Bryophyllum pinnatum* Resurrection plant (and probably also *Kalanchoe integra* var. *crenata* Never die) besides a house, makes the intentions of all persons approaching it friendly (5). Among some tribes in its growing area, *Costus afer* Ginger lily, a perennial ginger-like forest herb, planted in front of a house serves a similar purpose in addition to spiritual protection from evil forces (27).

Cola acuminata Commercial cola nut tree, C *nitida* Bitter cola (symbol of welcome and hospitality, love, friendship and reconciliation; associated with Islam and Oriental religion)

In N Nigeria, the seeds of *Cola acuminata* Commercial cola nut tree and *C. nitida* Bitter cola are regarded as the greatest symbol

or gesture of welcome and hospitality among many tribes (30). Most social functions start with 'offering of kola-nuts' by the host, quickly followed by the elaborate 'ceremony of breaking the kola'. Without observing this traditional procedure, the function may be deemed to be ill-fated and uncustomary and may be invalidated (30). Similarly, in many southern regions of Cameroon, the kola-nut is seen as a symbol of love and friendship (29). Kola-nut features in many social ceremonies; funerals, weddings and initiations; and it is an important symbol of reconciliation and friendship (29).

In Nigeria, the fruits of *Dacryodes edulis* Native pear, which is popularly eaten in combination with roasted or boiled fresh maize, *Zea mays*, is also extensively used as a mark of hospitality (30). However, the tree itself may be used to signify disapproval in some circumstances (30). For instance, a bride whose conduct or behaviour is unbecoming is sent back to the parents with the leaves of Native pear. Similarly, according to the custom of some people in Anambra State, if after a visit, the host escorts his guest and deliberately stops under a Native pear tree, it warns him not to visit again (30). In S Nigeria, *Clerodendrum capitatum,* an erect or scrambling shrub of both savanna and forest, is invoked by the Yoruba in an Odu incantation under the name 'funmo'mi' to have one's house full of people - the name meaning 'come close' (38).

According to Chinese mythology, *Salix babylonica* Weeping willow represents the sentiments of friendship and mercy, as well as grace (17). *Prunus domestica* Plum and *P. persica* Peach, mostly northern temperate fruit trees, represent brotherliness and cordial relationship - the blossoms of both trees symbolizing physical charm and loveliness of women, as well as of the spirit (17). In Korea, *Magnolia* species Magnolia flower is symbolic of love and happiness, hope and future, and firm faith (11). Among Christians, *Myrtus communis* Myrtle represents compassion; and *Reseda odorata* Mignonette represents mildness (11). In years gone by, *Lonicera periclymenum* Honeysuckle was traditionally regarded as a sign of true devotion (3). *Elettaria cardamomum* Cardamon is an important ingredient of curry powder, and also used for various flavourings. In Arabic countries, Cardamon coffee is a common beverage and a symbol of hospitality (33); and in Ancient Greece, *Malva silvestris* High mallow, a biennial plant of the Malvaceae family, symbolizes calmness and sweetness (32). (See also CHARM; and MARRIAGE).

References.

1. Ainslie, 1937; 2. Ankoma-Ayew, (pers. comm.); 3. Back, 1987; 4. Beloved, (pers. comm.); 5. Bouquet, 1969; 6. Bubuwa Tubrar, 1985; 7. Burkill, 1935; 8a. Burkill, 1985; b. idem, 1997; c. idem, 2000; 9. Cofie, (pers. comm.); 10. Commeh-Sowah, (pers. comm.); 11. Crow, 1969; 12. Cudjoe, (pers. comm.); 13. Dalziel, 1937; 14. De Feo, 1992; 15. Everard & Morley; 1970; 16. Ferry & al., 1974; 17. Graf, 1987; 18. Hutchinson & Dalziel, 1958; 19. Irvine, 1930; 20. Jackson, 1973; 21. Keay, 1989; 22a. Kerharo & Adam, 1964; b. idem, 1974; 23. Kerharo & Bouquet, 1950; 24. Klah, (pers. comm.); 25. Lawla, 1984; 26. Lotschert & Beese, 1983; 27. Mensah, (pers. comm.); 28. Moro Ibrahim, (pers. comm.); 29. Nkongmeneck, 1985; 30. Okafor, 1979; 31. Oso, 1977; 32. Pamplona-Roger, 2001; 33. Perry & Greenwood, 1973; 34. Pitkanen & Prevost, 1970; 35. Rayner, 1977; 36. Siti, (pers. comm.); 37. Sofar, (pers. comm.); 38. Verger, 1967; 39. Walker, 1953; 40. Walker & Sillans, 1961; 41.Watt & Breyer-Brandwijk, 1962.

31 LUCK

And unto one he gave five talents,
to another two, and to another one;
to every man according to his several ability...
Matthew 25:15

Anything happening by chance is often labeled as luck. Should it be favourable, it is good luck; otherwise it is, unfortunately, bad luck or hard luck or ill-luck. A variety of factors are believed to influence luck among many cultures worldwide. These include the environment; colours; persons; clothing; pets; numbers; the calendar (either some days, or some months, or some years). Other factors are the heavenly bodies - the stars; the planets and other celestial bodies; phases of the moon; the zodiac; and so on. Traditionally, it is the belief that some plants do also influence or control events - and are either accordingly named or find use as charms for good luck or against ill-luck.

GOOD LUCK CHARMS

In N Nigeria, *Leucas martinicensis,* an aromatic annual labiate, is regarded as a charm for luck. The seeds are rubbed along with butter in the hands while the person says 'arziki jaka', meaning 'good fortune come'; and *Ipomoea argentaurata,* a woody climber, may be burned as a fumigant for clothing and as a charm for luck (5). In this country, *Phragmanthera* and *Tapinanthus* species Mistletoe, epiphytic parasites, growing on *Stereospermum kunthianum,* a savanna woodland tree ,are considered a lucky find (5). In S Nigeria, the Igbo at Tjele consider *Momordica cissoides,* a trailing cucurbit, a charm against ill-luck; and the Igbo at Uburubu rub *Microglossa pyrifolia,* a composite, on the skin to bring good luck (N.W. Thomas fide 2a). In this country, a plant locally called 'afagbara', meaning 'buy with sixpence', is invoked by the Yoruba to save someone's good luck (20). In Ghana and Ivory Coast, the brown-stemmed type of *Musa sapientum* Banana, known as 'kwadupa' in Akan, enters into rituals to drive away ill-luck - the fruits are placed at cross-roads in the night while naked with incantations for luck (3).

In Zimbabwe, the Manyika use the root of an *Anthospermum* species

as a good luck charm - a small amount of the chewed root being put on the head, and the root mixed with vaseline being smeared on the face every morning (22). Other plants used by the tribe as good luck charms include *Protea roupelliae* Sugarbush with other plants; the roots of *Ipomoea crassipes* Wild patata; *Dianthus mooiensis,* a herbaceous plant; and the composites *Berkheya seminivea* and *Lopholaena disticha* (22). The tribe also grow *Boophone disticha* Candelabra flower, a lily, outside their hut as a charm to bring good luck (22). In Mexico, *Oxalis tetraphylla* and *O. deppei,* prostrate plants with violet and rose-red flowers respectively, have four-parted leaves which are sometimes sold as lucky charms (16). In Europe, *Illicium verum* Star anise is carried by the superstitious folks as a good luck charm (17); and the five petals of *Prunus mume* Japanese apricot, an introduction from China, denoted good luck (7).

In S Nigeria, Yoruba traditional doctors use *Termitomyces robustus,* an edible mushroom, in the preparation of charm food for luck (18). One way in which the charm is made is by roasting the mushroom together with chalk and the wood of *Pterocarpus osun,* a forest tree legume, and divining on it. This is mixed with African black soap and used for washing. It is supposed to bring good luck to everyone using it (18). This usage is based on the folklore that a man called Ogogo was dogged by ill-luck. He consulted 'Orunmila', God's deputy on earth, who was asked for certain sacrifices to be made at the base of *Annona senegalensis* var. *senegalensis,* a savanna shrub. On the ninth day, there were mushrooms around the shrub, and Ogogo found them good to eat - hence the Yoruba name for the mushrooms 'ewe'. The charm is also prepared by roasting the mushroom with the bark of *Ceiba pentandra* Silk cotton tree and that of *Adansonia digitata* Baobab and divining on it. This is eaten periodically (18).

In southern Cameroon, the bark of *Distemonanthus benthamianus* African satinwood is used by the Beti as a talisman for good luck (13). The root of *Mimusops kummel,* a tree, is deemed to hold magical powers to confer good luck and success in business for which it is prepared as 'fortune baths' (Ruffo fide 2c). A gum exudes from the bark of *Piliostigma thonningii,* a tree legume in savanna, and in native folklore a propitious effect is attained by chewing the bark or gum, which also enters into some magical prescriptions, being considered lucky (5). In S Africa, the Southern Sotho use *Eriospermum* species, a lily, as a charm for good luck; and the Zulu regard *Capparis tomentosa,* a thorny savanna climber or straggling shrub, as a general charm against misfortune (20). In this country, the Sotho-Tswana group use *Aloe saponaria* Soap aloe, another lily, or *A. tenuior* for ensuring good luck (20). In Zambia, *Lonchocarpus*

capassa Lance tree or Mbandu, a tree legume, is regarded as a lucky charm among the Luvale (22); and in Tanzania, *Commelina bracteosa,* a herb, is put to magic nostrum for coping with bad luck in the Tanga Province (2a).

In Senegal, the root of *Neocarya (Parinari) macrophylla* Ginger plum, a forest tree with edible fruits, enters into charms against bad luck (11); and the branches of the tree are placed before new huts about to be occupied for the first time as a fetish to bring luck (12). In Ghana, the leaves of *Lawsonia inermis* Henna with palm kernels are immersed in a mixture of shea-butter and palm kernel oil and left in the sun to melt away the evil and diffuse to a red colour. The preparation is applied as charms for good luck (3). In this country, the edible fruits of *Drypetes floribunda,* a small tree in dry forest and savanna regions, is believed by some tribal groups on the Accra Plains to give protection against misfortune in the forest (9). In S Nigeria, the Yoruba invoke *Amorphophallus abyssinicus,* an aroid, in an Odu incantation to 'get good luck' (20). In this country, the leaves of *Lophira alata* Red ironwood, a large forest tree with hard timber utilized for railway sleepers, are used as charms to bring luck by semi-Bantu tribes (14). In N Nigeria, the planting of *Boswellia dalzielii* Frankincense tree as a live fence is believed to prevent bad luck, since the Hausa name 'hano' is supposed to be derived from 'bana', meaning to prevent or to obstruct (5).

In Liberia, *Memecylon memoratum,* a shrub or small forest tree, has magical attributes. A witch-doctor can produce good luck for those who bring the tree (himself) a present before he cuts a piece of the bark to make a sort of snuff (4). This is compounded with a white substance gathered from the bush; and when the supplicant requires aid of the juju, the mixture is sprinkled on the forehead while at the same time crying out the magic word *'jaywree'* (4). In N Nigeria, the seeds of *Garcinia kola* Bitter cola, popularly called 'minchingoro' in Hausa, meaning 'male kola-nut', are traditionally valued and attributed with magical powers among the tribal groups that use it (see WEALTH: CHARM PRESCRIPTIONS). For instance, when the seed is chewed with a prayer first thing in the morning, it is believed to be a good luck charm throughout the day (3). Bitter cola is a rare, slow-growing forest tree valued in its growing area for the wood, but not the seeds. The tree is a first-class chew-stick, locally called 'tweapea' in Akan. Ironically, like kola-nuts, the seeds of *Cola acuminata* Commercial cola nut tree and *C. nitida* Bitter cola, have more value both socially and commercially and are also better appreciated as a stimulant by tribes in the savanna and Sahel regions than those in the forest zone where the plants grow.

Children's Cases

In the folklore of W Africa, *Portulaca oleracea* Purslane or Pigweed is a children's charm for luck, and an emblem of goodwill in its distribution area (5). In N Nigeria, necklaces made of the seeds of *Afzelia africana* African teak, a tree legume of savanna and fringing forests, are worn by children against bad luck (5). In the Region *Vernonia cinerea* Little ironweed is a good luck charm (3).

In the northern Peruvian Andes, *Huperzia* species Huaminga chica, a fern, is one of the plants used to promote good luck in children (6). The plant is put in a bottle with other magical plants, *Arcytophyllum nitidum* (hierba de la estralla), a mountainous plant and *Cyperus articulatus* Piri piri, a sedge; some perfume with seven drops of lagoon water and honey is added and magical words are spoken while naming the person that the ritual is made for (6). Different colours of the seeds of *Zea mays* Maize or Corn are also believed to be very potent amulet for good luck when carried in the pocket (6). In England, the seeds of *Entada gigas* or *E. rheedei (pursaetha)* Sea bean or Sea heart, tropical drift dissemules (for the meaning, see RELIGION: SYMBOLIC PLANTS: Sea-Beans), were given as good luck pieces to children likely to go down to sea, because the seeds had been washed ashore in sound condition (8). British children say *Trifolium* species, a four-leaf clover, is lucky or a love-charm (10).

Entada rheedei Sea bean or Sea heart 1/3 natural size

Games of Chance

Gambling, popularly called *chacha* in the Region, is one area where good luck is evidenced. In N Nigeria, *Biophytum petersianum* African sensitive plant, an erect annual weed, is well known in Kano the use of the roots, which are chewed with red natron (native sesquicarbonate of soda) and rubbed on the cowry shells before staking them in the gambling game *chacha* (5). In a medico-magical usage, the people believe that *Strophanthus gratus*, a liane with milky latex, brings luck;

Chrysanthellum indicum var. *afroamericanum*

so in Sierra Leone, a leaf is rubbed between the hands to be fortunate in gambling; similarly, when in danger of the ordeal or other trial by superstition (Lane-Poole fide 5). In the Region *Chrysanthellum indicum* var. *afroamericanum,* a savanna composite herb, enters rituals for luck in gambling (3). In Congo (Brazzaville), the leaf-sap of *Ageratum conyzoides* Billy goat weed, a composite, rubbed on the hands of card players 'improves' their luck (1). In S Nigeria, the seeds of *Canna indica* Indian shot, which are called 'ido', are used as counters in a game of chance also called 'ido', the name being taken from the Yoruba name of the plant (15). In Central African Republic, a gambler in Ubangi carries a piece of *Eleusine indica*, an annual grass, for good luck (21).

BAD LUCK CHARMS

Some plants could, however, be unlucky charms. For instance, *Sempervivum tectorum* Houseleek, a temperate plant, is often planted on cottages to keep slates in position; and traditionally it was considered unlucky to uproot them or to allow them to flower (16). In Japan, the flower heads of *Camellia japonica,* a wild forest shrub locally called 'tsubaki' drop off whole, not petal by petal. This made the samurai consider them an unlucky emblem, seeing in them too much resemblance to human heads falling (19). (See also CHARM).

ASSOCIATED PLANTS

In Senegal, *Cercestis afzelii,* a climbing forest aroid, is known as 'ka lonk' in Diola, meaning luck; and in Benin Republic, *Elaeis guineensis* var. *idolatrica* King oil palm is called 'fa-de', a name referring to the god of fate (5). In N Nigeria, the Hausa name for *Polycarpaea linearifolia*

and *P. corymbosa,* herbaceous annuals or perennials, 'mai-nasara', means luck bringer - as such the plants are used as 'good luck' charms (5). In S Nigeria, the Yoruba name of *Leonotis nepetifolia,* a robust herb, 'oka', has reference to good luck (2b). In Japan, *Ardisia crenata* Coral berry, a low evergreen pot plant locally called 'manryo', meaning 'ten thousand' has very lucky associations, especially for New Year - a 'ryo' was a golden coin (19). The common name of *Cordyline fruticosa* Good-luck plant or Tree-of-Kings, an introduced plant whose natural distribution extends from India, Indo-Malayan archipelago, northern Australia to New Zealand and Papua, New Guinea suggests its use for, or presence as a lucky charm.

References.

1. Bouquet, 1969; 2a. Burkill, 1985; b. idem, 1995; c. idem, 2000; 3. Cofie, (pers. comm.); 4. Cooper & Record, 1931; 5. Dalziel, 1937; 6. De Feo, 1992; 7. Everard & Morley, 1970; 8. Gunn & Dennis, 1979; 9. Hall, 1978; 10. Hepper, (pers. comm.); 11. Kerharo, 1966; 12. Kerharo & Adam, 1963; 13. Laburthe-Tolra, 1981; 14. Meek, 1931; 15. Newbury, 1938; 16. Perry & Greenwood, 1973; 17. Pitkanen & Prevost, 1970; 18. Oso, 1977; 19. Richards & Kaneko, 1988; 20. Verger, 1967; 21. Vergiat, 1970; 22. Watt & Beyer-Brandwijk, 1962.

32 MANHOOD

....but when I became a man,
I put away childish things.
1 Corinthians 13:11b

Strength and bravery or courage are the physical requirements of normal male development. In addition, potency, good looks and, above all, the ability to have an issue further testify to manliness. Impotency and sterility (both mishaps often blamed on witchcraft in traditional Africa) are despised in society. From the initiation of boys, through virility charms for adult male, to invigorating and longevity prescriptions for the aged, plants and their products play a significant role.

INITIATION

Among some tribal groups in the Region and elsewhere, youths are initiated into manhood as part of the normal male development. In N Nigeria for example, in a superstitious practice based on the durability of split stems and the bond of tribalism, *Loeseneriella (Hippocratea) africana,* a savanna liane in forest fringes, enters into prescriptions for decoctions drunk by Fula youths preparatory to the ordeal of manhood known as *sharo* (9). The seeds of *Datura stramonium* Trumpet stramonium, a plant with hallucinogenic properties, is added to a drink given to Fulfulde youths to incite them on entering the *sharo* contest, and at the ordeal of manhood (20). Similarly, Trumpet stramonium and *Datura innoxia,* a shrubby plant also with hallucinogenic properties, were at one time used by various Indian tribes during the initiation ceremony for boys entering manhood (22). In NE and N America, an intoxicating medicine is given to Algonkian Indians at the manhood initiation ceremony. Boys experience a form of mental disorders for 20 days, forgetting entirely their former life, to start adulthood completely oblivious that they had ever been children (26). (In this respect, the plant has application as a drug for 'brain-washing').

In Equatorial Africa, *Tabernanthe iboga* Iboga, an apocynaceous forest shrub with hallucinogenic properties, has for centuries been used by the Gabonese and other Africans in religious rituals relating to rites of passage from adolescence to adulthood, and also for healing purposes (8). In the ordeal of manhood of the Lalas (Lola area), fronds of *Asparagus africanus*, a climbing annual with numerous very spiny branches, are used to make a large head-dress for youths (9); and in Lesotho, the root-powder is rubbed into incisions on the body of boys in the initiation ceremonies to make them strong and brave (14, 31). In S Africa, milk in which the root of *Holarrhena febrifuga*, a tree with milky latex, has been boiled, is used to wash a boy entering puberty; or the seeds of *Datura* species, known as 'langboontjie' among schoolboys in country schools, is used in a kind of initiation ceremony, the aim being to administer sufficient to produce the inebriant effects (31). In Botswana, *Scilla saturata*, a lily, is used by the Sotho in the initiation ceremony of boys - it is thought to inebriate them, and for this purpose the plant is mixed with *Phygelius capensis*, locally called 'mafifi-matso', in kaffir corn porridge to which has been added a little of the flesh from an enemy who has been killed in battle (31). (For the initiation of girls, see preamble to FERTILITY).

STRENGTH

In S Nigeria, the Yoruba invoke *Dioscorea dumetorum* Bitter yam, a poisonous wild yam with prickly bulbils, in a sinister incantation to bestow virile power (28). In this country, two unidentified plants locally called 'otakiti' and 'masa' in Yoruba, meaning 'somersault' and 'don't run', respectively, are invoked to make an old man become strong, and to induce bravery (28); and the Igbo rub themselves with *Leucas martinicensis*, a labiate, for strength (7b). In N Nigeria, the root of *Pericopsis (Afrormosia) laxiflora* Satinwood, a savanna tree legume, is one of the constituents of a virility charm (19). In Gabon, *Combretum paniculatum* ssp. *paniculatum*, a shrub, is a fetish plant for muscular strength (30); and in Cameroon, the bark of *Baillonella toxisperma* Djave nut, a large forest tree, has superstitious uses in native medicine to ensure strength (9). This is likely an instance of sympathetic magic referred to the size of the tree. Similarly, in Sierra Leone, the Temnes call *Adansonia digitata* Baobab, 'anderabai', meaning 'the chief's body', alluding to the belief that a root decoction taken with food makes a man stout (7).

In Kenya, a bark and root decoction of *Acacia nilotica* ssp. *subulata*, a shrubby savanna legume, is drunk by the Masai youth to acquire strength

and courage; and in Botswana, *Cymbopogon dieterlenii*, a grass locally called 'lebatjana', is used by the Southern Sotho, with *Elyonurus argenteus* Lemon-scented grass, called 'hloko', among other things, as 'moditola', a medicine to make a young man strong and true (31). In Zambia, the Luvale use a cold infusion of the root of *Psydrax (Canthium) venosa,* a straggling shrub, to 'make a man strong' (31). In W Africa, some *asuman* (fetish) carry names and come close in importance to *abosom* (deities). Among these is 'Makekamboso' which gives physical strength (12).

LONG LIFE

The Baobab is one of the longest-lived trees. The Temne name 'anderabai' could alternatively be a reference to long life. The age of a Baobab cut down for the Kariba Dam project was determined by C^{14} dating as 1010+- 100 years old, with an inference that really large individuals could indeed, be several thousand years old (27). Some trees are reckoned to be 4,000 years old (13, 21). This is probably an over-estimation since the oldest living trees in the world, *Sequoiadendron giganteum* Great sequoia or Wellingtonia in the Sierra Nevadas of C California are reckoned to be 4,000 years - with heights reaching 360 feet or more. The oldest tree in the world is *Pinus longaeva* Bristle-cone pine also called the Methuselah tree, and it is estimated to be over 4,700 years old. However, *Taxodium mexicana* Mexican cypress, a relative of the sequoia, holds the record for circumference. The most famous specimen is located near the city of Oaxaca, Mexico, in the town of Santa Maria del Tule, and its base measures an impressive 150 feet. It is supposed to be from 4,000 to 5,000 years old (21).

A few other trees are known to be long-lived. The curious *Xanthorrhoea preissii* 'Blackboys' or Grass-trees, the long-lived xerophytic perennials of south-western Australia, with thick woody palm-like trunks are also believed by tradition to grow one inch in 100 years (13). *Dracaena draco* Dragon tree from the Canary Islands has the reputation of being exceedingly long-lived (22). It is an exceptionally long-lived tree, and two specimens on the island of Tenerife are believed to be among the oldest trees in the world (18) . Similarly, *Welwitschia mirabilis* Welwitschia, the unique xerophytic plant of the Namib Desert in Namibia, have been proved to be centuries old. In venerable old age, the trunk expands into extraordinary mounds of pleats and folds like a giant fasciation; and from radiocarbon dating the biggest, which are said to attain four metres across, must be in excess of 2,000 years (25). 'Welwitschia plants grow slowly and often live 1,000 to 2,000 years' confirms *The World*

Book Multimedia Encyclopedia. The plant actually has only two leaves. The original pair continue growing for the entire life of the plant. An Africaans name for it, 'tweeblaar-kanniedood' literally means 'two-leaf-cannot-die'.

'In the book, *The Mystic Mandrake*, the author refers to a Russian gentleman who looked like a man of forty, but who really was seventy years of age. When asked about his secret of youth, he replied that it was his constant use of *Mandragora officinarum* Mandrake, a herb renowned since ancient times for its supposed magical and rejuvenating properties' (23). Tibetan yogis in the snowy heights of the Himalayas are claimed to achieve remarkable longevity on a diet composed of *Hordeum vulgare* Barley flour, plus wild roots (23). The blossom and fruit of *Prunus persica* Peach, an introduction to the Region, were symbols of longevity in prehistoric China, and were traditionally given as presents on birthdays (11). Since the Bamboo becomes stronger as it grows, it promises healthy long life (13). The Bamboo includes many genera of plants. Among these are *Arundinaria, Bambusa, Dendrocalamus, Oxytenanthera, Phyllostachys, Sasa* and *Shibataea.*

In W Africa, *Pandanus* species Screw pine is a symbol of longevity. The tree is planted with rites by an elder of the family for the long life of the members. The belief is that the tree does not normally die unless deliberately uprooted or burnt (4). The species in the Region include *P. abbiwii* and *P. aggregate (candelabrum)*. *Tilia europaea* Linden is a majestic tree which lives for several centuries. In Europe, the tree symbolizes the family and homelife (21). In S Nigeria, the Yoruba invoke the leaves of *Dioscorea hirtiflora,* a wild yam, in an incantation 'to become old and to stay long on earth (28, Verger fide 7c). In Senegal, *Polycarpaea linearifolia,* a short erect annual or perennial, is an ingredient of numerous potions beneficial to mental health and longevity of old people (16a, b); and in Ghana, a stem bark decoction of *Khaya senegalensis* Dry-zone mahogany taken regularly by draught as 'okum-adadaa', meaning 'removing aged features', is attributed with similar properties among numerous others. The stem bark is sold in medicine markets for this and other ailments. Where the tree is planted as an avenue, it is constantly debarked. This practice renders the suitability of the tree for avenues questionable.

In Europe, the roots of *Eryngium maritimum* Sea hollies preserved with sugar are believed to be exceedingly good when given to old and aged people that are consumed and withered with age - they have the property of nourishing and restoring the aged and amending the defects of nature in the younger (22); and *Melissa officinalis* Lemon balm tea was believed

Ginkgo biloba Ginkgo or Maidenhair tree (legendary and mythical plant; symbol of longevity; associated with Islam and Oriental religion)

to ensure a long and trouble-free life (3). In Japan, China and some other countries in the East, a potion preparation from *Ginkgo biloba* Ginkgo or Maidenhair-tree, a living fossil and perhaps the oldest tree available, or symbol of longevity, is drunk for longevity - an instance of sympathetic magic. Ginkgo has been described as a well known vegetal fossil which has been growing on the earth crust for over 160 million years (5). The extract represents one of the most important and valuable therapeutic tools for the treatment of...headaches and ageing-related problems in general (5). In Japan, the evergreen Pine symbolizes long life and constancy; and *Prunus pendula* Weeping cherry is known to be more than 1000 years old (24). Another fossil plant is *Metasequoia glyptostroboides* Fossil tree which was thought to be extinct, until it was found growing in China (24). For 5,000 years, the entire population of China and Korea, over a billion, that is 1000,000,000 people, have placed unlimited faith in *Panax schinseng (gingseng)* and *P. quinquefolium* Gingseng as a herbal elixir of life for the restoration of youth to the aged (23). Among the attributes of *Centella asiatica* Indian navelwort or Gotu kola, is the saying that 'two leaves a day will keep old age away', this is the claim made by the ancient Sinhalese for Gotu kola (23). The plant is widely distributed in tropical Africa, Asia and Australia.

POTENCY

Another method of invigorating the system is by administering an aphrodisiac - a popular item of medicinal plant peddlers. The Greeks dedicated *Punica granatum* Pomegranate to the god Aphrodite since its fruit was alleged an aphrodisiac (21). It is any food, drink or drug that arouses sexual desire. Potency is the true mark of manhood among many cultures worldwide. A man with erectal disorder is classified as 'a woman'. However, the potent man is not necessarily fertile.

In S Nigeria, the Yoruba invoke *Stachytarpheta cayennensis,* Rat tail verveine, a semi-woody herb of wastelands, in an Odu incantation to be granted virile power; and the Igbo people invoke *Solanum incanum* Egg plant or Garden egg to show virile power superior to that of 200 hunting dogs, *Canis familiaris,* unable in chasing a leopard, *Panthera pardus,* to catch it (28). An unidentified plant locally called 'atabese' in Yoruba, meaning 'pepper on muddy floor of bathroom' is invoked to obtain virile power (28). In Ivory Coast, *Oldfieldia africana,* a large tree, is considered a powerful fetish tree with efficacious medicinal virtues. The bark is added by medicine-men to prescriptions to increase potency; and the leaves of *Phyllanthus muellerianus,* a shrub or climber, are eaten in the belief that they promote male fertility (1). In Congo (Brazzaville), there is a fanciful attribute of the leaf of *Icacina mannii,* a scandent forest shrub. The belief is that if a leaf is held in the hand during sexual intercourse, ejaculation will be delayed, and will only happen when the leaf is placed on the head (6). Among others, *Inula helenium* Elecampane, a shrubby temperate composite, awakens genital virtues when eaten (21).

In the Eastern Province of Zambia, Nyanja boys of twelve or thirteen take a root decoction of *Ochna mossambicensis,* a shrub, in the form of a porridge 'to make them men when they meet a girl'; or apply the sap of *Vitex mombassae,* locally called 'mfutu' to make the penis grow bigger (31). In S Sudan, the Azande tie *gbaga* (the fruit of a palm-tree) to their girdles so that they be very potent sexually and have powerful erections. It is medicine for masculinity (10). Another charm for potency worn by the tribe appears to be the seeds of *Entada gigas* or *E. rheedei (pursaetha)* Sea bean or Sea heart (10). In this country, the root of *Pericopsis (Afrormosia) laxiflora,* a medium-sized savanna legume, is recorded as a virility charm (Meek fide 9). In Congo (Brazzaville), *Pachyelasma tessmannii,* a forest tree, is full of strong juju. It is considered to be the most potent aphrodisiac known (6). In this country, some people believe that if the fruit pulp of *Dracaena arborea,* a palm-like tree, is rubbed on the penis, erection in intercourse is prolonged (6).

In W Africa, the Ewe-speaking tribe of Ghana, Togo and Benin Republic however, do believe that deliberately urinating on the leaves of *Sida linifolia,* a malvaceous annual believed to be fetish (4), and *Bryophyllum pinnatum* Resurrection plant (2) results in erectal disorder. In Congo (Brazzaville), *Olax latifolia,* a shrub or small tree, is deemed to have aphrogenic and magical properties. It is said that women may eat the leaves to make men impotent (6). The Peul of Senegal ascribe medico-magical powers to *Pennisetum polystachion,* a polymorphic grass,

annual or perennial, particularly with a corn cob and *Tapinanthus* species Mistletoe, epiphytic parasites, in treating erectal disorders (16a, b).

Sympathetic Magic

Some prescriptions for potency, however, are based on sympathetic magic - like the use of a rhino horn as a popular aphrodisiac, because the male rhinoceros, *Diceros bicornis*, is sexually active. In Tanzania, the dried and pulverized stems of *Sarcostemma viminale*, a leafless climbing shrub with milky latex, are mixed with the dried powdered penis of a ram, *Ovis* species, and taken with milk as an aphrodisiac (Koritschoner fide 7a). In W Africa, the use of the bitter yellow roots of the shrubby forest menisperm, *Penianthus zenkeri* and *Sphenocentrum jollyanum*, as aphrodisiac, either as chewing sticks or in 'bitters', could most probably have initially been deduced from the pulp of the edible orange fruit which is slimy like semen. The convoluted roots of *Strophanthus hispidus* Arrow poison are similarly peddled as an aphrodisiac, because they resemble the male organ of the duck, *Anas* species.

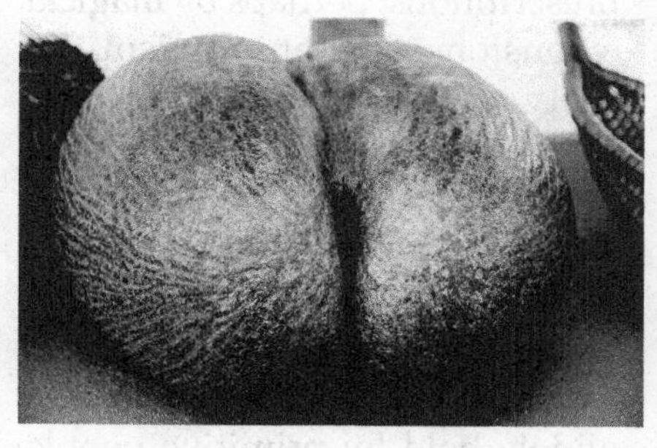

Lodoicea maldivica Coco-de-mer -1/20 natural size (legendary and mythical plant; associated with traditional religion)

The traditional belief by owners of *Lodoicea maldivica* Coco-de-mer that the endoderm is an aphrodisiac (15) is probably purely sympathetic magic based on the undeniable resemblance of the nut to certain portions of a woman's anatomy. It could also be based sympathetically on the germinating seed - which to all decent people appears most vulgar. The mysterious sexual powers attributed to *Mandragora officinarum* Mandrake (11), was also probably simply associated with the man-shaped roots. 'A mandrake root, in forked female form was considered a powerful aphrodisiac and still is considered as such in many parts of the world today. The same virtues

Lodoicea maldivica Coco-de-mer - (Habit)

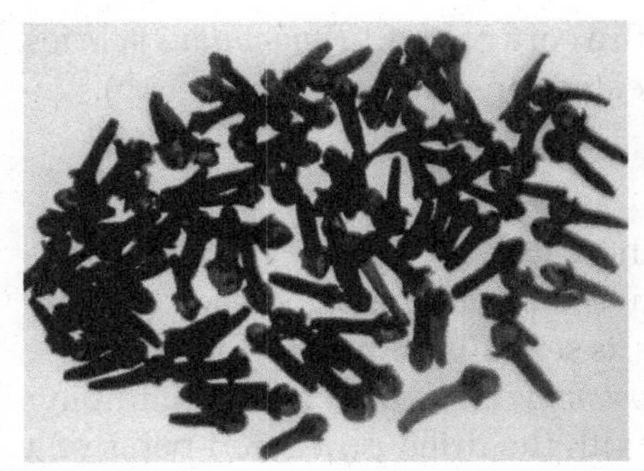
Eugenia caryophyllus Clove 1/2 natural size

were attributed to Ginseng by the Chinese' (23). In the Middle Ages and the Renaissance, herbalists and apothecaries saw in *Eugenia caryophyllus (caryophyllata)* Clove the representation of an erect penis, with the testicles at the base; therefore, it was supposed to act on the male genitalia (21). Similarly, the flower of *Clitoria ternatea* Blue pea with small keel relative to the large standard suggests mammalian clitoris, as such the plant is used as an aphrodisiac.

In the Region, the stem of *Listrostachys pertula*, an epiphytic orchid, is reported, but without detail, to be an ingredient of aphrodisiac prescriptions, perhaps by magical extension of its epiphytism (9). In Ivory Coast, the mucilage produced from macerated whole plant of *Hybanthus enneaspermus*, an erect savanna bush, is eaten by the Bété in treating male sterility. This is perhaps based on the Doctrine of Signatures for it is widely held by primitive people that male sterility - as distinct from erectal disorder - is due to excessive fluidity of the semen which the mucilage can rectify (17). In Central African Republic, the root of *Becium obovatum* var. *hians*, an annual herb labiate with a hard perennial rootstock, is used in an unexplained manner 'to harden the penis'. This seems to be a phytopromorphic erotic fantasy to simulate the hard woody and knobby root-stock (7b).

Clitorea ternatea Blue pea.

In S Nigeria, *Eriosema pulcherrimum*, an erect savanna leguminous herb, is held to have superstitious uses (9). The root is somewhat fleshy and forked, and contains a blood-red sap which coagulates in air. On an anthropomorphic basis it is endowed with aphrodisiac properties in Ubangi, Central African Republic (29). In Senegal, a fanciful aphrodisial application by Pulaar folk is recorded in which the gum of *Faidherbia (Acacia) albida*, a large savanna tree legume, is taken with the meat of a bull, *Bos indicus* and *B. taurus*, preferably the testicles, for erectal disorder;

or cooked with various things of which the viscera of a porcupine, *Artherurus africanus,* is the most esteemed because the animal's diet consists of plants considered to be beneficially active (16a, b). In this country, *Tapinanthus bangwensis* Mistletoe, an epiphytic parasite, is used by the Basari in medico-magical clinical treatment for erectal disorder and to break spells (16a, b). (For further instances of plants believed to have aphrodisiac properties, see LOVE: ASSOCIATED PLANTS).

References.

1. Adjanohoun and Aké Assi, 1972; 2. Ashieboye-Mensah, (pers. comm.); 3. Back, 1987; 4. Badu, (pers. comm.); 5. Bonati, 1992; 6. Bouquet, 1969; 7a. Burkill, 1985; b. idem, 1995; c. idem, 2000; 8. Cantor, 1990; 9. Dalziel, 1937; 10. Evans-Pritchard, 1937; 11. Everard & Morley, 1970; 12. Fink, 1989; 13. Graf, 1978; 14. Guillarmod, 1971; 15. Gunn & Dennis, 1979; 16a. Kerharo & Adam, 1964; b. idem, 1974; 17. Kerharo & Bouquet, 1950; 18. Lanzara & Pizzetti, 1977; 19. Meek, 1931; 20. Oliver Bever, 1983; 21. Pamplona-Roger, 2001; 22. Perry & Greenwood, 1973; 23. Pitkanen & Prevost, 1970; 24. Richards & Kaneko, 1988; 25. Rowley, 1972; 26. Schultes, 1972; 27. Swart, 1963; 28. Verger, 1967; 29. Vergiat, 1970; 30. Walker & Sillans, 1961; 31. Watt & Breyer-Brandwijk, 1962.

33 MARKETING

The wicked one is making false wages,
but the one sowing righteousness, true earnings
Proverbs 11:18

The open display of goods and services for sale is a competitive one. As such it is the practice among some traders, big and small, to acquire a fetish, juju prescriptions or other charms for success in marketing. Prosperous traders may therefore be the centre of gossip as resorting to a fetish, witchcraft or some local deity for the secret behind their success. It is the belief that failure in marketing is equally to be blamed squarely on evil influences, therefore the necessary precautionary measures are usually taken to counteract these forces. Plants do feature in some of these charms, rituals and prescriptions for marketing, or to prevent money from being snatched spiritually. Some plants are also said to be used against marketing and for bargaining.

CHARM PRESCRIPTIONS

In N Nigeria, the root of *Cyperus articulatus,* a sedge with aromatic tubers, enters into prescriptions to ensure success in trade (13); and traders and travellers in this country carry *Polycarpaea linearifolia,* an erect annual or perennial herb, for protection (Thornewill fide 4). The flowers of *Ipomoea argentaurata,* a trailing convolvulus, open in the morning, and the Hausa name 'farin gamo', means 'luck in trading early in the morning' (6). In Ghana, the fruit of *Solanum americanum (nigrum)* Black nightshade, an erect common weedy annual herb, with the flowers of *Ageratum conyzoides* Billy goat weed, both collected with rituals, ground with frankincense and dusted on wares, enter into prescriptions as a charm for marketing by the Ga-speaking tribe (12). In this country, the tiny dark fruits of Black nightshade with the leaves of *Phyllanthus niruri* var. *amarus (amarus),* an annual weedy euphorb herb, are similarly used with rites for the same purpose (12) - (but see below).

In its distribution area, *Triumfetta cordifolia* Bur weed is reputed in

Euphorbia hirta Australian asthma herb

trade charms by fetish priests and other practitioners, but the methods employed in its collection and preparation are ceremonial (14). This is likely an instance of sympathetic magic hinging on the hooks of the burs. To attract customers, *Euphorbia hirta* Australian asthma herb is approached and offered a hen, *Gallus domestica,* (the type with frizzle feathers) by slaughtering it, and sprinkling the blood on the herb. Other species of the plant nearby are then collected and placed over the doorway of the shop to complete the charm (14). In the Region, *Euphorbia prostrata,* a trailing weed with milky latex, dried then ground in lavender, is applied as cosmetic by traders to attract customers (14).

In its distribution area in the Region, the dried prickly petiole of *Lasimorpha senegalensis (Cyrtosperma senegalense)* Swamp arum, an aroid often growing in swampy areas in the forest, charred with the seeds of *Aframomum melegueta* Guinea grains or Melegueta and salt is applied as cosmetic to the skin either in lavender or pomade as a vehicle; or the charred powder rubbed in incisions as a charm for success in marketing among the Akan-speaking people of Ghana and Ivory Coast, the Nzima-speaking people of Ghana and some other ethnic groups (7). In N Nigeria, 'sarka' means 'attract market' in Hausa, thus the leaves of *Asparagus* species, climbing lilies in savanna, locally called 'sarka', collected with the necessary rituals, are dried and ground in lavender which is applied as a cosmetic, and sprinkled on wares to attract customers (15). The whole dried plant may also be burnt as incense in a ritual fumigation within the shop for the same purpose (15). The species in the Region are listed in Table 1 below:

Species	Plant Family	Habit and Habitat
Asparagus aethiopicus	Liliaceae (Asparagaceae)	a scandent or climbing plant in savanna
A. africanus	Liliaceae (Asparagaceae)	an erect plant in wooded savanna
A. falcatus	Liliaceae (Asparagaceae)	a climbing or erect plant in savanna
A. flagellaris	Liliaceae (Asparagaceae)	a scandent plant in woodland savanna
A. pubescens	Liliaceae (Asparagaceae)	a trailing or branching plant in savanna
A. racemosus	Liliaceae (Asparagaceae)	a creeping or climbing plant in savanna

A. schroederi	Liliaceae (Asparagaceae)	an erect spiny herb of savanna
A. setaceus	Liliaceae (Asparagaceae)	a trailer-climber, cultivated ornamental
A. sprengeri	Liliaceae (Asparagaceae)	a much-branched climber, cultivated
A. warneckei	Liliaceae (Asparagaceae)	a climber over shrubs in savanna

Table 1. Asparagus species in the Region

Among the Hausa of N Nigeria, the Zarma of Mali and Niger and the Fula the leaves of *Calotropis procera* Sodom apple, sometimes with the flowers, collected on Fridays, dried and tied in a white calico among wares, enters into prescriptions by medicine-men for marketing (15). In these countries, the flowers and leaves of *Crescentia cujete* Calabash tree or the leaves alone; or either *Pancratium trianthum* or *P. hirtum* Pancratium lily, herbaceous annual lilies, collected with rituals, dried and tied in a white calico among wares is reported to promote marketing (15). The dried powdered whole plant of *Scoparia dulcis* Sweet broomweed, a herbaceous annual weed of cultivated land, with local soap used for a ceremonial bath early in the morning is a charm for success in marketing; and *Eragrostis tremula,* a tufted annual grass, enters into rituals for marketing or for prospects in business among some tribal groups in its distribution area (14). (For the full rituals see CEREMONIAL OCCASIONS: PURIFICATION & SACRIFICE: Purification).

Crescentia cujete Calabash tree.

Among many tribes in the Region, the dried powdered leaves of *Lawsonia inermis* Henna, a shrub in open savanna often cultivated for hair- or nail- or skin-dyeing, with the powdered seeds of the white variety of either *Cola acuminata* Commercial cola nut tree or *C. nitida* Bitter cola in water for a ritual bath, is a charm to attract customers (5). It must be the white variety of kola-nuts, because the nuts of the red variety are rather believed to repel customers (5). Similarly, the leaves of *Phyllanthus niruri* var. *amarus (amarus)*, a weedy herb with sub-woody stem, chewed together with the tubers of *Manihot esculenta* Cassava and sprinkled on an enemy's wares with a spell, is a sinister charm to prevent or inhibit sales of this rival's goods (8). In Ivory Coast, medicine-men use the leaves of *P. niruroides,* a semi-woody euphorb herb, which has

been macerated and kept for 7 days, to wash their faces to improve business (9).

In W Africa, a talisman or an arm bangle stuffed with dried powdered *Oldenlandia corymbosa*, a prostrate weedy herb sometimes tended or even potted by fetish priests and medicine-men, previously collected with rituals, then libation offered during the preparation, is worn as a charm for bargaining (14). The ware is taped by the one wearing the amulet or talisman, who then calls any price, and the seller would invariably accept the offer (14). In Tanzania, witch-doctors put the leaves of *Oldenlandia herbacea*, an annual weedy herb, in a shop to attract customers (11). In Congo (Brazzaville), a few leaves of *Boerhaavia diffusa*, a weedy annual herb, carried on the person will ensure success in commerce or any enterprise involving risk (2). In this country, medicine-men hang a piece of the liane of *Synclisia scabrida*, a forest menisperm, at the door of their house to secure good patronage (3).

With some of the tribal groups in the Region, the dried ground whole plant of *Solenostemon monostachyus*, a herbaceous labiate, or *Vernonia cinerea* Little ironweed, an annual composite weed, in lavender sprinkled on the body as cosmetic and over the wares, serves as charms by medicine-men for success in marketing (14). *Abrus precatorius* Prayer beads enters into charms to attract customers. The red and black shinny seeds ground together with sixteen seeds of *Aframomum melegueta* Guinea grains or Melegueta, then rolled into balls and dried like poultice are placed among the wares. It is important to remove the charm from the wares as soon as customers flock round (14). A young seedling of *Bryophyllum pinnatum* Resurrection plant which had been raised on one's compound with assorted ingredients - including a few seeds of Guinea grains, one white kola-nut and one seed of *Cola verticillata*, a forest tree with slimy three-cotyledonous seeds, are all placed in a hole, with a bottle of mineral water or soft drink. The items are all covered up with soil while libation is being poured. The ritual is a charm to attract customers - and for other requests that may be demanded (14). The bottle of soft drink is a symbol of gratitude in advance.

In S Africa, articles for sale by an African, such as wooden pillows, spoons, bowls, sticks, earthenware pots and mats are sprinkled with a decoction of a *Pelargonium* species Wild geranium to ensure a quick sale; and in Botswana, if cattle are washed with an infusion of *Anthospermum pumilum*, locally called 'masopolohane', a quick and profitable sale will ensue (19). In Central African Republic, birds are said to eat the seeds of *Ficus asperifolia* Sandpaper tree whereby there is a belief that mistletoe parasitizing this plant brings good luck in commerce; that is

customers will come as eager for business as the birds are for the seeds (4b). Similarly, in S Nigeria, Yoruba wishing to establish a new market, invokes *F. sur (capensis),* a fig tree with abundant fruits, since the trees attract wild animals, and people, and they will come to do trade (17).

In S Nigeria, *Termitomyces microcarpus,* an agaric mushroom, is used medicinally by Yoruba native doctors as an ingredient in preparation of a charm supposed to bring good luck, particularly to traders (16). To prepare this charm, 200 fruiting bodies of the mushroom are pulverized by roasting in a pot together with ripe banana, pawpaw, salt and some herbal ingredients. The preparation is collected on a piece of white cloth, and certain magical verses are recited upon it (16). This is then tied with thread and hung above the door of the trader's shop. It is strongly believed that this has the power of promoting the sale of articles by drawing the buyers into the shop (16). In this country, Yoruba invoke *Sesamum orientale (indicum)* Sesame as an unfailing sales agent, attracting customers to a market; and the Yoruba name for *Phaseolus lunatus* Lima bean or Broad bean, 'ere', meaning 'profit', derives from an incantation 'to find something lost' (17).

Fetish

Describing the silent trade, it has been recorded that 'against each class of articles, so many cowry shells or beans are placed, and, always hanging from a branch above, or sedatively sitting in the middle of the shop, a little fetish' (10). 'The number of cowry shells or beans indicate the price of the individual articles in the various heaps, and the little fetish is there to see that anyone who does not place in the stead of the articles removed their proper price, or meddles with the till, shall swell up and burst' (10). In fact, the practice - or a slight modification of it - exists, and it is still to be seen among certain tribal groups in W Africa to date. For instance, in the sale of sugarcane and other farm produce in parts of the Nzema area of Ghana, a few coins are displayed in a calabash container by the roadside with the produce, to indicate the price per unit article; and openly attached to a nearby post - a fetish. The settlement is usually linked by a narrow footpath some 100 and 250 metres off the road.

The plants which enter into prescriptions for marketing, are on the whole also generally preventive of witchcraft and evil forces, or are symbolic of luck. It is, therefore, apparent that these forces do inhibit sales.

PREVENTING MONEY SNATCHING

It is traditionally believed among many cultures worldwide that evil spirits, bad juju, black magic or witchcraft and dwarfs are capable of snatching money from individuals, traders, merchants and banks. The snatching could either be by 'remote control' or it could be by physical contact. In the latter instance a customer buys some goods and when his or her money is added on to the rest, it effects the snatching. In W Africa, one occasionally finds a bank note (especially notes in the bigger denominations) with one corner torn off. The general belief again is that the practice counteracts spiritual snatching.

As a further preventive of the practice, the stone (excrescences of calcium malate) that is sometimes found in the heartwood of *Milicia (Chlorophora) excelsa* or *M. regia* Iroko with the fruit of *Capsicum frutescens* Chillies and a millipede, *Iulus terrestris*, are burnt together into black powder and sprinkled in the container in which the sales are kept (14). The leaves of Iroko rubbed in the palm of the hands before receiving the money is an additional measure of protecting it from being snatched by these forces (14). On the contrary, it is the belief that money saved in an Iroko chest could be mysteriously halved or totally spirited away. Among both commercial traders and some market women in the Region, the clawed seeds of *Martynia annua*, a decorative foetid weed sometimes cultivated, when kept together with money, is believed to prevent it from being spiritually snatched away by evil forces (1). The seeds are, consequently sold by medicine peddlers for this purpose, among others.

With the Yoruba of S Nigeria, to relief oneself from money which is obtained but which apparently disappears without having been spent on anything tangible, the prescription consists of a leaf of *Spondias mombin* Hog plum or Ashanti plum with the gland which produces a smelly fluid in the civet cat, *Viverra civetta*, and palm oil as ingredients. "Use it to wash hands, drop palm oil in the hands everyday, the hands will become tough and the medicine will start to hurt the hands "(18). Alternatively, a type of spinach, a rat, *Cricetomys gambianus*, and hailstone are pounded together with soap and used to wash the hands; or the whole pod of *Aframomum melegueta* Guinea grains or Melegueta with a particular leaf (unspecified), the head of a type of snake (unspecified) and the tail feather of the grey parrot, *Psittachus erithacus*, as ingredients are pounded and rubbed into the cuts made around the wrist of both hands (18).

References.

1. Afenu, (pers. comm.); 2. Bouquet, 1969; 3. Bouquet & Debray, 1974; 4a. Burkill, 1985; b. idem, 1997; 5. Cofie, (pers. comm.); 6. Dalziel, 1937; 7. Eshun, (per. comm.); 8. Gamadi, (pers. comm.); 9. Kerharo & Bouquet, 1950; 10. Kingsley, 1901; 11. Kokwaro, 1976; 12. Lamptey, (pers. comm.); 13. Meek, 1931; 14. Mensah, (pers. comm.); 15. Moro Ibrahim, (pers. comm.); 16. Oso, 1977; 17.Verger, 1967; 18. Warren, Buckley & Ayandokun, 1973; 19. Watt & Breyer-Brandwijk, 1962.

34 MARRIAGE

What therefore God hath joined together,
Let not man put asunder.
Matthew 19:6b

Traditional marriage is a bond of friendship between an adult male and female as husband and wife - after the payment of a fixed dowry and performance of the necessary customary rites (both varying from region to region, or from tribe to tribe). Like all institutions, marriage is bound by laws and regulations – also varying according to the culture. The main difference between traditional and western marriage is that the former may be polygamous - that is allowing more than one wife per husband, while the latter is strictly monogamous - that is one husband to one wife. Cases of polyandry where a woman is permitted to have, and many do have more than one husband, occurs in Tibet and parts of the Far East.

In traditional Africa and elsewhere, plants or plant prescriptions feature in the various aspects of this union. Among these are the timing, the selection of a partner, payment of the dowry, stability in marriage, faithfulness of the partners (especially the wife) and rivalry in marriage. There are prescriptive charms both for and against seduction and rape. There are also plants associated with western marriage.

TIMING

With some of the tribal groups, the fruiting of the 'birthright' tree signifies time for marriage age. For instance, in Central African Republic where a tree is planted for every newborn child by the Oubangui tribe (see CHILDBIRTH: POST-NATAL), it is the belief that when the tree begins to fruit, the time will have come for the child to marry (31).

SELECTION

Some plants are also believed to assist in selecting a suitable partner. For example, some Hausa names of *Bauhinia rufescens,* a shrub or small savanna tree legume, are derived from *tsage,* meaning 'to split'; thus a youth will tear the stem apart to find an omen when he is thinking of choosing a bride (11). In S Africa, the Zulu use a root infusion of *Dianthus crenatus* Wild pink with *Tephrosia lurida,* a legume, as an emetic, accompanied by washing the face with the froth which rises from the infusion. The treatment is used where an unmarried person of either sex fails to find a mate (35). In this country, the Zulu and the Thonga use a decoction of the bark of *Sclerocarya caffra* Cider tree as ritual cleansing emetic before marriage (35). In Japan, an Ainu legend tells the story of a Goddess turned into *Adonis amurensis,* a garden plant with yellow flowers, when she refused to marry the god of her father's choice (29).

In S Nigeria, a number of plants are invoked by the Yoruba in an Odu incantation to obtain a wife (32). Table 1 below lists some of these plants:

Species	Plant Family	Habit and Habitat
Cochlospermum planchonii	Cochlospermaceae	a shrub in savanna woodland
Dracaena fragrans	Agavaceae	a shrub or small tree often planted
Sarcophrynium brachystachys	Marantaceae	a semi-woody rhizomatous herb in swamps
Sesbania pachycarpa	Papilionaceae	an erect legume herb of muddy swamps
Spigelia anthelmia	Loganiaceae	an annual herb in waste places

Table 1. Plants Invoked by the Yoruba to Get a Wife

Spigelia anthelmia

In Europe, Bachelor's button was a name given to flowers of a common genus *Lychnis,* also sometimes to other flowers like *Gomphrena globosa* Bachelor's button. They were so called because they were said to be carried about by bachelors in the hope that if they flowered the carrier would be successful in love (10). In Congo (Brazzaville), a young man out courting washes his face with the sap of *Cyathula prostrata,* a herb, then chews a piece

of twig used as a chew-stick to advance his courtship (27). In Europe, *Pimpinella anisum* Anise, a plant in the Umbelliferae family, was called 'husbands are back' since those husbands who leave home due to the bad breath of their wives, when the latter take Anise, the former are back (25). There is a belief that when an orange or an apple peel is thrown over one's shoulder or when apples were paired and thrown down on the day of Saints Simon and Jude, they are supposed to represent the initials of the future partner. This is a form of divination. Divination by means of plants is called *botanomancy,* and it is largely used for revealing the future of one's husband or bride, and the character of the same (10). (For more instances of divining with plants see FORTUNE-TELLING AND DIVINATION).

DOWRY

In Hausaland, 'a calabash of kola-nuts (the seeds of *Cola acuminata* Commercial cola nut tree or *C. nitida* Bitter cola) and 15,000 cowry shells are always sent to the female when the suitor proposes marriage or otherwise, and their acceptance or rejection signify her gratification or displeasure with the offer. As kola-nuts are said to be aphrodisiacs, there may be something symbolical in this gift (30). In the Upper East area of Ghana, the dowry was paid in three portions. The first portion, at the time of asking for the girl in marriage, consisted of kola-nuts, *Pennisetum glaucum (americanum)* Pearl, Bulrush or Spiked millet, and Guinea fowl, *Numida meleagris.* The second, at the time of marriage, consisted of four cows, *Bos* species; and the third was a goat, *Capra* species, to be delivered if the woman gave birth to a boy (20). In S Nigeria, payment of dowries and various ceremonies among the Igbo also involve specific numbers of yam tubers (22). In this country, a plant locally called 'musaya' in Yoruba, meaning

Pennisetum glaucum Pearl, Bulrush or Spiked millet

'take a wife', is used to obtain a wife without paying the bride-price (9).

Among some of the tribal groups of the Guinea-savanna where *Nicotiana tabacum* Tobacco is cultivated, the dried leaves form part of the marriage dowry. In N Nigeria, *Lawsonia inermis* Henna, a popular skin dyeing plant features prominently in Hausa marriage rites - both for the bride and the bride-groom (30). With the exception of two other instances - one recorded in Senegal (11) (see CHILDBIRTH: POST-NATAL), there appears to be no other record of the use of raw plant product as dowry in W Africa. An indirect method is the 'money to buy a wife' among the Fan tribe in Congo (Brazzaville) (18). 'This he does by collecting ebony and rubber and selling it to the men who have been allotted goods by the chief of the village. It takes a good time to get enough rubber to buy a lady' (18). The ebony was then obtained from *Diospyros* species, and the rubber from *Landolphia owariensis* Vine rubber. In China, *Paeonia officinalis* Common paeony was offered in marriage dowries during the Tang dynasty (A.D. 618-906) (26).

Bartering with plant products was, however, common in ancient times. In addition, kola-nuts were employed as money for local trading. In fact, so precious was the commodity that 'to eat kola-nut with, or present some kola-nuts to, a Mandingo or Wolof, places a stranger on the same footing as the tasting of salt does with an Arab; and after such a ceremony one is entitled to protection and assistance. A kola-nut is a good kind of passport and *viséd* for any Moslem town' (13a). Similarly, in Peru, the seeds of *Theobroma cacao* Cocoa served as coinage among the Incas (see WEALTH: ASSOCIATED PLANT NAMES).

STABILITY IN MARRIAGE

Some plants are believed to assist in a successful marriage life. For instance, in Ivory Coast, the Maninka prepare a leaf macerate of *Santaloides afzelii*, a straggling shrub by streams and in the forest, and use the liquid as a wash in order to have a stable matrimonial home (2). In Burkina Faso, to drive away an evil force or spirit believed to destabilize and break up marriage or cause the foetus to abort, *Lippia multiflora* Gambian tea bush and the aromatic rhizomes of *Cyperus* or *Kylinga* species called 'turare-wuta' or 'kaajiiji' in Hausa both collected with the necessary ceremony, are burnt with a chameleon, *Chamaeleon vulgaris,* a common ingredient in fetish rituals, in fumigation on her birthday (23). In S Cameroon, the 'oven' tree, *Didelotia africana,* a legume, is approached by the Beti for help with difficult problems (such as broken

marriages), but can only be used by healers who have the power to communicate with it (3).

In S Africa, a Zulu couple who suspect that the discords in their life are produced by someone desirous of separating them, use *Hermannia depressa* Rooi opslag as a protective charm (35). In this country, a root decoction of *Acridocarpus natalitius,* a climbing shrub, is used by the African as a charm, applied by sprinkling, to prevent a wife from absconding (35). As charms to prevent a bride from running away from the husband among some of the ethnic groups of the Guinea-savanna, a piece of meat which, unknown to her, has been sprinkled with the powdered root-bark of *Annona senegalensis* var. *senegalensis* Wild custard apple, then tucked under the left armpit of the husband is cut up, roasted on skewers as *khebab* or *kyikyinga* and offered her (21). This is likely an instance of sympathetic magic. The powdered root-bark of *Gardenia ternifolia* ssp. *jovis-tonantis,* a savanna shrub traditionally believed to conduct lightning, enters into similar charms.

In W Africa, *Sida linifolia,* a little erect malvaceous weed of cultivated land popularly believed to be fetish, potted on either side of the doorway or the main entrance to the house with ceremonial rites, is invoked by the Ewe-speaking tribe in an incantation with a spell to postpone indefinitely a wife's intention or threat of running away from the husband (12). The local name of the herb, 'woduowugbugbo' literally means 'they planned but retreated'. In S Nigeria, *Melinis repens,* an annual grass, enters a Yoruba incantation 'to make a wife come back to her first husband' (32); and in N Nigeria, *Pistia stratiotes* Water lettuce is an ingredient of a medicine to stop a woman running away from her husband (8). In Senegal, the Tenda consider *Baissea multiflora,* a climbing shrub with white latex, with the Mistletoe growing on it a charm to be used to induce a woman to return to her husband (15). In Gabon, *Leonotis nepetifolia,* a robust bush, enters into a compact made between husband and wife and between lovers in a magical sense (33).

In Nigeria, *Crotalaria pallida (mucronata),* a legume, is used in a sort of love potion to induce the voluntary return of a wife, if she can be got to drink or eat it unknowingly (11). Similarly, *Tridax procumbens* Coat buttons, a composite, is used in rituals to induce a runaway wife to return home (see JOURNEY: COMPULSORY RETURN HOME). Among some of the tribal groups in the Region, the leafy juice of *Triumfetta cordifolia* Burweed is instilled into each other's eye by both partners in a ritual to secure and sustain the marriage (19) - likely an example of sympathetic magic hinging on the hooked burs of the plant; and in Hausaland, the fruit of the 'begeyi' tree (botanical identity unknown) will reconcile husband and wife, if eaten (30). *Erythrina senegalensis* Coral flower or Bead tree together with *Newbouldia laevis,* a popular fetish tree, are symbols of reconciliation (19). In S Nigeria, *Tragia benthamii* Climbing nettle, a plant with irritating hairs, enters a Yoruba Odu incantation against death separating husband and wife (32).

On the contrary, among many of the coastal tribes generally to sweep dirt onto someone is believed to prevent him or her from either being married or of having a successful marriage (19). The sweeping broom which is often made with the mid-ribs of *Elaeis guineensis* Oil palm is associated with many supernatural attributes. In S Nigeria, *Imperata cylindrica* Lalang grass, a notorious weed of cultivated land, is invoked in an Yoruba incantation 'to make a husband fight with his wife' (32).

In Ghana, firewood is the symbol of submission, and among the Ga tribe, all women on good terms with their mother-in-law vie with each other in sending the most impressive tokens, some women send four or six logs the first year of their marriage, eight or ten the second year, and a dozen logs the third year (16). Among some other tribal groups in the Guinea-savanna zone however, the husbands rather present firewood to their in-laws.

Rivalry in Marriage

To displace a rival in marriage among the Akan-speaking tribe of Ghana and Ivory Coast, the juice of *Hilleria latifolia,* a perennial herb commonly found around villages in the forest regions, is squeezed with a spell onto two individual adjacent leaves of *Kalanchoe integra* var. *crenata* Never die, a fleshly herb, which are then collected, crushed and placed where this rival would walk over. The ritual is usually started and ended with libation (1). To excel a rival in any field - such as in job opportunities, in sports, in singing and in marriage, among some of the coastal tribes of the Region, the whole plant of *Alternanthera sessilis* Sessile joyweed, a herbaceous amaranth associated with water, collected with the necessary rites, is ground into Florida Water, a type of lavender, and added to one's bath water (19). During an incantation while bathing, the names of all the days of the week are recited except the day on which the subject was born, for excellence over other persons born on those days (19).

To cause divorce - either of partners or of a rival among some of the tribal groups in its distribution area, two leaves of *Uraria picta,* a savanna legume woody herb, collected with rites, are placed with the upper surfaces facing each other, among incantations and a repetitive

spell for divorce as it unrolls. The leaves are first rolled together lengthwise with the upper surfaces facing each other, then allowed to unroll, in which process they turn with the lower surfaces now facing each other (5). This is likely an instance of sympathetic magic hinged on the leaves first facing each other, then turning away from each other. In S Africa, *Psorospermum febrifugum* Rhodesian holly is used as a charm by the Nyanja to entice someone else's husband by putting some of the plant on the wife's fire-place (35).

Polygamous marriage is practised in Africa, and in traditional marriage, the husband unquestionably dominates the partnership in all respects - being the undisputed leader. In Ivory Coast, *Datura metel* Hairy thorn-apple or Metel is used by women in exercising patience to tolerate the foolishness of their husband, while at the same time keeping them under their domination (17). 'A high level of reasoning is shown where a woman seeks a charm to give her the power of ruling the husband in Hausaland. The malam tells her that she must bring him some milk of a buffalo-cow, *Syncerus caffer,* and she gets this after having gradually made the beast accustomed to her presence. When at last she brings the milk, the malam asks 'how did you get it?'. 'By strength of will, by luck, and by coaxing', she replied. 'Good', says the malam, by the same means that you obtained the buffalo's milk will you be able to rule your husband' (29).

FAITHFULNESS IN MARRIAGE

In the marriage union, faithfulness among the partners should be a matter of course - being one of the important moral laws in the institution of matrimony. According to the Greek historian, Pliny, (first century C.E.) berries of *Vitex agnus castus* Chaste tree strewn on the beds of soldiers' wives was a testimony of the wives' faithfulness while their husbands were at battle; and Greek priests gave newlyweds wreaths of *Hedera helix* Ivy to signify fidelity (6). In S Nigeria, 'adosusu' is the local name of a plant in Yoruba meaning 'we stand together', and it is used by the tribe in respectful

Vitex agnus castus Chaste tree.

recognition of marital fidelity (9). Traditional marriage is polygamous, so unfaithfulness on the part of the husband is often dismissed as gossip. The offence often attracts little attention and hardly any punishment. At best a symbolic compensation or an atonement in addition to traditional rites ends the matter. In S Africa, a pregnant Chewa woman wears a part of *Syzygium cordatum* Waterberry, a tree, as a charm when she learns that her husband has committed adultery, so as to prevent her baby from being born with deformity (35).

Adultery

There are three options open to an adulterous wife. She may either confess voluntarily or be compelled to, and bear the consequences (this is usually followed by some compensation and then forgiveness); or deny the accusation and be subjected to superstitious ordeal to prove her innocence or guilt (see ORDEAL: PLANT TESTS: Adultery and Lesser Crime); or opt for a divorce. There are also charms to make an adulterous wife mad (see DISEASE: MENTAL DISORDERS).

To compel an unfaithful wife to reveal her secret affairs verbally while asleep, among some of the tribal groups in the Region, a single leaf of *Bryophyllum pinnatum* Resurrection plant is picked with ceremonial rituals, dusted with gun powder and parcelled (as the purpose of the charm is recited) under her pillow without her knowledge (19). In Gabon, *Mussaenda tenuiflora,* a shrub of montane locations, is a fetish plant for certain tribes to test whether married women have committed adultery (33). The spiritual uses of *Sida acuta* Broomweed, a common malvaceous weed, include charms to make an unfaithful husband or wife change his or her immoral habit (19). In S Nigeria, Yoruba invoke *Aframomum sceptrum,* a perennial forest herb in the ginger family, in an orgastic soliloquy by the husband of an unfaithful wife; and *Sida rhombifolia*, a semi-woody plant of open waste places, enters into Yoruba incantation to force an unfaithful woman to return home (32). Among the Hausa of N Nigeria, the Zarma of Mali and Niger, the Fula and some other tribes the aerial roots of Ficus *dekdekena (thonningii)*, a common shade tree, are employed in charms to convert an adulterous wife and win her love again. The root infusion is substituted for her bath water to be used in washing the genitalia (21).

Seduction. Besides voluntary sexual relations outside marriage, there are also instances of seduction, where the woman is either coerced or influenced to yield to the advances of the man. There are sinister prescriptions by medicine-men to compel a married woman to yield to the advances of another man. For example, among the tribal groups in the Guinea-savanna, the unopened leaves of *Piliostigma thonningii,* a decorative shrub or small savanna tree legume, are used with rites as a love-charm to seduce someone else's wife and she will not tell anyone about it (21). The leaf is collected early in the morning with a spell and before speaking to anyone, and smuggled under her pillow, and the charm is on her (21). The sympathetic magic appears to be the unopened leaf and the fact that it is collected early in the morning before speaking to anyone.

In Senegal, the Tenda consider *Baissea multiflora,* a climbing shrub with white latex, with the Mistletoe growing on it a charm to be used to seduce a woman (15). In SE Senegal, a Tenda man will smear ointment containing the ash of *Desmodium gangeticum,* an under-shrub, over himself to promote sexual success with the woman he loves; or in more complicated preparation, the burnt ash compounded with shea-butter is smeared over the glans of his penis to further seduction (15). In this country, the Tenda dry and burn the leaves of *Costus spectabilis,* a herb in the Ginger family. The leaf ash is mixed with *karité* butter and rubbed on the chest by youths to ensure seduction of any girl of their choice (15). In Ivory Coast, the Koulango and the Brong use *Biophytum petersianum* African sensitive plant, a herb, to make preparations for seduction. The plant is approched slowly, touched lightly when the leaves close; then saying *yabagdio toa* ('old woman, close yourself') as it is plucked and powdering it up is put in the food of the person to be seduced or held (17). In Congo (Brazzaville), a man wishing to seduce a woman rubs his body with some palm-wine in which the leaves of *Iodes africana,* a forest liane, have been macerated (7).

Punitive and Preventive Measures

Whether unfaithfulness is through voluntary adultery or through seduction, the wife is under obligation to confess to the husband. If voluntary confession fails, then measures could be taken to expose and disgrace her in public. Various methods were also adopted to prevent unfaithfulness and, at the same time, punish or even kill her paramour. Sometimes the woman was also killed.

Death or Mutilation. For example, the penalty inflicted on anyone who had a *liaison* with any of the numerous wives of the *Ashantihene* was a terrible one. Not only (it is alleged) were the woman and her paramour killed, ... but the mother, father, and maternal uncle of both parties also suffered death (28). The condemned man is dragged by *Smilax anceps (kraussiana)* West African sarsparilla, a thorny creeper locally called 'kokora' in Akan, threaded through the nasal septum and gradually but systematically dismembered until finally the head is chopped off (28). Similarly, among the Yoruba-speaking tribe of S Nigeria, 'adultery in a wife is punishable by death or divorce, but as a rule the injured husband beats this erring wife, and recovers damages *(oji)* from the adulterer. In extreme cases, where the husband is a man of rank, and discovers the couple in the act, they are sometimes both put to death'(13b). In Bioko (Fernando Po), the left hand of an adulterous wife is cut off at the wrist as a punishment among the Bubi tribe (13c,18). In *Witchcraft among the Azande* (14) a photograph of 'a man who has been mutilated for adultery' is shown - the right hand has been cut off at the wrist.

Among the Ashantis of Ghana, if any wood-carver's wife was unfaithful to her husband, and the latter, being unaware of this, went to work, then 'his tools will cut him severely' (28). In the event the adulterous wife pays a fine for sacrifice of atonement and purification upon the wood-carver's tools as a prelude to any settlement (28). Among the Azande tribe of S Sudan, it is said that a man may be killed in warfare or in a hunting accident as a result of his wife's infidelities, therefore, before going to war or on a large-scale hunting expedition, a man might ask his wife to divulge the names of her lovers (14).

Maagun. In Yoruba mythology, *maagun* is a magic drug put on a woman, unknown to herself, so that when she commits adultery it may cause her paramour to fall over three times and die. The effect of *maagun* on the paramour may assume different forms; for example constant coughing, somersaulting, extreme lassitude and haemorrhage (24). *Termitomyces robustus,* an edible mushroom, is used medicinally as a remedy for *maagun* (24). This remedy is prepared by pounding the mushroom with *Tapinanthus* species Mistletoe, epiphytic parasites growing on *Jatropha curcas* Physic nut, with the peppery fruits of *Piper guineense* West African black pepper and some fresh pork. The compound is collected in a bottle and thoroughly mixed with lime juice or schnapps and administered orally (24). Alternatively, the

mushroom is replaced with the seeds of *Parkia biglobosa (clappertoniana)* West African locust bean and shea-butter - the ingredients being used to prepare soup which is taken when affected by *maagun*, or used as a preventive (34); or a giant rat, *Thryonomys swinderianus*, and *Allium sativum* Onion is prescribed - 'burn them together, put it into the blood of sheep, *Ovis* species, and mix it with shea-butter, then rub it on the body' (34).

Infection with Venereal Disease. In S Africa, a Zulu youth suspecting his lady-love of unfaithfulness, makes an infusion of *Bulbine natalensis,* a lily, with various magical powders of animal origin, and drinks the mixture which is thought to act as a sedative diuretic. The 'evil' properties are supposed to affect the girl at intercourse without injuring her, but they are subsequently absorbed into the bladder of the rival youth should he have intercourse with her (35).

Unable to Withdraw. With some ethnic groups of the Guinea-savanna, the black powder from a burnt shrub called 'kankadafau' in Hausa (botanical identity unknown) enters into juju prescriptions to prevent the man from withdrawing after sex with another man's wife (21). Some of the powder is placed in the hole of an axe in an incantation mentioning the wife's name, then the axe head is fitted in and hit firmly in place (21). Alternatively, the powder is applied four times to the eye of an opened padlock with a similar incantation or a spell, then locked with a key (21). To reverse the juju, the victims must first be exposed to public ridicule and hooting as necessary primaries. After this the husband hits the naked buttocks of the rival with some of the black powder, in an invocation to effect the withdrawal, before the victim can withdraw (21). It is traditionally believed that only the husband can reverse the charm, and that any undue delay, deliberate or inadvertent, could result in the death of both victims (21). The sympathetic magic appears to lie with the hole/axe-head, and the key/padlock, respectively.

In S Nigeria, the shape of the bicarpelate fruit of *Hedranthera barteri,* a decorative shrub with milky latex, evokes the bawdy Ijo name 'okotitaku', meaning the testicles of a goat, *Capra* species, and also the Yoruba name 'oko aja', meaning the penis of a dog, *Canis familiaris* - the Yoruba in this context also call upon the plant (probably the fruit) in an Odu incantation to prevent adultery in women on the simile that a dog in coitus finds it difficult to withdraw (26). Various forms of this above juju are said to be practised in the Region - varying from place to place, and even within the same area.

Sustained Erection - the medical term is Priapism. There is also a juju prescription to induce 'sustained erection' or 'erection non-stop' in a man who commits adultery with another man's wife. It is the exact opposite of *xala* (see below). Selected assorted plants including *Ocimum gratissimum* Fever plant, a perennial labiate often cultivated on compounds, *Amaranthus lividus* or *A. viridis,* common weeds and *Eclipta alba,* a common composite herb growing by water, are all collected with the necessary rites (19). The plants are placed on the ground and covered with some earth on which is stood the first egg of a maiden hen, *Gallus domestica,* and around which is coiled the penis of a goat, *Capra* species - the whole being baptised with the blood of this goat among incantations and libation (19). Some of the herbs are charred and sprinkled where the wife would walk on to activate the juju (19).

To reverse this juju, the culprit must first confess, and be exposed to public ridicule and jeered at. The whole plant of *Portulaca oleracea* Purslane or Pigweed is then charred into black powder, rubbed on the erected penis and administered in food and drink. It is only after these traditional rituals that the juju may be evoked in an incantation by the husband to reverse the erection - else the victim dies (19). The sympathetic magic appears to be the first egg of the maiden hen, *Gallus domestica* and the penis of the goat, *Capra* species. There are various forms of this juju in the Region - varying from place to place, and even within the same area.

Xala. As a further precautional measure to prevent adultery and rape, there are juju prescriptions known as *xala* in Senegal. The charm is effected through a series of knots on a rope amid incantations, recitations and a spell. The purpose of this juju is to deny a man an erection altogether. This juju may be contracted by parents to protect their daughters before marriage; or by husbands to prevent their wives from committing adultery or being seduced or raped. The primary objective of the prescription is to counteract seduction and rape or charms to this effect. (For the consequences of unfaithfulness during pregnancy, see CHILDBIRTH: PARTURITION). (See also ORDEAL: PLANTS TEST: Adultery and Lesser Crime).

WESTERN MARRIAGE

In addition to traditional marriage western marriage or wedding (dubbed for better for worse) is practised in the Region - especially among the *élite.* Some plants are associated with such wedding ceremonies. For example, *Antigonon leptopus* Corallita or Wedding flower, an introduced decorative climbing perennial, and *Rosa* species Rose (both for its beauty and as an emblem of love) feature in bridal bouquets. *Rosmarinus officinalis* Rosemary stood for fidelity and was included in bridal bouquets (4). In S Africa, *Dombeya natalensis,* a decorative plant, on account of its large white fragrant clusters is also called Wedding flower; and *Serruria florida*, a winter-blooming plant, is in great demand in this country for wedding and corsage sprays - the popular name, Blushing bride, referring to this practice as well as its delicate colouring (26).

In Europe, the sweetly fragrant waxy-white flowers of *Citrus sinensis* Sweet orange were in great demand as bridal wreaths; *Spiraea x arguta* Bridal wreath is so called because of its dainty white flowers; and *Mesua ferrea* Ceylon iron wood, whose persistently fragrant china-white blossoms are used for perfumery, are often stuffed into pillows and cushions for bridal beds in the Orient (26). In Japan, one method of retaining the fleeting moment of the cherry blossom is to salt them; these salted flowers are then infused to make the tea which is given to a bride before she goes to the wedding ceremony (29). (See also LOVE).

References.

I. Abban, (pers. comm.); 2. Adjanohoun & Aké Assi, 1972; 3. Amat & Cortidellas, 1972; 4. Back, 1987; 5. Beloved, (pers. comm.); 6. Blumenthal & al., 2000; 7. Bouquet, 1969; 8. Bubuwa Tubrar, 1985; 9. Burkill, 2000; 10. Crow, 1969; 11. Dalziel, 1937; 12. Dogbe, (pers. comm.); 13a. Ellis, 1883; b. idem, 1885; c. idem, 1894; 14. Evans-Pritchard, 1937; 15. Ferry & al., 1974; 16. Field, 1937; 17. Kerharo & Bouquet, 1950; 18. Kingsley, 1897; 19. Mensah, (pers. comm.); 20. Maja Naur, 1999; 21. Moro Ibrahim, (pers. comm.); 22. Okigbo, 1980; 23. Osmanu, (pers. comm.); 24. Oso, 1977; 25. Pamplona-Glover, 2001; 26. Perry & Greenwood, 1973; 27. Portères, 1974; 28. Rattray, 1927; 29. Richards & Kaneko, 1988; 30. Tremearne, 1913; 31. Vergiat, 1969; 32. Verger, 1967; 33. Walker & Sillans, 1961; 34. Warren, Buckley & Ayandokun, 1973; 35. Watt & Breyer-Brandwijk. 1962

35 MEDICINE

There is a plant for every illness.
Abbe Kneipp

Traditional medicine is generally associated with fetish, juju, the supernatural, the spirit world (such as dwarfs, ancestral spirits and local deities), the occult and other invisible forces and powers. Much native medical practice cannot easily be separated from magic and superstition (17). In African culture, traditional medical practitioners are always considered to be influential spiritual leaders as well, using magic and religion along with medicine (5). For instance, a priest-healer *(okomfo)* combines the roles of religious leader and healer; and in the treatment of diseases he uses not only medicinal plants, animal parts and minerals, but also religious healing rituals (26). This figure is the Shaman, substantially the same in all cultures: a connoisseur of phytotherapy, who combines elements of botany, knowledge of herbs, toxicology with religious and ritual elements based on magic, ancestral beliefs and often superstition (18).

Nevertheless, herbal healing, as practiced by a section of simple native herbalists, may involve neither magic nor superstition; but is based entirely on the curative properties of the herbs. 'Herb remedies or folk remedies and folk-lore are associated together, which is a mistake. Folklore can be amusing in its superstitions in the cure of the sick. On the other hand, folk medicine or herbalism is rich in herbal wisdom' (8). A herbalist *(dunsini)* normally confines the treatment of patients to the application of plant medicine, with the use of healing rituals being an exception (26). The strong conviction of some traditional healers that evil spirits, bad juju, sorcery, witchcraft and other dark forces are responsible for a host of ailments, and the belief that the administration of any medication would be ineffective unless these forces are first driven away, explains why some herbalists are also fetish priests, witch-doctors or members of a cult. Usually, where the collection of the herbs and the preparation of the medicine are accompanied with rituals, and where

the healing is completed with purification and sacrificial offerings, it is the likelihood that the disease is not a natural one. Certain customs are linked with such a healing system and the person healed by this kind of society is forbidden to eat certain kinds of food, use certain types of plants as firewood, and so on; either for a certain duration of time or throughout his lifetime (20). The practice is normally with the full approval of the patient, but if he or she is too sick to consent to - the relatives.

DEFINITION

Traditional medical practice involves the formulation of plant material, animal parts or mineral products in prescriptions (with or without ingredients) either for their biologically active parts or for their supernatural powers or both - for the prevention of, the protection from or the treatment of disease (physical, mental or spiritual), an injury, a curse or a spell. The medication is prescribed as infusions, concoctions, decoctions, in enemas, as suppository placed in the rectum or vagina to dissolve, or applied to incisions cut on the affected parts; or effected spiritually through charms, amulets or talisman. The practice may or may not be preceded by a ritual, and similarly, completed with or without purification and sacrificial offerings. Such practitioner is termed medicine-man, native-doctor, herbalist, fetish-man, witch-doctor or juju-man; and acquires this talent by patrimony or by initiative training. 'The profession of herbalist is a male domain. It can be inherited or self-acquired. Many *nnunsifo* (herbalists) have acquired their knowledge on their own. Inheritance is usually from father to son, that is patrilineally; daughters are normally excluded' (26). It has been observed that if a woman becomes prominent as a healer she may be suspected of witchcraft (44). Some practitioners claim to have acquired their knowledge from dwarfs or from the spirit world (see DWARFS AND FAIRIES: DWARFS: Curative Powers).

INSIGHT

How the curative properties of herbs were associated with specific ailments is a subject of controversy. There are several schools of thought. These include trial and error, observing animal and plant behaviour, interpreting signs and symbols, dialogue with the plants, revelation, witchcraft and devilry. Another school of thought is that they were contributed severally by all these various means listed above.

Trial and Error

Some medicine-men suggest trial-and-error. After series of disappointments and failures, the right plant or plant part, with the preparation and dosages, is eventually found from experience. In fact, all the other methods listed below are first tried before they are proven.

Animal Behaviour

Other medicine-men claim to have acquired the knowledge by observing the feeding habits of animals - cats, *Felis domestica;* dogs, *Canis familiaris;* livestock, poultry and snakes. For example, writing on the hallucinogenic effects of *Tabernanthe iboga* Iboga, an apocynaceous shrub, it has been observed, 'the discovery of the plant, indeed, may not have been by man but boars, *Potamochoerus porcus,* in the jungle. Several accounts mention that the natives saw boars dig up and eat the roots of the plant, only to go into a wild frenzy, jumping around and perhaps fleeing from frightening visions. Porcupines, *Artherurus africanus;* and gorillas, *Gorilla gorilla;* according to the natives, occasionally did the same thing' (53). In addition to medicinal plants, the knowledge of food plants could be similarly explained. 'It is not difficult to imagine how man discovered that grains were good for food. Probably he saw birds and small animals eating them and decided to see what they were like' (47).

Plant Behaviour

The observance of plant behaviour could also account for, and explain their healing properties. For instance, it has been recorded from observation that 'all xeroids have a special chemical make-up which closes any wound almost immediately so the plant will not lose its precious water. The wound then heals with almost miraculous rapidity and the plant begins to grow in another direction. Ancient man observed this, reasoned that if it worked for the plant it would work for him, and thus, from earliest written history, we find references to *Aloe vera'* (50, 52).

Aloe vera (legendary and mythical plant)

Signs and Symbols

In nature, the structure and size, the branching system, the leaf shape, the colour and shape of the flower, the presence of an exudate and its colour, the features and characteristics of the bark or any signs and symbols on other parts of the plant, are all believed to have a meaning and to serve a purpose. That is, others propose that there are natural signs on plants indicating which ailment they cure most effectively. Certain parts of a plant might resemble in form or colour some part of the human body, and it was believed that diseases of the said organ could be cured by the application of the corresponding plant (16). This was the *Doctrine of Signatures,* which stated that every plant was signed, as it were, with its own use, and it was only necessary to look for and understand the signature (16). According to 'doctrine of sympathetic resemblances' all growing things reveal through their structure, form, colour and aroma, their peculiar usefulness to man (58). The proponents of the natural uses, and of sympathetic magic, link the V-shaped signs on *Annona* fruits with those on poisonous snakes (Vipers, species of *Bitis and Echis*) - implying that the two neutralize when put together - that is, *Annona* species do cure snake-bite. (See also DISEASE: HEART DISEASE; INTELLIGENCE: PRESCRIPTIVE CHARMS: Sympathetic Magic; LOVE: CHARM PRESCRIPTIONS: Sympathetic Magic; MANHOOD: POTENCY: Sympathetic Magic; POPULARITY AND SUCCESS: GENERAL CHARMS: Sympathetic Magic.

Dialogue

Some medicine-men and fetish priests do claim to understand the language of herbs, and therefore, to obtain the curative properties through this dialogue. For instance, among the initiative training of fetish priests, it has been observed, 'he is taught the art of foretelling, of interpreting animal voices and understanding the language of plants and trees' (26).

Revelation

By far revelation - either by celestial or ancestral spirits, or local deities (including dwarfs - see DWARFS AND FAIRIES: DWARFS), and during spirit possession (see TWINS: CEREMONIES: Spirit Possession) - appears the widest source of information on the curative properties of plants. The supporters of revelation by ancestral spirits claim the information may be communicated in a dream or in a vision; or the spirit may appear in human form, disclose the secret and vanish again. In the dreams of the *nnunsifo* (herbalists), these spiritual forces also reveal plants and their medicinal applications (26). In the case of revelation by dwarfs, the novice might be spirited away for a period. Many priests and priestesses claim to have lived among the *mmoatia* (dwarfs) in the forest for a certain period of time and to have been taught the art of healing by these (26). Herbalists in particular are believed to attain medicines for the treatment of rare diseases from *mmoatia* (26). The revelation by gods or deities is through a possessed medium (see TWINS: CEREMONIES: Spirit Possession). A medium will give information with respect to the cause, the type and treatment of a disease or an accident as well as to theft, approaching misfortune, and how to avoid these, or on causes of conflicts in the community (26).

Witchcraft, Devilry and the Occult

There is widespread traditional belief that witchcraft, devilry and the occult are revelatory sources into the insight of the secrets of medicinal plants - their collection, preparation, healing properties and any rituals that may be associated with the medication.

SECRECY

The healing property acquired, becomes the copyright of the medicine-man - a property he jealously guards. To prevent others from infringing on these rights, herbal prescriptions are often mixed with black powder, or crushed and crumbled, or collected in the night to hide the identity. The active ingredients of many traditional folk medicines are often deliberately obscured by the presence of non-active materials in order to conceal the identity of the former and preserve the reputation of the medicine-man or witch-doctor (65). Some traditional healers die with this wealth of knowledge - leaving posterity the poorer.

RITUALS

Traditional medicine-men normally perform certain rites before collecting medicinal plants or revere the plants in some other ways. For instance, the mixing of plants during collection is normally avoided. Not only does the practice affect the efficacy of the preparation, but it

Medicine market

is believed that this could actually render the preparation toxic. In addition, plants dug for their roots are as a rule, covered up again with soil, allegedly to ensure the efficacy of the medicine - but it appears the practice is, in effect, a traditional means of conserving biodiversity for sustainable yield. (Collecting medicinal plants from the wild, endangers the species population. See HUNTING :Table 3 for other activities that threaten the survival of the species). Some plants are said to be collected on certain days only, or during a specific phase of the moon. Among the psychic healers it was common belief that the medicine will not work if the herbs are collected by a woman. Indeed if a female is sent into the bush other people might think she is a witch, so it is safer to send boys (Bob Loggah fide 44). Sometimes an egg, a cowry shell, some powdered corn dough, white clay or a coin is offered to the plant before a part is removed - signifying 'payment' or 'compensation', and to ensure its potency. Before a *dunsini* (herbalist) picks a plant, he will make a sacrifice (either some eggs, a fowl, *Gallus domestica;* or a bottle of schnapps) to the *sasa* which may dwell in it; this is to prevent the spiritual forces from turning against the *dunsini* and/ or his patients by withholding the plant's curative power (26).

In Sierra Leone, before a Mende medicine-man of the Pujehun District collects specimens of *Diospyros thomasii* or *D. heudelotii,* small forest trees, or *Psychotria rufipilis,* a shrub in forest, for healing purposes, the first line of the Koran is recited – this is because God is regarded as supreme. When collecting *Anthonotha macrophylla*, a forest tree legume, which is used for yellow fever the patient has to go round the tree first (42). In Ghana, a ritual is performed before cutting a piece of *Spiropetalum heterophyllum,* a lofty undulating forest liane believed to be fetish, for medicinal purpose (55) - (see LACTATION: RITUAL INDUCEMENT). In its distribution area, *Eragrostis tremula,* a tufted annual grass, enters into rituals by traditional herbalists for healing powers (45) - (see CEREMONIAL OCCASIONS: PURIFICATION CEREMONIES: Purification). In Ivory Coast and Burkina Faso the importance of *Strophanthus hispidus* Arrow poison for medicinal use is, out of difference, accompanied by particular rite (36). In Congo (Brazzaville), *Quassia (Simaba) africana,* a forest shrub, is endowed with magic properties. In collecting its leaf or bark it is essential to offer a piece of money and make a prayer to the plant's puissance (10). 'I have seen a witch-doctor who was treating a patient for nothing place a piastre of his own on the ground, and when I asked him what he was doing he explained that it would be a bad thing if the medicine did not observe a fee, for it might lose its potency' (23). There are contradictory views in the remuneration of psychic and spiritual healers. While one claims that the medicine cannot be used without payment of a fee - however small; another view is that 'my father told me not to charge for medicine, otherwise it would not work' (44).

Instances of rites in disposing of used medicinal plants also do occur. For example, among the Bulsa tribe of NW Ghana, if medicinal roots, bark, branches, and so on, have been used, they must not be burnt; otherwise the sickness will come back; so the *tiim-nyono* (medicine-man) comes to the house of his recovered patient and buries these used parts in an anthill so that the white ants, *Macrotermes* species, carry them away - the fee for performing this ritual is a small fowl, *Gallus domestica* (38). In traditional practice used medicinal plants are more usually discarded with rites. For instance in Senegal, the leaves of a Mistletoe, an epiphytic parasite, is steeped in water with *Bombax costatum,* a savanna tree, for a week, during which time the body of a sick person is washed with the water morning and evening. At the end of the week, the Mistletoe is thrown to the four-winds at a crossroads (24). (See PURIFICATION below).

Sun-worship

Some traditional medicinal practices are suggestive of sun-worship. For instance, in Ivory Coast and Burkina Faso, when the bark of *Ceiba pentandra* Silk cotton tree is required for medical treatment it is common to take it from either the east or the west side, a mark of sun-worship evinced by all races, even those Islamised; and in the former country, a bark decoction of *Trichoscypha patens,* a forest tree, is used to bathe a patient ill without diagnosable symptoms – taken from the east and the west sides of the tree (36). In Ivory Coast, the bark of *Erythroxylum mannii,* a tree, taken from the east and west sides of the stem in a cryptic relic of sun-worship, is pulped with citron and melegueta pepper for use in frictions for intercostal pains and pleurisy (13b).

CHILDREN'S CASES

In Ivory Coast and Burkina Faso, *Bidens bipinnata* Spanish needles is considered a fetish plant *par excellence* for children: a bath containing the plant gives strength and protection from illness and afflictions resulting from a crowned crane, *Balearica pavonina,* flying overhead. This can be averted by rubbing the limbs with its sap and scattering some of the lees on the ground so that passers-by take the ill-effects away with them (36). In Liberia, an infusion of Spanish needles is given to babies (13a) – perhaps superstitiously; and in Ghana, the whole plant in water serves to purify the life of the sick in transit (1). In S Nigeria, the Ijo use the leafy stems of *Momordica cissoides,* a cucurbit, in bath preparations to cure a child suffering from *endeé* during a pregnancy of its mother (13a).

In this country, *Crinum natans,* an aquatic lily in perennial streams, is used in a superstitious way to ward off ill-health in new-born infants (13a). In N Nigeria, the root of *Bergia suffruticosa,* a heath-like undershrub, is used by the Hausa as ingredient in *dauri* prescriptions at weaning time to prevent sickness; and the ground bark of *Combretum molle,* a savanna tree, along with cereal foods is used in ceremonial preparation given to young children as preventive of sickness or other troubles (17). Among the Hausa of this country, the fruit with the leaves of *Guiera senegalensis* Moshi medicine, a shrub in sandy wastes and semi-desert areas, are common ingredients in more or less ceremonial prescriptions for strengthening children or preventing disease *(dauri)* in young children (13a). In Zambia, the Lamba tie strips of the inner root-bark of *Euphorbia inaequilatera* Melbossie round the loins of infants to relieve constipation; and in Botswana, the Southern Sotho use *Anacampseros arachnoides,* locally called 'serelile', as a charm and a medicine for a sick orphan child (64). In Europe, magical powers ascribed to *Viscum album* Mistletoe, a parasitic epiphyte in temperate regions, included a panacea for children's ailments and a universal remedy against poisons and diseases (51). The parasite's equivalent in W Africa are species of *Phragmanthera* and *Tapinanthus.* (For a list of Mistletoes in the Region, see DISEASE: LEPROSY: Table 1).

MEDICO-MAGICAL TREATMENT

Instances of medico-magical treatment of disease is practised in Africa and elsewhere. This is the application of magical means against disease, afflictions, dangerous persons, charms, spells, and so on. Usually these 'medicines' do not work without a continuous renewal of their magical power by sacrifices on these objects (that is part of the blood drips on the medicine) (38). In W Africa, many of the medicinal uses of *Achyranthes aspera* Devil's horsewhip, a weedy amaranth, may have arrived through magical attributes. For instance, in Sanskritic medicine it was a sorcerer's plant for driving away demons, removing enchantments, curses and illness. In Tanzania, the Zigua add the plant to medicine to remove spirit possession (Tanner fide 13a).

In Ghana, when the leaves of *Distemonanthus benthamianus* African satinwood, a forest tree legume with characteristic red bark, are required for medicinal purposes among the Ashantis, libation is poured under the tree with incantations and a request to collect some parts of the tree to cure a specified ailment. If the offer is accepted, meaning that the medication would work, some of the young leaves drop by themselves and may be collected with the bark for the purpose; but if no young leaves drop, it is an indication that the offer has been rejected and that the medication would not work (55). The Akan name 'bonsamdua', means 'Devil's tree' – most likely in reference to the red colour of the bark. In this country, the peeled stems of *Hibiscus lunariifolius* Rama fibre, a tall shrub, enter into magic, and are an ingredient of medicines for treating anaemia, lassitude and fatigue (32).

The large vines of *Saba senegalensis* var. *senegalensis*, and var. *glabriflora,* forest lianes with milky latex, have medico-magical attributes; and as such the plants are sometimes conserved for superstitious reasons in the vicinity of villages and also for their edible fruits (17). In S Nigeria, the pulverized wood of *Pterocarpus soyauxii* Redwood is used in medico-magical rites; and the fruit of *Elaeis guineensis* var. *communis* Oil palm, with thin kernel shells, is used in Ibibio magico-medical practice (17). In Ivory Coast and Burkina Faso, some of the medicinal applications of *Sterculia tragacantha* African tragacanth, a forest tree, have magical implications where a multiplicity of indications suggest a single line of action is preferable (11). In Congo (Brazzaville), *Pentadiplandra brazzeana,* a lianous bush, is believed to hold medico-magical properties; and its relative commonness along roadsides and cultivated sites may be related to this (10).

In Senegal, the Fula and the Wolof use *Diospyros mespiliformis* West African ebony tree in a medico-magical treatment for psychosis (35a). In Liberia, Sierra Leone and other parts of the Region, *Bersama abyssinica* ssp. *paullinioides*, a medium-sized forest tree, has medico-magical uses (see JUJU OR MAGIC: MEDICO-MAGICAL TREATMENT). Similarly, *Olax gambecola,* a forest shrub usually by streams, and *Borassus aethiopum* Fan palm both have various medico-magical applications (32). In W Africa,

Annona senegalensis var. *senegalensis* Wild custard apple is in general considered a panacea for most illnesses, acquiring thereby not a little medico-magical attribute; and *Capparis tomentosa,* a thorny scrambling savanna shrub, enters widely into local pharmacopeias throughout Africa, often with magico-medical attributes (13a). The shrub is used in Senegal for protection and for the treatment of poisoning, spells, and so on (35a, b). In Gabon, *Barteria fistulosa,* a slender forest tree, is associated with fierce black ants, *Crematogaster* species, capable of inflicting painful bites. As a result of these attributes, the tree enters medico-magical usage, principally in connection with pain (63).

In S Nigeria, the fruit of *Pleioceras barteri* var. *barteri,* an erect or climbing shrub with milky latex, enters into gynaeco-medico-magical applications (see CHILDBIRTH: PRE-NATAL). In Ghana, the fruits of this shrub are used superstitiously in local medicine as well (32). In Senegal, the resin of *Daniellia ogea* Gum copal tree is burnt as incense in medico-magical treatment; and *Burkea africana,* a savanna tree, is ascribed with undisclosed medico-magical attributes (35b). In this country, *Cenchrus biflorus,* an annual grass, is used in a medico-magical way as a vulnerary on damage sustained in wrestling contests (35b). In the savanna region of Ivory Coast and Burkina Faso, *Blighia sapida* Akee apple is fetish with an important place in medico-magic practice (36); and in Ivory Coast, *Spermacoce (Borreria) octodon,* a savanna herb, enters into magical preparation for the removal of interdictions during the course of an illness, and to protect the sick (11, 36). In The Gambia, the seed of *Sesamum radiatum* Wild beniseed, an erect savanna herb, is put to magical use. Placed in a small bottle and tied onto the rope of a juju, it is believed to prevent spread of transmittable diseases (Hallam fide 13d).

In S Africa, the latex of *Ficus carica* Common fig is used as an application to a wart to cause its disappearance, a usage also recorded from India. Another procedure, with a more magical flavour, is to transfer the wart to a fig tree by grafting a drop of blood from it on to a young twig (64). In this country, an infusion of the root of *Ranunculus* species is used in a magical way by the Mpondo. When sickness is believed to be caused by ancestors, the patient is washed before the kraal gate, the ancestors being meanwhile called upon; or is washed with a cold infusion of the root of *Senecio (Crassocephalum) latifolius,* a composite, and occasionally the sufferer is made to drink some of the infusion (64). In E Africa, the Lobedu use the burnt end of a piece of the root of *Clematis oweniae,* locally called 'kiobaobe' to draw a circle round a sprained joint, which procedure has some magical significance (64).

In N Nigeria, the rhizomes of *Cyperus articulatus*, a sedge by water (often cultivated for the aromatic rhizomes), is added to water which has been used to wipe Koran text from a writing-board, and after evaporation, the residue is used to fumigate the body in sickness in a medico-magical application; also the seeds of *Monodora myristica* Calabash nutmeg are endowed with magical attributes for which they are valued in many medical preparations (13a). In S Nigeria, the seeds, under the name of *arino,* enter into a Yoruba Odu incantation against disease (61). Among this ethnic group 'Osanhin' is the god of medicine (21). The Yoruba tribe prescribe the leaf of *Vernonia amygdalina* Bitter leaf and *V. colorata,* common composite shrubs or small trees in derived savanna, and invoke it in an incantation against *kono* disease (4). In Senegal, *Caralluma dalzielii* and *C. decaisneana,* succulent herbs with white latex, are used in a number of medico-magical treatments for various ailments including perhaps also Parkinson's disease (34a, b). In this country, Pulaar medicine-men have a medico-magical use of a decoction of *Waltheria indica,* a shrub, in baths and massages for treating roseola (German measles) (24).

In Europe, *Tamarix gallica, T. tetrandra, T. ramossisima (pentandra)* and *T. hispida* Tamarisk, temperate shrub-like plants, were believed to have magical powers and Pliny recommended the leaves as an ointment for 'chilblaines' or 'night-foes'; and Indian medicine-men smoked the leaves of *Nicotiana tabacum* Tobacco for the purpose of magic and healing (51). Long ago, *Chamomile matricaria* Chamomile was grown all round herb gardens in the belief that it would keep other plants growing near free from disease, and it was known as the 'Plant's Physician'(6). There is a magical sickness on the Island of Pantelleria off Sicily, called *scantu* ('fright sickness') and infusions of *Ruta chalepensis,* a shrub, and *Marrubium vulgare,* a labiate, is prescribed by the women who cure it after they have performed the magical cure (28).

CHARMS AND TALISMAN

The treatment of disease or its prevention or protection from attack, could also be in the form of a charm or a talisman. For instance, in East, West and S Africa, the seed of *Mucuna* species, probably *M. sloanei* Horse-eye bean or True sea-bean, with a thong attached, is carried as an amulet or a charm to protect the owner from sickness (Muir fide 29). Similarly, the hard black seeds of *Ensete gillettii* Wild banana are used as necklaces, rosaries, wristlets, and so on, worn to prevent sickness; and the fruit of *Elaeis guineensis* var. *communis* Oil palm, with thin kernel

shells, has superstitious uses as charms against poisoning (17).

In Botswana, the dry bulb of *Nerine angustifolia*, a lily, locally called 'lematlana', mixed with parts of other plants, is used as a charm against illness; or the Southern Sotho use *Anthospermum pumilum,* locally called 'masopolohane', as a charm to hasten convalescence (64). In this country, *Pimpinella caffra,* an umbel, is used by the Southern Sotho as a charm to drive away *nohana* (intestinal worms) and *thokolose* (a mysterious spirit which is supposed to appear at night in certain places and during certain illnesses). For the latter purpose the hut is fumigated with the smoke from the burning root and the inmates wash themselves with a decoction made from the plant; however, the process must be done at night at a distance from the hut (64). The tribe also use *Eragrostis plana*, a grass, as one of the charm ingredients in a preparation used for treating fractures (64). In Zimbabwe, the Manyika use the root of *Lopholaena disticha,* a composite, as a general charm to aid the sick - however, it is never used alone but added to other medicines (64). In S Nigeria, *Eragrostis tenella,* a delicate tufted annual grass, is used by the Igbo tribe as 'medicine' against 'bad medicine' - a sort of re-insurance (13d).

In Ivory Coast, *Floscopa africana* ssp. *africana,* and two other subspecies *petrophila* and *majuscula*, erect or straggling herbs usually by streams, are reported to be endowed with magical properties of protection from all illnesses. The plant is pounded with a leafy twig of *Microglossa pyrifolia,* an erect or straggling shrubby composite, and a little water and the skin of a house-spider, *Nephilengys* species - rolled into a pellet and dried and stored in the horn of an antelope, *Adenota kob kob,* or in a fold of leather (talisman) about the house (36). In the western part of the Region, seeds of *Cassia sieberiana* African laburnum hidden under the floor of the house will render inconstant women living there (4a). In S Nigeria, the Yoruba invoke *Typha domingensis* Bulrush or Cat's tail to encourage a sick person to eat (61). In Zimbabwe, the Manyika have a charm use of *Solanum linnaeanum (sodomeum)* Apple of Sodom: the root is tied to the foot of a patient to prevent him from becoming worse if a menstruating woman comes into the hut or room; and the twig of *Salix woodii* Wild willow is a constituent of a charm medicine carried by the African doctor (64). In the W Indies, *Cestrum nocturnum* Night blooming cestrum is used as a stupefying charm medicine (64). In Hawaii, a necklace of the seeds of *Coix lacryma-jobi* Job's tears, a perennial grass by streams, is endowed with curative charms (64); and in England, *Daldinia concentrica,* a fungus, was, until quite recently, carried by old men in Surrey and Sussex as a charm against cramp (Rolfe & Rolfe fide 48).

SCORPION STING

There are plant prescriptions for the treatment of scorpion sting, or for immunity from the sting. The practice is, sometimes, based on sympathetic magic. For instance, in Ethiopia (17), and in Nigeria (13a), *Heliotropium subulatum*, an erect branched perennial, is applied for scorpion stings; the herbalist's indication for this being the shape of the tail-tipped petals, resembling a scorpion's sting. In Sudan (33), there is a similar use of *H. strigosum*. In W Africa, *Sesbania sesban,* a herbaceous legume, has superstitious association with reference to scorpion stings; and also in practice the fresh root, applied as a paste, has been described as an excellent remedy (17). In N Nigeria, *Crotalaria atrorubens,* a sub-erect or erect savanna herb, is used by the Hausa on scorpion stings - hence their name 'maganin kunama', meaning 'medicine for scorpions', *Buthus* species (17); and the root of *Merremia tridentata* ssp. *angustifolia,* a prostrate and twining annual in savanna, eaten with bran is supposed to confer a prolonged immunity (a year or more) to scorpion-sting, provided one does not deliberately, apart from in food, eat salt during that period. The plant is thus loosely known by the same Hausa name 'maganin kunama'(17).

In Senegal/Guinea area, there is a superstition that when one eats *sumbala,* the fruits of *Parkia biglobosa (clappertoniana)* West African locust bean one must wash one's hands or a scorpion will kill one (3). The old English name for *Myosotis* species, Scorpion grass, refers to the curled flower heads which were thought to resemble scorpion tail and provide an antidote for their sting (51). The Ancient Egyptians believed that *Delphinium* species had the property or power to keep off scorpions (51). In Hausaland, 'twins are supposed to have a special power of picking up scorpions without injury, but I have seen others do it who are not twins' (59). On the contrary, in Jabel Marra, it is believed that the fruit of *Sclerocarya birrea,* a savanna tree, when eaten attracts scorpions (13a) to the eater.

VETERINARY CASES

The treatment of sick animals may be given a superstitious touch in traditional medicine. For instance, in N Nigeria, the bark of *Erythrina senegalensis* Coral flower or Bead tree is used superstitiously as a diuretic for horses, *Equus equus,* with copious draughts of water (32). *Cucumis figarei (pustulatus)* and *C. prophetarum,* trailing cucurbits, are medicine for fowls, *Gallus domestica,* placed in their drinking water to help growth and

prevent disease, render them immune to predatory hawks, *Stephanoaetus coronatus,* increase egg-laying, and so on - a usage which is probably wholly superstitious (17). In Ethiopia, *Cyperus alternifolius,* a robust leafless cultivated sedge, enters ritual for the cure of sickness. Culms are laid across the kraal to rid people and cattle of illness - the green and cool appearance is held to cool the malady (Strecker fide 13a). In Europe, the embryo of *Entada gigas* or *E. rheedei (pursaetha)* Sea bean or Sea heart, tropical drift dissemules (for the meaning, see RELIGION: SYMBOLIC PLANTS: Sea-Beans), which is bitter like quinine, was used among the Norse as a purgative and a medicine for cattle, *Bos* species (Gunnerus fide 29) - probably superstitiously.

WITCH AND FETISH

Witchcraft, witch-doctors and fetish do feature prominently in traditional medicinal practices throughout Africa and elsewhere. In fact, the practitioner could be a member of a cult, because certain categories of diseases said to be caused by witchcraft can only be cured by witch-doctors, juju-men or fetish priests who perform the necessary rituals before the patient is cured with the herb (42). Whenever witchcraft is suspected, it is the traditional belief that patients are cured, not by the therapeutic treatment of the witch-doctor, but by a bargain struck between the witch-doctor and the witch (23). It is also traditionally believed that *bayie* (witchcraft) does not only cause illness on the part of the victim, but is also directed against the *obayifo* (witch) herself (25); and evil thoughts and intentions, whether you are aware of them or not, make yourself and other persons ill (26).

Similarly, instances of fetish practice in traditional healing are common. For most of their illnesses, the natives take various baths; most of these are connected with fetish and work only along superstitious faith (30). At times *Tapinanthus bangwensis* Mistletoe is used in various illnesses to consecrate water, with which the sick person may be washed (30). In Ghana, *Cyperus articulatus,* a perennial sedge that is often tended or cultivated for the aromatic rhizomes, has similar uses. In Gabon, fetish-priests sprinkle the debris of the sedge over the body of their clients to give them strength (63). In Benin Republic, the fruit of *Kigelia africana* Sausage tree is much used by witch-doctors (Laffittes fide 13a) and also in Tanzania (64). In its distribution area in W Africa, *Loesenera kalantha,* a medium-sized tree legume, is regarded as a councillor, and people pay large sums to witch-doctors to sit by it and tell it their trouble, plucking a few leaves (17).

In S Africa, every Mpondo nursing mother wears two pieces of the root of *Asparagus* species, climbing lilies, locally called 'kubaol', and nibbles a tiny piece each time before feeding her baby in case someone may have placed bewitching matter on the path along which she has been walking (64). In this country, when a Xhosa is supposed to be bewitched, his herbalist administers an emetic made of the root of *Capparis citrifolia* Cape caper (64). In Botswana, the Sotho use a root infusion of *Pittosporum viridiflorum* Stinkbas to keep the evil influences of witchcraft from their patients; and the Karanga use the root of *Wormskioldia longepedunculata* Cucuto for the relief of sudden pains thought to arouse from witchcraft (64). In S Cameroon, healing rituals among the Beti for illnesses caused by sorcery are always held under *Distemonanthus benthamianus* African satinwood, a large forest emergent tree legume with characteristic red bark (43). In this country, afflictions 'caused' by sorcery are treated with *Copaifera religiosa,* the 'oven tree', among the Evuzok people (43).

In Kenya, the Masai use *Osyris tennuifolium* or *O. wightiana* African sandalwood as an ingredient in one of their secret witchcraft medicine, but the significance of this is not known; and in Botswana, *Harveya speciosa,* locally called 'lekxolela-la-basoth', is used by the Southern Sotho in treating a person who has been bewitched by his relatives (64). For the treatment of the climacteric the Tswana of this country administer a root decoction of *Delosperma herbeum,* locally called 'lomalanthufe', and rub the powdered plant into scarifications over the vertebral joints, or rub *Scilla natalensis,* a lily, into the back, the joints and other parts of the body (64). In Botswana, the Southern Sotho rub the powdered bulb of the lily into scarifications over sprains and fractures to make them strong and resistant to witchcraft (64). In Botswana, *Massonia bowkeri* Abraham's book, a lily, is similarly used by the Sotho witch-doctor for certain illnesses - if the plant is indicated by the divining dice it is powdered and placed in incisions on the patient's body (64).

In Peru, *Trichocereus pachanoi,* a cactus, is used as with other hallucinogenic plants, as a revelatory agent to make known the source of bewitchment deemed responsible for illness and misfortune; and specialized healers, called *maestros,* use it to treat illnesses believed to be caused by witchcraft (19). In this country, *Nicotiana rustica* Wild tobacco, is used in massive doses by the Campa Shaman, and is credited as the general source of his power to see and communicate with the spirits and to cure or diagnose illness (56). In Iquitos, S America, a drink called *ayahuasca* - prepared from the bark of *Banisteriopsis caapi* or *B. inebrians* or *B. rusbyana,* forest lianes, with various other plant ingredients - is administered by a witch-doctor or brujo in a ceremony to induce various

levels of hallucination for many reasons. Among others, it is possible for him to locate a missing object, identify the cause of an allergy, and often *ayahuasca* may be used by the brujo to treat psychological disorders among his patients (27).

In W Africa, *Ocimum canum* American basil is used by the natives in various illnesses, mainly such as are attributed to witchcraft or the deceased; and the leaves of *Allophylus spicatus,* a savanna shrub, are used as a fetish in cases of illness - in a complex method (Thonning fide 30). In Ivory Coast and Burkina Faso, *Khaya senegalensis* Dry-zone mahogany, a savanna tree, is a fetish, and its use is subject to certain rituals such as removal of the bark for medicine only from the east and west side of the trunk, and dosage is regulated by a law of numbers (36). In this country, *Geophila obvallata,* a herbaceous prostrate plant, is used by the fetish priest to analyze sickness that cannot be easily diagnosed (36). In Senegal, a general fetish of the Socé is that no medicine is any good unless seven leaves of *Terminalia avicennioides,* a savanna tree, have been added to it (35a, b). In Ghana, the fronds of *Elaeis guineensis* Oil palm are sometimes hung in front of houses where there is a sickness to warn people from going inside and offending the medicine of the native doctor (32).

In Ivory Coast, *Lannea welwitschii,* a big forest tree with the typical pitted bark like gun-shot wounds, is a fetish to some tribes - the leaf-sap being given to the sick to hasten recovery (36). For the same reason, the drug obtained from *Afzelia africana* African teak, a large tree legume of savanna and fringing forest, follow certain rules: a precise dose has to be taken on a precise hour - three doses in three days, or three times daily for males and four times for females - the calabash used for the drug may either be held with the right or left hand (36). The tree is usually used with *Tamarindus indica* Indian tamarind and *Ficus sur (capensis),* a fig tree (36). In S Nigeria, the Yoruba invoke *Capsicum annuum* Pepper and *Cola caricifolia,* a forest tree, in an incantation against chest disease and disease of the bones, respectively (61). The tribe also invokes *Hibiscus subdariffa* var. *sabdariffa* Sorrel in an incantation against insect-bites and to cure leg disease (61). In this country, the Ekoi people believe that *Croton zambesicus,* a small tree often planted in villages, is a fetish and a powerful medicine to restore to health an important person (Talbot fide 17). In Senegal, the Diola put *Piper guineense* West African black pepper into a fetish medicine (34).

In S Africa, the Zulu use the leaf of *Ekebergia meyeri* Dog plum as an anthelmintic, and one of the ingredients in a medicine taken for the expulsion of 'intestinal beetles' (64). In this country, medicine to prevent

'insects' being introduced into the stomach by witchcraft is prepared from *Pisosperma capense*, a cucurbit, and *Geum capense*, a plant in the Rose family (64). In Botswana, the Southern Sotho use *Cussonia paniculata* Cabbage tree as an enema, or use *Withania somnifera* Geneesblaar as remedy to rid the intestines of parasites, especially 'beetles' introduced by witchcraft (64). In this country, the Southern Sotho also use the leaf and bark of *Psydrax (Canthium) ciliata*, locally called 'kheoha', as an enema for the relief of abdominal pains caused by the presence in the intestines of a small dung beetle, *Scarabaeus sacer*, which has been introduced by witchcraft (64).

Some prophylactic medicines have to stand on special stands, the idea being almost exactly that of insulation and conduction. For instance, a pot of medicine to keep off witches must be insulated from the earth - it must not stand on a clay or earthen pedestal - for witches power can travel through the earth (25). On the other hand, it must not stand on wood unless the wood is thoroughly dead and dry, for witches have power over fluids such as blood and sap (25).

SYMPATHETIC MAGIC

The Doctrine of Signatures and sympathetic magic are apparent in some traditional medical practices. In some cases the use of plants is based on wishful thinking, the Doctrine of Signatures, or some other fantasy (65). Some of the herbs collected conform with the Doctrine of Signatures, that is the observable qualities of a plant indicate its utility - for example, the red liquid from the fruit and leaves of *Sabicea vogelii*, a slender climbing shrub, suggests its use in stopping persistent menstruation (42). In S Sudan, it has been observed that the homeopathic element is so evident in many magical rites and much of the *materia medica*....(that) it is recognized by the Azande themselves (23). They say, we use such-and-such a plant because it is like such-and-such a thing...; we do so-and-so in order that so-and-so may happen (23). 'Based on superstition, heaven had indicated to mankind the purpose for which each plant had been placed on earth marked clearly with signs that resemble parts of the human body for which it was to be used' (52). As such the leaves of *Anemone hepatica* Windflower which look like the anatomical lobules of the liver, were to inspire Renaissance physicians to use it for hepatic disorders (50).

In W Africa, an unspecified part (probably the fruit) of *Hedranthera barteri*, an apocynaceous shrub with white flowers, on the Doctrine of Signatures, is used by the Igbo of S Nigeria to prevent miscarriage (N.W.

Thomas fide 13a) (see MARRIAGE: FAITHFULNESS IN MARRIAGE: Punitive Measures: Unable to Withdraw); and in SE Asia, the flowers of *Celosia argentea,* an erect annual amaranth with silvery-white pink-tinged yellow or red flowers – often cultivated, are considered medicinal for conditions whose symptoms include discharge of blood such as dysentery, haemophthysis and menstruation (13a). In this connection it is probably the red-flowered form of the species that apply on the Doctrine of Signatures (54). In S Nigeria, *Harungana madagascariensis* Dragon's blood tree, a shrub or small tree, features in a Yoruba incantation under the name 'amuje', arrester of blood, for the purification of the blood (35).

In Congo (Brazzaville), the application of various parts of *Combretum recemosum*, a scandent liane in forest, to the following ailments may be on the Doctrine of Signatures because of the red colour of the flowers: the whole plant for all genito-urinary and gastro-intestinal affections accompanied by bleeding; the root macerate or decoction in draught for dysentery; the leaf-sap for haemorrhoids; the bark-pulp for bleeding during pregnancy; the powdered leaves or roots for haematuria, convulsive coughing and tuberculosis; the sap as a haemostatic and cicatricant; and the powdered bark or leaves to circumcision wounds (10). In this country, *Selaginella myosorus,* a herbaceous forest fern, is credited with magical power as one of the select remedies used for treating illnesses of diabolical origin, the patient being fumigated over an abundant quantity incinerated on hot embers (10). The leaves are alternate on the stem, and the sophistry in Ghana of treating with this plant is that the disease is deceived missing the opposite leaves and thus sparing the patient. In S Africa, *Cassytha filiformis* Dodder, a slender parasite with leafless stems, is used as a wash in the Cape area to stimulate hair growth (64). The application is also found in Asia and arising on the Doctrine of Signatures because of the vigorous matted hair-like growth of the plant (13c). Similarly, *Adiantum capillus-veneris* Venus's hair is used to promote hair growth (50).

Throughout the Region, *Manniophyton fulvum,* a straggling bush or liane, on the Doctrine of Signatures, is widely used for affections in which blood is manifest. In Congo (Brazzaville), the plant is used as a haemostatic and cicatricant on wounds, and for treating dysentery, piles, haemophthysis and dysmenorrhea (10). In Ivory Coast, it is used for painful menstruation (31). In this country and Burkina Faso, the red stem-sap is used topically on herpes and other dermal infection (36); and in Congo (Brazzaville) the leaf-sap is similarly applied even to areas of leprosy (10). In Liberia, the powdered dried leaves are sprinkled on sores; and in Sierra Leone, a stem decoction is drunk for blennorhoea or gonorrhoea (13d). In Java, *Pycnoporus sanguineus* and *Polystictus sanguineus,* polypore fungi, are used for treating haemophthysis on account of their red colour (64).

In Sierra Leone and Liberia, *Bersama abyssinica* Winged bersama, a savanna tree, has medico-magical uses with important juju practices in the latter country (17). The Basa name meaning 'broken bones' arises from a curious treatment to mend broken limbs. When someone breaks a leg, the patient is placed on a mat with the leg set straight and held in place with stones. A fowl, *Gallus domestica,* is caught and has a leg broken in the same place. The bark, wood and some leaves are pulped up with palm-oil to make a salve. This is applied to the fowl, not to the patient whose healing is derived by sympathetic transference from the unfortunate bird. When the fowl begins to walk again, it is time for the patient to do likewise (15). In Liberia, magical power is said to reside in the bark of *Pentaclethra macrophylla* Oil bean tree. Because of the crooked twisting of the trunk, the bark is used to treat hunchback (62). In SE Asia, the jointed nature of the stems of *Euphorbia tirucalli,* a shrub, lead to the use of the plant, on the Doctrine of Signatures, to assist in the healing of broken bones (12, 54). In Malaya, leaves of *Pedilanthus tithymaloides* Redbird-cactus or Jew-bush or Slipper-flower, a cactus-like shrub, arranged in rows like the legs of a centipede, *Lithobius* species, so the sap is used for the treatment of centipede bite, scorpion sting and skin cancer (13b).

In Ivory Coast and Burkina Faso, *Vernonia nigritiana,* a composite herb in savanna, is credited as being an emmenagogue, probably because of its blood-red flowers (36). The application of the dried crushed finely ground haustorium of *Phragmanthera* and *Tapinanthus* species Mistletoe, epiphytic parasites, in prescriptons to heal wounds is hinged on the haustorium being a healed wound at the point of attachment of the parasite to the host. In Botswana, the Southern Sotho administer a root decoction of *Rubus ludwigii* Blackberry or *R. rigidus* Bramble when there is acute pain during illness, the belief being that the prickly plant will struggle with the pain and overcome it (64). In Europe, *Symphytum officinale* Comfrey had a reputation in Medieval times for knitting broken bones and reflected in its name which comes from the Latin *conferre,* meaning to bring together (6). (See also DISEASE: HEART DISEASE).

OTHER SUPERSTITIOUS USES

Besides instances of medico-magical, charms and talisman, witch and fetish and sympathetic magic in traditional medicinal practice, there are also instances based on superstition. For instance, in W Africa, *Ficus* species have various superstitious medicinal practices; and many of the medicinal uses of *Phragmanthera* and *Tapinanthus* species Mistletoe, epiphytic parasites, are based on superstition and vary according to the nature of the host (32). (For a list of Mistletoes in the Region, see DISEASE: LEPROSY: Table 1). In Senegal, Mistletoe leaves are steeped in water with *Bombax costatum*, a tree, and *Costus spectabilis*, a perennial herb in the ginger family, for a week, during which a sick person is washed with the water morning and evening. At the end of the week the Mistletoe is thrown to the four-winds at crossroads (24). There are also many superstitious uses of *Lophira alata* Red ironwood in traditional African medicine (32). In parts of the Region, *Euphorbia balsamifera* Balsam spurge and *E. lateriflora*, tree euphorbias, are grown in some villages and used superstitiously for medicinal purposes (17).

In S Nigeria, Igbos at Asaba are reported to use *Elytraria marginata*, a small rosette-shaped acanth, perhaps superstitiously in medicine against poisoning caused by bats, species of *Cynopterus, Eidolon, Epomophorus, Glossophaga* and *Pteropus* (N.W. Thomas fide 13a). In this country, the Yoruba name of *Bombax buonopozense* Red-flowered silk cotton tree, 'esho', is used in a superstitious sense when referring to its medicinal uses, in the belief that the remedy would be ineffective if called by its proper name (46). In this country, the fruits of *Pachyelasma tessmannii*, a forest tree, with the seeds and the bark, are used superstitiously by the Ekoi people in native medicine and other practices (17); and *Desmodium ramosissimum*, an erect semi-woody forest perennial, has superstitious uses (13d). The Ijo of this country claim that for an antidote against serious poisoning, 'pick four leaves of *Struchium sparganophora*, a semi-succulent annual composite herb, and chew them with seven 'lark' (? small) peppers and some salt and you will soon feel well' (Williamson fide 13a). In this country, the Yoruba call an unidentified plant which causes sedation of fish, 'iromi', meaning 'my thoughts', and invoke it to calm the body of a sick person (Verger fide 13d).

In Senegal, young girls suffering from ailments of fatigue connected with puberty are thought to benefit by lying on a bed of fresh leaves of *Carica papaya* Pawpaw for the duration of the illness (35a, b). In this country, a herbalist is reported to cure earache by prescribing the patient to wear a collar of *Sanseviera senegambica* African bowstring hemp (7,

35); and in Sierra Leone, the leaf decoction of *Glyphaea brevis,* a small forest tree, is an ingredient in medicine used by Secret Societies (Thomas, Herb. Oxf. fide 17). In S Africa, the European ties slices of raw potato, *Solanum tuberosum,* behind the ears with a red cloth for delirium, until the potato turns black (64); and in Europe, the leaves and flowers of *Borago officinalis* Borage, a Mediterranean garden herb, put into wine is believed to make men and women glad and merry, driving away all sadness, dullness and melancholy (51).

Bad Dreams

In S Africa, the Southern Sotho make a steam bath of *Helichrysum nodifolium* var. *leiopodium,* a perennial composite herb, by pouring an infusion on hot stones, which is applied to patients suffering from fever or bad dreams; or a lotion made from *Crassula transvaalensis,* a fleshy herb, is used by the tribe as a charm against disturbing dreams about one's ancestors - in making use of this it is essential that the body be bathed by an older person and not the patient himself (64). In Botswana, the Sotho inhale the smoke from burning *Lasiosiphon anthylloides,* a shrub locally called 'moomang', for fevers and bad dreams (64). In S Africa, the Zulu either use *Popowia caffra,* a scandent shrub, as a charm against bad dreams; or take a bark infusion of *Sideroxylon inerme* Milkweed to dispel bad dreams (64). The Zulu also use the root of *Hibiscus pusillus,* locally called 'uguqukile', as an emetic in the treatment of bad dreams; or take the bark infusion of *Ansellia humilis,* an epiphytic orchid, as an antidote for bad dreams; or the head is held in the smoke from the burning fibrous root for the same purpose (64).

In Ghana, the heartwood of *Daniellia oliveri* African copaiba balsam tree, locally called 'sanya', is ground and burnt to drive away forces believed to be responsible for bad dreams (22); or the egg-like seeds of *Okoubaka aubrevillei,* a tree believed to be fetish, is placed under the pillow or the bed amidst prayers for a similar purpose (45). In Sierra Leone, herbalists give a preparation of *Spilanthes filicaulis* Brazil cress, a herb, to 'clean' the

Spilanthes filicaulis Brazil cress

tummy of someone who fancies he has been given bad food in a dream (Boboh fide 13a). In Congo (Brazzaville), to stop nightmares and erotic dreams a macerate of *Cogniauxia* species, a composite climber, and *Rauvolfia vomitoria* Swizzle-stick, a small tree, in palm-wine is drunk before retiring (10).

In S Africa, when a person suffers from nightmares and headaches, he is likely to be advised by his African doctor to plant a cutting of *Vangueriopsis lanciflora* Wild medlar, locally called 'mpemba', or a meal being scattered around the cutting; or a heavy twig of *Acokanthera venenata*, Poison tree is burned in the hut to keep away *impundulu* which is a type of visiting evil spirit which chokes people in their sleep and causes bad dreams (64). In this country, the Southern Sotho use *Senecio asperulus*, a composite, as a charm for bad dreams in children; and in Zimbabwe, the Manyika grow *Boophone disticha* Candelabra flower, a lily, outside their huts as a charm to ward off evil dreams (64). In Europe, *Vinca* species Periwinkle has been used for diverse complaints including nightmare; and in Australia it is believed that a twig of *Viscum album* Mistletoe, an epiphyte, laid on the door-step safeguards inmates from nightmares (51). In Japan, *Nandina domestica* Heavenly bamboo is planted close to a door leading into the home for the convenience of members of the household who may suffer a bad dream or nightmare - if he confides details of this to the Heavenly bamboo, it is believed that he will suffer no harmful after effect (51).

Drift Dissemules

The obscure origin of sea-beans (see RELIGION: SYMBOLIC PLANTS: Sea-Beans for the meaning) gave them an aura of mystery and magical power of healing. When the Gulf Stream was unknown, the origin of tropical drift dissemules or sea-beans was subject of speculation, superstition and lore. This further added up to the belief in their curative properties. The prevailing view was that sea-beans grew on under-water trees or tangles. (For a list of the plants, see RELIGION: Table 1). In S Nigeria, the seeds of *Entada* are used medicinally by the Yoruba tribe - the uses being probably superstitious (18). And in many parts of Africa, *Mucuna* species enter into charms for protection form sickness (see CHARMS AND TALISMAN above).

By far the most speculative dissemule in folklore is *Lodoicea maldivica* Coco-de-mer. Until the discovery of the parent plants on Praslin and Curieuse Islands in the Seychelles by Mahé de la

Bourdonnais in 1743, the only source of the nuts was the drifting or stranded nut; and claims concerning its powers as a cure-all became ever more grandiose. The sick became immune to other diseases if they drank water that had been kept in the shell for some time; and to which had been added a piece of the endoderm (29).

Even today Coco-de-mer seeds are precious as a plant curiousity (being the biggest seed in the world). It costs at least £50.00 (fifty pounds sterling) to take one of these seeds out of the Seychelles.

All-in-One

The practice of professing a single plant as a miracle to cure every imaginable ailment that can possibly afflict mankind is not new. It has been the usual boast of some medicine-men, a common slogan of many drug peddlers and equally the desire, hopes and expectations of countless patients. In addition to Coco-de-mer, several other plants are branded as a panacea for a host of ailments. For instance, *Aloe vera*, a lily, and *Sassafras officinale* Sassafras are included in the cure-all plants (52). For more than 200 years Sassafras was exploited in disease-ridden Europe as a panacea for many ills (52); and *Cnictus benedictus* Blessed thistle was mentioned as heal-all in treatises on the Plague from whence the specific epithet benedictus was derived (9). Others are *Arctium lappa* Burdock and *Stachys officinalis* Betony (50). In the Middle Ages, *Ruta graveolens* Common rue was considered a cure for countless ills from indigestion to bee stings and was often called 'Herb of Grace' (51). The roots of *Anemopsis (Houttuynia) californica,* a bog plant from California, were strung into necklaces - in the form of beads - as a precaution against malaria and other diseases; and *Digitalis purpurea* Fox-glove, a temperate decorative weed, was employed for a miscellany of ailments including the 'Kings's Evil' (51).

American Indians considered *Nicotiana tabacum* Tobacco a magical herb and a cure-all for countless ailments (51). Tobacco was regarded by the aboriginal people of America as sacred and a special gift of the gods to man. When the plant reached the Old World this attitude was adopted by Europeans along the Indians' ideas of the efficacy of tobacco as a remedy against almost all bodily ills (64). Two thousand years ago, the plant was one of the great curers of all ills; and today it is one of the great killers of all times (60). Tobacco has become the drug that has produced the most diseases, the most deaths, and the highest expenditure world-wide (50). Ginseng's genus name *Panax* is

Mandragora officinarum Mandrake (legendary and mythical plant)

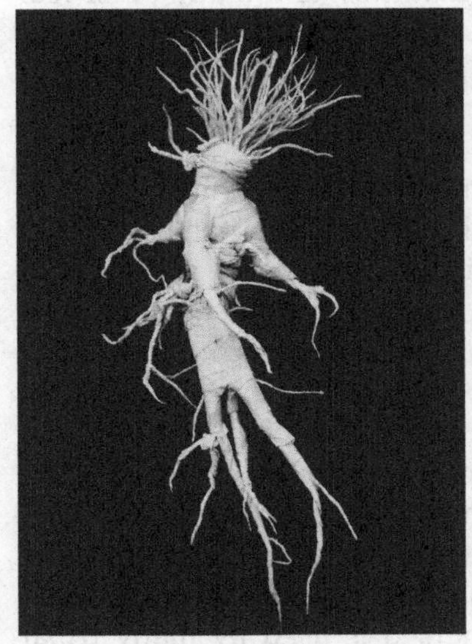
Mandragora officinarum (root)

Senna occidentalis Negro coffee

derived from the Greek word pan (all) akos (cure), meaning cure-all (9). The forked roots of *Panax quinquefolium* and *P. schinseng* Gingseng, were believed to be a panacea for countless ills. Human-shaped specimens were particularly valuable (50), and quite literally worth their weight in gold (51). The roots of *Mandragora officinarum* Mandrake, often shaped in human form, were supposed to cure a host of maladies (51). The popularity of both Gingseng and Mandrake as panacea for all ills is confirmed (52).

At the peak of their maturity which is also usually around a full moon, cut flowers have been observed to transfer any healing properties to water from a clear stream with miraculous results. 'After the flowers had been standing in the sunlight for several hours, he found that the water had become strongly impregnated with the vibrations and power of the plant... Thousands of patients throughout England and the world were to vouch for their efficacy, and many thousands still depend on this elixir of flowers to cure them of innumerable ailments' (58).

Traditional medicinal practice in W Africa is no exception. Some plants are claimed to cure several ailments. Among these include *Annona senegalensis* var. *senegalensis* Wild custard apple, *Balanites aegyptiaca* Desert date or Soap berry, *Senna (Cassia) occidentalis* Negro coffee, and *Ocimum gratissimum* Fever plant. The egg-like seed of *Okoubaka aubrevillei*, a rare forest tree occurring mainly in Ivory Coast and Ghana and believed

to be fetish, placed in bath water for ritual washing is said to heal many ailments (14). In Liberia, *Scottellia coriacea,* a forest tree, is an infallible panacea used by all herbalists (15). In Congo (Brazzaville), *Oncoba spinosa* Snuff-box tree, a thorny savanna shrub, has a great reputation as a panacea for all sickness and protector against evil influences and spirits - no part of the plant must be cut without making an offering and explaining to the plant the reason and expectations (10). On the traditional usage of orchids, it has been observed, 'In many parts of Africa, the tubers and fleshy roots are used as sources of healing remedies, not just to ward off witchcraft or to ensure a happy outcome for a courting episode, but also to treat, apparently successfully, such diverse maladies as loss of speech, flatulence, worm infections, difficulties in pregnancy, madness, and many other real and imagined sorrows' (57).

PURIFICATION

This is a traditional ceremony summoning the beneficial forces in a protective, religious, cleansing ritual to strengthen one spiritually against sin, evil, witchcraft, bad juju, disease, and so on; and to expel these forces or their influence. Among many of the coastal and forest tribes in the Region, the whole plant of *Portulaca oleracea* Pigweed or Purslane is used as a ritual bath, or dipped in water which is sprinkled around by fetish priests as a sign of completion of healing, or for general purification (2). In Ghana, Togo and Benin Republic it is not uncommon among the Ewe-speaking tribe to wash patients with water in which Purslane has been macerated before the healing as well, to drive away any evil prior to administering treatment (39). After a juju cure, a cold bath in which Purslane has been immersed is used by many tribal groups in the forest area to stop the operations of the spiritual forces that had been employed, as well as other forces that may cause harm (45). *Costus afer* Ginger lily, a tropical perennial herb, and *Hyssopus officinalis* Hyssop, a temperate labiate, have both featured in purification rituals in their respective distribution areas and beyond since biblical times. *Lavandula angustifolia* Lavender,

Costus afer Ginger lily (symbol of purification from sin)

Hyssopus officinalis Hyssop (legendary and mythical plant; symbol of purification from sin; associated with The Christ).

another temperate labiate is used for purification of body and spirit (9).

As a precautionary measure among some coastal tribes, seven leaves of *Plumeria rubra* var. *acutifolia* Frangipani or Temple Flower and seven leaflets of *Elaeis guineensis* Oil palm in water, enter into rituals used by medicine-men to wash themselves after curing a patient of an ailment as a preventive of its occurrence on them (37). Similarly, in the treatment of an epileptic among some of the ethnic groups in the Region, the children and all members of the household of the fetish priest bathe with the medicine before the sick person is treated, to prevent its recurrence on them (22). In Botswana, *Acanthospermum hedyotideum* or *A. rigidum* Granaatbossie is one of the ingredients in a Southern Sotho charm medicine for a person who has assisted in the nursing of a sick person or taken an active part in a funeral - failure to take the medicine results in a failure or a partial failure of the person's crops (64).

In W Africa, there is the general belief among many of the tribal groups that convalescents regain their strength more rapidly after an early morning walk in the forest or a visit to the farm - the explanation and popular belief being that as the leaves along the path brush against the sick, they carry the illness away and cleanse the sick (49). It could also probably be the more salubrious forest atmosphere coupled with the exercise (see also DISEASE; JUJU OR MAGIC; and WITCHCRAFT).

References.
1. Ankoma-Ayew, (pers. comm.); 2. Anobah, (pers. comm.); 3. Appia, 1940; 4. Aubréville, 1950; 5. Ayensu, 1981; 6. Back, 1987; 7. Berhaut, 1988; 8. Bethel, 1968; 9. Blumenthal & al., 2000; 10. Bouquet, 1969; 11. Bouquet & Debray, 1974; 12. Burkill, 1935; 13a. Burkill, 1985; b. idem, 1994; c. idem, 1995; d. idem, 1997; e. idem, 2000; 14. Commeh-Sowah, (pers. comm.); 15. Cooper & Record, 1931; 16. Crow, 1969; 17. Dalziel, 1937; 18. De Feo, 1992; 19. Dobkin de Rios, 1997; 20. Dokosi, 1969; 21. Ellis, 1894; 22. Eshun, (pers. comm.); 23. Evans-Pritchard, 1937; 24. Ferry & al, 1974; 25. Field, 1937; 26. Fink, 1989; 27. Flores & Lewis, 1978; 28. Galt & Galt, 1978; 29. Gunn & Dennis, 1979; 30. Hepper, 1976; 31. Holland, 1908-22; 32. Irvine, 1961; 33. Jackson, 1973; 34. Kerharo, 1969; 35a. Kerharo & Adam, 1964; b. idem, 1974; 36. Kerharo & Bouquet, 1950; 37. Klah, (pers. comm.); 38. Kroger, (pers. comm.); 39. Kulefianu, (pers. comm.); 40. Labulthe Tora, 1981; 41. Lawler, 1984; 42. Macfoy & Sama, 1983; 43. Mallart Guimera, 1969; 44. Maja Naur, 1999; 45. Mensah, (pers. comm.); 46. Millson, 1891; 47. Moore, 1960; 48. Oso, 1977; 49. Otu, (pers. comm.); 50. Pamplona-Roger, 2001; 51. Perry & Greenwood, 1973; 52. Pitkanen & Prevost, 1970; 53. Pope, 1990; 54. Quisumbing, 1951; 55. Sereboe, (pers. comm.); 56. Siegel, Collings & Diaz, 1977; 57. Stewart & Hennessy, 1981; 58. Tompkins & Bird, 1974; 59. Tremearne, 1913; 60. Van Sertima, 1976; 61. Verger, 1969; 62. Voorhoeve, 1965; 63. Walker & Sillans, 1961; 64. Watt & Breyer-Brandwijk, 1962; 65. Wickens, 1987.

36 OCCULT

Woe unto thee that call evil good, and good evil; that put darkness for light, and light for darkness; that put bitter for sweet, and sweet for bitter!
Isaiah 5:20

The practice of occultism is an ancient and universal one. It involves healing, divination, prophecy, visionary and other forms of services and advice in a mysterious or supernatural way which is beyond the range of scientifically known or recognizable phenomenon. The term literally means 'hidden', 'covered over' or 'concealed'. Occultism, like witchcraft, could be influenced by the forces of the Devil. In the folklore of traditional Africa and elsewhere, some plants are associated with the practice - either to promote it or to counteract its influence.

PLANTS FOR THE PRACTICE

Hallucinogenic plants are usually associated with occultism. It has been observed that magicians, witches and practitioners of ancient, occult science have often utilized psychedelic drugs to induce altered states of consciousness (2). The plants with psychedelic properties referred to include those listed in Table 1 below:

Species	Plant Family	Common Name	Remarks
Amanita muscaria	Basidiomycetes	Amanita	A poisonous mushroom
Artemisia absinthium	Compositae	Wormwood	Temperate plant
Artemisia vulgaris	Compositae	-	Temperate plant
Cannabis sativa	Cannabinaceae	Marijuana or Indian hemp	A perennial tropical weed
Erythroxylum coca	Erythroxylaceae	Cocaine	A tropical shrub
Nicotiana tabacum	Solanaceae	Tobacco	A cultivated annual
Salvia divinorum	Labiatae	-	A herbaceous annual
Tabernanthe iboga	Apocynaceae	Iboga	A perennial tropical shrub
Tagetes species	Compositae	Marigold	An annual decorative

Table 1. Some Psychedelic Plants.

Tagetes species Marigold

Datura species, plants with decorative flowers often tended around villages, are traditionally believed to be associated with occult practice - very likely for their hallucinogenic properties. The species in the Region are listed in Table 2 below:

Species	Plant Family	Common Name	Habit
Brugmansia (Datura) suaveolens	Solanaceae	Angel's trumpet or Moon flower	Foetid annual or perennial
Datura candida	Solanaceae	-	Foetid annual or perennial
Datura innoxia	Solanaceae	Hairy thorn-apple	Foetid annual or perennial
Datura metel	Solanaceae	Hairy thorn-apple or Metel	Foetid annual or perennial
Datura stramonium	Solanaceae	Trumpet stramonium or Jamestown weed	Foetid annual or perennial

Table 2. Datura species in the Region

Datura metel Hairy thorn-apple or Metel

In Ghana, two sisters - aged three and five - from Aburi who, mistakenly, thought the fruits of Metel were those of *Solanum incanum* Egg plant or Garden egg and used them to prepare a sauce on the farm to eat in 1987 were unconscious for twenty-four hours - regaining consciousness at the Tetteh Quarshie Memorial Hospital, Mampong with shouts and yells. In southern N America, *Datura innoxia* Hairy thorn apple is of importance in rites (5b). It was taken during rituals to gain occult power, and in a spring ceremony to ensure good luck and long life of initiates, who, dancing like wild animals, would finally fall into a stupor to attain adulthood (5b). Throughout its growing area in the Region, the leaves of *Ipomoea pes-caprae* Beach convolvulus, a perennial creeper on the sea shore, collected with the necessary rites are chewed with the seeds of *Aframomum melegueta* Guinea grains or Melegueta by many of the coastal tribes, then washed down with one or two tots of local gin *(akpeteshi* or *ogogoro)* in a ceremony to induce deep

sleep during which, it is believed, visions of both one's enemies and true friends alike will be clearly seen (10). In Gabon, a root-extract of *Mimosa pigra*, a sensitive scrambling shrub, together with other certain plants, is dipped into the eyes of novitiates in the Bwiti free-masonry to give them the ability to see that which is hidden (12).

In C Sumatra, *Celosia argentea,* a weedy annual amaranth often cultivated as a vegetable, enters into most occult ceremonies to propitiate the protective spirit (5a). Similarly, for occult reasons, *Calotropis procera* Sodom apple is added, along with *Tamarindus indica* Indian tamarind, a savanna tree legume often cultivated for the edible fruits, *Allium cepa* var. *aggregatum (ascalonicum)* Shallot and resins in gunpowder in Chad, as in India (6). In N Nigeria, the peeled stem of *Hibiscus lunariifolius* Hemp-leaved hibiscus has superstitious uses in medical and occult practice by the Hausas (6). In Ivory Coast, the Guéré name of *Inhambanella guereensis,*'kantu', a forest tree, meaning 'tree of spirits', suggests that the tree has some mystic significance, but no explanation is given (3, 4). In Ghana, *Pancovia bijuga,* a small forest tree, is reported occurring in fetish forest, but whether it is deemed to hold occult powers is not stated (5b).

In Ivory Coast and Burkina Faso, a few drops of the latex of *Elaephorbia drupifera,* a savanna tree with fleshy branches containing abundant very caustic white latex (the high forest equivalent is *E. grandifolia*), in occult way can make the criminal lose his sight; but when he confesses, an antidote comprising a virgin young girl's urine and a preparation of the dried leaves of the tree cooked together as drink is given to reverse it (9). The practice is a form of trial by ordeal. (For more about the practice, see ORDEAL). In practice a drop or two of the milky latex in the eye is known to cause blindness unless quick remedial measures are taken. The antidote is immediate application of mucilaginous plants or a wash with plenty of water (1).

PLANTS AGAINST THE PRACTICE

Some other plants are however, protective of occult forces or neutralize the action of these forces. In N Nigeria, the leaves of *Rhynchosia sublobata,* a twining legume, are prepared along with natron (native sesquicarbonate of soda) - and probably other medicines - for washing the whole body to prevent poisoning or other injury by occult means (6). Similarly, a decoction of *Indigofera hirsuta* Hairy indigo, collected and prepared with ceremony, is taken by draught and used for a ritual bath by the Hausa of N Nigeria, the Zarma of Mali and Niger and the Fula as preventive

Indigofera hirsuta Hairy indigo.

of any spiritual injury by occult means (11). A ritual bath with the bitter decoction of *Phyllanthus niruri* var. *amarus (amarus)*, an annual euphorb (7); or a whole plant macerate of *Croton lobatus*, an euphorb weed of cultivated land (10); could also be used to counteract these forces. Whilst bathing, some of the decoction or maceration is sipped as a definite request is meditated upon (7, 10).

A leaf of *Pergularia daemia*, a trailing annual asclepiad, collected and chewed with prayer (10), is believed protective and preventive of any injury by occult forces. The whole plant of *Pergularia* may be buried with rituals in front of the house with prayers for spiritual protection from occult forces (10); or the dried plant may be charred in an earthenware and either sprinkled alone around the household, or mixed with the fruit fibre of *Elaeis guineensis* Oil palm, invoking the plant during the ceremony in an incantation to neutralize any destructive intentions (8). 'Being impervious to occult influences, *Leea guineensis*, a soft-wooded shrub, is used to strengthen an oath and therefore, give it assurance of purity of motive' (Cooper fide 6). (See also FETISH; JUJU OR MAGIC; and WITCHCRAFT).

References.

1. Abbiw, 1990; 2. Albert-Puleo, 1978; 3. Aubréville, 1959; 4. Aubréville & Pellegrin, 1957; 5a. Burkill, 1985; b. idem, 2000; 6. Dalziel, 1937; 7. Fabianu, (pers. comm.); 8. Gamadi, (pers. comm.); 9. Kerharo & Bouquet, 1950; 10. Mensah, (pers. comm.); 11. Moro Ibrahim, (pers. comm.); 12. Walker & Sillans, 1961.

37 ORDEAL, TRIAL BY

.... and others had trial of cruel mockings and scourgings,...they were stoned, they were sawn asunder, were tempted, were slain with the sword...
Heb 11:36

Some plants were used in Africa and elsewhere for ordeal trials because of their poisonous properties. In this trial serious offences such as alleged witchcraft and murder, or less serious ones like adultery and theft were judged by superstitious ordeal. Suspected persons were made to drink a poisonous decoction to test their guilt or innocence. The traditional belief being that the innocent would vomit out the poison, but that the guilty would be poisoned and possibly killed; unless an antidote was immediately given to neutralize the poison. The test may also be applied externally. By this system of trial, the investigation, sentence and punishment (unless acquitted) were carried out simultaneously. There was, of course, no provision for appealing. The practice is now illegal in law.

It is one of the mysteries of traditional practice that any suspect ever survived, because extremely poisonous principles like erythrophleine, physostigmine or eserine and strychnine have been isolated from some of the ordeal plants. As further proof of the poisonous nature of ordeal preparations, it has been recorded, 'a guilty man does not dare to drink *orisha,* but the innocent will submit to the ordeal without fear, and, indeed, frequently demand it in order to prove their innocence; whence it follows that it is the guiltless who ordinarily perish' (9). There is certainly more to the practice of administering ordeal prescriptions as a means of trial and judgement than appears. It is also an open secret that witch-doctors, fetish priests and medicine-men are capable of many tricks, and there were various means of manipulating the test. Besides plant-based trial, other forms of trial were practised.

PLANTS TEST

Plants that traditionally featured prominently in ordeal-trials were the bark of *Erythrophleum suaveolens* Ordeal tree, a savanna legume tree; the seeds of *Physostigma venenosum* Calabar bean, the product of a legume

Erythrophleum suaveolens Ordeal tree

Physostigma venenosum Calabar bean 1/2 normal size

vine - sometimes cultivated for the purpose - (see Witch Trial below) and the root-bark of *Strychnos* species, forest lianes noted for their poisonous properties. Sometimes stem-bark is used as an ordeal (3). The use of *Strychnos* species in ordeal trial has not been recorded for W Africa.

Adultery and Lesser Crime

In Cameroon (23b), Gabon (3, 23a, 33), Congo (Brazzaville) (3, 4), Central African Republic (3, 5, 25), and the Democratic Republic of Congo (Zaire) (3, 23a) - all in the Central African area - the root-bark of *Strychnos icaja*, a large forest liane known as *mboundou, mbomdo* or *mbondo*, is the source of a frequently used ordeal-poison. This applied especially to young plants, the older lianes being known as *kpo* in Cameroon. *Strychnos densiflora,* a large liane, is said to be the *mboundou* or *mbomdou* ordeal tree of Gabon, but these names are widely applied to *S. icaja* in southern E Cameroon, Gabon and Congo (Brazzaville). The root-bark is an ordeal-poison in E Cameroon (2). In the Democratic Republic of Congo (Zaire), *Strychnos angolensis*, a liane, is recorded as being used as an ordeal-poison in the Equator Province (2, 3); and *Strychnos samba*, a forest liane, is used as an ordeal-poison in Central Africa (6c).

It is the root-bark of young plants that have not yet reached 2 cm high that is taken. This is perhaps on the premise that the bark of young plants has a lower toxicity which, in the case of an ordeal trial,

gives the accused a sporting chance to prove innocence. To prepare the judicial dose, chips of the root-bark are macerated and fermented. The liquor turns brown and when effervescence ceases, the accused is given a cupful to drink (32, 33). In limited quantity, the concoction is not fatal, causing intoxication and diuresis. An antidote is said to be human faeces with a chicken's egg and sugarcane juice (33). Strong dosing causes vomiting, or death with strong convulsions (3).

In Liberia, *Vismia guineensis,* a small tree in secondary forest, known as 'ge-an' in Basa, is regarded as a fetish with magical properties - the local name partly meaning 'the spirit world'. A pot containing the bark, along with palm oil (or palm kernel oil) and some of the oily resin of the plant, is placed on a tripod formed of the three branches of the tree; and suspected criminals immerse the hands in the boiling mixture as a superstitious ordeal (Cooper fide 8). This method of ordeal trial is also applied to lesser offences like adultery. In this country, the Basa use *Mareya micrantha,* a shrub or under-storey tree, as an ordeal-poison (6b). In Gabon, a purple-leaved form of *Canna indica* Indian shot is used in ordeal trials in cases of alleged adultery; and the roots of *Chasmanthera welwitschii,* a woody forest liane, are used as an ordeal-poison - judgement resting on constipation or diarrhoea, for the innocent or guilty, respectively (33).

In Ivory Coast, the latex of *Elaephorbia grandifolia,* a forest tree, is a common ordeal-poison applied to the eye. Anyone admitting their crimes quickly receives an antidote - a decoction of dried leaves of this plant made in a young virgin's urine (21, 28). (The savanna equivalent is *E. drupifera*). In Central African Republic, the root and stem-bark of *Hymenocardia acida,* a savanna tree, are used as an emetic antidote to ordeal-poison in Ubangi (28). In Senegal, the latex of *Euphorbia balsamifera,* an erect shrub has been used in ordeal-poison (20a, 20b); and in N Nigeria, *E. kamerunica,* a cactiform tree, has been recorded as one of the commonest ordeal-poisons amongst the Pabir and Busa tribes of Bornu (24). The latex is smeared on a straw which is licked saying the proverbial oath. The priest then administers a series of three draughts of the juice with water (to which he may have added an emetic as he thinks fit) which the accused then drinks (24). Taken in excess death may swiftly follow, as has been recorded by the use, apparently of the latex of this species, taken in trial by ordeal on the Benue River (8).

In the Democratic Republic of Congo (Zaire), sap from the bark and cambial area of *Entada abyssinica,* a savanna tree, are used as an ordeal-poison by introducing it under the eyelid (18). In this country (18) and Congo (Brazzaville) (4), the bark-sap of *Piptadeniastrum africanum,* a forest timber tree, is used as an ordeal-poison by dripping into the eye. The application is so painful that the very threat is said to extract a 'confession' (4). In Congo (Brazzaville), sap of *Indigofera capitata,* a herbaceous or semi-woody savanna legume, is administered as an eye-instillation in an ordeal for married women accused of infidelity; and *Barteria fistulosa,* a slender forest tree, is invoked to chastise a guilty person from a distance (4). In places *Andira inermis,* a savanna tree, is regarded as evil. The Hausa name, *gwaska,* means 'ordeal' reflecting that in N Nigeria it is, or has been, used as an ordeal-poison. The tree's use in states of dementia perhaps adds to this sinister regard (8, 19, 26, 27).

In Congo (Brazzaville), *Scoparia dulcis* Sweetbroom weed, an annual herb of wastelands, features as an ordeal in a trial known as 'Hand washing' or 'Locked hands'. A cup containing crushed leaves in water is placed before the contending parties who then wash their hands in the liquid, beating their palms together, each stating his case. The guilty one of the parties is the one who becomes incapable of moving his hands which are locked together, palms facing outwards. This manifestation is explained by the tantalizing action of the liquid on the arm muscles (4). In the Democratic Republic of Congo (Zaire), *Celtis durandii* and *C. wightii,* forest trees, enter into ordeal-poisons or are ingredients in the practice; and are subject to superstitious beliefs and observances (17).

In Zambia and Zimbabwe, the African uses *Elaeodendron* species for trial by ordeal. If the accused becomes rapidly unconscious, without vomiting, he is guilty (34). In Mozambique, a species of *Datura,* probably *D. metel* Metel or Hairy thorn-apple is administered by the Tonga for trial by ordeal. If the accused person rapidly becomes unconscious and dies without vomiting, then he is held to be guilty (34). In The Gambia, a plant-infusion is used as an ordeal truth poison: suspects of a crime such as stealing would be given by the local religious teacher an infusion of *Brugmansia (Datura) suaveolens* Angel's trumpet or Moon flower, a shrub with long white flowers, to exhort the truth (16). In Cameroon, the seeds of *Baillonella toxisperma* African pearwood or Djave nut are reported used as an ordeal-poison (8, 26).

In W Africa, the sweeping broom is employed in ordeal trial by the Akan-speaking tribe of Ghana and Ivory Coast, and some other tribal groups. Two brooms are pushed into each other end to end, and placed in a loose loop round the neck of a suspect with incantations

and a spell by the fetish priest or medicine-man performing the test. The brooms tighten up and start to strangle the guilty until he or she confesses. However, with the innocent one, the brooms remain slackened round the neck (10). Brooms are usually made with the mid-ribs of *Elaeis guineensis* Oil palm, and in Akan culture they are revered as sacred objects (see TABOO: GENERAL FOR ALL AND SUNDRY Sweeping Brooms).

In S Nigeria, there is a method of ordeal practised by the Yoruba speaking people called *gogo* - drawing of lots. 'A certain number of grass stalks, one of which is bent, are held in the hand or wrapped in a piece of cloth, so that the ends only show, and each person in turn draws one - the bent stalk indicating the one who is at fault' (9c). Another form of ordeal involves the empty fruit kernel of *Borassus aethiopum* Fan palm. This is previously filled with *Sida linifolia*, a malvaceous weed popularly believed to be fetish, and other assorted herbs, then suspended round the neck by a string, and buried by the suspect or suspects with a message. The belief is that if the suspect is guilty, then the kernel will dig deeper and deeper into the soil pulling him or her along with it. However, if the suspect is innocent, the seed will pop out of the earth again (10).

Borassus aethiopum Fan palm (legendary and mythical plant).

There is also boiling palm-oil method of ordeal in which suspects (in case of adultery) plunge their hands into boiling oil for a moment. Those found blistered the next day are adjudged guilty and punished accordingly (22a). A husband who suspects his wife of having been unfaithful to him, but is unable to prove it, while the wife strenuously denies her guilt, subjects her to an ordeal (9b). He obtains from a priest, to whom he states the case, certain leaves which the priest informs him possess medicinal and magical qualities. These leaves he mixes with water in a calabash, in the presence of his wife, while an earthen pot containing palm-oil is placed over a fire. When

the oil is boiling the wife has to dip her hand in the water in which are the leaves, and then at once plunge it into the boiling oil. If the hand should sustain no injury, she is guiltless; but if it is scalded, she is guilty (9b). Few wives knowing themselves to be guilty will stand this test. Generally they confess at the last moments, knowing that the beating that the husband will inflict will be more bearable than the pains of a scalded hand (9b).

For theft, boiling water ordeal may be used along the lines of the above. In the former Gold Coast (now Ghana), there was fire ordeal for priest or priestess which was supposed to show whether they have remained pure, and refrained from sexual intercourse, during the period of retirement; and so were worthy of inspiration by the gods (9b). In S Africa, the Zulu make a torch of the bark of *Ptaeroxylon utile (obliquum)* Sneezewood or Cape mahogany, a timber tree, and the fat of a python, *Python regius,* to discover an evil-doer in the household by thrusting its burning end against the bare body of each inmate. The guilty one only is burnt (34). In this country, the Southern Sotho believe that if a portion of *Aster muricatum,* a composite, is put into the food of an accused person, he readily confesses this guilt (34); and in Zimbabwe, *Zanha golungensis,* a tree of fringing forest in the savanna zone, has use in ordeal trials in the Nkata Bay area (6d).

Witch Trial

In folklore, witch trials are almost always by superstitious ordeal. Among pagan tribes, the most picturesque usage of *Erythrophleum suaveolens* Ordeal tree is an ordeal brew for persons suspected of witchcraft or serious crimes (8). Similarly, *Physostigma venenosum* Calabar bean or Ordeal bean, the product of a legume vine, and known as 'esere' in Ibibio, featured prominently in ordeal trials in S Nigeria in particular, and throughout the Region generally. In Sierra Leone, a bark decoction of *Agelaea trifolia,* a forest climber with red fruits, is reported to be used as a witch ordeal; and a bark decoction of *Lasimorpha senegalensis (Cyrtosperma senegalense)* Swamp arum, an aroid in swamps with prickly petioles, is dropped in the eyes of a fowl or a witch as ordeal (N.W. Thomas fide 6a). In S Africa, the root of *Cassine croceum* Saffraan is a Xhosa emetic and has been used by their witch-doctors for trial by ordeal (34).

In certain sorcery ceremonies in Gabon, especially ordeal-poison trials, the sorcerer will sprinkle his assistants with the leaf-water of *Leea guineensis,* an erect or sub-woody shrub (33). In S Sudan, there is

a witch trial known as *makama* among the Azande (11). 'The *makama* consists of a cone of wood and conical sheath, also of wood. As the cone is thrust into the sheath it is easy or almost impossible to pull it out. Witch-doctors hand it to members in their audience at séances to discover whether they are witches. They hold the sheath and present the cone to the people they want to test. Those who fail to pull out the cone are suspected of witchcraft' (11).

Witch trials have been common rituals in many villages - almost invariably the victims being old women. The climax of the ceremony which the people anxiously awaited was the vomiting of the poison. Those who failed to vomit out all the liquid are jeered and booed out of the town and ostracized (1). On the other hand, should the person accused of being a witch vomit the poison and recover, the accuser would be fined such an amount as to ruin the whole family (2a). 'Witches are torn to bits, destroyed in every savage way, when the ordeal has conclusively proved their guilt - mind you, never before (22b). Some of these savage ways of destruction are 'slow roasting alive; mutilation by degrees before the throat is mercifully cut; tying to a stake at low tide that the high tide may come and drown; and any other death human ingenuity and hate can devise' (22a). In S Sudan, there is a 'bad medicine', *ngbasu Mani*, among the Azande, which kills witches and adulterers and other wrongdoers but not innocent men (11). 'One cooks the medicine on top of a termite mound and stirs it, uttering spells until an eye appears in the middle of the medicine. One pierces the eye with a spear, smashing the pot to pieces' (11).

In addition to the exposure of the witch, it is the belief that the spirit of the ordeal is held to be able to manage to suppress the bad spirit trained by the witch to destruction. Human beings alone can collar the witch and destroy him in an exemplary manner, but spiritual aid is required to collar the witch's devil, or it would get adrift and carry on after its owner's death (22b). (For the consequences of this, see WITCHCRAFT: EXTERMINATION).

Besides the use of plants, there were other methods of witch ordeals. 'When David Livingstone came across the ordeal in the Zambezi, he told his African followers about the water-test formerly in use in Scotland: the supposed witch, being bound hand and foot, was thrown into a pond. If she floated, she was considered guilty, taken out and burned; but if she sank and drowned, she was pronounced innocent. Livingstone, with his usual dry humour, added: 'The wisdom of my ancestors excited as much wonder in their minds as their custom did in mine' (15). In Britain, there was a test, often used in witchcraft

trials, where suspects were made to touch the bones (of the victim) (31). There is a similar practice - 'carrying the body' in Anomabu (9b), and in Ashanti (29) - both in the then Gold Coast (now Ghana). These practices also occurred among the Ga-speaking tribe. 'The rite consists in imploring the spirit of the dead man or woman to assist the living in pointing out the witch who, by his or her black magic, has compassed the death. This the dead person does by causing those who are 'carrying the body' to push or knock against the guilty party' (12).

The practice also occurred among the Dormaa people of Ghana. 'In former times it was a common practice to carry dead bodies in order to expose *abayifo*. The corpse or hairs and fingernails from a deceased person were carried through the village until the *sunsum* of this person took possession of the spirits of the carriers and led them to the person responsible for the death. Since colonial times, this method of trial - like the poison oracle - has been prohibited' (13). In N Nigeria, this method was employed among the Hausa in selecting a new chief (30). (See KINGSHIP: QUESTION OF SUCCESSION); and also in India. 'An Indian custom seems to support this view. In the case of a suspicious death among the Gonds, the relations solemnly call upon the corpse to point out the delinquent, the theory being that if there had been foul play of any kind, the body, on being taken up, would force the bearers to convey it to the house of the person by whom the spell had been cast' (30).

Sometimes animals may also be used in witch trials. Among the Ga-speaking tribe witch trials consist of the accused taking a fowl and cutting partly through its neck, not completely severing the head, but allowing the fowl to run about in death-throes for some minutes before finally lying down to die. If the fowl dies with the breast upwards, the accused is not guilty. However, if the fowl dies breast-downwards, the sacrifice has been rejected as the offering of a guilty person (12). In Sudan, there is record of fighting cocks in a curious ordeal on the granite rock of Kobshi. 'When two are litigating about a matter, each of them takes a cock which he thinks the best for fighting, and they go together to Kobshi. Having arrived at the holy rock, they set their birds fighting, and he whose cock prevails in the combat is also the winner in the point of litigation. But more than that, the master of the defeated cock is punished by the divinity, whose anger he has provoked; on retiring to his village, he finds his hut in flames (30). (For the use of fowls in divination, see FORTUNE-

TELLING AND DIVINATION: PREDICTING THE FUTURE: Animals in Divination).

Anyone shown by the ordeal to be innocent is given a small piece of white clay as a token, and is rubbed with powdered clay to proclaim his innocence. Anyone found guilty has a mouthful of chewed kola-nut spattered upon him (12). (For the penalty, see WITCHCRAFT: EXTERMINATION).

OTHER TESTS.

Another method of ordeal trial is with aggry beads. (For the probable origin of the beads, see INFANTS: PHYSICAL DEVELOPMENT: Other Products). In this test, a bead is placed in a bowl of water, and one takes a mouthful of water and the bead into his mouth, invoking, at the same time, the power of the bead to kill him if he lies. This test generally has the desired effects, as the natives are usually too superstitious to take the bead and make the required invocation, if they know they are not speaking the truth (9a). In Congo (Brazzaville), *Ageratum conyzoides* Billy goat weed, a common weedy composite, is applied in a mild form of trial by ordeal. The sap is spread on the hands of the accused which is then pricked with a needle. Only if guilty will any pain be felt (4).

A needle may also be thrust through the tongue of each member of a household in succession to discover a thief, it being believed that it will fail to pierce the tongue of the person who committed the theft (9b). Before needles came into general use, the sharp thorn of *Citrus aurantiifolia* Lime were similarly employed. Though not explicitly stated, it is assumed , of course, that no pains whatsoever would be felt by the innocent, and that assurance to that effect would be given before members submit themselves to the ordeal.

Another method is to cut a hole about the size of a six penny-piece in a small gourd and fill it with ink. Each of the suspected persons then dips his forefinger into the ink, and those who are innocent would be able to withdraw again without trouble. But directly the finger of the guilty person enters, the gourd closes on it, and will not release it - not even if pulled or struck - until a malam has recited a portion of the Koran over it (30).

Fire could also be used in ordeal trials. 'The suspected persons are made to sit round a fire as close as possible. If a person shivers he is guilty, but should no member of the party do so within a certain time - say an hour - all are innocent, and another party is called up' (30). (See also WITCHCRAFT).

References.

1. Abruquah, 1971; 2. Biset, 1970; 3. Bisset & Leeuwenberg, 1961; 4. Bouquet, 1969; 5. Bouquet & Debray, 1974; 6a. Burkill, 1985, b. idem, 1994; c. idem, 1995; d. idem, 2000; 7. Cooper 7 Record, 1931; 8. Dalziel, 1937; 9a. Ellis, 1881; b. idem, 1887; c. idem, 1894; 10. Eshun, (pers. comm.); 11. Evans-Pritchard, 1937; 12. Field, 1937; 13. Fink, 1989; 14. Gilbert & Boutique, 1952; 15. Gluckman, 1963; 16. Hallam, 1979; 17. Hauman, 1948; 18. Hart & Biney, 1962; 19. Irvine, 1961; 20. Kerharo & Adam, 1964; 21. Kerharo & Bouquet, 1950; 22a. Kingsley, 1897; b. idem, 1901; 23a. Leeuwenberg, 1972; b. idem, 1979; 24. Meek, 1931; 25. Motte, 1980; 26. Oliver, 1960; 27. Oliver-Bever,1983; 28. Porteres, 1935; 29. Rattray, 1927; 30. Tremearne, 1913; 31. Underwood, 1971; 32. Walker, 1953; 33. Walker & Sillans, 1961; 34. Watt & Breyer-Brandwijk, 1962

38 POPULARITY AND SUCCESS

And the fame of David went out into all lands; and the Lord brought the fear of him upon all nations.
1 Chronicles 14:17

The traditional uses of plants include prescriptions and charms believed to make one popular (generally liked or well-known in the society) or successful (prosperous in attaining one's objective or wealth or position). In addition to charms, some of these prescriptions appear to be based on sympathetic magic.

GENERAL CHARMS

In N Nigeria, Fula youths wear *Ectadiopsis oblongifolia,* a shrub in savanna woodland, around the waist as a charm to secure popularity; or the tribesmen use *Kohautia senegalensis,* an annual weed, as a magic medicine to ensure popularity (3). In this country, the Fula add *Evolvulus alsinoides,* a spreading perennial convolvulaceous herb with light blue flowers, to milk to bring success (3). As charms for a prosperous day, *Evolvulus* or *Scoparia dulcis,* Sweet broomweed, collected with ceremony, and macerated in water, is used for a ritual bath after prayers in the morning and before talking to anyone (1). A piece of *Holarrhena floribunda* False rubber tree carried in the hand when going to the bush, or on business, acts as a favourable charm, both to ensure success or to escape punishment for evil-doing (3). For the same purpose a piece of False rubber tree may be burnt in the house (3).

Among the Ewe-speaking tribe of Ghana, Togo and Benin Republic, and other coastal tribes in the Region, a piece of the hide of an animal of prey with the dried

Sida linifolia

leaves of *Sida linifolia*, an erect hairy malvaceous weed of cultivated land believed to be fetish, collected with traditional rites, covered in leather as talisman with the necessary affirmations, and carried on ones body, enters into charm prescriptions to command respect even among one's seniors (6). In S Nigeria, when a new title is conferred on a man among the Yoruba-speaking people, a leaf of *Newbouldia laevis,* a tree similarly believed to be fetish and called '*akoko*-tree', is given to the recipient as a sign of honour (4). In Gabon, a leaf-macerate of *Leea guineensis,* a shrub, used as a body-ointment will make ordinary people acceptable to their superiors (8); and in S Africa, the Sotho-Tswana group use *Aloe saponaria* Soap aloe, a lily, for ritual purification to ensure success (9).

Sympathetic Magic

Some of the charm prescriptions appear to be based on sympathetic magic. For instance, in N Nigeria, the leaves of *Ceiba pentandra* Silk cotton tree, a popular huge tree in both forest and savanna vegetation, along with other herbs, enter into prescriptions to ensure popularity (3); and in S Nigeria, *Launaea (Lactuca) taraxacifolia* Wild lettuce, a common weed of cultivated land, is so ubiquitously appreciated that it features in a Yoruba invocation for someone to be well-known in the community (8). Both prescriptions appear to be based on sympathetic magic. Among many ethnic groups in the Region, *Phragmanthera* and *Tapinanthus* species Mistletoe, epiphytic parasites, especially those growing on some specific hosts, enter into prescriptions by medicine-men for high social position (6). In this instance the sympathetic magic appears to be referred to the unusual and unique habit of the epiphyte. (For a list of species of the epiphyte growing in the Region, see DISEASE: LEPROSY: Table 1).

Portulaca oleracea Purslane or Pigweed (symbol of purification, of goodwill and of peace)

Among many of the ethnic groups in its distribution area, the whole plant of *Portulaca oleracea* Purslane or Pigweed or *Chrysanthellum indicum* var. *afroamericanum*, a composite savanna herb, collected with the necessary rites, is used in a ceremonious bath with the

traditional white sponge and any sweet-scented soap for seven or seventeen consecutive days as prescriptions for success or progress or popularity within the community (6). The sponge is made exclusively from the fibres of *Momordica angustisepala,* a cucurbit forest climber. This is likely an instance of sympathetic magic because the Akan names of this sponge - *'sapow-pa'* or *'ahensaw'* - has reference to superiority and kingship, respectively. The involvement of a sweet-scented soap further supports the suggestion. In Gabon, *Streptogyna crinita,* a perennial grass, is commonly used to catch mice, *Mus musculus;* rats, *Cricetomys gambianus,* rodents and water birds; and entangled animals are unable to free themselves (9). Sophistry arising from such practical applications leads to the use of the plants in a charm mixture to capture the confidence of influential personages (9).

As charms for general success at interview and at appointment, a stem scraping or the poultice of ground leaves of *Abrus precatorius* Prayer beads, a climbing perennial legume, previously collected and prepared with rituals, mixed with water and some honey is applied to the forehead, and also licked with prayers and a definite request. The addition of honey is likely an instance of sympathetic magic. *Cuscuta australis, Cassytha filiformis* Dodder, epiphytic parasites, and *Vernonia cinerea* Little ironweed, a composite, enter into charms by medicine-men for success at an interview, or for a specific request, or general success, or for prosperity and popularity (6). The plant macerate is left in the open for the night. At dawn, the macerate is used for a ritual bath on a rubbish heap, or at crossroads with a new sponge and local soap. If the demand is specific and requested of a person, then the name is mentioned while bathing (6).

In N Nigeria, the adhesive property of the leaves and of the fruit burs of *Desmodium velutinum,* a half-woody shrub legume, gives rise to superstitious practices in prescriptions and potions for popularity (3). The sympathetic magic probably referable to the clinging burs of the fruits. Among the coastal tribes and other ethnic groups in its distribution area in the Region, *Euphorbia hirta* Australian asthma herb, an annual with milky latex, either macerated in water and used for bathing; or dried, ground and applied in lavender; or chewed with honey; enters into charms by medicine-men for good speech (6). The addition of honey suggests an instance of sympathetic magic. *Sophora japonica* Pagoda tree is a legume and a native of China where it is known as the 'Tree of Success in Life' (7).

References.

1. Ammah-Attoh, (pers. comm.); 2. Ankomah-Ayew, (pers. comm.); 3. Dalziel, 1937; 4. Ellis, 1894; 5. Meek, 1967; 6. Mensah, (pers. comm.); 7. Richards & Kaneko, 1988; 8. Verger, 1967; 9. Watt & Breyer-Brandwijk, 1962.

39 PROTECTION, PHYSICAL

The name of the Lord is a strong tower,
The righteous runneth into it, and is safe.
Proverbs 18:10

A number of plants enter into prescriptions for charms to protect the skin - like an invisible impenetrable sheath - against cutting instruments, the arrow, the spear and firearms. There are equally charms for general protection against fire, motor accidents, and household accidents, for immunity in fighting, for boxing and for protecting a house from destruction.

CUTTING AND PIERCING WEAPONS

In W Africa charms against cutting instruments - popularly known in Hausa as *kaskaifi* or *maganin k'arfe,* meaning 'edge destroyer' and 'medicine against iron' respectively - are more common among the tribes of the woodland savanna and Sahel region to the north, than they are among the tribes of the forest region to the south. A display by a 'knife-proof' man cutting through the bare tummy or trying to gouge out the eyes with a previously tested sharpened knife, without a drop of blood nor a scratch beats the imagination. In Ghana, an incident involving a student who could not be injected (the only means of saving his life) is reported from the Okomfo Anokye Teaching Hospital in Kumasi. Perhaps, unknown to him, the parents had protected him permanently many years earlier from cutting weapons - a practice which ironically cost him his life. Some medicine-men, however, claim that with the performance of the necessary rites, this juju or charm may be reversed.

In N Nigeria, the leaf and pod of *Bauhinia rufescens,* a savanna tree, are boiled and mixed with corn porridge and butter by the Hausa to prepare *maganin k'arfe* for protection against spear, knife or arrow (8). The fruit of *Combretum sericeum*, an under shrub (8), and *C. racemosum,* a liane in forest (3), have superstitious uses in prescriptions to ensure immunity against cutting weapons. Other plants with hard roots like *Uraria picta,* a woody legume in grassland, is used as a medicinal charm together with *Sida rhombifolia* Wireweed, a weed, and *Rogeria adenophylla*, a savanna herb, (especially used by hunters) mixed with plants known as *kask'arfe,* to rub or wash on the body as a magical preventive of injury by cutting weapons (8). If the prescription is applied to the sword or knife of a possible adversary, it will blunt its edges (8).

In Nigeria, the root of *Polygala arenaria,* an annual, is worn by some Benue tribes alongside others with similar properties as a charm against cutting weapons (8). The roots of *Stylosanthes erecta* and *S. mucronata,* prostrate leguminous plants, are equally worn as a charm, or as ingredients with others used as a wash, also in the belief of protection from sharp weapons (8). The plants smoked like tobacco or boiled as a decoction and drunk, is also a charm against injury by a cudgel, and will cause the weapon to break (8). In parts of W Africa, especially among the Hausa of N Nigeria and the Zarma of Mali and Niger and the Fula, the leaf and pod of *Piliostigma recticulatum* and sometimes also *P. thonningii,* shrub legumes or small trees in dry and moist savanna regions, respectively, often planted or tended in villages, are boiled, mixed with corn porridge and shea-butter and taken as magic against spear, knife or arrow (17). This infers that the protection may also be acquired from within.

In N Nigeria, the Hausas use ninety-nine fruits or seeds of each of seven different plants of which *Dichrostachys cinerea* Marabou thorn, locally called 'dundu' in Hausa, a thorny legume, is one, to make into a charm against cutting weapons - either worn around the waist or used as a wash or as drink (8). This usage is confirmed (23). 'I was told that the fruit of the small 'dundu' tree if ground up and drunk with water will make it impossible for the drinker to be wounded by a sword'. Other decoctions are of more use against arrows and clubs (23). This is another example, and a confirmation of protection from within. The leaf or pulverized leaf of Marabou thorn, collected and prepared with rites, and taken in food, or the root scrapings of *Gardenia ternifolia* ssp. *jovis-tonantis,* a savanna shrub believed protective of lightning, used to cook rice, millet or guinea corn, or the powdered root-bark of *Cassia sieberiana* African laburnum with milk and shea-butter eaten in a ceremonial corn or *gari* meal, enter into prescriptions for protection against cutting objects (17). The prescription further supports the suggestion of acquiring the protection from within.

In N Nigeria, the seeds of *Senna (Cassia) obtusifolia* Foetid cassia, an annual herb legume, are often an ingredient in charms against cutting weapons, and the seeds of *Afzelia africana* African teak, a tree legume in

savanna or fringing forests, are sold in Hausa markets and used to avert injury by weapons (8). The leaves and roots of *Tephrosia bracteolata*, a small under-shrub legume with pink and purple flowers, are ingredients in charms used against injury by hunters and warriors; and in Liberia, the Mano people use *Olax gambicola,* a shrub by streams, along with various other ingredients to prepare a magic medicine protecting against spear and firearms (8). In The Gambia, water in which the roots of *Desmodium velutinum,* a semi-woody erect savanna shrub, have been boiled is held to have powerful juju against knives. After imbibing the water, a symptom of the juju's power is a constant need to scratch (13).

In N Nigeria, a mixture of the dried, pulverized leaves of *Paullinia pinnata,* a woody or sub-woody climber with tendrils in both forest and savanna regions, with gunpowder is rubbed on the body to confer immunity from wounds (8). The Hausa name of *Mitracarpus hirtus (scaber),* a common annual weedy herb, 'goga masu', means 'smear spear', in reference to its magical use as a protection against injury (effective for one day only). The body is rubbed with the plant or washed with an infusion of it before going to war (8). In S Nigeria, *Dracaena fragrans* and *D. smithii,* palm-like shrubs in forest - both with fragrant flowers, have a superstitious use to render one knife-proof; and in a superstitious way the crushed fruits of *Crinum zeylanicum (ornatum),* a lily besides streams in savanna , are rubbed on the feet of farmers apparently as a charm or preventive of injury by the hoe (8). Among the Hausa of N Nigeria, the Zarma of Mali and Niger and the Fula the leaves and roots of *Sida acuta* Broomweed, a very common perennial malvaceous weed, serves as an ingredient in a prescription to render immunity from cutting objects (17).

Sida acuta Broomweed

Counter-Charms

It is traditionally believed that there are sinister charms that counteract immunity against cutting instruments. The direct result of this is that the knife deeply cuts and wounds the performer. As a precautionary measure against possible counter-charms, those who exhibit knife-proof charms in public places, do look out for, and are said to be capable of finding out whether, any person with such powers and intention is hiding among the spectators. Among the Ewe-speaking tribe of Ghana, Togo and Benin Republic and other tribes in its distribution area, *Imperata cylindrica* Lalang grass, collected with rites and a spell enters into sinister prescriptions to counteract any immunity from cutting instruments. A blade of the grass previously 'treated' is either clandestinely placed on the cutlass or knife before the performance, or is hidden in the enemy's pocket during the display. (5). In N Nigeria, a stick cut from *Commiphora africana* African bdellium, a gum-copal tree, is held to have power to neutralize protective charms (6).

FIREARMS

In traditional medicinal practice, certain plants are used in prescriptions, often with ingredients for immunity against firearms. From the collection of the plants and the ingredients, the preparation of the prescription, its administration to the completion of the 'fortification' are all shrouded in secrecy and accompanied by prayers, rituals, ceremonies, taboos, sacrificial offerings, libation and incantations. For instance, the bark of *Securidaca longepedunculata,* Rhode's violet, a shrub or small tree in savanna vegetation, is traditionally valued for its magical and spiritual properties. The shrub is called 'kyirituo' in Akan, meaning 'against gunshot', and enters into charms to acquire immunity from fire-arms (19). Since the plant is believed to be 'against metal' however, it has to be debarked with a wooden implement or a stone, but never with a metal tool, else the charm will be rendered ineffective (15). Samples of the bark or what is often claimed to be the bark of 'kyirituo' are sold in medicine markets for the purpose. The seeds of the shrub equally enter into prescriptions as charms against bullets (1). Similarly, the small finely granulated seeds of *Milicicia (Chlorophora) excelsa* and *M. regia* Iroko, huge forest timber trees, collected and prepared with rituals, enter into charms worn by some tribal groups in the Region against gunshot

(11). In S Africa, the Southern Sotho use *Aloe ecklonis*, a perennial lily, as a charm to turn the bullets of the enemy into drops of water (26).

Commenting on the use of bullet-proof charms in N Nigeria, it has been observed that 'there was a special kind called *sha bara* which had a great vogue when the European first began to conquer the country. Its virtue being that by its means the white man's bullets would not only cause no harm to the wearer but would even rebound and wound the one who had fired the rifle. Considering the number of casualties, it is strange to think that the trade still flourishes' (23). The observer adds, 'my cook had fought against our troops at Kano, and been defeated, but his faith in native charms was as strong as ever' (23). In S Nigeria, the leaf of *Oxyanthus subpunctatus*, a forest shrub, is deemed to make one bullet-proof (7c).

Sympathetic Magic

Some of the prescriptions appear to be based on sympathetic magic. The unusually curious habit of *Biophytum petersianum* African sensitive plant, an annual weed with rosette leaves which are sensitive to weather and to touch, attracts traditional uses among many cultures in the Region. For instance, the whole plant, collected with rites and a pledge, is dusted with gunpowder, then enclosed in a finger of banana which is consumed peels and all, while one is totally submerged in a river or a lake as a protective charm against gunshot (15). The inclusion of gunpowder suggests sympathetic magic. In N Ghana, the plant is an important ingredient in many juju preparations among the Dagombas. Among many tribal groups, African sensitive plant, with a leather bangle, the meat of a tortoise, *Kinixys* species, and other ingredients, is used to cook a ritual meal consumed within a circle surrounding an anthill. The bangle, so 'fortified', is supposed to protect the wearer from bullets (15). The sympathetic magic seems to lie in the shell of the tortoise.

An Anthill

Thevetia peruviana Exile oil plant or Milk bush or Yellow oleander (associated with Islam and Oriental religion)

The seeds of *Thevetia peruviana* Exile oil plant or Milk bush or Yellow oleander, collected with rituals, invoked in an incantation, and then covered with leather as talisman, is believed to protect the wearer against gunshot (15).

In Liberia, the Mano people believe that *Olax viridis,* a small tree in forest, when put to magical concoction, can confer protection against spear and firearms (8, 16). The prescription - usually of assorted ingredients - includes, among others, the whole plant and the flesh or dried heart of an enemy slain in war. As part of the ritual, no knife nor iron implement must be used in the preparation, nor must the preparation come into contact with palm kernel. The preparation is dried in a covered pot over a fire till charred, then ground into powder and put in the horn of a sheep, *Ovis* species, decorated with cotton thread and cowry shells. This horn is rubbed with fresh blood of a decapitated war victim and slung round the neck. The horn needs to be 'fed' from time to time by rubbing well with blood flowing from war victims (Harley fide 7c). The inclusion of the heart of an enemy and blood of war victims suggests an instance of sympathetic magic.

The *Kunkuma* fetish, the greatest in Ashanti, among others, prevents the owner from being shot (22a, b). It is made with the leaf ribs of *Raphia vinifera* Wine palm, with the springs of a flint-lock gun and other ingredients tied with a rope of twisted fibre of *Sansevieria liberica, S. senegambica* or *S. trifasciata* African bowstring hemp, and the whole stained and clotted with *esono* (elephant, *Loxodonta africana*) dye, eggs and the blood of sheep, *Ovis* species; and fowls, *Gallus domestica;* that have been sacrificed upon it. (22a, b) The inclusion of the springs of a flint-lock gun suggests an instance of sympathetic magic. Some prescriptions for charms against gunshot are in effect, vanishing charms, and triggered by fear. The unfortunate disadvantage is that the owner could be shot from behind if he or she does not see the assailant first. (For spiritual protection from firearms see TALISMAN: FUNCTIONS: Firearms).

Bullet Extraction

Some plants are used in traditional medicine to effect the extraction of bullets from the body. In Ghana and Ivory Coast, the leaves of *Rourea (Byrsocarpus) coccinea,* a shrub or climber in secondary forest or savanna thickets, collected and chewed with magical incantations, finds use as charms to extract bullets from the wounded to save his life (18). The local name of the shrub in Akan 'awendade', means ' to chew metal'. Alternatively, the ground leaves of *Boerhaavia* species Hogweed or that of *Heliotropium indicum* Indian heliotrope or Cock's comb, collected and prepared with specific rituals then mixed with black soap is applied to the gun-shot wound by medicine-men to extract the bullet (15). There are also plant prescriptions for the extraction of arrows. In Europe, *Origanum (Amaracus) dictamnus* Dittany of Crete, a northern temperate labiate, is a plant of proverbial virtues including the power of drawing out splinters. The Ancients believed that if an arrow hit a goat, *Capra* species, and the beast fed on Dittany, the arrow would fall out again – hence the human application (20). (See also MEDICINE: INSIGHT: Animal Behaviour).

FIRE

The names of some plants suggest that they are protective against fire and are accordingly used as such. Some other plants enter into charms and prescriptions for the purpose. In N Nigeria, the evergreen bark of *Commiphora kerstingii*, a small tree often planted, called 'garkuwar-wuta', seems to suggest the idea that the tree is little likely to burn, so it has acquired vernacular names suggestive of protection against fire, and survival of property and therefore of inheritance (8). The local name in Hausa means 'shield against fire'. Similarly the Hausa believe that *C. africana* Africa bdellium has seven lives, because it is cut and not killed, and they therefore, plant it to form a hedge in arid districts (8).

In some villages and settlements, *Jatropha gossypiifolia*, Red physic nut, *J. curcas* Physic nut and *Craterispermum laurinum*, a shrub or small tree besides streams, are traditionally used for live fences in the belief that they are little likely to burn. For a similar reason, *Thevetia peruviana* Exile oil plant or Milk bush or Yellow oleander is planted around a fetish grove near Agbogba on the Accra Plains, in Ghana. Among some of the tribal groups in the Region, *Bryophyllum pinnatum* Resurrection plant is equally believed to withstand burning – many of the traditional uses reflecting this property (See also MOTOR ACCIDENT below).

At Kamboma, Sierra Leone, people use *Polystachya microbambusa*, an epiphytic or terrestrial orchid to decorate the apices of their hut roofs, because the plant has fire-resisting capacity and seems to survive bush fires. Could there be some thought of protection against fire ? (Macdonald fide 7c).

There is a folklore about *Parquetina nigrescens*, a climbing shrub with white latex, locally called 'nsurogya', in which a turkey laid its eggs, nested and hatched them in an enclosure of the shrub at a period when all the surrounding vegetation was burnt down by bush fire. The local name of the climber in Akan, which is the same for *Gongronema latifolium* since the two plants are very similar, means 'does not fear fire'- either because it appears quickly after burning the bush, or because the green stems are unaffected in binding quality by fire. The climber is traditionally used by palm-wine tappers to bind palm branches when making fire (14). The plant is ascribed with magical powers, with the strong conviction that even if one holds a single leaf he or she could walk through fire unhurt (12). Other climbers like *Pergularia daemia,* and *Telosma africanum*, trailing asclepiads, and *Adenia rumicifolia* var. *miegei (lobata)* a woody climber in the Passion flower family, go by the same local name and are believed to have similar properties. In N Nigeria, *Blepharis linariifolia,* a herbaceous acanth, called 'faskara tooyi' in Hausa, means resistant to burning.

In Benin Republic the Azoon fetish protects streets, houses and buildings of every description, and in addition the people place around the house a country rope, that is one made of grass, festooned with dead leaves, which is a fetish to prevent the building from catching fire (9a). When a large fire occurs, they frequently kill the owner of the habitation in which the fire broke out, considering that it originated through some sacrilege or omission of fetish worship (9a).

MOTOR ACCIDENTS

Some plants are ascribed with the property of protection from motor accidents. For instance, the leaves of *Bryophyllum pinnatum* Resurrection plant, enter into charms as a warning against an impending motor accident. The leaves are collected with the necessary rituals, and burnt with black ants, *Crematogaster* species, and a live chicken, *Gallus domestica,* in a ceremony with incantations and a prayer of request. The charred residue is then enclosed in a talisman and suspended on the vehicle. It is the belief that in the event of an impending accident the talisman gives an advance warning by chirping like a live chicken (4). The sympathetic

magic apparently lies in the plants natural tendency to resist burning.

Among many tribes in its distribution area and beyond, it is the belief that the roots of *Securidaca longepedunculata* Rhode's violet, a savanna shrub or small tree, collected with the necessary ceremonial rituals, and carried on one's person, among its numerous attributes, warns and protects the wearer from all motor accidents, including even plane crash (2). *Pedilanthus tithymaloides* Redbird-cactus or Jew-bush or Slipper-flower, a cactus-like shrub introduced from tropical America, is credited with the same properties (15). The Hausa of N Nigeria, the Zarma of Mali and Niger and the Fula believe that a piece of the dried stem of *Calotropis procera* Sodom apple, collected with prayers and carried on one's person warns and protects the owner from motor accidents and all evil forces (17). In Gabon, *Mammea africana* African mammee-apple, a forest tree, is held to have magical power for warding off accidents (25). In S Nigeria, the Yoruba believe *Hallea (Mitragyna) stipulosa* African linden, a tree of swamp forest, has supernatural power for protection and imprecate it in an incantation to avoid misfortune on the road. The local name 'ewe obi', literally meaning push leaf, infers that 'please don't send (push) me to heaven' (24).

In its distribution area, the fine granulated seeds of *Milicia (Chlorophora) excelsa* and *M. regia* Iroko, giant forest timber tree, and *Uraria picta,* a woody fibrous legume in savanna traditionally believed inimical to evil forces, enter into prescriptions by medicine-men as preventive and protective of motor accidents when carried on one's person (11). The whole plant of *Sida acuta* Broomweed, a very common weed, and that of *Costus* species, perennial forest herbs in the Ginger family, are used as preventive or protection from motor accident by many tribal groups (15). The dried whole plant of *Tridax procumbens* Coat buttons, a common weedy composite, with a grain of maize that has escaped the mill (likely an instance of sympathetic magic) wrapped with a prayer of request and worn as talisman is protective of motor accidents (15); or *Tradescantia* and *Zebrina* species Wandering Jew, both introduced decorative houseplants, collected with the necessary ceremony, then dried and powdered with kaolin and a drop of the blood of a pigeon, *Columba unicincta* or *C. guinea,* all covered with a piece of white cloth, with a prayerful request , and worn as talisman is prescribed by medicine-men to protect the owner from motor accidents (15).

OTHER INCIDENTS

In folklore, *Chamaecrista (Cassia) mimosoides* Tea senna, a herb or low shrub legume, is supposed to confer invulnerability; and in N Nigeria, *Vetiveria nigritana,* a robust perennial grass on stream sides, is tied up in charms in leather armlets called *kambu* in Hausa and worn by hunters to prevent injury in fight and other attacks (8). In this country, the fruit of *Balanites aegyptiaca* Desert date or Soap berry, a savanna tree, is an ingredient in a superstitious prescription called *guba* in Hausa to ensure immunity in fight, and protection against defeat in boxing (8) ; and a slave who does wrong rub his body with *Ludwigia octovalvis* Primrose-willow, an erect herb with yellow flowers - sometimes in marshy areas, or sucks it with red natron (native sesquicarbonate of soda) to avoid pain if beaten for his fault or to escape detection (8). However, in S Nigeria, *Dalbergia lactea,* a scandent legume shrub, enters into an invocation in an Yoruba Odu incantation 'to get an enemy beaten wherever he goes' (24, Verger fide 7b). The leaves and stems of *Hypselodelphys* species, scandent marantaceous shrubs in forest, are employed by medicine-men in the preparation of charms for strength and immunity in fighting, and to withstand severe beating without a scathe (27). Among the Akan-speaking tribe of Ghana and Ivory Coast, the plant, locally called 'babadua', is a symbol of strength and durability. For instance, mud houses built in pre-colonial days in Kumasi, Ghana, with the stems of 'babadua' as lateral supports, still stand to today (27). The species are *H. poggeana* and *H. violacea.* *Abrus precatorius* Prayer beads has similar uses for physical strength (15).

In Guinea, a leaf decoction of *Aphania senegalensis,* a forest tree, is taken in a superstition against falls and accidents (21); and an infusion of the leaves along with butter is given to anyone who falls from a tree to restore the circulation. The leaves and fruits of *Citrus aurantiifolia* Lime enter into similar prescriptions against accidents. In Gabon, a variety of *Saccharum officinarum* Sugar cane with red leaves is credited with magical powers of protection when planted alongside village paths (25). The root of a plant, (botanical identity unknown) scraped, dried, and threaded on a string with white beads, protects the wearer from injury, and also makes him bold and outspoken. In emergency the wearer bites off and chews a small piece of the root (9b) ; and the tubers of *Cyperus articulatus,* a sedge often cultivated for the aromatic rhizomes, strung into a necklace and a waist-girdle and worn by pagan tribes is said to keep insects away (8). In Senegal, a leaf of *Ceiba pentandra* Silk cotton tree stuck in a calabash is considered by Tenda people to have

a charm protective of a house, and to ensure peace within family and between women (10).

In S Africa, the Southern Sotho use *Linum thunbergii* Wax flax as a charm to prevent accidents to huts; and the Sotho-Tswana group use *Aloe saponaria* Soap aloe or *A. tenuior*, perennial lilies, as a ritual purification for escaping from danger (26). In SW Nigeria, the rhizomes of *Eulophia angolensis*, a ground orchid, is worn by little girls in Oyo town as a protective charm (Sanford fide 7c); and in S Nigeria, *Amaranthus lividus* and *A. viridis* Wild or Green amaranth are laid under the first course of a new house, or sprinkled over the ground before building to 'cool' it (N.W. Thomas fide 7a). In W Africa, *Cissus quadrangularis* Edible-stemmed vine commonly grows from an anthill at the foot of a tree, and the latter being uninjured the vine is sometimes planted beside houses where termites, *Macrotermes* species, appear (8). (For spiritual protection, see EVIL: GHOST AND SPIRIT WORLD; and TALISMAN).

References.

1.Agyeman, (pers. comm..); 2. Akligo-Zomadi, (pers. comm..); Akpabla, (pers. comm..); 4. Ankomah-Ayew, (pers. comm..); 5. Beloved, (pers. comm..); 6. Bubuwa Tubra, 1985; 7a. Burkill, 1985; b. idem, 1995; c. idem, 1997; 8. Dalziel, 1937; 9a. Ellis, 1883; b. idem, 1887; 10. Ferry & al., 1974; 11. Gamadi, (pers. comm..); 12. Gyedu, (pers. comm..); 13. Hallam, 1979; 14. Irvine, 1930; 15. Mensah, (pers. comm..); 16. Michaud, 1966; 17. Moro Ibrahim, (pers. comm..); 18. Nkwantabisa, (pers. comm..); 19. Noamesi, (pers. comm..); 20. Perry & Greenwood, 1973; 21. Pobeguin, 1912; 22a. Rattray, 1923; b. idem, 1927; 23. Tremearne, 1913; 24a. Verger, 1967; b. idem, 1986; 25. Walker & Sillans, 1961; 26. Watt & Breyer-Brandwijk, 1962; 27. Wofa Yaw, (pers. comm.).

40 RELIGION, INDIGENOUS

Thou shalt worship the lord thy God,
And him only shalt thou serve.
Matthew 4.10

By nature, man tends to worship spiritual beings - the identity of this being depending entirely on the individual's belief, aspirations and level of spiritual development. It could also be significantly influenced either by environmental factors or by one's concept of the universe and its origin, or by both of these factors. A collection of individuals with similar beliefs constitutes a religion. There are various types of religions throughout the world. In Africa, there are many local traditional religions that differ considerably from one another and worship deities of many names (Anon).

NATURE AND DEITIES

Traditional or indigenous religion is the worship of ancestral spirits, or man-made objects like idols (see FETISH) , or non man-made objects identified with relief objects like hills, rivers, streams, lagoons, lakes and the ocean; or rocks, trees and other natural objects; or celestial bodies like the sun, the moon and the stars - in the belief that these are guardian spirits. For instance, lagoon-worship has been recorded among the Ga-speaking tribe of Ghana (18); and sun-worship among the Egyptians (50). Natural occurrences like lightning and thunder may be mysteriously interpreted, deified and worshipped. There is also fire-worship. For instance, it is believed that the Akus of Sierra Leone are fire-worshippers (14a). Fire-worshipping is not peculiar to W Africa for the practice is reported in India (39). Like water bodies and fire, the earth may also be worshipped. In W Africa men worship the Earth, and in this worship groups who are otherwise in hostile relations annually unite in celebration (21).

The worship of animals like the leopard, *Panthera pardus;* the crocodile,

Crocodylus species; and the python, *Python regius,* similarly occurs. The Dahomians (now Benin Republic) worship the sun, the moon, fire, the leopard and the crocodile (14b). It is recorded that in Ghana, everyone in Nungua used to worship the leopard and the hyena, *Hyaena hyaena,* but now only the Wulomei do so; nobody, however, may use leopard skin on his sandals (18). Ophiolatory (snake-worship) takes precedence of all other forms of Dahomian religion, and its priests and followers are most numerous. The python is regarded as emblem of bliss and prosperity, and to kill one of these sacred boas is strictly speaking, a capital offence (14b). Now the full penalty of the crime is seldom inflicted, and the sacrilegious culprit is allowed to escape after being mulcted of his worldly goods, and having 'run-a-muck' through a crowd of snake-worshippers armed with sticks and fire-brands (14b). It is also recorded that in the eastern states tributary to Ashanti, for instance, in Kwahu, the python is in some few villages reverenced or worshipped; but this *culte* does not appear indigenous, and is borrowed from the neighbouring Dahomian people, who, in turn, appear to have derived it from Whydah (14c).

Snake worship is also reported among the Ga-speaking tribe at Tema and Nungua in former days (18). The worship of snakes is also recorded among the Mani Secret Society in Congo (Brazzaville) (15). 'There is a special *Mani* word for snake, and members are forbidden to eat snakes. New members are told that they must not be alarmed if at any time they meet snakes, but that if they greet them with one of the formulae of the association, the snakes will glide harmlessly away' (15). Snake-worship is also mentioned among the Hausa of N Nigeria, and the Kikuyu of E Africa (49). There are also instances of individuals who keep snakes (usually pythons) not as pets in the strict sense of the word, nor as deities, but for some other spiritual purpose - likely a religio-magico one, which is not clearly defined.

In S Nigeria, the guana or iguana or monitor, *Varanus niloticus,* another reptile, is sacred or fetish at Bonny in Old Calabar; and in Ghana, the crocodile, *Crocodylus* species, is sacred in Paga, Upper East Region. Sacred crocodiles were also recorded in R Prah, in the then Gold Coast (now Ghana) (14c). In this country, crocodile-worship was recorded at Dixcove (29). The practice also occurs in the lower Congo (Brazzaville) (29).

Some plants may also be worshipped. For example, *Adansonia digitata* Baobab is worshipped in Senegambia (22a). Trees believed to be fetish or sacred are, in fact, all deified and worshipped by various tribal or religious groups in the Region. Such trees include *Ceiba pentandra* Silk cotton tree, *Okoubaka aubrevillei*, a rare forest tree, *Milicia (Chlorophora) excelsa* and *M. regia* Iroko, *Blighia sapida* Akee apple and many species of *Ficus* (see FETISH: FETISH PLANTS; and SACRED PLANTS: FIG TREES). Tree-worshipping is not peculiar to W Africa, for in N Europe the Druids were known to worship trees - the practice still exists in a mild form. There is also iron-worship. For example, in S Nigeria, one sees a *Ficus* or *Dracaena* tree in many a compound among the Yoruba, to the side of which are tied iron rods, which indicates that the god of iron (Ogun) is worshipped there (38). The worshipped object (the tangible one, such as a carved figure, a stone, a lump of earth, a relief object, an animal, or a big tree) is only symbolic - the emphasis is rather what is believed to dwell within it (the intangible one or the indwelling spirit) (14a).

THE SUPREME GOD

A good many scholars have argued that in effect these minor gods were worshipped, prayed to and feared to the exclusion of the Supreme God, while a new crop of researchers and writers are saying that these minor gods are mere intermediaries (27). In reality, the above deities are lesser gods only - for traditional religion recognises the existence of the Supreme God. Most of the people firmly believe in a Supreme Being who is 'unique' and 'the absolute controller of the universe'. Several factors testify to this assertion - folk-proverbs or sayings and folklore, beliefs, proper nouns of the Almighty and plant names, among others. One of the most fundamental of the differences between people must be the question whether they believe in God or not; for on that depends their whole interpretation of the universe and history - on that depends their answer to so many other questions (10).

The Evidence

Among others, some of the folk-proverbs or sayings and some plant names support the saying that traditional Africans believed in God the Almighty long before the advent of both Christianity and Islam.

Folk-proverbs. The folk-proverbs of Africans, for instance, bear testimony to this belief in God. It has been recorded that in the collection of 3,600 proverbs made by J.C. Christaller among the Ashanti, the name of God occurs frequently (45). Writing on

traditional Igbo religious beliefs, it was recorded (27) that the great God (Chi-nkwu) is believed to be the author of heaven and earth who makes animals and plants grow. It is similarly recorded (38) that among the many gods, said to number 401, in Yoruba mythology, the Creator God (Ilorin) is recognised clearly as chief and ruler over all. It is also recorded (19) that in Ghana, Onyame, the Creator God, is in the centre of the belief of Christians, Muslims and the adherents of traditional Dormaa religion alike, and that they all believe in the same God. In Africa, some of the traditional names of the Supreme Being are Allah in Hausa, Chi-nkwu in Igbo, Ilorin in Yoruba, Mawu in Ewe, Mbori in Azande, Molimo in Basuto, Nkulunkulu in Zulu, Nyankopon or Onyame in Akan, Nyonmo in Ga, Thixo in Xhosa and Zamba in Ntumu.

Plant Names. Some plant names further support the belief in the existence of God. The Akans call *Myrianthus arboreus, M. libericus* and *M. serratus,* under-storey forest trees, 'nyankoma', meaning 'God's heart'; the Nzema name of *Vitex grandifolia,* a small forest tree, is 'nyamele-kukwe', meaning 'God's coconut' (28b) - perhaps from the acorn-like fruits and calyx. In S Nigeria, *Crinum jagus* and *C. natans,* bulbous lilies in forest swamps, are called 'ede-chukwu' and 'ede-obasi' respectively in Igbo, meaning 'God's cocoyam' (13). *Mariscus alternifolius (umbellatus),* a tufted sedge, is known as 'oru angi' in Ijo, meaning 'God's angi' (9a). In Congo (Brazzaville), the people endow *Desmodium adscendens,* a straggling legume in forest, with magical properties, using vernaculars for it which link the concept of God with the groundnut, *Arachis hypogaea* (6). In Ghana, the Fantis call *Anchomanes difformis* and *A. welwitschii,* herbaceous forest aroids with one enormous divided leaf, 'nyame kyim', meaning 'God's umbrella' (28a); and the Awunas call *Elaeis guineensis* var. *idolatrica* King oil palm 'sede', meaning 'God's palm' (28b).

Christopher Columbus learnt to know *Dioscorea alata* Water yam or Greater yam (in the New World) in 1492 under the name of 'nyame' (31). In Ghana and Ivory Coast, the Akan name for God is also Nyame - probably a nomenclatural coincidence, but nevertheless, a coincidence of religious significance. Many such names common to both African and American cultures have been enumerated (50) as proof of a pre-Columbian contact. *Entada gigas* and *E. rheedei (pursaetha)* Sea bean or Sea heart, popular tropical drift dissemules (see below under SYMBOLIC PLANTS: Sea Beans), are also called Columbus-bean because according to tradition the seeds provided

inspiration to Columbus and led him to set forth in search of lands to the west (23).

FOREIGN RELIGION

Islam or Moslem and Christianity are the main foreign religions to the Region. Despite the fact that these are the dominant religion, traditional religion has stood its ground. With the exception of N and S Africa which are Sunni Muslim and Christian respectively, almost all the remaining parts of the continent are tribal religion or animist (20). Islam was introduced earlier by land from the north and the east by the Arabs. Christianity however, was introduced by sea from the south by the Portuguese, the Spanish, the Danes, the Dutch, the British, the French and the Germans.

RELIGIOUS CEREMONIES

In Africa and elsewhere, plants are used in ceremonies associated with the practices and beliefs of traditional religion - such as in sacrificial offerings and libation, and in ritual purification or cleansing.

Sacrifices

Indigenous religion involves sacrifices and libation to the spirits and ancestors on occasions or anniversaries, or through these lesser gods to the Almighty God. Libation or 'ritual drink' to the spirits of the ancestors does not only occur at the individual and family level, or at traditional festivals and ceremonies only, but also forms part of the official programme at state functions in Ghana at least - both at home and abroad.

Libation at State Function

Dioscorea **species Yam Sacrifice.** Yam is the basis of important religious cults and the New Yam Festival is the crowning ceremony of the year (Talbot fide12). In S Nigeria, the festival usually

Libation at Individual or Family level

represents the major event among the Igbo tribe very much like Christmas in the annual calendar of socio-religious functions (37). Not only are special religious offerings made during the time that yams are harvested, but in various occasions yams constitute a component of offerings used in sacrifices to various gods (37). In Ghana, the production of large yams is part of the religious cult of the Vhe people's New Yam Festival (24). These are preserved at the abode of the yam spirit from one season to the next (24). Similarly in Western Polynesia, complex taboos and magical or religious rites surround yam cultivation in Uvea, (30). (For the religious prohibition, see TABOO: FOOD: Yams *Dioscorea* species).

Incense Burning and Smoking. Incense burning and tobacco smoking have long association with religious sacrifices. (For a list of incense-producing trees, see EVIL: INFLUENCE ON DISEASE: Children's Cases). The burning of incense and spices has had a place in worship from time immemorial and there is little doubt that the custom of smoking developed from the inhalation of the fumes from tobacco burnt as an offering to the Great Spirit (53). Besides being an ancient practice, incense burning occurs among many religious groups. 'The burning of incense has from earliest ages been intimately connected with the religious sentiments of man - being practised by Pagans, Jews and Christians'. For instance, *Commiphora molmol* Myrrh is familiar to many cultures as incense used in religious rituals (6). In churches, various kinds of aromatic gum-resins are used. 'In the Pagan temples *Santalum album* Sandalwood holds the highest rank, pieces of the wood being burned before the images of their deities' (40).

Rituals

The use of plants like *'Nyame-dua'*, kola-nuts and others in religious rites occurs throughout Africa. The practice is also recorded from other parts of the world.

Alstonia boonei Pagoda tree (symbol of Dormaa's attachment to God; associated with traditional religion)

'Nyame-dua'. In Ghana and Ivory Coast, *Alstonia boonei,* 'nyame-dua', has religious association for the Akan races as shown by the local name, meaning 'Sky-God's Tree' (28b, 47). The name arises from the whorled branches being used to support fetish bowls holding food for spirits at domestic shrines (see also FETISH: FETISH PLANTS: Trees). 'Only the *Onyame-dua* (God's Tree), a three-topped tree (*Alstonia boonei* Pagoda tree), or a fork of this tree, serve as symbols of the Dormaa's attachment to God. Before the spreading of the Christian faith, the *Onyame-dua* could be found in palaces, shrines and in the inner courts of most houses. Although the *Onyame-dua* has in the meantime disappeared from the royal palace of the Dormaahene, it can still occasionally be seen, especially in the homes of traditional healers in the villages. In the fork of the tree a clay pot is placed which is filled with water and herbs, *nyankonsuo* (divine water)' (19).

This symbol of traditional religion has been adopted by Christianity. For instance, at the Emmanuel Methodist Church in Labadi, a suburb of Accra, a cement model of the whorled branches holds a bowl of consecrated water used for baptism. On the contrary, in Indomalaya *A. scholaris* is called 'Devil Tree' (2).

Kola-nuts. In W Africa, *Cola acuminata* Commerical cola nut tree and *C. nitida* Bitter cola are classed according to colour - red, white or pink. The white nuts are preferred and command a much higher price, besides having a special significance in social and religious ritual (13). In Nigeria, the kola-nut is, in fact, used in saying prayers by both Christians and pagans alike (36). Kola-nut has entered so deeply into African tradition that to the Muslim races it is considered divine, brought to Africa by the Prophet Mohammed (2). Similarly, an infusion of *Gongronema latifolium,* a climbing shrub with milky latex, is taken as a cleansing purge by Moslems during Ramadan (9a). In Sudan, a leaf infusion of *Dregea abyssinica*, an asclepiad, is taken in Kordofan in ceremonies of purification during the month of Ramadan (4). In Somalia, there is belief that when one eats the fruit of *Ziziphus mauritiana* Jujube tree it will remain in the stomach for 14 days; and should a Muslim believer die during that period, ascent direct to Paradise is assured (9c).

Other Plants. In Liberia, the bark of *Homalium letestui*, a forest tree, is used in local religious practice (Cooper fide13); and in S Nigeria, the Yorubas use *Dissotis rotundifolia,* a common decumbent forest herb in damp places, in religious rites in the consecration of idols and purification ceremonies (13). In Nigeria, *Sesamun orientale (indicum)* Beniseed or Sesame is prominent in all religious rites among the Bagarmi of the Benue region (13); and in ancient India, Sesame was central to certain Hindu religious ceremonies and was called *homadhanya* or sacrificial grain (40).

Sesamum orientale Beniseed or Sesame (associated with Islam and Oriental religion)

In the Plateau State of Nigeria, *Tacca leontopetaloides* African arrowroot has an important place in rituals - it enters ancestor worship (52). In the Region, religious teachers (malams) cut the stems of *Cymbopogon giganteus,* a loosely tufted perennial grass (with erect culms), for use as pens (13). In Nigeria and N Ghana, the Hausa use *Evolvulus alsinoides,* a tufted or spreading convolvulaceous hairy herb with small blue flowers, in religious practices; and the leaves of *Philoxerus vermicularis,* a decumbent herb and coastal sand-binder, have religious use for preparing water for ceremonial bathing (28a). In Ghana, *Costus afer* Ginger lily, a robust forest herb, is used in religious ceremonies in Ashanti (42a, b); and in N Nigeria, the stem of *Pennisetum purpureum* Elephant grass, cut to a sharp point, is used as a spear in mimic battles of religious rites among the Jukun-speaking peoples (33). In Ghana, the false fruits (beads) of *Coix lacryma-jobi* Job's tears, a perennial grass growing near rivers and wet places, are worn on religious or ritual occasions, and the crude fibre of *Sansevieria liberica, S. senegambica* and *S. trifasciata* African bowstring hemp is

used for threading ornaments and religious objects to wear (13). In Nigeria, *Albizia gummifera* Flat-crowned tree bears reference to religious and mystical usage among the Yoruba-speaking people (13). In this country, the articulated branches of *Detarium microcarpum* and *D. senegalense* Tallow tree, tree legumes, are of special significance in heathen worship (36).

In Zambia *Lannea stuhlmannii*, a tree, figures a great deal in spirit worship in the Balovale District; and in E Africa the juice of *Sclerocarya caffra* Cida tree is used in certain of the Shangana ceremonies (53). In India the flower of *Gloriosa superba* Climbing lily is used in religious ceremonies (53). As the lotus of the Nile, *Nymphaea lotus* Water lily played an important part in the lives of Ancient Egyptians from the Fourth Dynasty (app. 4,000 B.C.) especially in religious observances (39). In Asian theological tradition, Water-lily flower is considered to have been the birth-place of both Brahma and Buddha. Nevertheless, the sacred lotus of Buddhism is not a *Nymphaea* species, but is *Nelumbium nelumbo (Nelumbo nucifera)* Nelumbo or Blue lotus (9c). The plant has close association with Buddha and the Buddhist faith (44). In Indo-China, the wood of *Artocarpus heterophyllus* Jack fruit was used for sacred Buddhist buildings because of the yellow colour; and the dye for vestments of Buddhist priests (8, 11, 26).

Ritual uses of plants do occur in the religions of other countries outside traditional Africa. For instance in Europe, *Verbena officinalis* Vervain, a small decorative plant, is said to have been used in the religious ceremonies of the Druids and Romans; and in the Far East the colour of the flowers of *Colquhouhia coccinea,* a high altitude decorative shrub of the Himalaya, Thailand and Yunnan, has significance for those of the Buddhist faith (16). Among the Hindus, *Thevetia peruviana* Exile oil plant or Milk bush or Yellow oleander is often chosen as an offering to the god Siva (31). The fruits of *Elaeocarpus angustifolius,* a large tree of the evergreen rain-forest, are particularly important as rosaries worn by Hindu mendicants, followers of Siva, as they afford assistance to the attainment of Heaven and Siva's company (32). Such rosaries contain 32, 64 or 101 stones, representing the number of eyes of Siva (8).

In C America, *Chiranthodendron pentadactylon* Hand-flower or Monkey's hand, a tree whose stamens resemble a red hand, was a focus or religious cult among the Mexican Indians who regard it with superstitious awe, believing that only one specimen of the tree existed and that to propagate it would offend the gods (16) . In addition to being eaten, *Capsicum annuum* Pepper played a role in the

religious observations of many Indians (25). In Peru, *Erythroxylum coca* Coca, a shrub, is 'The Divine Plant of the Incas' used for royal and religious occasions and chewed by all to impart prolonged endurance (35). In Lower Dahomey, *Zea mays* Maize or Corn, like *Pennisetum glaucum (americanum)* Pearl, Bulrush or Spiked millet, is said to have magical origin and powers, and is used for religious purposes (1). On the various uses of the products obtained from Maize, it has been recorded that the Aztecs used popcorn in religious rites (34).

ASSOCIATED PLANT NAMES

Some plant names have reference to religion or to worship. For example, the specific name of *Cochlospermum religiosum (gossypium),* an introduced decorative tree from the drier parts of India, suggests religious uses. *C. planchoni* and *C. tinctorium,* savanna shrubs, are indigenous to the Region. *Bignonia capreolata* Cross vine, an ornamental introduction, and related species have religious and superstitious uses in the New World because of the cross-like arrangement of the tissue in the lianes (9a). In Japan, *Nandina domestica,* a stiff graceful evergreen cane, is called heavenly bamboo (44). In N Nigeria, *Aerva javanica,* an erect or sub-erect under-shrub amaranth, is called 'alhaji' in Hausa, from its appearance in allusion to the habit of Mecca pilgrims. *Abrus precatorius* Prayer beads, a climbing leguminous plant, is so called because the black and red shining seeds were formerly threaded as rosaries and used in prayers; similarly, *Maranta leuconeura* and other species in the genus are sometimes called Prayer plant because of the manner in which the young leaves fold together.

RELIGIO-HALLUCINOGENIC PLANTS

Some intoxicating plants that induce illusions when chewed or inhaled or injected, have been employed in religious ceremonies and meditation purposely for their physical influence on the mind - a phenomenon which the priests craftly display as spiritual. For instance, historians believe that the priests of Delphos took *Datura stramonium* Jamestown weed, *D. metel* Metel or Hairy thorn-apple, or *D. innoxia,* shrubby plants with hallucinogenic properties, in order to produce the paraxysms attributed to Divine power, a ruse also employed by Peruvians for similar purpose (39). Similarly, *Ipomoea tricolor,* a cultivated introduction to the

Region, was used by the Aztecs as an hallucinogen for their religious ceremonies and medicine. There is also convincing evidence that widespread religious practices have been based on the controlled use of hallucinogenic fungi.

It was believed that through the use of hallucinogenic mushrooms such as *Amanita muscaria* Death cap, it was possible to gain knowledge of God's will or to prognosticate the future; and in some religions the basic concept of a particular deity or pantheon may have arisen directly from fungus-induced visions (12). In C America and Mexico, several groups of Indians ingest certain Basidiomycetes, especially *Psilocybe mexicana*, a sacred mushroom, for their hallucinogenic qualities or colourful visions; and mushrooms figure prominently in their religious ceremonies (43). This usage is also reported in the extreme west and extreme north-east of Siberia and probably the European Lapps (43).

Writing on hallucinogenic mushrooms, it has been recorded that 'the toxins psilocybin and psilocin occur mainly in species of *Conocybe, Panaeolus* and *Psilocybe* (46). The intentional use of these so-called 'magic mushrooms' for their hallucinatory effects is well known in C America, where their ritual exploitation continues today, and as an illicit pursuit in certain Western cultures. When accidentally ingested they cause alarming symptoms of intoxication, but these subside within a few hours and only supportive therapy is normally called for' (46).

There are a few recorded instances of the use of hallucinogenic plants in traditional African religion. In Tanzania, the sun-dried leaf and flower of *Brassica juncea* Indian mustard are smoked like hemp in order to get in touch with the spirits (51). In S Cameroon, initiates of the Fang tribe consume the roots of *Tabernanthe iboga* Iboga, an apocynaceous shrub with hallucinogenic properties, so that they can see god; and the shrub is believed to be vehicle through which the people communicate with their god (5). This shrub remains a central feature of local religion, and its spectacular effects have hampered native acceptance of Christianity in Gabon, backed by the proverb, 'Iboga and baptism are not compatible' (41). In this country, the root of *Mostuea hirsuta,* an undershrub, chewed alone or with other plants like *Pausinystalia yohimbe,* a forest tree, as hallucinogenics, alters the mind of participants to enable them to pass into that 'other world' of mystery (17, 51).

SYMBOLIC PLANTS

Legendary and superstitious statements in Europe, Britain and America accredit *Passiflora glabra (foetida)* Stinking passion flower, *P. edulis*

Passion fruit, *P. quadrangularis* Giant granadilla and *P. laurifolia* Water-lemon (the last three being introduced fruit plants from C America) and other species in the genus with religious beliefs based mainly on symbols on the floral structure, the root, the stem and even the stem anatomy (3). Passion flower, refers to the symbolism associated with the plant by early Spanish friars and missionaries in tropical America, who 'made it an epitome of our Saviour's Passion'. The corona was the crown of thorns; the five stamens were the wounds, and the three styles, broadened at the ends into club-shaped stigmas, represented the nails (31). The ovary was the hammer (6).

Passiflora glabra Stinking passion flower (associated with The Christ)

In the central receptacle one can detect the pillars of the cross; and the calyx was supposed to resemble the nimbus, or glory, with which the sacred head is regarded as being surrounded (16). The five-parted leaves recalled the clutching hands of the soldiers; the curling tendrils were a reminder of the whips with which the Saviour was scourged; and the ten sepals of its blooms represented the ten faithful apostles (Judas and doubting Thomas - some versions say Peter being omitted) (39). Writing on the harmonic life of plants, another interpretation to the peculiar floral structure of the Passion flower is given (48). 'When one considers... that a Passion flower contains two ratios, a five-part petal and stamen arrangement and a three-part pistil, even if one rejects a logically reasoning intelligence, one must admit that the soul of plants are certain form-carrying prototypes. In the Passion flower's case musical thirds and fifths - which work, as in music, to shape the blossom forms as intervals (48).

After the rather extravagant use of *Rosa* species, Rose by the Romans which threatened its acceptance by the church, the flower assumed a venerable status in the religious life of the Middle Ages (39). Early Christians saw in it the emblem of martyrs - the petals representing the five wounds of Christ and the white Rose the virginity of Mary (39). The rosary commemorates a chaplet of Roses supposed to have been bestowed on St. Dominic by the Virgin Mary, and were originally strings of beads made from tightly pressed Rose petals which gave out a pleasing fragrance (39). In Medieval times, a Sunday in mid-lent was known as Rose Sunday – and even today, a golden Rose blessed by the Pope may be sent as a mark of outstanding pontifical favour to special personages (39).

In W Africa, *Croton zambesicus,* a small tree euphorb planted in villages, is a symbolic tree among the Ekoi people of S Nigeria, known at least in its religious significance, as *afam* (the name of a juju cult) (Talbot fide 13). In C Africa, *Kigelia africana* Sausage tree is regarded as holy and religious meetings are held in the shade of the tree (53).

Sea-Beans

Tropical drift dissemules or sea-beans are fruits or seeds that are carried ashore by ocean currents - very often from distant lands. Many of them are treasured for the belief in their spiritual potentials. Among collectors of sea-beans, *Merremia discoidesperma* Mary's-bean, also called Crucifixion-bean because of the cross stamped on the surface, is perhaps the most precious stranded seed (Martin fide 23). To pious people who found these stranded seeds, the cross gave the seeds special meaning; the seeds had obviously survived the ocean and would now extend their protection to anyone lucky enough to own one (Martin fide 23).

Tropical beaches are the main sources of drift dissemules. Some of the precious samples from W Africa and elsewhere are listed in Table 1:

Species	Family	Common Name	Habit
Aleurites moluccana	Euphobiaceae	Candle nut	Tree
Andira inermis	Papilionaceae	Dog almond	Tree
Caesalpinia bonduc	Caesalpiniaceae	Bonduc or Grey nickernut	Climbing shrub
Calophyllum inophyllum	Guttiferae	Alexander laurel	Tree
Cycas circinalis	Cycadaceae	Fern-palm or Queen sago	Shrub
Dioclea reflexa	Papilionaceae	Marble vine or Sea purse	Climbing shrub
Entada gigas	Mimosaceae	Sea bean or Sea heart	Liane
Entada rheedei (pursaetha)	Mimosaceae	Sea bean or Sea heart	Liane
Enterobium cyclocarpum	Mimosaceae	-	Tree
Ipomoea alba	Convolvulaveae	White moonflower	Climbing shrub
I. pes-caprae	Convolvulaceae	Beach convolvulus	Climbing shrub
I. tuba	Convolvulaceae	Sea moonflower	Perennial twiner
Lodoicea maldivica	Palmae	Coco-de-mer	Tree
Merremia discoidesperma	Convolvulaceae	Mary's-bean or Crucifixion-bean	Climbing shrub

Mucuna species	Papilionaceae	True sea-bean	Climbing shrub
Pterocarpus species	Papilionaceae	African kino	Tree
Spondias mombin	Anacardiaceae	Hog plum or Ashanti plum	Tree

Table 1. Some Tropical Drift Dissemules

The religious significance of *Lodoicea maldivica* Coco-de-mer, the largest of the world's drift dissemules (and the largest seed) was first observed by General Gordon, a British soldier administrator, in 1882 when he discovered on Praslin Island in the Seychelles, the *Valle de Mai* - the valley of the Giants. The General became convinced that this beautiful valley was the biblical Garden of Eden, and that the Coco-de-mer trees were the Tree of Knowledge of Good and Evil (23). (For a list of religious plants, see APPENDIX - Table III. See also CEREMONIAL OCCASIONS; FETISH; JUJU OR MAGIC; MEDICINE; SACRED PLANTS; SHRINE; and TABOO).

References.

1.Adande, 1953; 2. Aubréville, 1959; 3. Ayensu, 1981; 4. Baumer, 1975; 5. Binet, 1974; 6. Blumenthal & al., 2000; 7. Bouquet, 1969; 8. Burkill, 1935; 9a. Burkill, 1985; b. Idem, 1994; c. Idem, 1997; 10. Butterfield, 1954; 11. Chadha, 1985; 12. Cook, 1977; 13. Dalziel, 1937; 14a. Ellis, 1881; b. idem, 1883; c. idem, 1887; 15. Evans-Pritchard, 1937; 16. Everard & Morley, 1970; 17. Fernandez, 1972; 18. Field, 1937; 19. Fink, 1989; 20. Garrett, 1986; 21. Gluckman, 1963; 22a. Graf, 1978; b. idem, 1986; 23. Gunn & Dennis, 1979; 24. Haynes & Coursey, 1969; 25. Heiser, 1980; 26. Holland, 1908-22; 27. Ilogu, 1974; 28a. Irvine, 1930; b. idem, 1961; 29. Kingsley, 1901; 30. Kirch, 1978; 31. Lotschert & Beese, 1983; 32. Macmunn, 1933; 33. Meek, 1931; 34. Moore, 1960; 35. Mortimer, 1901; 36. Okafor, 1979; 37. Okigbo, 1980; 38. Parrinder, 1953; 39. Perry & Greenwood, 1973; 40. Pitkamen & Prevost, 1970; 41. Pope, 1990; 42a. Rattray, 1923; b. idem, 1927; 43. Raven, Evert & Curtis, 1976; 44. Richard & Kaneko, 1988; 45. Smith, 1950; 46. Storrs & Piearce, 1982; 47. Taylor, 1960; 48. Tompkins & Bird, 1974, 49. Tremearne, 1913; 50. Van sertima, 1976; 51. Walker & Sillans, 1961, 52. Walu, 1987, 53. Watt & Breyer-brandwijk, 1962.

41 SACRED PLANTS

She is a tree of life to them that lay hold upon her:
And happy is every one that retaineth her.
Proverbs 3:18

In Africa and elsewhere, plants that are respected and feared either for their association with fetish or juju, or are objects of worship and sacrificial offerings, or are symbols of traditional values - such as trees for carving devil masks and images or trees forbidden to be felled or to be used for fuel-wood or medicinally important plants are all believed to be sacred in folklore. Naturally, all such plants have, in one way or the other, been protected from generations. This is an effective traditional method of conserving biodiversity for sustainable yield. Some plants may be sacred to a religious group only, or to a society such as witches, or medicine-men, or a cult, or to the whole tribe. Some other plants may be sacred throughout the Region.

HUGE TREES

Huge trees more often tend to be classified as sacred, and generally believed to be fetish. For instance, in pagan folkore, *Ceiba pentandra* Silk cotton tree is said to be sacred by the Tera of N Nigeria (35), by the Ashantis of Ghana (48a, b), by the Ga folk of Nungua in Ghana (18), and to be more or less sacred by the Igbo tribe of S Nigeria (13), and the Yoruba tribe of this country (15b). It has been recorded from S Nigeria that: 'the temples of the chief gods are usually situated in groves of fine trees, among which one or two large silk-cotton trees which seem to be regarded with veneration throughout all W Africa, towers above the rest' (15b). There is a similar observation from Benin Republic. 'The reverence which is paid to usually tall and fine trees forms a contrast to the foregoing barbarous beliefs. The silk-cotton tree and the well-known poison-tree of W Africa are those most normally selected. Libation in honour of these trees are poured into perforated calabashes placed round their roots' (15a). (The 'well-known poison-tree' is *Erythrophleum*

suaveolens Ordeal tree). In Britain, *Quercus robur* Oak was so respected that a branch of the tree was stamped onto British coins for centuries (7). Other species are *Q. petraea* and *Q. alba* (7).

Similarly, the Baobab is revered by many tribal groups. Because of its horizontally spreading branches, the Africans believe that when God created the world he turned *Adansonia digitata* Baobab upside down with the roots into the air (20) - (but see DEVIL: PLANTS ASSOCIATED WITH THE DEVIL). In Ghana, the 'shadjo tso' or the 'kpledzoo shito' in front of the Meridian Hotel, the sacred and monumental tree believed to contain the souls of the living in Tema, is a Baobab. In N Nigeria, some tribes reverence Baobab by cutting symbols on the bark (35). In present day Malawi, David Livingstone, the British missionary and explorer, is reported to have carved his initials onto a Baobab in 1858 (20) - probably along R Shire on the banks of L Shirwa. Earlier, in 1855 an incident involving another big tree is recorded. 'Livingstone had often heard the Makololo speak of a place they called *Mosi-oa-tunya,* 'the place of sounding smoke', which he was to name, less attractively, Victoria Falls (37). When he first set eyes on this stunning spectacle (the largest waterfall in the world), he did not show much enthusiasm, although he did surrender to an uncharacteristic impulse and carved his initials on a tree (unspecified) growing on an island near the falls' (37).

In W Africa, the Ewe-speaking tribe of Ghana, Togo and Benin Republic (1) and some tribes in Ivory Coast and Burkina Faso (27) regard *Antiaris toxicaria* Bark cloth tree as sacred. In Benin Republic, the people regard *Cylicodiscus gabunensis* African greenheart, another huge forest tree, as more or less sacred (13); and in S Nigeria, *Pentaclethra macrophylla* Oil-bean tree is often a sacred tree among the Igbo (55b). Other big trees regarded as sacred in Igboland are *Distemonanthus benthamianus* African satinwood, a forest tree with characteristic red bark; *Bombax buonopozense* Red-flowered silk cotton tree, with deep pink or red petals - flowering when leafless - and *B. brevicuspe,* a tall tree of rain forest (40). African satinwood is also sacred among the Yoruba because it is associated with the mythical god, 'Shango', King of Oyo. 'He usually goes armed with a club called *oshe,* made of the wood of the *ayan*-tree, which is so hard that a proverb says, 'the *ayan*-tree resists the axe'. In consequence of his club being made of this wood, the tree is sacred, to him' (15b). (See EVIL: PLANTS FOR EVIL for yet another story about 'Shango' and the *ayan*-tree). In Ivory Coast, the Red-flowered silk cotton tree is sacred to the 'Ando', giving a general association with the ancestors (57); and a man and woman coming to live in a village make a ritual planting (10d).

In Ivory Coast and Burkina Faso, *Milicia (Chlorophora) excelsa* and *M.*

regia Iroko, forest timber trees, are reported sacred where offerings by the fetish priests may be found in containers under the trees (27). It is under the Iroko that the fetish receives his revelation, and due to the sacred nature of the tree, fragments of the bark are added to black medicine to reinforce the treatment (27). In a study comparing the religious significance of different forest trees for various W African cultures, it was found that Iroko was a sacred tree throughout the Region - it was often protected, and sacrifices and gifts were given to it (53). In S Nigeria, the Igbos and others in the Region often regard Iroko as a sacred tree (55a); and the tribe believe that from the tree come souls for the newborn (13, 41). In N Nigeria, the Igala believe the Iroko to be the first tree to be shown to man by God (13, 41). In its distribution area, the stone in the heart-wood of Iroko (excrescences of calcium malate) also serve as a token of superstition among many traditional groups (1) (see also EVIL: SPIRITUAL PROTECTION; and WITCHCRAFT: PLANTS FOR WITCHCRAFT: Plants Associated with Witch Meetings). The Iroko is used chiefly for building, whence probably it comes to be the emblem of refuge (15b). It is claimed that murmurings which only the initiated can interpret may be heard from furniture made of Iroko. In addition the furniture may move by itself, and doors made of Iroko may also open by themselves. (See also MARKETING: PREVENTING MONEY SNATCHING).

In S Nigeria, there was at one time a very large specimen of *Schrebera arborea,* a deciduous forest tree, in Wana Town, which was considered sacred (10c). In Ghana, the famous 'Big Tree' - a popular tourist attraction - is *Tieghemella heckelii* Makore, locally called 'baku' in Akan; and it is believed by the local people to be sacred. The tree is situated near the main Agona Swedru-Akim Oda road in the Esen Epam Forest Reserve. Sacrificial offerings and coins are usually found under the tree, or pieces of white calico attached to the trunk.

FIG TREES

Many *Ficus* species are believed to be sacred in the Region and elsewhere. It is, in fact, the genus with the largest number of sacred trees. In Ghana, *Ficus vogeliana* and *F. vogelii* are regarded as sacred trees in Akropong Akwapim; *F. sycomorus (gnaphalocarpa)* is sacred among the Ewe-speaking tribe (1); *F. lyrata* is sacred in the forest areas; and *F. sur (capensis)* is sacred in Ivory Coast (25). In N Nigeria, *F. dekdekena (thonningii)* is sacred and emblematic tree among several tribes (13). In this country, the tree is also sacred to the Angas (Kerang), the Birom, the

Aten (Ganawuri) and many other tribes (35). In S Africa, it is a sacred tree of the Luvale and the Lunda (58); and people in the Babati District of Tanzania (10c). In Kenya, the tree is sacred to the Somalis and the Boran who make offerings to it (10c). In E Africa, particularly among the Kikuyu, many *Ficus* species have been used from generation to generation as sacred shrines or as places of sacrifice to ancestral spirits (58). 'In many of the South-western Asian countries, figs were considered sacred. The Romans considered them the fruit of the god Bacchus. In Hebrew life and the Bible, the significance of the fig began with the story of the Garden of Eden' (45).

In W Africa, a condemned man who escapes to touch a fig tree in the sacred grove at Santemanso in Ashanti, in Ghana, is pardoned. The chief who insists that he should be killed will himself die before the end of the year (48). In an incident in C Africa, a condemned man was tied up and thrown into a river, but he managed to untie the ropes and to escape. He was caught and tied up again - this time with a weight, but he again untied the ropes and escaped. He was again arrested, tied up and knocked unconscious on the head before he was drowned in the river. The Chief responsible for his execution died himself before the end of the year (28). Compared with the observation in Ashanti (48), it is apparent the professed death of the chief is a manifestation of the wrath of the god of pardon - and not necessarily a spiritual force within the fig tree. Furthermore, in Africa it is traditionally believed that deliberate and serious offences that incur the wrath of the gods or ancestral spirits is punishable by death before the end of the year. It is apparent that this judgement has no respect for persons nor office. Nevertheless, the reason why only the fig tree but no other tree in a sacred grove has this property of pardoning condemned men needs an explanation.

In India, *Ficus religiosa* Pepul or Bo-tree is a sacred tree (5). The tree is also sacred in Tibet - where it is carried in the silver amulet box worn by all Tibetian women (20). 'Inside this charm box, she carries a lamaist prayer sheet or supplication, and a leaf of the venerated Bo-tree, jealously guarded. For greatest merit, this treasure must not be picked, but worshippers wait patiently by a tree watching for a falling leaf' (20). The specific name comes from the Latin *'religiosus'*, meaning sacred. According to legend it was under this tree that Buddha received the enlightenment in 500 B.C. (33), which was to form the basis of his teaching (30). The history of India mentions that a grove of *Mangifera indica* Mango was presented to Buddha for his use as a place of repose as the fruit was food for the gods (45).

OTHER TREES

In addition to large trees and some species of *Ficus*, many other trees are believed to be sacred in the Region. In N Nigeria for example, *Holarrhena floribunda* False rubber tree has been seen as a sacred tree in a village of the Arago tribe in the Benue Region - the tree is reported to have produced no fruit (Hepburn fide 13); and in both Ghana and Ivory Coast, another latex-producing tree, *Elaephorbia drupifera*, a savanna tree euphorb, is considered sacred (25). (*E. grandifolia* is the forest equivalent). Among some coastal tribes in the Region, *Croton zambesicus*, a small tree often planted in villages, is believed to be sacred. In S Nigeria, *Raphia hookeri* var. *planifolia* Wine palm and *R. vinifera* King bamboo-palm, occurring in evergreen forest and freshwater swamp-forest of the Niger Delta, are locally regarded as sacred and worshipped, and are therefore in no way exploited (42).

At Maradi in Niger, *Diospyros mespiliformis* West African ebony tree is held to be a sacred tree of the village (32). In S Nigeria, *Newbouldia laevis*, a popular fetish tree, is more or less sacred or symbolic tree among the Igbo, and often planted in small groves in front of the chief's house. The tree is a symbol of deities among the Efik, the Ekoi and the Ibibio, and a cutting or sapling is always brought from the old town to the site of a new one (13). (The inference, in this instance, is that the survival or death of the cutting is indicative of whether the new town should be inhabited or abandoned). (See also FORTUNE-TELLING AND DIVINATION: THE FUTURE: The Founding of Kumasi). *Newbouldia* is also sacred in Ghana - especially among the Ashanti in the forest zone and the coastal tribes in the savanna. The tree occurs in fetish places in Ivory Coast, Benin Republic, Liberia, Sierra Leone and in Togo.

In S Nigeria, *Markhamia tomentosa*, a small tree commonly found in fringing forest, is probably sacred among the Ekoi and other tribes in the same way as *Newbouldia* (13). Throughout the Region, *Erythrina senegalensis* Coral flower or Bead tree is sometimes grown for sacred purposes (25). Sacred properties are ascribed to *Okoubaka aubrevillei*, a rare tree, in both Ivory Coast and Ghana because of the associated fetish, juju and mystical powers and the taboos attributed to it (see FETISH: FETISH PLANTS: Trees). The genus name *Okoubaka*, locally means 'tree of death' in Anyi (3). In Niger, *Tamarindus indica* Indian tamarind, a savanna tree legume, is considered sacred in Soumarana village (32); and in Senegal, as the dwelling site of djinns (26).

In Ivory Coast (25) and throughout its distribution area (10b), *Afzelia africana* African teak, a tree of savanna and fringing forest, is widely

considered sacred because of the magical properties ascribed to the seeds. In parts of W Africa, *Elaeis guineensis* var. *idolatrica* King oil palm is regarded as sacred; and *Baphia nitida* Camwood, a shrub or small tree legume; and probably also *Pterocarpus osun* and *P. soyauxii* Redwood, tree legumes in forest, are in some degree, regarded as sacred (25). In Nigeria, *P. mildbraedii*, a medium-sized tree legume, is reported sacred (40). Camwood is used for dying sacred objects. In Cameroon, the 'oven' tree, *Didelotia africana*, a tree legume in forest, is sacred among the Beti (2). Among Moslems, *Phoenix dactylifera* Date palm is sacred (45). Mohammed said, 'there is among the trees one that is preeminently blessed, as is the Moslem among men, it is the (Date) palm' (45). In The Gambia, some villages consider *Cola cordifolia,* a savanna tree, sacred, and will consult it before any big social event (such as a circumcision ceremony) is undertaken (23).

In certain parts of the Region, *Gardenia nitida,* a shrub or small tree sometimes planted in villages, and *Octoknema borealis,* a forest tree, are regarded as sacred (25). In Ghana, *Oxyanthus unilocularis,* a shrub or small tree in forest or fringing forest in savanna, is a sacred tree in Akwapim and in Ashanti (25). In S Nigeria, *Cordia millenii* Drum tree, a forest tree, is sometimes regarded as a sacred tree in Yoruba villages (25). In Ivory Coast, *Trichilia prieuriana,* a small tree, is sacred in the forest of Bondoukou (27). In Gabon, *Tabernanthe iboga* Iboga, an apocynaceous shrub with hallucinogenic properties, is regarded as sacred (46). The sacred Iboga appears at several other points in the life of the Bwiti (46). In this country, followers of the Bwiti masonry consider *Cissus aralioides,* a forest liane, sacred (8); and a dwarf form of *Musa sapientum* Banana and *M. sapientum* var. *paradisiaca* Plantain are sacred to followers of the Bwiti Secret Society (10c).

In E Africa, branches of *Clausena anisata* Mosquito plant and *C. inaequalis* are carried as sacred leaves in ceremonies; and *Vangueriopsis lanciflora* Wild medlar, locally called 'mpemba', is sacred to the Luvale (13). In Zimbabwe, *Kirkia acuminata* White seringa, a tree, is a sacred tree to the African; and in S Africa, *Faidherbia (Acacia) albida,* a tree legume, is held sacred by the African of the Transvaal Springbok Flats (58). In this country, the Shangana medicine-men use the wood of *Annona chrysophylla,* a shrub, for a sacred fire - the wood must not be used by ordinary folk (58). In Gabon, a long-branched willow-like shrub (botanical identity unknown) is a sacred plant. 'Adooma from Kembe Island especially drew my attention to this shrub, telling me his people who worked the rapids always regard it with an affectionate reservation. He said it was the only thing that helped a man when his canoe got thrown over in the dreaded Alemba, for its long tough branches swimming in, or close to the water, are veritable life lines, and his best chance' (28).

In W Africa, *Olax subscorpioidea*, a shrub or small tree in forest or savanna region, has sacred uses in Ivory Coast (3) - probably because of the garlic scent in the bark. *Leea guineensis*, a soft-wooded shrub; *Blighia sapida* Akee apple, a medium-sized tree usually planted near dwellings; and *Hoslundia opposita*, a shrubby labiate; are all considered sacred by various tribal groups in this country. In places, *Glyphaea brevis*, a shrub of secondary jungle, has sacred import (47) and is cut for wands. In the Niger River delta it is planted around shrines (10d). In Niger, *Acacia seyal*, a shrub or small tree in savanna, is said to be sacred in a village of Touroumboudi, but the significance is not stated (32).

HERBS

Some herbaceous plants are traditionally believed to be sacred among the various tribes in the Region. Such herbs are often seen to be tended in and around villages, or sometimes cultivated by medicine-men and fetish priest. Table 1 below lists some of these herbs:

Species	Family	Common Name	Habit
Kalanchoe integra var. crenata	Crassulaceae	Never die	a thick fleshy perennial
Leptadenia hastata	Asclepiadaceae	-	a trailing perennial
Momordica balsamina	Cucurbitaceae	Balsam apple	a trailing annual
M. charantia	Cucurbitaceae	African cucumber	a trailing annual
Oldenlandia corymbosa	Rubiaceae	-	a trailing annual weed
Pergularia daemia	Asclepiadaceae	-	a trailing perennial
Plumbago zeylanica	Plumbaginaceae	Ceylon leadwort	a climbing shrub
Scoparia dulcis	Scrophulariaceae	Sweet broomweed	an erect annual weed
Sida linifolia	Malvaceae	-	an erect annual weed
Tradescantia species	Commelinaceae	Tradescantia	a trailing perennial
Zebrina species	Commelinaceae	Zebrina	a trailing perennial

Table 1. Some Herbs Believed to be Sacred

In N Nigeria, a plant which is perhaps *Cyanotis lanata*, an annual herb with bulb-like base like an onion, is grown in masses on thatched roofs as a juju or sacred object (13). In S Africa, *Verbena officinalis* Vervain is regarded as a holy object (58); and in the Middle Ages, holy powers were attributed to *Cnictus benedictus* Blessed thistle, a handsome member of

the thistle family, and it thus came to be known as the Holly or Blessed thistle (4). In China, the Emperor Shen-nong, who compiled the *Medical Bible of the Yellow Emperor (Huang-di nei jing)* sometime between 2967 and 2597 B.C.E., counted *Glycine max* Soybean among the five sacred crops (7); and *Filipendula ulmaria* Meadowsweet, a perennial herb and native to Europe and Asia, was one of the three most sacred herbs used by ancient Celtic Druid priests (7).

GROVES

Sacred or fetish groves are relics of the forest, and were formerly burial grounds for important people. Such forests have never been cut by man, but been preserved. This is contrary to the observation that fetish groves have usually sprang up on sites of abandoned habitations where the bones of the village ancestors lie buried (51). As custom demands, offerings are made and libation offered to the ancestral spirits on occasions. Although many cultural traditions are disappearing with the rapidly changing social and physical environment, sacred groves often remain as valued elements of cultural heritage (29). However, as traditional beliefs weaken in the face of western education and religion, these groves will be increasingly endangered, and it is to be hoped that ways may be found to ensure their continued conservation (22).

Sacred grove

This forest type is found only in Ghana (21); but similar forest types are reported in Ivory Coast and Burkina Faso (10a). The presence of sacred groves in Ivory Coast (52) and in Cameroon (29) has been described as places where moral values are taught and passed on from one generation to the next. Certain plants are associated with these groves, or are specially planted in them. In Ghana (13) and in Nigeria (40) *Costus afer* Ginger lily is planted in sacred groves. In Ghana, *Combretum collinum* ssp. *hypopilinum,* a savanna shrub or small tree, is common in Tengani sacred groves (10a). In both Ivory Coast and Burkina Faso, *Strophanthus hispidus* Arrow poison is planted in sacred groves and considered an important fetish plant (Lynn fide 10a). Similarly, *Cyperus articulatus,* a sedge famous for its aromatic rhizomes, is commonly grown in sacred groves, and the scented rhizomes offered to the Shades of the Departed (10a).

In Nigeria, *Treculia africana* African breadfruit; *Detarium microcarpum* and *D. senegalense* Tallow tree, tree legumes in forest and forest-savanna region respectively; and *Brachystegia* species, forest tree legumes, are associated with sacred groves (39). The species in the Region are listed in Table 2, below:

Species	Plant Family	Habit and Distribution
Brachysegia eurycoma	Caesalpiniaceae	a river-bank tree from Nigeria to Cameroon
B. kennedyi	Caesalpiniaceae	a tree of evergreen forests in S Nigeria
B. laurentii	Caesalpiniaceae	a forest tree from W Cameroon and Dem. Rep. of Congo (Zaire)
B. leonensis	Caesalpiniaceae	a forest tree from Sierra Leone, Liberia and Ivory Coast
B. nigerica	Caesalpiniaceae	a rain-forest tree of S Nigeria

Table 2. Brachystegia species in the Region and Their Distribution.

(For the importance of groves in the conservation of plants, see JUJU OR MAGIC: CONSERVATION).

LEGENDARY PLANTS

In Europe, *Laurus nobilis* Laurel or Sweet bay, an introduction from the Mediterranean region, features in legends with various attributes. For instance, it is called *daphne* by the Greeks after the legend that the nymph Daphne was turned into a Laurel to escape from Apollo (30). The tree became sacred to that god and was used in festivals in his honour, and when the cult passed to Rome, conquerors and poets were crowned with Laurel, a custom that survived for a very long time. Even Napoleon's crown featured a golden Laurel wreath (30).

In Africa, legendary plants include *Adansonia digitata* Baobab in the savanna countries generally; *Ceiba pentandra* Silk cotton tree throughout its distribution area in both the savanna and forest countries; *Distemonanthus benthamianus* African satinwood (see EVIL: PLANTS FOR EVIL); *Okoubaka aubrevillei*, a fairly rare forest tree believed to 'resurrect' after felling; *Milicia (Chlorophora) excelsa* and *M. regia* Iroko, and also *Morus mesozygia*, a forest tree - all three believed to haunt or to be haunted (see GHOST AND SPIRIT WORLD: HAUNTING TREES). Other legendary plants are *Ficus* species; *Borassus aethiopum* Fan palm; *Lodoicea maldivica* Coco-de-mer, the largest seed in the world; and

Welwitschia mirabilis, a curious xerophyte of the Namib desert.

Mushrooms are generally surrounded by legends; and the mushroom is an object of lore. Recently it has been claimed that a sacred mushroom cult existed among the Maya, the principal indigenous civilization of America (34). In Japan, there are many legends about *Cinnamomum camphora* Camphor tree, and many of the trees are sacred and venerated (49). (See APPENDIX-Table II for Legendary and Mythical Plants).

NATIONAL AND STATE PLANTS

In many countries throughout the world, the traditional values of plants reflect as national or state symbols, emblems or arms for various reasons. It may be a reflection of the cultural inheritance, or of legendary and mythological significance, or commemorative of an event - and therefore of historic importance, or symbolizing spiritual and religious aspirations and other values. In W Africa, *Elaeis guineensis* Oil palm was the emblem, of all the English-speaking countries before independence (see ANIMALS: SPIRITUAL SIGNIFICANCE : Emblem). In Ghana, *Hyphaene thebaica* Dum palm or Gingerbread palm, a branching palm, is the emblem of the people of Agbosome. According to legend the tree sprouted from the buried horns of a ram, *Ovis* species, which was sacrificed to the gods for peace in the area (14). The palm is still to be seen growing in the town to this day.

Quercus species Oak (with some five hundred species) has played a large role in the fortunes of the British Isles, and the tree is often regarded as symbolic of England (16); however, *Rosa gallica* Red Tudor rose, is the National Flower of England (44). *Onopordum acanthium* Scotch thistle, a prickly composite with striking sculptured leaves, constitutes the badge of the Stuarts and the national emblem of Scotland (see WAR: ASSOCIATED PLANTS: Lore). The national emblems of Ireland and Wales are *Trifolium repens minus* Irish shamrock, and *Narcissus pseudo-narcissus* Daffodil or Trumpet narcissus, respectively (19).

The order of the star of India comprises the flower of the Rose of England and two crossed Palm branches - as the Hindus compare their country to the Lotus, the petals suggesting central India and the leaves the surrounding province (44). Lotus is known as 'padma' in India, and venerated as a symbol of the Ganges; while in Egypt it is a sacred plant (33). *Cynodon dactylon* Bermuda grass is celebrated in the ancient Vedas as the 'Preserver of Nation' and 'Shield of India' (36) - (but see DEVIL: PLANTS NAMED AFTER THE DEVIL). In Chinese mythology, *Nymphaea lotus* Water-lily or Sacred-lotus, the symbol of truth and purity, is the favourite flower of Kuan-yin, the Goddess of Mercy (20) - (but see EVIL: PLANTS FOR EVIL). To other authorities the Sacred-lotus is *Nelumbium nelumbo (Nelumbo nucifera)* Nelumbo or Blue lotus (24).

The Rising sun in the Japanese flag represents *Chrysanthemum* species Chrysanthemum (not the sun) with a central disc and sixteen flaring petals (44). In this country, *Paulownia tomentosa (imperialis)*, a tree with violet-blue flowers, is one of the imperial plants of old Japan (like the Chrysanthemum) and was used in the crests of the Mikado (44). The 16-petalled Chrysanthemum has been the official crest of the Imperial Family since Meiji days, and used by them as a personal motif and decoration long before that (49). *Helianthus annuus* Sunflower, a native of S America and Peru, was once the emblem of the Sun God of the Incas; but *Cantua buxifolia*, from the Bolivian and Peruvian Andes, is the National Flower of Peru (44). *Reseda odorata* Mignonette is the National Flower of Saxony (12). *Convallaria majalis* European lily of the valley is the National Flower of Finland; and *Lapageria rosea* Chilean bell flower is the National Flower of Chile (44). *Caesalpinia echinata* Pau Brasil is the national emblem of Brazil, and the plant from which the country gets its name. Puerto Rico uses *Delonix regia* Flamboyant or Flame tree as its national emblem; and Nicaragua, a country of mestizos, includes *Ceiba pentandra* Silk cotton tree in its national arms (33). *Telopea speciosissima* Waratah, a bushy plant with coral-red flowers, is the National Flower of New South Wales; Protea species is the National Flower of S Africa (where there are over 100 species); and *Tilia americana* Linden is the National Flower of Prussia.

Eschscholzia californica California poppy is the State Flower of California; and *Lewisia rediviva* Bittercup, a western N American plant with attractive rose-purple flower, is the State Flower of Montana; while *Aquilegia caerulea*, with soft lavender-blue and creamy-white flowers, is the State Flower of Colorado (44). *Acer platanoides* Maple is the National Tree of Canada, and its leaf shows on the National flag. Some other species of *Acer* have similar leaves. The people of Israel (in biblical times) were symbolized by *Vitis vinifera* Grape vine (24). *Myosotis alpestris* Alpine forget-me-not is the floral

Vitis vinifera Grape vine 1/5 natural size (legendary and mythical plant; symbolizes people of Israel in biblical times)

emblem of the State of Alaska (16); *Magnolia grandiflora* Laurel magnolia, a tree with globular creamy-white thick-petalled flowers with a strong lemon aroma, is the State Flower of Louisiana and Mississippi; *Magnolia* species Magnolia flower is the National Flower of Korea; *Kalmia latifolia* Mountain laurel, a showy shrub forming dense thickets, is the State Flower of Pennsylvania; while *Strelitzia reginae* Bird of paradise flower, a native of Transkei Republic in S Africa, is the emblem flower of Los Angeles; and *Epacris impressa* Common heath, an Australian shrubby plant with tubular pink white or red flowers, is the State Flower of Victoria (44).

Punica granatum Pomegranate (legendary and mythical plant; symbol of fertility, fecundity, longevity and love; biblical plant; National Plant of Libya and Spain)

The fruit and flowers of *Punica granatum* Pomegranate appear frequently as a motif in ancient Middle Eastern art, especially in embroidery (16); and was regarded by the Egyptians and Greeks as sacred (33). However, Pomegranate is the National Flower of Spain; and *Viola odorata* is the National Flower of Greece (12). The leaf of *Acanthus spinosus* Spiny bear's breech or *A. mollis* is similarly, reputed to have inspired the traditional design on Corinthian columns in architecture (44). *Olea europaea* Olive was sacred to the Greeks. So venerated was the Olive tree that it was Athena's legendary gift to the ancient Greeks who named the city Athens after her (50). The oil so frequently mentioned in the Bible was olive oil - symbolic of the Holy Spirit (24). (See APPENDIX - Table IV for a full list of National Plants).

Olea europaea Olive (legendary and mythical plant; symbol of peace, freedom and purity; Emblem of The United Nations).

RELIGIOUS PLANTS

Plants associated with traditional religion, Islam and Oriental religion, other religions, the Christ and the Crucifixion, the Virgin Mary, biblical plants and plants associated with temples generally, are traditionally believed to be sacred.

Traditional Religion

In C Africa, *Kigelia africana* Sausage tree is deemed holy, and religious meetings may be held in its shade (58). Sausage tree is also sacred in Nigeria (40). *Theobroma cacao* Cocoa comes from the Greek *theos,* god, and *broma*, food; and means food of the gods (30). In W Africa the plant is, however, such an important cash crop and foreign exchange earner - especially in Ivory Coast and Ghana - that the economic importance overrides any other traditional values. In Ghana, *Alstonia boonei* Pagoda tree, has sacred association in Ashanti - the Akan name 'nyame-dua', meaning 'Sky-God's Tree' (see RELIGION: RELIGIOUS CEREMONIES: Rituals). In S Nigeria, the Igbos regard *Chrysophyllum albidum* White star apple as a sacred tree, worshipped by women in hope of maternity (55). It has been suggested, albeit with some doubt, that *Sarcostemma viminale,* a scrambling bush with latex in semi-arid conditions, was the Soma, the Divine Plant of the Ancient Ayrians sanctified in the Rigveda (56).

Islam and Oriental Religion

Moslems consider *Cola acuminata* Commercial cola nut tree and *C. nitida* Bitter cola as sacred and brought by the prophets (25). However, in Ivory Coast and Burkina Faso when black doctors are preparing their medicine they add one or two nuts of kola to it since they consider the trees sacred (27). In the Orient, *Nelumbium nelumbo (Nelumbo nucifera)* Nelumbo or Blue lotus, an aquatic introduction, is held sacred to Buddha who

Nelumbium nelumbo Nelumbo or Blue lotus (symbol of The Ganges; National Plant of India and Nepal; associated with Islam and Oriental religion)

is thought to have been born in the heart of a Lotus blossom (44). In Japan, the young leaves of *Hydrangea macrophylla* var. *serrata* was used to make 'amacha', the sweet tea used on April 8th, the birthday of the Buddha (49). In India, *Ixora grandifolia,* from Malay Archipelago with white flowers, is held sacred to Shiva and Vishnu, and offered to the god 'Ixora' - a local deity; and *Dictamnus albus* Gas plant or Burning bush or Candle plant is considered a sacred plant by the Fire-worshippers of India because of the inflammable nature of a volatile oil which the plant secrets (44).

Other Religions

Nicotiana rustica Wild tobacco, is considered to be the tobacco of the fire god 'Tatewari' of the Mexican Indians (54). According to Huichol folklore and mythology, this god was reportedly once a hawk, *Sephanoaetus coronatus,* and is said to give one visions (54). *Viscum album* Mistletoe, an epiphytic parasite, was the object of superstitious veneration in British folklore, especially among the Druids; and *Dianthus caryophyllus* Carnation was also known as the divine Flower by the Ancients (44).

The Christ.

Some plants are named after Jesus Christ for one reason or another, and accordingly venerated. For instance, the common name of *Caladium bicolor* Heart of Jesus, likely in reference to the red-leaved variety; and *Porana paniculata* Christ plant or Christmas-vine suggests the plants are sacred. *P. volubilis* has been introduced to the Region as a decorative climber. In Europe, *Anemone coronaria* De caen or Giant flowered French is also called 'Blood drops of Christ' from the legend that the plants sprang up from the soil brought from the Holy Land, and spread over Campo Sancto at Pisa to bury the honoured dead (44).

Caladium bicolor Heart of Jesus (associated with The Christ)

Ziziphus spina-christi Christ thorn, a shrub with paired thorns - one straight, one curved - as the name suggests could be the plant used for the Crown of Thorns placed on Jesus' head at his crucifixion. *Tragia* species, climbers with irritating hairs, are also traditionally believed to have composed the Crown of Thorns. Other plants that some writers consider could have been used are *Paliurus spina-christi* Christ's thorn or *Poterium spinosum* Spiny burnet (24). According to some legends *Acanthus spinosus* Spiny bear's breach is one (of many plants) credited with furnishing the material for Christ's Crown of Thorns (44). *Crataegus oxyacantha* Hawthorn was similarly thought to have been used for the Crown of Thorns, and country folk believed it could be a sacred plant. They would carry prigs of the shrub to protect themselves against sickness and plant Hawthorn trees near houses to ward off evil and bring happiness (4).

Euphorbia millii (splendens), a somewhat climbing woody plant armed with stout spines and native of Madagascar, is also known as Crown of thorns or Christ thorn. Similarly, *Koeberlinia* species, *Canotia* species and *Holacantha emoryi*, shrubs or trees restricted to the Mexico deserts, are all known as Desert crucifixion-thorn (6). *Hyssopus officinalis* Hyssop, a herbaceous labiate, is remembered as the herb dissolved in vinegar and used as a sponge at the Crucifixion (45). *Ailanthus altissima* Chinese tree of heaven is an ornamental from China (31).

The Virgin Mary

Plants associated with the Virgin Mary are equally believed to be sacred. For instance, *Anastatica hierochuntica* Rose of Jericho or Mary's flower or Resurrection flower, an introduction from N America, is a symbol of resurrection because it is supposed to have blossomed at the birth of Christ, closed at His Crucifixion and blossomed again at Easter. *Nerium oleander* Oleander, a decorative exotic believed to be the biblical willows, is also called Rose of Jericho (24). However, death has been caused by the use of oleander wood for meat skewers (44) - even the ingestion of two Oleander leaves may cause heart failure and death (43). In addition, the flowers...are deadly poisonous to dogs, *Canis familiaris*; *Equus species* such as donkeys and

Nerium oleander Oleander(Biblical plant)

mules; and many four-legged animals... and also poisonous to human beings (43). The bloom of *Nymphaea lotus* Water lily were similarly, living symbols of regeneration and purification - for their manner of rising with the return of the rains, pure and undefiled from the mud and slime of dried up water-courses, suggests immortality, purity and resurrection (44). *Bryophyllum pinnatum* Resurrection plant from Madagascar is so called probably in reference to the vegetatively reproductive leaves. Other plants with reproductive leaves are *Cystopteris bulbifera* Bulblet fern, which produces new plants from bulblets on the leaves; and *Camptosorus rhizophyllus* Walking fern, that sprouts new ones at the tip of each long pointed leaf. Among Christians, resurrection was represented by the foliage of *Buxus sempervirens* Box (12).

Buxus sempervirens Box (symbol of Palm Sunday and of resurrection)

Tagetes species Marigold, herbaceous composite annuals, refer to the Virgin Mary, and the flowers are traditionally used in Catholic events concerning the Virgin Mary (7). *Silybum marianum* Blessed

Silybum marianum Blessed Mary thistle (associated with The Virgin Mary)

Lilium candidum Madona lily (associated with The Virgin Mary)

Mary thistle is a dark-leaved shrubby composite; and *Lilium candidum* Madona lily with immaculate white flowers, golden-yellow anthers and a delightful fragrance is probably the oldest cultivated lily - many beautiful paintings of the Renaissance times linking the species with the Virgin Mary (44). According to tradition, when the Virgin and the infant Christ were fleeing from Herod into Egypt, they took refuge behind *Juniperus communis* Juniper bush (45).

Biblical Plants

Plants associated with the *Bible* are by tradition sacred - not only in their native lands (mainly the Mediterranean region), but also in many other lands far and near where its message - peace on earth and goodwill towards men - has influenced and inspired the lives of generations past and present. Among such plants are *Sternbergia lutea,* a Mediterranean bulbous plant, thought by some authorities to be the 'Lilies of the field' in the *Bible;* and *Ornithogalum arabicum, O. narbonense* and *O. umbellatum* Star of Bethlehem - native of Europe and N Africa - lilies with cluster of white flowers (44). The *Bible* mentions in Exodus 28 that the Jewish high priests wore garments made from *Linum usitatissimum* Flax (7). *Prunus armeniaca* Apricot, a Mediterranean fruit tree, is believed by some authors to be the apples of gold in Proverbs 25:11 (24). Biblical researchers have now shown more than likely, that Eve tempted Adam with Apricot instead of *Pyrus malus* Apple, and no wonder, for they are delicious (45). According to scientific study, the manna God provided to feed the wondering tribes of Israel was algae on the water, which is *Chlorella* species Chlorella (45). Some herbs and spices of the Bible are listed in Table 3 below:

Species	Plant Family	Common Name	Bible Reference/Remarks
Aloe vera	Liliaceae	Aloe	John 19
Artemisia absinthium	Compositae	Wormwood	Revelation 8:11
Brassica nigra	Cruciferae	Black mustard	Matthew 13:37
Cinnamomum zeylanicum	Lauraceae	Cinnamon	Song of Solomon 4:12-15; Exodus 30:23
Coriandrum sativum	Umbelliferae	Coriander	Exodus 16:31
Crocus sativus	Iridaceae	Saffron	Isaiah 35:1
Nandostachys jatamonsi	Valerianaceae	Spikenard	Song of Solomon 1:12
Pimpinella anisum	Umbelliferae	Anise	Matthew 23:23
Ruta graveolens	Rutaceae	Common rue or Herb of grace	Luke 11:42
Tanacetum vulgare	Compositae	Tansy	Symbol of bitter herbs eaten at Passover

Table 3. Some Herbs and Spices of the Bible

Temples.

By tradition, the timber for temple construction, or plants grown in and around such places of worship are believed to be sacred. Plants used in religious offerings are equally sacred. Perhaps the largest and most significant building in biblical times was the Temple of Jerusalem - constructed with the choicest timbers, and with gold-plated pillars, walls, rafters, doors; and floored with precious stones. The fragrant wood of *Cedrus libani* Cedar of Lebanon, the emblem of Lebanon, a beautiful evergreen tree, is the timber which King Solomon used to build the Temple of Jerusalem about 950 B.C. Other coniferous trees were used, such as *Pinus halepensis* Aleppo Pine, *Juniperus drupaccea, J. excelsa* Juniper and *Cupressus sempervirens* Cypress (24).

Earlier, Ramses II (ca.1290-1223 B.C.) used Cedar of Lebanon for the construction of the Temple of Amen at Thebes (today Karnak) (Baikie fide 11). The timber of *Dipterocarpus trinervis,* native of central and western Jain and Bali, is reserved for sacred purposes, such as the construction of temple (16). (See TABOO: FELLING TREES for another tree useful for seats and pulpit in a church).

In addition to constructional purposes, plants have also been associated with mosques, temples and other places of worship since time immemorial, both for decoration and for their traditional values and beliefs. For instance, in W Africa, *Phoenix dactylifera* Date palm is sometimes planted as an ornamental at mosques (13); and *Caralluma dalzielii,* an erect succulent asclepiad, called 'karan masallaachii' in Hausa, meaning mosque read, suggests that the plant is probably also planted at mosques. In India, *Plumeria rubra* var. *acutifolia* Frangipani or Temple flower, an introduced decorative to the Region, is the Temple tree of the Hindus (20). In the Old World, Frangipani is also recorded often associated with Buddhist temples (16); in order that the blooms may be readily available as temple flowers and offerings to the gods (44). *Paeonia suffruticosa* The tree paeony or The moutan is a very ancient plant in China, and was brought to the temples of Japan in the second half of the 8th century, at the same time as Buddhism (49).

In Peru and Mexico, *Helianthus annuus* Sunflower was a plant highly prized by the people, who adorned their temples with sun flowers made of gold (4). It is recorded (20) that in Peru, *Cantua buxifolia,* the Magic flower of the Incas, has been used since ancient times to decorate their temples. Similarly in Japan, *Cryptomeria japonica* Sacred 'Japanese cedar' is reported (20) as planted in temple compounds in Kamakura; and in parts of N Europe, *Savastana odorata* Holy grass is spread before the doors

of churches and in the path of religious procession during festivals (36). In Japan, there are often beds of *Nelumbium nelumbo (Nelumbo nucifera)* Nelumbo or Blue lotus near temples, and frequent carvings of both flowers and leaves within temples (49). To sit on the Throne of Hasu (The Lotus Throne), means rebirth in paradise after death, and it is the goal of the Buddhist faith (49).

In China and Japan, *Ginkgo biloba* Ginkgo or Maidenhair-tree, a living fossil and symbol of longevity, has been planted since earliest times, particularly near palaces and temples (38). But for the attention given it by Buddhist priests, it would probably now be extinct (38). However, it has also been observed that the tree has survived because it was cultivated by Chinese Monks (30). The great Shwe Dagon Pagoda, near Rangoon, is the most venerable place of worship in all Indochina, and its golden spire, shining in the sun, can be seen for miles. A seed of the sacred *Ficus religiosa* Pepul or Bo-tree was carried by a bird to the top of one of the stupas surrounding the central spire, and from it grew a little tree (20).

In China, *Nandina domestica* Heavenly bamboo, a decorative plant, is used for alter and temple decorations (44). In India, *Michelia champaca,* an introduction to the Region, is used as an ornamental on the grounds of Hindu and Jain temples; and *Thespesia populnea* Portia tree, a naturalised malvaceous coastal shrub or small tree, was believed to be sacred by the Tahitians (*thespesia* meaning 'divinely decreed') who planted it around places of worship (16). In biblical times, *Hyssopus officinalis* Hyssop, a herbaceous labiate, was regarded as the herbal symbol of purification from sin, and used in the cleansing of holy places (4). In W Africa, *Costus afer* Ginger lily, a perennial forest herb, and *Portulaca oleracea* Purslane or Pigweed, a weedy annual, are similarly used for purification by many of the tribal groups.

Finally, some plants serve as sacrificial offerings or are burnt as incense during religious ceremonies in places of worship. For instance, in Mexico a bundle of *Tagetes lucida* Marigold, a composite, is used by the Huichol Indians as religious offerings in temples, government benches and sacred shrines (54). In Japan, *Illicium anisatum (religiosum),* a shrub or small tree from this country and Formosa, is frequently cultivated around Buddhist temples, and its bark was formerly burned as incense (44). In Europe, branches of *Rosmarinus officinalis* Rosemary, a fragrant evergreen labiate, were burned in churches as a substitute for incense; and an aromatic resin extract from *Styrax officinalis* Storax, native from France to Israel and W Turkey, is used (among other things) as incense in churches (44). In W Africa there are several gum copal trees used as

incense in churches and places of worship - among other things. (For a list of gum copal trees, see EVIL: INFLUENCE ON DISEASE: Children's Cases Table 1). (See APPENDIX - Table III for a list of Religious Plants. See also RELIGION; SHRINES; and TABOO).

References.

1. Akpabla, (pers. comm.); 2. Amat & Cortadellas, 1972; 3. Aubréville, 1959; 4. Back, 1987; 5. Bailey, 1954; 6. Benson, 1959; 7. Blumenthal & al., 2000; 8. Bouquet & Debray, 1974; 9. Bouscayrol, 1949; 10a. Burkill, 1985; b. idem, 1995; c. idem, 1997; d. idem, 2000; 11. Chaney & Basbous, 1978; 12. Crow, 1969; 13. Dalziel, 1937; 14. Doe-Lawson, (pers. comm.); 15a. Ellis, 1883; b. idem, 1894; 16. Everard & Morley, 1970; 17. Ferry & al., 1974; 18. Field, 1937; 19. Fowler & Fowler, 1953; 20. Graf, 1978; 21. Hall, 1978; 22. Hall & Swaine, 1981; 23. Hallam, 1979; 24. Hepper, 1982; 25. Irvine, 1961; 26. Kerharo & Adam, 1974; 27. Kerharo & Bouquet, 1950; 28. Kingsley, 1897; 29. Koagne, 1986; 30. Lanzara & Pizzetti, 1977; 31. Lawrence, 1963; 32. Leroux, 1948; 33. Lotschert & Beese, 1983; 34. Lowy, 1972; 35. Meek, 1925; 36. Moore, 1960; 37. Mountfield, 1976; 38. Novak, 1963; 39. Okafor, 1979; 40. Okigbo, 1980; 41. Oliver, 1960; 42. Otedoh, 1982; 43. Pamplona-Roger, 2001; 44. Perry & Greenwood, 1973; 45. Pitkanen & Prevost, 1970; 46. Pope, 1990; 47. Portères, 1974; 48a. Rattray, 1923; b. idem, 1927; 49. Richard & Kaneko, 1988; 50. Roberts, 1990; 51. Rose Innes, 1971; 52. Sanago, 1983; 53. Schnell, 1946; 54. Siegel, Collings & Diaz, 1977; 55a. Talbot, 1926; b. idem, 1932; 56. Tyler, 1966; 57. Visser, 1975; 58. Watt & Breyer-Brandwijk, 1962.

42 SECRETS

Fear them not therefore, for there is nothing covered, that shall not be revealed; and hid, that shall not be known.
Matthew 10:26

It is often required of employees, members of clubs, organizations or societies and individuals to undertake an oath of secrecy as a condition to an employment, or a binding requirement of the group. Various methods are practised. It could be the signing of a declaration, or swearing by either the Bible, the Koran, or a local deity, or a fetish, or an ancestral spirit. It could even involve the use of the individual's blood in a ritual or juju ceremony. Plants may also be used.

SOCIETIES

Some of the most powerful secret societies on the W African coast are: the Poro of Sierra Leone, the Oru of Lagos, the Egbo of Calabar, the Isyogo of the Igakwa, the Ukuku of the Benga, the Okukwe of the M'pongwe, the Ikun of the Bakele, and the Lukulu of the Bachilangi Baluba (12). In Sierra Leone, the Society of Bonn, another Secret Society, is associated with infamous rites among the Sherbros - the ritual murder of a virgin annually (5a). This country is famous for its secret societies. For instance, there is also a Bundu women's society and 'Humo' women's society (4). In S Nigeria, there is the Ogboni society among the Yoruba people (5b). In Gabon, there is the Bwiti and the Byeri Secret Societies among the Fang and the Eshira, respectively (14); and also the 'Ndjembe' and the 'Nyemba' societies (17). In Angola, two examples of cannibalistic societies occur - the Maghena Secret Society of the Bonyala, and the Quissama

Local deity

Secret Society (9). In S Sudan, there is the Mani, the Biri, the Nando, the Kpira, the Siba and the Wangara societies among the Azande, which have been described as closed associations (6).

ASSOCIATED PLANTS

In African tradition, some plants enter prescriptions to prevent the disclosure of secrets. In S Nigeria for instance, the Yoruba name of *Cuscuta australis* Dodder, a leafless parasitic twiner, 'omonigedegede', meaning 'lonely child' implies that such a child keeps secrets, and thus they use the plant in an invocation to conceal a secret (16). Similarly, the Yoruba name of *Senecio abyssinicus*, an annual composite herb, 'amunimuye', meaning 'bringing life' or 'producing life' or 'catching a person's thoughts', suggests magical uses, and enters in prescriptions to prevent someone knowing about a matter (4). *Senecio baberka* 'Ragwort' is reported to have the same uses (1).

In S Nigeria, the Edo name of *Pararistolochia goldieana*, a forest climber with large purplish-red flowers, 'ugbogielimi', is also the name of the black hat of the mother of Ovia, the founder of a secret society among the Edo people (15). (See also DEATH AND DYING: PLANT NAMES AND DEATH).

In Sierra Leone, the Mende name of *Strophanthus gratus*, a lofty liane with white latex is 'sawa' (3a). This is a general term for medicine - made of a decoction of leaves - that is used to cure violations of Secret Society rules (3a). More specifically it applies to *Gouania longipetala*, a scandent forest shrub or liane; and an infusion of mashed leaves mixed with other herbs goes into a cleansing 'medicine' used by certain Secret Societies (3a). In this country, *Strophanthus* is used in the ritual of the women's society 'Humo' (Lane-Poole fide 4). The liane is reported to have secret and juju uses in Nigeria as well (1).

Coix lacryma- jobi Job's tears

In Sierra Leone, *Psydrax (Canthium) subcordata*, a forest tree always inhabited by black ants, *Crematogaster* species, is made use of by the Sowisia of the women's society called Bundu (4). In this country, the oil of *Licania (Afrolicania) elaeosperma*, a tree in the Rose family with drooping branches and foliage, is mixed with white clay to smear as body scent - also in the Bundu Secret Society ceremonial by women (4). The false fruits (seeds) of *Coix lacryma-jobi* Job's tears, a perennial grass on river banks, have a place at meetings of the Bundu society (3b). In this country, *Glyphaea brevis,* a shrub or small forest tree, has relevance to medicine used by Secret Societies (11). In N Nigeria, *Monocymbium ceresiiforme,* a tufted perennial grass, is used in pagan forms of oath in Bornu (4). In Congo (Brazzaville), *Plagiostyles africana,* a shrub or small forest tree, is endowed with powerful fetish properties for Secret Societies; and the bark is put into all manner of love-potions and charms - benign, malignant, protective or offensive (2). In Senegal, the sticky buds of *Gardenia sokotensis,* a shrub or small savanna tree, are thought by the Basari to carry portents. A young girl undergoing initiation into a Secret Society will put one on the bottom of a calabash when she is given beer. If the beer bubbles and froths over, someone is trying to bewitch her (8).

In Gabon, the young shoots of *Scleria boivinii (barteri),* a scandent perennial sedge; or the leaves of *Cissus producta,* a large herbaceous forest climber; together with other herbs enter into a ritual dish given to the Bwiti Secret Society initiates to eat when the intoxicating effects of taking *Tabernanthe iboga* Iboga, an apocynaceous shrub, are wearing off (17). *Alchornia floribunda,* an euphorb shrub, is used in almost precisely the same way as Iboga in the initiation to another Secret Society, the Byeri, which is prevalent among the Fang and the Eshira (14). In this country, the root of *Mostuea hirsuta,* an under-shrub or shrub of open savanna, is considered to be a strong aphrodisiac and to be capable of staving off sleep. Its action seems to be comparable to Iboga and *Pausinystalia johimbe,* a forest tree, which is taken by followers of the Bwiti Secret Society on dance nights. The root of *Mostuea* may be chewed alone or taken in admixture with the other two species. The function of these plants is primarily as excitant, so that dancers will last out the night long, and as hallucinogens altering the mind of the participants to enable them to pass into that 'Other World' of mystery (Fernandéz fide 17). The roots of Iboga are also used in the initiation rites to a number of Secret Societies, of which the Bwiti is the most famous (14).

In Gabon, the leaves of *Urera oblongifolia,* a glabrous liane in forest, enter into a complicated formulation in the secret masonry 'Ndjembe' for applying to the body of novitiates as protection from evil spirits (17). In this country, *Lapportea aestuans,* a herbaceous plant with irritating hairs, is a fetish in the masonic rites of the 'Ndjembe' and the 'Nyemba' societies. During the initiation, young novitiates are rubbed with the leaves and stems, and are then put in a bath which causes intolerable pain

(17). The twigs of *Barteria fistulosa,* a slender forest tree, are hollowed out and colonized by powerful fierce black ants, *Crematogaster* species, capable of inflicting painful bites (17). Such is the painfulness of the ants' bite that in Gabon, initiates to the female sect of masonry known as 'Ndjembe' or 'Nyemba' have to prove their capacity of endurance by collecting so many ants from the tree (17).

In Gabon, the latex of *Elaephorbia drupifera,* an euphorb tree, is also used along with others to drip into the eyes of novitiates in Secret Society initiation; and by Fang mediums in trances (17). If the *Alchornia* concoction taken to induce trance is slow to act, some latex mixed with oil is wiped over the eyeball with a feather. This affects the optic nerve to produce bizarre visual effects and a general dizziness (7, 10). *Lygodium microphyllum,* a climbing or scandent fern, is used by the Fang and other races in initiation dances of Secret Societies (17). In this country, *Thaumatococcus daniellii* Katemfe has importance in the ritual of Secret Societies, where clumps are specially grown at meeting places, and the participants of dance ceremonies tie strips of the petiole around their waist (17). In this country, *Platycerium stemaria* and *Drynaria laurentii,* decorative epiphytic forest ferns, are planted for decoration on Bwiti Secret Society huts or houses (17).

The phrase *sub rosa* was linked with Cupid's gift of *Rosa* species Rose (the emblem of love and charity and of discretional secrecy) to Harpocrates, the God of Silence, as a bride not to reveal the armour of Venus. Accordingly, whenever secret matters were discussed a Rose was suspended from the ceiling, and what took place beneath it was strictly *sub rosa* (under the Rose) (13).

References.

1. Ainslie, 1937; 2. Bouquet, 1960; 3a. Burkill, 1985; b. idem, 1994; 4. Dalziel, 1937; 5a. Ellis, 1883; b. idem, 1894; 6. Evans-Pritchard, 1937; 7. Fernandéz, 1972; 8. Ferry & al., 1974; 9. Freyberg, 1935; 10. Hargreaves, 1978; 11. Irvine, 1961; 12. Kingsley, 1901; 13. Perry & Greenwood, 1973; 14. Pope, 1990; 15. N.W. Thomas, 1910; 16. Verger, 1967; 17. Walker & Sillans, 1961

43 SHRINES, PLANTS IN AND AROUND

In all thy ways acknowledge Him,
and He shall direct thy paths.
Proverbs 3:6

A shrine - the sacred house of a fetish or a god - offers physical shelter to the supernatural forces. It is the meeting ground between humans and their spiritual overseers (11); or indicates any type of potential abode for a spiritual element, entity or force (12). In the Region, this 'physical shelter' is usually erected with simple local materials. It may be of mud walls with grass thatch; a wooden structure; a hut of woven palm fronds, or of *Pennisetum purpureum* Elephant grass. Sometimes the abode is a bamboo fence enclosing one or more fetish plants. In the savanna country, it is not uncommon to erect a shrine under the shade of *Adansonia digitata* Baobab. In the Region, shrines are more often confined to rural areas or the outskirts in urban centres. On the contrary, elsewhere shrines may be constructed and furnished like temples, and situated in the cities.

Tongo shrine

ASSOCIATED PLANTS

Some plants are, by tradition, associated with shrines - either planted around them for protection and for decoration, or used in ceremonies to make new ones, or used in ritual purification or in sacrificial offerings to the deity believed to occupy the shrine. Some plants also enter into charms to destroy the spiritual force within the shrine.

Throughout the Region and particularly in their distribution area, *Newbouldia laevis,* a popular fetish tree; *Jatropha curcas* Physic nut;

Jatropha gossypiifolia Red physic nut

Jatropha gossypiifolia Red physic nut; the large species of *Crinum*, such as *C. jagus, C. distichum* and *C. zeylanicum (ornatum)*, bulbous lilies in swamps; and *Dracaena arborea*, another popular fetish tree, are planted before shrines. In N Nigeria, *D. fragrans* and *D. smithii*, shrubs or small palm-like trees with fragrant flowers, (or other species of similar habit) may be seen at pagan shrines along with *Euphorbia* species (1). *Sansevieria liberica, S. senegambica* and *S. trifasciata* African bowstring hemp are usually planted at pagan shrines (1). In S Nigeria, *Albizia ferruginea* and *Tetrapleura tetraptera*, medium-sized tree legumes, and *Euphorbia kamerunensis*, a tree or shrub with candelabriform branching system often planted in towns and villages, are sacred plants associated with shrines in Igboland (7). In this country, *Garcinia kola* Bitter cola; *Glyphaea brevis*, a medium-sized tree in regrowth forest; and *Aframomum melegueta* Guinea grains or Melegueta are found around shrines or used in religious ceremonies and sacrifice in Igboland (7); and *Oxyanthus subpunctatus*, a shrub or small forest tree, is used by the Igbo tribe with chalk to fasten to 'ikenga' (image) at a shrine (N.W. Thomas fide 3).

In S Africa, a stick of *Terminalia sericea* Assegei wood is stuck, apparently with magical significance, in the floor of a shrine in which homage is paid to the ancestral spirits at planting and harvesting times; and in Mozambique, of the four sticks planted round a low circular shrine by the Tonga, in which the sacred stones are embedded, two are from *Annona chrysophylla*, a savanna shrub (13). In E Africa, *Trochomeria macrocarpa (macroura)*, a climbing cucurbit in open country, and *Ficus sur (capensis)*, a fetish tree, are reported common at shrines in some villages of Shari-Chad (1). In Congo (Brazzaville), some plants are associated with shrines. 'In some parts of the long single street of most villages, there is a low hut in which charms are hung, and by which grows a consecrated plant, a lily, an euphorb, or a fig' (6).

In W Africa, the whorled forks of a branch of *Alstonia boonei* Pagoda tree, a tree with white latex and associated with traditional religion, is used as support for food in domestic shrines for spirits among the Akan-speaking tribe of Ghana and Ivory Coast (1). In Ghana, a large branch of *Costus afer* Ginger lily, a robust perennial forest herb, is carried and used

to sprinkle water on shrines during the Apo ceremony at Tekyiman (1) - probably for purification. The larger varieties of *Urginea* species, such as *U. indica* and *U. altissima*, bulbous lilies in savanna, are sometimes seen in villages and shrines or over doorways, or planted for superstitious use (1). In this country, two sticks of the well known fetish plant of the Ada people - *Ochna afzelii* and *O. membranacea*, savanna shrubs, are used by the fetish priests in this traditional area to cross the entrance to every shrine. This is called *kpangme* (2). In this country, most shrine rooms and houses in Ashanti and Brong Ahafo have dried plantain or banana leaves used to cover pots of palm-wine hanging from rafters; and at every Yam Festival these are torn down and thrown away to begin a new year. Some people grind these leaves and use them for medicine (12). In the Region, *Zebrina pendula* Wandering Jew and *Tradescantia* species Spiderwort, decorative introductions to the Region, are occasional plants cultivated or carefully tended at shrines and spiritual places.

In the Far East, the fallen flowers of *Amherstia nobilis*, one of the world's loveliest flowering tree and native of SE Asia are used as an offering at Buddhist shrines (4); and in the Middle Ages, *Rosa* species Rose was commonly used in England for decorating shrines (8). In Japan, *Cleyera japonica* Sasaki, an evergreen tree, is traditionally believed to be the first tree that grew out of the chaos of creation, and a pair of Sasaki are often planted in front of a shrine as they are the sacred trees of Shinto; and the branches are offered at alters and at ceremonies (10). Similarly, *Eurya japonica* Hisakaki, an evergreen shrub, is also used in Shinto rites (10); and *Eriobotrya japonica* Biwa, an evergreen tree, is usually planted in temples and shrines rather than in private gardens (10). In this country, the magnificent 20 mile avenue of *Cryptomeria japonica* Japanese cedars leading to the Toshogu Shrine in Nikko is a lasting memorial to Masatsuna Matsudaira who rebuilt the shrine in 1652. Japanese cedars are often seen in shrine precincts with ropes round their trunks to mark them as sacred (10). In this country, there are old and respected specimens of *Ginkgo biloba* Maidenhair tree in many Shinto shrines (marked by their encircling ropes and/or white papers) held to be the residence of sacred spirits. The ancient *Ginkgo* still standing by the steps of Tsurugaoka Hachiman Shrine in Kamakura has a special place in history - it provided cover for Shogun Sanetomo's assassin in 1219 (10).

Construction

In Ghana, several plants have religious use in being ceremonial ingredients in the brass pan when making a new shrine in Ashanti. For instance, *Piliostigma thonningii,* a shrub or small tree legume in savanna regions, and *Clausena anisat*a Mosquito plant are used among the tribe in religious rites for making a new shrine, being supposed themselves to possess spirits *(saman)* (9). A full list of such plants (9) is given in Table 1:

Species	Plant Family	Common Name	Part Used	Habit
Abrus precatorius	Papilionaceae	Prayer beads	Seeds	Climber
Acacia kamerunensis	Mimosaceae	-	Bark	Liane
Ageratum conyzoides	Compositae	Billy goat weed	Whole plant	Herb
Alternanthera pungens	Amaranthaceae	Khaki bur or weed	Whole plant	Herb
Cardiospermum grandiflorum	Sapindaceae	Balloon vine	Whole plant	Climber
C. halicacabum	Sapindaceae	Balloon vine	Whole plant	Climber
Clausena anisata	Meliaceae	Mosquito plant	Leaves	Shrub
Costus afer	Costaceae	Ginger lily	Whole plant	Herb
Distemonanthus benthamianus	Caesalpiniaceae	African satinwood	Bark and roots	Tree
Erythrina mildbraedii	Papilionaceae	Bead tree	Bark and roots	Tree
E. senegalensis	Papilionaceae	Coral flower	Bark and roots	Tree
Erythrophleum suaveolens	Caesalpiniaceae	Ordeal tree	Bark and leaves	Tree
Holarrhena floribunda	Apocynaceae	False rubber tree	Bark	Tree
Hyptis pectinata	Labiatae	-	Whole plant	Herb
H. suaveolens	Labiatae	Bush tea-bush	Whole plant	Herb
Justicia flava	Acanthaceae	-	Whole plant	Herb
Milicia excelsa	Moraceae	Iroko	Bark	Tree
M. regia	Moraceae	Iroko	Bark	Tree
Piliostigma thonningii	Caesalpiniaceae	-	Bark and roots	Shrub
Platostoma africanum	Labiatae	-	Whole plant	Herb
Salacia debilis	Celastraceae	-	Piece of stem	Liane
Spiropetalum heterophyllum	Connaraceae	-	Piece of stem	Liane

Table 1. List of Plants Used in Making a New Shrine in Ashanti.

The ingredients include any root that crosses a path, a projecting stump in a path over which passers-by would be likely to trip, and roots and stumps from under water. The rest are a nugget of virgin gold and aggry beads. All the ingredients are pounded and placed in the brass pan amid incantations and prayers (9).

Destruction

The practice of destroying or driving away the spiritual force within a shrine does occur among rival fetish priests in the Region. For instance, to neutralize or destroy the spiritual power believed to occupy any shrine, among the Ewe-speaking tribe of Ghana, Togo and Benin Republic and some of the coastal tribes, the leaves of *Phyllanthus niruri* var. *amarus (amarus)*, a weedy euphorb herb are collected with the necessary ceremonial rituals, and then chewed together with the tubers of *Manihot esculenta* Cassava. This is clandestinely sprinkled on the shrine with incantations and a spell (5). (See also SACRED PLANTS).

References.

1. Dalziel, 1937; 2. Doku Korle, (pers. comm.); 3. Burkill, 1997; 4. Everard & Morley, 1970; 5. Gamadi, (pers. comm.); 6. Kingsley, 1897; 7. Okigbo, 1980; 8. Perry & Greenwood, 1973; 9. Rattray, 1923; 10. Richard & Kaneko, 1988; 11. Twumasi, 1975; 12. Warren, 1974; 13. Watt & Breyer-Brandwijk, 1962.

SLAVERY

Stand fast therefore in the liberty wherewith Christ hath made us free, and be not entangled again with the yoke of bondage.
Galatians 5:1

Slavery existed in biblical times - it is an old institution. There are over thirty references to the practice in the New Testament of the *Holy Bible*; and three references to slavery in the Old Testament - in Exodus 2:23; Psalms 86:16 and Psalms 116:16. However, it is the Trans-Atlantic Slave Trade from 1760-1810 that has been of concern in modern times. Based explicitly on religion, justice and humanity, the entire British slave trade was abolished with effect from 1 May, 1807 (1). Today, slavery is officially illegal in law throughout the civilized world.

In the days of the slave trade, captives either awaiting shipment to the New World or on the march from the hinterlands to the coast, were tied up to each other with a loose chain by the waist or the ankle; or were yoked by the neck to prevent them from escaping. In traditional practice, plant charms also entered into prescriptions believed either to prevent the escape of captured slaves or to make them forget their kinsmen. There are equally plant prescriptions to counteract this charm.

PLANT PRESCRIPTIONS

In N Nigeria, the Hausa term 'manta uwa', meaning epiphytic orchids like species of *Ancistrorhynchus, Calyptrorchilum, Cyrtorchis, Diaphananthe, Plectrelminthus, Podangis, Stolzia, Tridactyle* and so on (see also INFANTS: PHYSICAL DEVELOPMENT: Plant Prescriptions: Sympathetic Magic), is also the name of a medicine (made sometimes of quite different plants) which is given to a captured slave to prevent his running away (4). *Crotalaria arenarea,* a legume, has been found as an occasional ingredient in superstitious practice to make a captured slave forget his kindred - hence the local name 'manta uwa', meaning 'forget mother' (4). Similarly, in Gabon, the fruit of *Massularia acuminata,* a forest shrub, is considered to be a charm for children who lose their mother;

Elaephorbia drupifera

it is supposed to help them to forget (6). In this country, stem shavings of *Elaephorbia drupifera,* an euphorb tree, were given to captives during slave-raids to subdue them, so that they lost any idea of escaping (6). Various euphorbiaceous plants were used, but principally this one (3b). In the Region *Pentodon pentandrus,* a herb, enters into rituals to free people's spirit from bondage (5). In the W Indies, *Dieffenbachia sequine,* an aroid, dubbed 'dumb cane' was formerly used in torturing slaves. It renders speechless a person who chews a piece of the stem (8).

COUNTER CHARMS

Among fetish priests, medicine-men, native doctors and other traditional practitioners in its distribution area however, *Paullinia pinnata,* a woody or sub-woody climbing shrub with tendrils, collected and prepared with ceremonial rites, a prayer and libation, enters into prescriptions as a counter measure, which nullifies the effects of any preparations mixed with a man's food, unknown to him, to make him subservient (5). A slave who does wrong rubs his body with *Ludwigia octovalvis* ssp. *brevisepala* Primrose-willow, an erect herb with yellow flowers - sometimes in marshy areas - collected with the necessary rites, or sucks it with red natron (native sesquicarbonate of soda) to avoid pain if beaten for his fault or to escape detection (4).

Paullinia pinnata

LORE

During the days that Samori and Babatu terrorized the people between Sierra Leone and Nigeria, killing and capturing slaves, robbing

and burning (7); the natives of Zoba, a village on the Ghana-Ivory Coast border, devised an ingenious method to outwit the raiders. As soon as scouts gave alarm of an advancing raid, the spiritually powerful chief turned himself and all the able-bodied men and women in the village into monkeys, species of *Cercocebus, Cercopithecus, Colobus, Erythrocebus, Papio, Piliocolobus* and *Procolobus* – leaving only the children and the aged – and back again into human beings after the raiders had left (2).

After several unsuccessful attempts - and in frustration - the raiders using counter-charms brutally shot and killed some of the unarmed and defenceless monkeys including the leader (the chief). With the demise of their chief and spiritual leader, the remaining monkey population could not be turned into human beings again. To date, the monkeys frequent the village during the day but return to the woods at night. Dead ones are retrieved from the wild by the people and given fitting burial in the village as kinsmen (2).

ASSOCIATED PLANT NAMES

In N Nigeria, *Merremia tridentata* ssp. *angustifolia,* a twining savanna annual, is called 'gammon bawa' in Hausa, meaning 'slave's head-pad (3a); and in Senegal, the Bambara/Malinke name of *Hyptis spicigera* Black sesame, an annual aromatic labiate often in damp places, 'bene-fing dion', has reference to a slave (4). (The Bambara tribe is also found in Benin Republic, Ivory Coast, Mali and Niger).

References.

1. Anstey, 1975; 2. Bonsra, (pers. comm..); 3a. Burkill, 1985; b. idem, 1994; 4. Dalziel, 1937; 5. Mensah, (pers. comm.); 6. Walker & Sillans, 1961; 7. Ward, 1956 8. Willis, 1931.

45 SNAKES & SNAKE-BITE

Ye serpents, ye generation of vipers,
how can ye escape the damnation of hell ?
Matthew 23:33

Poisonous snakes occur in W Africa. There are at least ten kinds whose bites may easily prove fatal (8). These include the Gaboon viper, *Bites gabonica;* Carpet viper, *Echis carinatus;* Green mamba, *Dendroaspis viridis;* and Black cobra or Spitting cobra, *Naja nigricollis.* Considering the frequency of deadly snakes, fatal bites are comparatively rare, even in the rural areas. It has been observed that 'though the deadly Black cobra and Niger jack are so common, one rarely hears of persons being bitten; and of those so bitten I never heard of one dying. It is well known that the natives have something specific against snake poison, the secret of which they guard carefully' (12). Some plants are either used for charming snakes or are used as anti-venom to snake-bite or as preventive of snakes or snake-bite or believed to repel snakes - some as charms, fetish or juju. Some other plants are named after snakes, or are associated with them for one reason or the other.

SNAKE CHARMING

The act of snake charming is reported to be common in India, where it probably originated. Since the late 1990s, when the Wildlife Protection Act (1972) was implemented, it is now illegal and snake charmers could be jailed if caught. In W Africa, it is mostly practiced by nationals of the Guinea-savanna to the north. Snakes charmers are simply fascinating, for snakes are probably the most feared reptiles, because of the venom. Though the fangs could be removed (as it is often the secret), nevertheless the mere handling of live, probably deadly cobras is a feat that few would dare attempt. The primary objective of snake charmers is to peddle in anti-venom to snake-bite and snake repellents.

In Ivory Coast, *Phyllanthus niruroides,* a weed of waste places, is believed to be a fetish used by snake charmers in the Man region

(19). In Ghana and Nigeria, the powdered leaves of *Hoslundia opposita,* an odorous labiate, with *Ocimum basilicum* Sweet basil and *O. canum* American basil are used to confer both immunity to snake-bite and the faculty of charming and handling poisonous snakes (10, 17). The labiates with assorted collection of the inflorescence of *Elaeis guineensis* Oil palm, gum copal and the head of a freshly-killed snake of a poisonous species, is baked black, powdered and mixed with palm-wine. The preparation is rubbed on the hands by snake-catchers – it stupefies the snake at once, and it will not bite (10, 17). In S Nigeria the chewed roots of *Scoparia dulcis* Sweet broomweed with the juice of *Nicotiana tabacum* Tobacco is used in Calabar to paralyze snakes (10) - the weed added for superstitious reasons.

Nicotiana tabacum Tobacco

The Hausa of N Nigeria, the Zarma of Mali and Niger and the Fula believe that the biblical rod which Moses turned to a serpent and back into a rod again in Exodus 4:2-4, is a branch of *Diospyros mespiliformis* West African ebony tree (23). It was the same rod that was lifted over the Red Sea to first split it apart, and then to return the sea to its original position in Exodus 14:16-29. In its distribution area, West African ebony tree is associated with many legends. (See WAR: ASSOCIATED PLANTS: Lore, and WEALTH: CHARM PRESCRIPTIONS).

SNAKE-BITE

The treatment of snake-bite usually starts with a ligature to prevent the venom from reaching the heart and therefore, the general circulation, followed by local incisions to suck out the venom before the necessary medication is administered. Traditionally, the practice may be given a fetish, medico-magical or general superstitious touch.

In Ghana, a superstitious remedy for snake-bite is the root of *Ocimum gratissimum* Fever plant, along with that of *O. canum* American basil and *Psidium guajava* Guava, mixed with the sap of the plantain stem (17). Among many coastal tribes in the Region, the leaves of *Pergularia daemia,* collected with rites, are immediately chewed and swallowed as a first aid for snake-bite, or the bitten spot is cut and a macerate of *Euphorbia*

hirta Australian asthma herb in water taken to induce the blood with the venom to gush out (22). The practice could probably be a superstitious one. In Senegal, the leaves of *Acacia macrostachya* are believed by the 'Socé' to be an effective antidote against snake-bite. Large quantities of the leaves must be eaten immediately as a prescription. This will stop the venom circulating in the blood stream (5, 18a).

In Ghana, there is a recorded incident in which the leaves of *Rourea (Byrsocarpus) coccinea,* a shrub or climber, was successfully used for snake-bite, but the natives were sure that it was not the healing power of the plant which had been effective here, but that it was a fetish cure, as the plant was a fetish of the healer (15). Another superstitious treatment of snake-bite (15) is with the whole plant of *Lantana camara* Wild sage, *Aframomum melegueta* Guinea grains or Melegueta and *Cymbopogon citratus* Lemon grass. In N Nigeria, the tuber of *Amorphophalus dracontioides,* a savanna aroid with underground tubers, is used by the Fula for snake-bite, but the prescription is apparently medico- magical, because to an infusion of the cut-up tubers are added scrapings and washings of a woman's hair oily with grease, probably along with other ingredients, and the concoction is given to the sufferer to drink (10).

Sympathetic Magic

The treatment of snake-bite could at times be based on sympathetic magic. For instance, in its distribution area in W Africa, the flowers of *Feretia apodanthera,* a bushy shrub in savanna, are said to be eaten by the snake, thus a person bitten by a snake will therefore get the flowers quickly, crush them in water and drink in the hope of vomiting to discharge the poison (10). In Ghana, the root of *Flabellaria paniculata,* a liane of woody savanna, ground up with *Piper guineense* West African black pepper and the dried head of either the cobra, *Naja nigricollis;* or the mamba, *Dendroaspis viridis*; or the viper, *Bitis gabonica* or *Echis carinatus*, is a vaccine used in the Mampong area for snake-bite (2). Seven cuts are made on the right side of the body and rubbed in with the preparation dissolved in spirit or lime-juice. The process is repeated after one week. This medication is said to confer immunization (2).

In E Africa, the root of *Cissus simulans,* a climber in the Vine family, mixed with portions of a snake, is a Lobedu snake-bite remedy; and in S Africa, a Zulu snake-bite remedy is made of equal amounts of powdered root of *Solanum capense* Nightshade and powdered snake

(31). In this country, the Luvale use *Gymnosporia senegalensis* Senegal pendoring as a snake-bite remedy either by burning the head of the snake with the root, mixing the ash with oil and applying the preparation to the tongue of the patient and to the bite, or if the head is not available, an infusion of the leaf is drank (31). In this country, the slimy leaf pulp of *Aloe davyana,* a lily, is an African snake-bite remedy, the material being rubbed into the wounds. Among the African there is a widely-held belief in this remedy, even for the bite of the deadly mamba, *Dendroaspis viridis* (31).

Wearing portions of the root-bark of *Securidaca logepedunculata* Rhode's violet, a savanna shrub or small tree, is a charm supposed to keep snakes away, and the use of the bark for snake-bite may be sympathetic magic (17). Among many African peoples, a bracelet of *Dichrostachys cinerea* Marabou thorn or a strand of the bark or a twig kept at the door of a hut is believed to keep away snakes (10); and in Liberia, an anklet of Marabou thorn is a charm against snake-bite, it is tied to prevent spread of the venom (10). The use of the root of Marabou thorn as an antidote to snake-bite could, therefore, be merely an instance of homeopathic magic. In Europe, *Echium vulgare* Viper's bugloss, a bee-visited British plant, dried and powdered, was once thought to be specific against the bites of vipers, species of *Bitis* and *Echis;* and mad dogs, *Canis familiaris* (26). In Trinidad and Tobago, *Ceropegia peltata,* an under-storey fast-growing tree, is prepared as a snake medicine by pounding a 3-inch piece of root and then boiling this with a piece of brown paper. It was claimed that the 'brown paper was a poison and that one needed a poison to kill another poison' (21).

PROTECTIVE CHARMS

Snake charms may be applied in various ways - either worn as a girdle, or rubbed on incisions, or taken internally, or kept in the house. In the Region, *Launaea (Lactuca) taraxacifolia* Wild lettuce is popularly used in juju to give protection against snake-bite, or the black powder prepared from the leaves of *Portulaca oleracea* Purslane or Pigweed and the head of a poisonous snake is taken in alcoholic drink (22) – likely an instance of sympathetic magic. Copper rings charred with certain herbs (unidentified) and the head of a poisonous snake are commonly sold in medicine markets in the Region as protective against, or preventive of snake-bite (28). *Euphorbia hirta* Australian asthma herb also enters juju preparation with assorted ingredients, which is taken as a spiritual protection against snakes. Even one's shadow over a snake kills it (22). In Senegal, *Bauhinia rufescens*, a scandent shrub or small tree legume in savanna, is credited with magical property to protect houses from snakes (18b).

In Ghana and Ivory Coast, *Parquetina nigrescens*, a twiner with milky latex in forest regrowth, 'nsurogya', meaning 'does not fear fire' in Akan, is also believed to be preventive and protective of snakes (14). Among many tribal groups in the Region, there is a general belief that the leaves or the whole plant of *Phyllanthus niruri var. amarus (amarus)*, a common weedy annual euphorb herb with sub-woody stem, carried on one's person is protective against snake-bite (24). In N Nigeria and Niger, Fula cowboys rub the plant on their legs for the same effects (22). It is the belief among many of the coastal tribes that it is suicidal for any snake to bite a person who possesses a fetish power controlled by *Alternanthera pungens (repens)* Khaki weed or bur, a common prostrate weedy amaranth (22).

The flowers of *Aristolochia albida* and *A. bracteolata* Dutchman's pipe are sometimes worn as juju or charm against snake-bite and scorpion-stings (1); similarly in S Nigeria, *Biophytum petersianum* African sensitive plant, a common weed - sensitive to both weather and to touch , is believed to be a charm against snakes (10). The young unopened leaves of *Piliostigma thonningii*, a savanna shrub legume or small tree, picked naked before sunrise, then charred with chameleon, *Chamaeleon vulgaris*, to a black powder and applied to incisions around the ankle, is a charm against snake-bite - the snake that attempts to bite will die instantly (22). In S Nigeria, *Harungana madagascariensis* Dragon's blood tree enters into rituals against meeting a snake while climbing a palm tree to tap (Boboh fide 7b). A young leaf from a terminal bud is carefully placed under the tongue. It must not touch the teeth. If the leaf remains unbitten, the tapper will be protected, but should he damage the leaf the protection will be lost (Boboh fide 7b).

The leaves of *Pergularia daemia*, a herbaceous semi-woody asclepiad believe to be fetish, pounded with china-clay and palm-wine and left to dry in the sun is tied round the ankle as a protection against snake-bite (19). In Ivory Coast, there is a popular belief in Dabakala that when the root of *Abelmoschus esculentus* Okro or Okra is thrown to the cobra, *Naja nigricollis*, it has the power to prevent it from spitting, and to render it inoffensive (19). In Botswana, before entering water to search for the body of a drowned person, the Sotho chew a portion of the raw root of *Kylinga alba*, a sedge, which proceeding is thought to protect them from a serpent which lives in the water (31). In the northern Peruvian

Andes if *Quassia (Simaba) cedron* Cedron is worn on the body, it has the magical virtue of protecting the wearer from the bite of the viper, species of *Bitis* and *Echis* (11).

Snake Repellents

Folktales about the snake-repellent properties of some plants are told in traditional Africa. For instance, among farming communities mothers place their babies under the shade of *Anchomanes difformis* or *A. welwitschii,* robust aroids, while working on the farm in the belief that the plant scares away snakes (21). (Children have sometimes been observed to play with poisonous snakes unharmed - or rather poisonous snakes are normally not known to bite innocent children). In S Africa, the Zulu often encircle a hut with *Tulbaghia violacea* Wild garlic, a lily, to keep snakes away, the supposition being that the odour repels them - the root of *T. alliacea* is also reported to be a charm to repel snakes (31). In this country, the tribe sprinkle a decoction of *Leonotis leonurus* Wild hemp about the kraal to keep snakes away, or think that the strong-scented rhizomes of *Kaempferia* species is good for driving away snakes (31). In Kenya, a piece of the root of *Carissa edulis* Carrisse, a thorny savanna shrub with edible fruits, is fixed by the Nandi to a hut-root as a snake-repellent (28) probably due to the strong smell of methyl salicylate from the crushed roots. In Zimbabwe and Tanzania, the root of *Acacia polyacantha* ssp. *campylacantha*, a savanna tree, is thought to be a snake-bite remedy and snake-repellent. It is put into the hut's rafters, and held also to keep crocodiles, *Crocodylus* species, away from fords (11).

Carissa edulis Carrisse. - 1/5 natural size

In Congo (Brazzaville), *Chenopodium ambrosioides* Indian wormseed or Sweet pigweed, a herbaceous annual or perennial, is said to repel snakes and to cure snake-bite (6); and in Cameroon, the scent of *Pentaclethra macrophylla* Oil-bean tree, a forest tree legume, is believed to be capable of driving away snakes in the Ewondo region of Yaoundé (9). The raw fruit is crushed or ground and rubbed on the feet before

leaving for the bush. It is even more effective if some seeds of Oil-bean are eaten. The snake will run away on your approach (9). Similarly, in some areas in Senegal, people going out into the bush, tie the fruit of Oil-bean tree on the feet as well as eating some grilled seeds (5). In S Nigeria, the Ijo believe that if *Senna (Cassia) occidentalis* Negro coffee is planted near their houses, it will drive snakes away. In Central African Republic, the Hausa and Senegalese expatriates in Bangui also plant Negro coffee nearby (29); and in Jabel Marra, there is a belief that the plant will actually kill snakes (16).

The leaves of *Manihot esculenta* Cassava are believed to have hypnotic influence on snakes. In W Africa the leaves are chewed by some of the coastal tribes before one looks for a snake to kill, to prevent it from biting (22). It is also the general belief that *Mucuna pruriens* var. *utilis,* the edible variety of *M. pruriens* var. *pruriens* Cow itch, repels snakes when planted around houses, and that the chewed fresh seeds neutralize any venom when one is bitten (20). In Cameroon, Cow itch is planted around houses to keep away snakes (30). In the Region, *Jatropha multifida,* a soft-wooded shrub is often planted as hedges and as an object of superstition. It is said, like *J. gossypiifolia* Red physic nut, to keep off snakes (19).

In Nigeria *Merremia dissecta,* a perennial twiner, is reported in Igala Province to keep snakes away from houses where it is grown (probably a superstitious belief), and the leaves in infusions are a major ingredient of snake-bite medicines for taking by draught and by topical application (Boston fide 7a). In SE Senegal, the Tenda lay the leaves of *Ocimum basilicum* Sweet basil around the house to keep snakes away (13). In this country, the Nyominka use *Calotropis procera* Sodom apple as a snake-repellent - perhaps in a superstitious way (7a). In Guinea, the roots of *Acridocarpus plagiopterus,* a scandent shrub, have a magical use against snakes (10). In western Sudan, folklore indicates that if one puts a fruit of *Citrus medica* Citron 'in one´s pocket and then goes to the bush, a snake will not kill one' (3). In India, *Allium sativum* Garlic is applied around windows and doors to repel snakes; and in Ceylon, the root-stock of *Gloriosa superba* Climbing lily is applied also around windows and doors to ward off snakes (31). It has been observed (25) that snakes fear *Fraximus excelsior* Common ash tree and they never dare go close to the tree.

It is equally believed that some plants do attract snakes. For example, in Ghana children are told that snakes will bite them if they eat the fruit of *Spondias mombin* Hog plum or Ashanti plum - probably because eating excess of the fruit is said to cause 'dysentery' (17).

Children enjoy the fruits of *Passiflora glabra (foetida)* Stinking passion flower, but are equally aware of the popular belief and saying that these fruits are also relished by snakes - with the result that the plant is, as expected, always approached with all the necessary caution and respect.

THE SNAKE SECT

Some plants are looked upon as symbol or fetish of Snake sects. For example, in Ivory Coast, *Ageratum conyzoides* Billy goat weed has protective fetish properties for the followers of the snake sect against snake-bite; and *Physalis angulata*, a robust herb, has undisclosed magical significance for the followers of the Snake sect in the Man area (19). In this country, the 'Serpent sect' of Man region consider *Rauvolfia vomitoria* Swizzle-stick, a shrub or small tree, and *R. serpentina*, an introduction to the Region, fetish; and *Erythrococca anomala*, a spiny shrub, is used by the Danané Snake sect to save themselves from bites of their dangerous snakes (19). The Yacouba tribe dominates this area.

In Sierra Leone, the Mende name of *Bertiera racemosa* var. *racemosa*, a shrub or small tree in high forest, 'kali-kuli', refers to the use of the plant in a charm used by snake society (kali = snake, kuli = to turn round dazed or dizzy) (10). Over most of the soudanian region, *Ceiba pentandra* Silk cotton tree is thought to be inhabited by the divine Python, symbol of maleness (18b). Indeed the very extensive surface spreading lateral root-system suggests long snakes around the base of the tree, and for some races, the roots especially are sacred as evoking a giant serpent (18b). (For snake-worship, see RELIGION: NATURE AND DEITIES).

ASSOCIATED PLANT NAMES

Some plants are named after snakes for various reasons. For example, *Achyranthes aspera* Devil's horsewhip or Snake's tail, a common amaranth weed, probably in reference to the tapering inflorescence, is known as 'hak' orin machiji' in Hausa, meaning 'snake' tooth', and *Trichosanthes cucumerina* Snake gourd , a cucurbit cultivated for the young snake-like fruits, is called 'snek-tamatis' in Krio, meaning 'snake tomato' – the taste of the ripe fruit (7a). The Fula name of *Citrullus colocynthis* Colocynth or Wild gourd or Bitter gourd, ' koron mboddi' , means 'snake's gourd' (7a); and the Madinka name of *Trianthema portulacastrum* and *Zaleya*

(Trianthema) pentandra Horse purslane, fleshy prostrated herbs, is 'sadjamba', meaning' snake leaf; - probably because the plants harbour snakes as suggested by the Hausa name, 'gadon machiji', meaning 'snake bed'. *Merremia tridentata* ssp. *angustifolia*, a prostrate twining savanna annual, goes by the same Hausa name.

In Ghana and Ivory Coast, *Dissotis rotundifolia,* a decumbent herb in damp places, is called 'bore da so' or 'bore kete' in Akan, meaning 'puff-adder lies on it, and 'puff-adder's mat, respectively (10). In these countries, the Akan names of *Momordica foetida (cordata),* a creeping herbaceous cucurbit, and *Salacia debilis,* a lofty woody liane in forest, are 'owoduane', meaning 'snake's food' (7a), and 'homa-kyereben', meaning 'green mamba climber' (17), respectively. In Senegal, the Diola name of *Strophanthus sarmentosus,* a scandent shrub or climber with milky latex, is 'fuladafo', meaning 'snake's skin' (7a). In Liberia, snakes frequent *Vitex micrantha,* a forest under-storey tree, and eat the fruits. The Basa name 'sah-sah' implies this, but it also means a person with an evil reputation who is a trouble maker (7d).

In N Nigeria, the Hausa name of *Chrozophora senegalensis,* an under shrub, is called 'walkin machiji' meaning 'snake's apron'- probably because of the leaf shape, and *Centaurea perrottetii* and *C. praecox* Star thistle, perennial and annual composite herbs respectively, are also known in Hausa as 'kububuwar makafi' , meaning 'blind man's cobra'. In S Nigeria, the Yoruba call *Oplimenus burmannii* and *O. hirtellus,* grass in forest shade, ' ite-oka' , meaning 'python's nest' (10). In Sierra Leone, *Melinis minutiflora* Stink grass is said to be a favourite sleeping place for pythons, hence the Kono name 'minan-sa-bine' meaning 'python's lair grass' (10); or 'boa-constrictor lie-down grass' (7b) - probably suggesting that pythons frequent these grasses.

In Japan, *Acer capillipes* with its green and white snake bark is known as Red snake bark maple, and *A. rufinerve* has the descriptive English name of Grey-budded snake bark maple (27). In the United States, *Darlingtonia californica* California swamp plant, an insectivorous plant, has hoods that terminated in serpent-tongued appendages, a circumstance which explains their common name of Cobra plants (26). (For more about snakes, see BURGLARY; CHARM; RELIGION; WEALTH and WITCHCRAFT).

References

1.Ainslie, 1937; 2. Ampofo, 1983; 3. Appia, 1940; 4. Bally, 1937; 5. Berhaut, 1975; 6. Bouquet, 1969; 7a. Burkill, 1983; b. idem, 1994; c. idem, 1997; d. idem, 2000; 6. Cansdale, 1961; 9. Cousteix, 1961; 10. Dalziel, 1937; 11. De Feo, 1992; 12. Ellis, 1881; 13. Ferry & al., 1874; 14. Gyedu, (pers. comm.); 15. Hepper, 1976; 16. Hunting Tech. Survey, 1968; 17. Irvine, 1961; 18a. Kerharo & Adam, 1964; b. idem, 1974; 19. Kerharo & Bouquet, 1950; 20. Kluga, (pers. comm..); 21. Lans, 2001; 22. Mensah, (pers. comm..); 23. Moro Ibrahim, (pers. comm..); 24. Nyarko, (pers. comm..); 25. Pamplona-Roger, 2001; 26. Perry & Greenwood, 1973; 27. Richard & Kaneko, 1988; 28. Sereboe, (pers. comm..); 29. Vergiat, 1970; 30. Visser, 1975; 31. Watt & Breyer-Brandwijk, 1962.

TABOO

But of the tree of knowledge of good and evil, thou shalt not eat of it: for the day thou eatest thereof thou shalt surely die.
Genesis 2:17

In Africa, as in other countries, certain practices are normally avoided in the community, because they are looked upon as being rather contrary to the expected norms and behaviour of the society. A taboo is therefore, a pattern of behaviour that is forbidden or banned by social custom. It is a medico-religious prohibition (39b). As means of enforcing these 'rules' or 'commands' or 'customary don'ts', disastrous and frightening consequences often shrouded in mystery and superstition are propounded as a deterrent. Some of the mishaps blamed on the consequences of deliberately breaking a taboo include failure of crops, loss of property, lost at sea, lost in the forest, abduction by dwarfs, affliction with boils or with leprosy, miscarriage and death of one's kinsmen or even oneself. Breaking cultural taboos and deviating from social norms is punished (35). The faithful are believed to be rewarded accordingly (35).

On the whole, taboos are fairly well complied with (especially among rural dwellers) - more or less as a habit - and rarely deliberately broken. Traditionally, it is believed that breaking a taboo offends the local gods and ancestral spirits who formulated them, and strictly obeyed these rules themselves. Taboos permeate all walks of life and aspects of cultural patterns of a community - such as diet, marriage, clothing, religion, the way and manner of greeting, trading, farming, fishing and so on. The 'rules' could either bind the whole community or tribe, or be aimed at target groups within these people only.

GENERAL FOR ALL AND SUNDRY

In some cases a taboo is general and binding on the whole society irrespective of age, class or sex. For instance, among the Nzima-speaking tribe of Ghana, the one-month collection of coconuts, an important cash

crop in the area, alternates with a period of three-months ban on the collection. Any one seen with coconuts during this prohibition period might have stolen them. There is large scale organized stealing of coconuts in the area, and the taboo is aimed at discouraging this practice. In this country, salt winning in the Songhor Lagoon is regulated in a similar manner. A stick of the traditional fetish plant of the Ada people, *Ochna afzelii* or *O. membranacea*, savanna shrubs, is fixed by the chief fetish priest in the middle of the lagoon to indicate the seasonal ban on this activity throughout the traditional area. Until the stick is removed by the same priest, salt-winning remains suspended (15). The lagoon is an important source of quality salt in the Region.

Like the birthright tree, the flourishing of certain trees are traditionally linked with the prosperity of some tribal groups and *vice versa*. For example, the people of Liberia will abandon a village in which a planted *Pterocarpus santalinoides*, a tree legume on river banks with floating fruits, wilts or dies; but remain as long as it flourishes (Cooper fide13). In S Nigeria, *Newbouldia laevis*, a popular fetish tree, has similar uses among some tribes. (See SACRED PLANTS: OTHER TREES). Since *Elaeis guineensis* Oil palm rarely branches, branched trees are regarded as living taboo properties (21); and such palms are taboo to the palm oil maker, and treated with fetish rites (13).

Branched *Elaeis guineensis* Oil palm

Among some tribes in Ghana, it is deemed an offence to carry whole bunches of oil palm fruits through the town (but not through the bush) (2). The bunches have to be cut up into individual nuts before entering the village or town. The justification of this prohibition is not clear. In Sierra Leone, *Phyllanthus muellerianus*, a shrub, is a taboo to the Ture clan (42); and in N Nigeria, *Prosopis africana*, a savanna tree legume, is a taboo to some of the Maguzawa people (Meek fide13).

Sweeping Brooms

In N Nigeria, Hausa prejudice exists against the use of *Hyparrhenia subplumosa,* a robust perennial grass, for brooms lest the user loses his goods and dies (13). The reeds of many other grasses may, however, be used for brooms. Mythical stories surround the sweeping broom in Ghana as well. In this country, the Fanti name 'ammbodzin', literally means 'not to be called by name' - a reference to its sacredness. The mortar, the pestle and whitlow (a festering finger or toe) have similar connotations. (The reason for the link with the latter is probably because it is a dreaded and painful disease which should better not be mentioned - for there is nothing sacred about it). 'Brooms, mortars, hearth, stool are treated with a measure of reverence, being believed to possess some sacred power over life' (4). For example, among the Akans of Ghana and Ivory Coast, it is believed that washing one's face in the morning with water kept overnight in a mortar results in seeing ghosts at night.

Sweeping broom

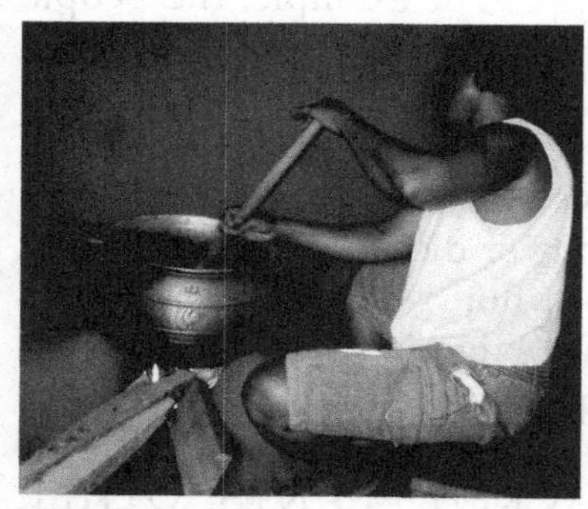
Hearth

Mortar

The deliberate throwing of sweeping brooms, especially those made with the leaflet midribs of *Elaeis guineensis* Oil palm, is a taboo. The penalty is a reciprocal throw from an unusually high altitude in a scaring nightmare. It is well known that anyone who experiences this nightmare - especially children (if they later get to know why) - do not ever throw sweeping brooms again. With many of the ethnic groups, to sweep onto someone is traditionally unacceptable (see MARRIAGE: STABILITY IN MARRIAGE). It is, likewise, a taboo among many tribal

Traditional stool

groups in the Region to hit someone deliberately with a sweeping broom or with a pestle. Such an act is traditionally considered to be highly sacrilegious. To hit someone with a stick of *Ochna afzelii* or *O. membranacea,* shrubs or small trees in savanna or forest margins, or even to point it at a person when annoyed is a taboo in the distribution area of the species, especially among the Ada people of Ghana (1). (See FETISH: FETISH PLANTS: Shrubs and Lianes).

Farming and Fishing

Among many tribes in Africa, farming and fishing activities are controlled by a string of rituals and taboos. The practice is, as a general rule, forbidden on some specific days. *Asase Yaa* ('mother earth') has a more direct influence on everyday life than *Onyame* (God) (20). (Among the Akan-, the Ewe- and the Nzima-speaking (Table 1a), and the Hausa (Table 1b), people are named - in part - after the day on which they are born. Within these tribal groups, whether literate or illiterate, people know the day on which they were born, but not necessarily the date. Females born on Thursday, for example, are called Yaa. Both the femininity and fertility of the land and its designation as a day of rest for farmers in the traditional area are

Days of the Week	Male Born		Female Born
	Traditional Name	Modernized Name	
Sunday	Kwasi, Kwesi	Yoosi	Akose, Akosuwa, Asi, Esi, Esiba
Monday	Kodwo, Kwadwo, Kwodwo	Dwoodwo	Adwoa, Adwoba, Adwowa, Adzo
Tuesday	Kobina, Komla, Kwabena	Kobi	Abena, Abla
Wednesday	Koku, Kwaku, Kweku	Abeeku, Kuuku, Yooku	Aku, Akua, Akuba, Ekuwa, Kuukuwa
Thursday	Kow, Kwao, Kwaw, Yaw	Yookow	Aba, Yaa, Yaba, Yawa
Friday	Kofi	Fiifi, Yoofi	Afi, Afiba, Afua, Efia, Efie, Efuwa
Saturday	Ato, Kwame, Kwamina	Yookwame	Ama, Ami, Amma

Table 1a. Names of Persons Born on Respective Days of the Week

Days of the Week	Male Born	Female Born
Sunday	Danlard	Ladi, Ladidi
Monday	Dantani	Lantana, Tani
Tuesday	Bature	Talata, Baturia
Wednesday	Balarabe	Balaraba, Laraba
Thursday	Danlamisi	Lami, Lamisi
Friday	Danjuma	Jummai
Saturday	Danasabe	Asabe, Asibi

Table 1b. Names of Persons Born on Respective Days of the Week

implied in the name *Asase Yaa*. Where the day of rest is Friday, 'mother earth' would be called *Asase Efuwa*). 'She demands a weekly day of rest on which farming is prohibited. Breaking this taboo can lead to serious consequences such as illness or crop failure' (20).

The taboo day for many tribal groups is either Thursday or Friday for farming; and Tuesday for sea-fishing. These days have been dedicated to the custodian of the land and the spirits of the sea, respectively. While the sea-fishing taboo day is more-or-less uniform throughout the Region, that for inland fishing varies from tribe to tribe, or according to the customary rites of the god of the river, lagoon or lake. Formerly offenders were put to death. 'For a follower of a god to violate the day sacred to that god is a serious offence among the Twi, the Ga, the Ewe and the Yoruba tribes, as to break the Sabbath was among the Jews (16b). As with the Jews, the offence is punishable with death, the notion being that if the honour of the god is not vindicated by his followers, they will suffer the neglect (16b). Nowadays, however, offenders are fined according to customary laws; but generally the fear of being abducted and spirited away by dwarfs or by *sasabonsam* or *ombuiri* (the Devil) in the forest, or enticed away by *mamewater* (mermaid) on the high seas, is sufficient to restrain any would-be offenders.

Lake Bosomtwe

Among the taboos about L Bosomtwe, a sacred lake in Ashanti in Ghana, two are concerned with fishing, and both relate to plants. Firstly, properly constructed rafts are a taboo, so logs of *Musanga cecropioides* Umbrella tree are used as rafts instead of dug-out canoes.

Secondly, it is a taboo among the fishing villages surrounding the lake (there were formerly thirty villages, but four have been submerged and not rebuilt (39a)) to use any other plants for fishing gear, like baskets and traps, except *Marantochloa* and allied species, slender rhizomatous forest herbs (39a). These include *M. congensis* Yoruba soft cane, *M. cuspidata, M. filipes* Yoruba soft cane, *M. leucantha* Yoruba soft cane, *M. mannii* Yoruba soft cane, *M. purpurea* Yoruba soft cane, and *M. ramosissima* Yoruba soft cane. The rest are *Megaphrynium macrostachyum* Yoruba soft cane, *M. distans, Sarcophrynium brachystachys* Yoruba soft cane, *S. prionogonium* Yoruba soft cane, *S. spicatum* and *Trachyphrynium braunianum*.

A controversial religious taboo surrounds the planting of the kola tree. Among some people to plant a kola tree is an impious act, and the person who commits it will die when the tree flowers. Among others, a kola tree is planted to commemorate a joyous event. Where superstition forbids the planting of kola seed, the naturally regenerated seedlings are transplanted (13). The two economic trees in the Region are *Cola acuminata* Commercial cola nut tree and *C. nitida* Bitter cola. A similar taboo surrounds *Vitellaria (Butyrospermum) paradoxa* Shea butter tree among some of the ethnic groups in its growing area - the woodland savanna. The belief is that the one who plants a Shea butter tree will die when the tree fruits. In S Sudan, the poison creeper, *benge,* used by the Azande in their poison oracle is associated with similar beliefs. 'If any one should dare to transplant the creeper his kinsmen would die'(17).

In S Africa, Zulu taboos do not permit of the fruit-pulp of *Oncoba spinosa* Snuff-box tree being extracted within the kraal vicinity in case the seed should grow there and possibly determine the death of the inmates (45). Instead, the pulp is extracted at the river so that the seed may be swept away (45). In Cameroon, it is the belief of the Ntumu tribe that if one blows across a bottle it will cause all the plants to die (41); and in N Nigeria, one may not eat what is saved of *Zea mays* Maize or Corn after the village has been destroyed by fire (43). The planting of some crops was also forbidden by tradition. For instance, *Arachis hypogaea* Groundnut has sociological and magical attributes. In parts of the Region it is not grown - for example in the Kirimi Valley of the Vogel Peak area where *Vigna (Voandzeia) subterranea* Bambara groundnut is preferred, and in some localities of Guinea-Bissau (26a). In Uvea, W Polynesia, the plucking of a single leaf of *Dioscorea alata* Water yam or Winged yam and dropping it in the swindden is believed to result in the spontaneous withering of all yam in that field (31). Water yam is associated with taboos in the Region as well - (see FOOD below).

SPECIFIC FOR TARGET GROUPS.

In other instances, only a section of the community is, by a taboo, forbidden from certain practices. That is, the taboo is specific, and aimed at a target group. Sex, for example, could be a delimiting factor. Thus a practice which is permissible among males may not necessarily be allowed among females. For instance, in Benin Republic, *Macrotyloma (Kestingiella) geocarpum* Kersting's groundnut and *Vigna (Voandzeia) subterranea* Bambara groundnut are forbidden to women, perhaps on the principle which in other tribes forbids women to eat the flesh of certain wild animals, lest they become strong-minded (12). In contrast in N Nigeria, there is no superstition inhibiting women from eating Kersting's groundnut in Nupeland (10b). In this country, kola nuts seem to have been forbidden to women in Bornu, in the period of Clapperton's first journey (13).

In Congo (Brazzaville), it is forbidden for women to touch the fruit of *Kigelia africana* Sausage tree under pain of suffering their breasts or womb falling, or if pregnant, of having a miscarriage (7) (but see FERTILITY: SYMBOLS OF FERTILITY: Sympathetic Magic). In S Nigeria, Sausage tree serves as boundary to mark territorial limits, amongst the Ijo of Kaiama, but their use is forbidden to the Yenagoa Ijo (Williamson fide10a). In Senegal, the Nyominka place a taboo on the fruit of *Ziziphus mauritiana* Jujube tree, cultivated for the edible fruits, which are reserved solely for Muslim priests (20a). In Ghana, the Ga tribe may not touch *Pennisetum purpureum* Elephant grass (19) probably because the grass has fetish implications. In this country, there are twelve family divisions among the Twi-speaking tribe. The plantain division or Abradzi-fo still abstain from *Musa sapientum* var. *paradisiaca* Plantain in the interior (44). However, in the south such an abstention is not now usual - this is due perhaps to the fact that the plantain is the principal article of food among the natives, and that a prohibition of it would entail considerable inconvenience, and in some cases hardship (16a). In Gabon, certain varieties of plantain are forbidden to some categories of persons (women, infants, uncircumcised males and persons with yaws) (44).

In Cameroon, many of the taboos which restrict or prohibit the consumption of certain foods among the Ntumu tribe are applied towards pregnant or virgin females (41). The proverb 'the herb the mother eats is the same the child eats' is a warning concerning this

(14). If such a female eats one of the tabooed foods it is said that the child will be deformed or mentally deficient. Thus when defects occur, breakage of such food taboos is blamed (14). In Ivory Coast, should a suckling baby fall ill while the mother is observing certain taboos, it is said to be cured by the mother taking a dose of the leaf-sap of *Agelaea obliqua,* a scrambling shrub in forest (8). In Nigeria, the Munshis plant *Erythrina senegalensis* Coral flower or Bead tree along with an *Euphorbia* species and other fetish plants, at the village shrine, quivers are hung on it, and this is a taboo to women (13). In the growing areas of *Vitellaria (Butyrospermum) paradoxa* Shea butter tree, on the contrary, it is a taboo for men to gather the seeds. The collection of the seeds is, therefore, done exclusively by women and girls. Another taboo specific to men is recorded. 'In Central Africa, a man is not allowed to cook when he is in the village, though he may cook on a journey. In the village, every man is dependent on some woman to cook for him' (22).

In Ghana, a woman might not carve a stool - because of the ban against menstruation (39b). A woman in this state was formerly not even allowed to approach a wood-carver while at work, on pain of death or a heavy fine (39b). There are also taboos on drums and drummers among the Ashanti tribe of Ghana. For instance, 'a drummer must on no account carry his own drums, lest he should become mad (39a). Women should not touch a drum and are not allowed to carry them; neither should a drummer teach his own son his art, but he has to engage some other drummer to do so (39a). Should a father teach his own son, it is thought the former would die as soon as the latter had become proficient (39a).

FUEL-WOOD

The use of some specific trees as fuel-wood is forbidden for a number of reasons among some tribal groups in the Region and elsewhere - either scarcity, or medicinal and other properties or uses, or fetish properties, or religious and other spiritual significance, or general superstition. This taboo, like the one that forbids specific plants from being cut (see FELLING TREES below) is, in effect, a means of conserving biodiversity for sustained yield. It is worth encouraging and support. (For other activities that threaten the survival of species, see HUNTING: Table 3). In N Nigeria, the Pabir tribe do not burn *Afzelia africana* African teak, locally called 'minta', lest their children be injured by fire (36); and in the Aku clan, it is a taboo to use the leaves or the wood of *Lophira alata* Red ironwood, a forest tree with very hard heavy timber, as fuel (36).

In Sierra Leone, there is a Mende superstition that it is permissible to cut *Anthocleista kerstingii* and *A. nobilis* Cabbage palm down, but if used as firewood, the people sitting round the fire will become sick (Deighton fide13).

In N Ghana, women do not use certain trees for fuel-wood among the Choggu tribe due to social and religious reasons (5). The trees forbidden as fuel-wood and for charcoal manufacture include *Sterculia setigera* Karaya gum tree, locally called 'pulumpun'. This tree is used for protection, thus to cut it is to cut a protector. In N Togo, the wood of this tree is avoided by people to make potash water for cooking as it is supposed to cause those who use it to swell up (13). It is perhaps a fanciful transference of the capacity of the gum to swell (10d). In N Ghana, the inhabitants around the Kpalevorgu Grove, near Tamale, are forbidden to use any wood from this grove as fuel-wood, else the whole hut will go up in flames. This grove consists of a pure stand of *Anogeissus leiocarpus,* a much preferred fuel-wood in the area. The grove is just by the roadside, and easily accessible. In this country, the Ga tribe may not burn *Premna quadrifolia* or the kernels of *Elaeis guineensis* Oil palm (19). In Gabon, *P. angolensis,* a forest tree, to some peoples is a taboo and its use in the kitchen is forbidden. It is however, permitted to throw leaves and twigs on the hearth fire in the evening as the bad smell generated is a precaution to keep away malignant spirits (44).

Sterculia setigera Karaya gum tree (forbidden to be used as fuel-wood or cut)

Local superstitious uses, as well as the medicinal uses, prevent *Stereospermum kunthianum*, a savanna tree, locally called 'zugubyetia' in Hausa, from being cut for firewood in parts of Africa - and in Sokoto, N Nigeria, the smoke is said to conduce to leprosy (13). The general belief is that you will not prosper if you use it for fuel-wood. In the Region, especially among the Ewe-speaking tribe of Ghana, Togo and Benin Republic there is a belief that trees whose fruits are inedible are not to be used for fuel-wood (3). In N Ghana, it is a taboo among the Grunshie tribe in the Navrongo District and other tribes in the Region, to touch a tree struck by lightning, much more use the wood for fuel-wood. (See LIGHTNING AND THUNDER: 'NEUTRALIZING'THE CHARGES

for the rituals performed to make such a tree safe again). However, the bark of such trees is often required as ingredient in fetish preparations and prescriptions. (See INFANTS: PHYSICAL DEVELOPMENT: Plant Prescriptions: Sympathetic Magic).

In N Nigeria the dry stem of *Annona senegalensis* var. *senegalensis* Wild custard apple and *A. senegalensis* var. *deltoides,* shrubs or small trees in savanna, are not burnt by the Fula as such action would show ungratefulness and unfaithfulness to nature. There is a strong belief that should these plants be burnt in a house there will be an outbreak of *nyau ladde* (McIntosh fide10a) - probably a dreadful disease. In Senegal, the Fula and the Fouladou never use the wood of *Combretum molle,* a savanna tree, to stove domestic hearths (28a, 29); and in Sierra Leone, the Mende say though *Anthocleista nobilis* Cabbage tree, may be felled, but if it is burnt for firewood, people sitting round the fire will become sick (10b). To some people there is an inhibition to burn *Cymbopogon schoenanthus,* a tufted perennial grass, on a fire (37). In Guinea, one does not put palm nuts, the fruits of *Elaeis guineensis* Oil palm on a fire because there is a cautious fear of being struck by lightning; nor does one throw in the air the pestle with which one pounds the nuts in the Toma country (Appia fide 10c).

In Central African Republic, *Indigofera conjugata,* an erect savanna herb, is used as a matter of expediency in the bush for firewood, but not by women during pregnancy or it may result in death at childbirth (10b). In Gabon, the wood of *Barteria nigritana* ssp. *nigritana,* a slender forest tree, is a taboo to certain races and is not permitted in the household kitchen (44). In Mozambique, Shangana medicine-men use the wood of *Annona chrysophylla* Wild custard apple for a sacred fire and the wood must not be used by ordinary folk (45). Among the Beti of S Cameroon, *Distemonanthus benthamianus* African satinwood is never burnt (32); and in Nigeria, the Yoruba and the Hausa never burn *Newbouldia laevis,* a tree popularly believed among many tribes in the Region to be fetish. (See KINGSHIP: CEREMONIAL RITUALS). (See also APPENDIX – Table V for a list of plants forbidden to be used as fuel-wood).

In some of the rural areas carrying a bundle of fuel-wood through the town (but not through the bush) is forbidden. The pieces of wood should not be tied together, but packed loosely on a pan, basket or any suitable container (2). Elders explain that the rational behind this taboo is to prevent the possible smuggling of firearms hidden within the bundle into town. There is also the possible risk of inadvertently conveying a snake or any dangerous reptile hidden in the bundle from the woods into town.

FELLING TREES.

The felling of some trees without certain religious rites, is avoided in the Region. (See also CARVINGS AND IMAGES: THE EVIDENCE). The belief is that such trees are either gods in themselves or the abode of such spirits or local deities. As mentioned earlier (see FUEL-WOOD above), the enforcement of this taboo is an effective conservation measure and worth encouraging and support. For example, among the Yoruba-speaking tribe of S Nigeria, 'the *Ashorin* tree is one which is inhabited by a spirit who, it is believed, would, if attention were not diverted, drive away anyone who attempted to fell the tree. The woodman therefore places a little palm-oil on the ground as a lure, and when the spirit leaves the tree to lick up the delicacy, proceeds to cut down its late abode' (16b).

In African mythology felling *Adansonia digitata* Baobab or *Ceiba pentandra* Silk cotton tree is regarded as a sacrilege. Similarly, *Stereospermum kunthianum*, a tree of savanna woodland, is held to have supernatural properties and this discourages felling. In many parts of W Africa, felling *Newbouldia laevis,* a tree believed to be fetish, is avoided (see KINGSHIP: CEREMONIAL RITUALS); and in Cameroon, *Distemonanthus benthamianus* African satinwood, a forest tree legume with reddish-brown bark, is never cut down among the Beti (32). In Ghana and Ivory Coast, tradition forbids the felling of *Okoubaka aubrevillei,* locally called 'odee' by the Akan tribe of both countries; or locally called 'okoubaka' by the Anyi of Ivory Coast, meaning 'tree of death'. Weird stories of felled trees that have allegedly 'resurrected' the next day are told by foresters who have felled these trees. It is also a taboo to go under an 'odee' or 'okoubaka' tree after a sexual intercourse without having had a bath - such an act being highly sacrilegious. One has to abstain from sex for at least seven days, as part of the rituals required, before venturing to go under the tree.

Milicia excelsa Iroko (symbol of refuge; legendary and mythical plant; associated with traditional religion; forbidden to be cut; National Plant of Ghana).

It is traditionally believed that *Milicia (Chlorophora) excelsa* and *M. regia* Iroko are also inhabited by a spirit, but it is not very powerful or malicious, and when a man

desires to fell such a tree it is sufficient protection for him to invoke the indwelling spirit of his own head by rubbing a little palm-oil on his forehead (16b). Normally, the trees are spared felling when clearing the forest for agriculture, but if there is need to fell, rites have often to be performed lest the tree be occupied by spirits (9). However, in Ghana, serial death has been recorded after the felling an Iroko. 'When the timber fellers arrived to fell the trees (there were three of them), the elders advised them to stop because the trees have been the protective deities of the village since its foundation, but they refused. As soon as the first tree was felled, the feller instantly collapsed and died. A few hours later, the supervisor also collapsed and died. After these two funerals, the contractor decided to convey the timber to the sawmill. The driver of the timber truck also collapsed and died. Since then, the remaining two trees still stand at Buokrom, a village near Tepa in Brong Ahafo to date (46). In a similar incident elsewhere, swarming bees, *Apis mellifera,* were used to drive away the fellers. '25 years after acquiring the piece of land, the village elders arrived with chainsaw operators to cut down the two Iroko trees because the trees belong to the stool land. After the operators have started the machines, I wore my bee-protecting suit and opened the hives. In a dramatic scene, the traditional rulers, their entourage and the two fellers fled - leaving the machines behind' (15).

In the past, owing to a strict taboo, Iroko was never cut down, nor was it allowed to be burned. Anyone who had to cut down the tree for Europeans feared to do so, since they might 'become mad or die' (14). It appears the sawyers had genuine reasons to fear. 'The *sasa* (vindictive spirit) is the invisible spiritual power of a person or animal, which disturbs the mind of the living, or works a spell or mischief upon them, so that they suffer in various ways. Persons who are always taking life have to be particularly careful to guard against *sasa* influence, and it is among them that its action is mainly seen. For example, executioners, hunters, butchers, and as a later development - among sawyers who cut down the great forest trees' (39b). A proverb, referring to the risks a man runs in cutting down trees inhabited by spirits, says, 'the axe that cuts down the tree is not afraid, but the woodman covers his head with etu' (a magic powder) (16b).

In Cameroon, there is a Yaoundé saying that if one cuts *Dialium guineense* Velvet tamarind one of his people will die during the year (Hédin fide13). However, the Aku of Sierra Leone who are of Yoruba descent say that their predecessors used to cut down Velvet tamarind in the belief that if it seeded and grew up around the village it brought fever (13). (The Ewes of Ghana, Togo and Benin Republic name girls born on Wednesday Aku (see Table 1a above); and among the Gas of Ghana, the Sempe family also name the first girl Aku). In N Nigeria, the Hausa and the Fula refrain from felling *Tamarindus indica* Indian tamarind, lest bad luck befalls the cutter (10b). In Liberia and Sierra Leone, there is superstition (not disclosed) about cutting *Piptadeniastrum africanum* African greenheart, a large forest tree (40). In S Nigeria, the tree is also left standing when the forest is cleared (11). In Ivory Coast, appropriate rites have to be observed before *Khaya senegalensis* Dry-zone mahogany is cut down lest it be occupied by spirits (9); and in SE Senegal, the Tenda carry out special sacrificial ritual at the foot of the tree for the benefit of any genies to which it might afford shelter (18). In this country, the Tenda will never cut down *Afzelia africana* African teak on cultivated land, and if the tree is very big it is thought to be the living place of spirits (18). (See APPENDIX – Table V for a list of plants that may not be cut down).

In Botswana, on no account must *Ziziphus mucronata* Buffalo thorn be cut by the African after the first summer rains have fallen, as this will cause a draught to ensue; and in S Africa, the Tlhaping believes that the cutting of *Acacia mellifera* var. *detinens* Hookthorn after the fall of the first rains results in adverse weather conditions (45). The African is warned against *Spirostachys africanus* African sandalwood or Tambootie tree from earliest youth and their fear of the living tree amounts practically to a taboo, stopping any desire to chop down or damage a living tree (45). Some Somali people will not cut *Withania somnifera* Winter cherry, a semi-woody perennial herb, lest bad luck befall (Hemming fide 10d). In Europe, bad luck was said to dog the footsteps of those who cut down *Sambucus nigra* Elder, a temperate tree, or burnt its branches (6). In S Nigeria, there is yet another tree that may not be felled. It has been observed 'to make the seats and pulpit the chief had the courage to use a magnificent tree which was regarded as principal juju of Akani Obio. The story goes that the people declared the juju would never permit it to be cut down. 'God is stronger than juju', said Chief Onoyom, and went out with a following to attack it. They did not succeed the first day, and the people were jubilant. Next morning they returned and knelt down and prayed that God would show Himself stronger than juju, and then hacking at the trunk with increased vigour, they soon brought it to earth' (34).

In Africa, the conviction that certain trees are supernatural and embody the souls of departed ancestors, and the consequent reverence paid to such trees with fear of adverse repercussions resulting from any violations dominate many cultures. It has been observed that, 'it

is the height of sacrilege to cut down bush or trees, or disturb the soil, in a spot where a local deity resides; and any such act is visited with the anger of the outraged god, who not unfrequently slays those who have violated the sanctity of the spot'(16a). For example, in Ghana and Ivory Coast, it is a taboo to intentionally cut *Spiropetalum heterophyllum*, a lofty undulating forest climber with red exudate, locally called 'homa-kyem' in Akan. The belief is certain death, if the liane is cut with ones shadow on it. The climber is held to be a god in these countries; local people will not cut it without carrying out rituals such as sacrificing fowls, *Gallus domestica*, or providing eggs (Enti fide10a). In Liberia, *Loesenera kalantha*, a tree legume, is held in superstitious regard and not allowed to be cut under severe penalty (Cooper fide13). Red sticky latex exudes from the cut phloem of many lianes in the plant family Connaraceae, Euphorbiaceae, and especially Papilionaceae (25). Other lianes with the red 'blood' exudate are, by wrong identification, included in the taboo. Table 2 lists some of these lianes:

Species	Plant Family	Remarks
Adenopodia sclerata	Mimosaceae	exudes red latex when cut
Dalbergia oblongifolia,	Papilionaceae	exudes red latex when cut
Dalbergiella welwitschii	Papilionaceae	exudes red latex when cut
Salacia miegei,	Celastraceae	exudes red latex when cut
Santaloides afzelii	Connaraceae	exudes red latex when cut
S. oliverana,	Connaraceae	exudes red latex when cut

Table 2. Forest Lianes with Blood-red Exudate

Similarly, *Dracaena draco* Dragon tree, which grows on the rocky precipices along the coast of Tenerife on the Canary Islands and Mandeira, is so called because reddish resin appear in cracks of the bark which created the legend that the sap turns to dragon's blood, with curious properties (23).

Another characteristic of 'homa-kyem' is its flat undulating habit; as such, similar lianes are for the same reasons also included in the taboo. Table 3 below lists some of the lianes with flat undulating habit.

Species	Plant Family	Remarks
Agelaea obliqua	Connaraceae	stem flat and undulating
Calycobolus africanus,	Convolvulaceae	stem flat and undulating
Combretum aphanopetalum	Combretaceae	stem flat and undulating
Dichapetalum albidum	Dichapetaliaceae	stem flat and undulating
Flabellaria paniculata	Malpighiaceae	stem flat and undulating
Hippocratea vignei	Celastraceae	stem flat and undulating
Loeseneriella (Hippocratea) africana	Celastraceae	stem flat and undulating
Neuropeltis acuminata	Convolvulaceae	stem flat and undulating
N. prevosteoides	Convolvulaceae	stem flat and undulating
Salacia alata	Celastraceae	stem flat and undulating
S. cerasifera	Celastraceae	stem flat and undulating
S. debilis	Celastraceae	stem flat and undulating
Securidaca welwitschii	Polygalaceae	stem flat and undulating

Table 3. Forest Lianes with Undulating Stem

FOOD

Like the biblical apple, folklore forbids the use of certain food plants. In S Nigeria for instance, certain food items associated with the Igbo cultural habits are taboos. These practices have both advantages and disadvantages and consequently significantly marked effects on nutritional status (38). In this country, *Cocos nucifera* Coconut palm is a taboo to some Igbos while some Yorubas say that drinking the milk makes men monkey-like (13). In Ondo Province, Coconut palm is associated with religious practices (10c). In the Region, the leaves and young shoots and flowers of *Leptadenia hastata,* a savanna twiner, is eaten as a vegetable; but in Burkina Faso its consumption by lepers is forbidden among the Kaya because of the Ydtaba fetish (29). Among the Ewe-speaking tribe of Ghana, Togo and Benin Republic, *Sorghum bicolor* Guinea corn and *Pennisetum glaucum (americanum)* Bulrush, Pearl or Spiked millet are a taboo (3).

In Liberia the inner bark of *Canarium schweinfurthii* Incense tree is pounded to rub on the skin for leprosy, and on ulcers. The patient on this treatment is under taboo not to eat catfish, species of *Clarias, Chrysichthys, Synodontis* or *Schilbe;* nor to have sexual intercourse, the bonus for observation being cure within one year (13). In Ghana, the fetish of Osu, a surburb of Accra eat *Jacquemontia ovalifolia,* a prostrate

perennial, as cabbage wherefore none of this tribe dare taste it, and the infringement of this command is regarded as a great crime (Thonning fide 26b). In this country, the Awuna name of *Lantana camara* Wild sage or Lantana, 'adelamanyi', means 'a hunter forbidden to eat'.

Yams (Dioscorea species)

In Ghana, it is a taboo to some Twi-speaking people to eat *Dioscorea alata* Water yam or Winged yam (13). *Dioscorea* species Yam is traditionally associated with several taboos due to its importance in the culture and history of the Region - especially within the Yam Zone (see CEREMONIAL OCCASIONS: FESTIVALS). For example, in S Nigeria the king and all his subjects after him can only eat new yam after the ceremonial sacrifice and sprinkling of water that constitute the most important element in the New Yam Festival (38). It is a serious offence to harvest or eat yam before this Festival; and when yams are eaten before the Festival special sacrifices have to be offered to the land or a locally designated god (38). In Ghana, a similar taboo pertains in Ashanti in connection with the Odwira (Yam) Festival. Only after ghosts, the local gods, and other non-human spiritual powers had partaken of the new crop, might the king, his chiefs and the nation eat of them (40b). It is equally a serious offence to steal yams. In ancient times, stealing of yams among the Igbos of S Nigeria was punishable by death, often accomplished not by hanging, but by dragging the culprit on the ground around the village until he was dead (38).

GROVES

Sacred groves are associated with a number of taboos. Groves may not be visited on certain days of the week. Visitors may have to remove their footwear and walk bare-footed. Taking photographs may not be permitted. Libation has to be poured to the gods and ancestral spirits, or payment for 'drink' or bottles of schnapps given before a visitor is allowed to enter. Cutting of any vegetation is strictly forbidden. It is a taboo for women in their menstruating period to enter sacred groves. The traditional belief is that deliberate offenders may bleed non-stop. One such well-known grove is the Boabeng-Fema Monkey Sanctury - the monkey grove near Nkoranza in Brong Ahafo Region of Ghana.

Fishing

Within some groves, it is a taboo to hunt or fish, so the animals are tame and abundant. The direct result of these religious sanctions, practices and beliefs is the preservation and conservation of biodiversity for sustained yield to the benefit of posterity. Another example, but less well-known is the Emuna grove near Kakumdo, Cape Coast, where the water in the pond practically boils with fish (24). Similarly, giant fish have been seen in the sacred stream near Nangodi in the north (24). Otweri, a sacred pond at Afwerase, near Aburi in Ghana, is the main source of drinking water for the inhabitants. The fear of the belief that should any of the fish be caught and eaten, the pond will dry up has preserved the single species of mud-fish, species of *Protepterus*.

The fetish of fish acts well because it gives a close season to river and lagoon fish. For instance, in Gabon, the Ajumba tribe around L Azingo believe that if the first fish that come up into the lake in the great dry season are killed, the rest of the shoal turn back, so on arrival of this vanguard they are treated most carefully, talked to with 'a sweet mouth' and given things (30). This tradition of protecting the stock is scientifically an important and effective means of conserving biodiversity for sustained yield. It is worthy of support and encouragement. In recent times, however, over-exploitation of fishing resources leading to depletion of stock and extinction of species has been of concern. Equally worse is catching fingerlings. In the Volta River at Kpong in Ghana, for example, *Sierrathrissa leonensis,* a small fresh-water fish hardly bigger than a driver-ant, *Dorylus* species, popularly dubbed 'one man thousand', is caught in large quantities despite a ban to this effect by the Fisheries Department. There is equally the wasteful and dangerous practice of using fish poison or explosives to kill whole populations of fish in a single operation.

PURIFICATION

After breaking a taboo the necessary customary rituals have to be performed to cleanse the victim and to pacify the gods and ancestral spirits. Among the people of Tekyiman in Ghana, a branch of *Costus afer* Ginger lily, a perennial forest herb, is laid up in front of the temple of Ta Kese to mitigate the evil should any one thoughtlessly break a taboo of this god (39a). To the Busummuru, a clan in Ghana, the whole plant of *Talinum triangulare* Water leaf is used in purifying ceremonies

after breaking a taboo (27). In Ivory Coast, *Boerhaavia diffusa* Hogweed, an annual herb, is an ingredient of a 'lotion of deliverance' used on convalescents to relieve them of taboos observed during their illness (29). In S Nigeria, the fronds of *Elaeis guineensis* Oil palm has the capacity to cleanse a priest or priestess who has become polluted by breaking a taboo (Williamson fide 10c). In Gabon, *Phyllanthus muellerianus*, a shrub or climber, is believed to have power to lift taboos and ritual interdiction (44).

In Ivory Coast, the Kru use *Elytraria marginata,* a small prostrate acanth herb with rosette leaves, to free a sick person from the effects of breaking a taboo - the plant is believed to be fetish (29). In Liberia, the extract of *Eulophia barteri,* a terrestrial orchid, is dripped into the eyes of a baby which cries without apparent reason, in a magical application, in a belief that the parents have broken a taboo during pregnancy (33). In SE Nigeria, the leaf of *Elaeis guineensis* Oil palm mixed with the shell of hatched eggs is used to purify the body. The leaf mixed with gin, local salt and three seeds of *Aframomum melegueta* Guinea grains or Melegueta are all put in a saucer, and either a girl who has not yet menstruated, or a woman in menopause holds a leaf in each hand in a ceremony to purify a priest or priestess who has become polluted or broken a taboo (10c). In its distribution area throughout the Region, *Portulaca oleracea* Purslane or Pigweed is used for purification. (See also JUJU OR MAGIC; and RELIGION).

References.

1. Adibuer, (pers. comm.); 2. Ampofo, (pers. comm.); 3. Amuzu, (pers. comm.); 4. Antubam, 1963; 5. Ardayfio-Schandorf, (pers. comm.); 6. Back, 1987; 7. Bouquet, 1969; 8. Bouquet & Debray, 1974; 9. Bouscayrol, 1949; 10a. Burkill, 1985; b. idem, 1995; c. idem, 1997; d. idem, 2000; 11. Burtt Davy & Hoyle, 1937; 12. Chevalier, 1913; 13. Dalziel, 1937; 14. Debrunner, 1959; 15. Doku Korle, (pers. comm.); 16a. Ellis, 1887; b. idem, 1894; 17. Evans-Pritchard, 1937; 18. Ferry & al., 1974; 19. Field, 1937; 20. Fink, 1989; 21. Gledhill, 1972; 22. Gluckman, 1963; 23. Graf, 1978; 24. Hall, 1978; 25. Hall & Lock, 1975; 26a. Hepper, 1956; b. idem, 1976; 27. Irvine, 1930; 28a. Kerharo & Adam, 1964; b. idem, 1974; 29. Kerharo & Bouquet, 1950; 30. Kingsley, 1901; 31. Kirch, 1978; 32. Laburthe-Tolra, 1981; 33. Lawler, 1984; 34. Livingstone, 1914; 35. Martin, 1995; 36. Meek, 1931; 37. Monteil, 1953; 38. Okigbo, 1980; 39a. Rattray, 1923; b. idem, 1927; 40. Savill & Fox, 1967; 41. Shepphered, 1988; 42. N.W. Thomas, 1916; 43. Tremearne, 1913; 44. Walker & Sillans, 1961; 45. Watt & Breyer-Brandwijk, 1962; 46. Wofa Yaw, (pers. comm.).

47 TALISMAN

Behold, God is my salvation;
I will trust, and not be afraid...
Isaiah 12:2b

A talisman is often a leather-covered pendant or charm or *gris-gris* worn around the neck, or as a waist band or girdle or as an arm bangle in the belief of offering specific or general service like personal protection - both physically and spiritually, or as a deity. A talisman may also be worn by the youth as a fashion or for adornment or as a craze - that is without any inherent charm whatsoever. It may also be a means of permanently carrying an identity or a valuable legacy.

The contends vary. They may consist of a line or verse from the Koran in Arabic characters, a portion of a leaf from the Bible, an animal or plant material, a bead, and so on. For instance, in Liberia, a piece of the bark of *Cassipourea firestoniana,* a forest tree, is reported to be a powerful talisman (2). In this country, *Pterocarpus santalinoides,* a tree legume usually along river banks, serves as a talisman; and for this purpose, it is planted in certain villages and also tended. So long as the tree flourishes, the villagers stay; but should it wilt or die, they abandon the place (Cooper fide 2). The talisman or sacred symbol *(shafa)* on which oats are sworn among some Nigerian tribes is composed of a bundle of leaves of *Combretum collinum* ssp. *hypopilinum,* a savanna tree or shrub, or the leaves of *C. micranthum,* a shrub or liane or small tree, (or presumably other similar species) enclosing various symbolic objects (9).

FUNCTIONS

The purpose for which a talisman is worn may be many and varied. These include success in hunting or fishing, fertility charms, fighting charms, physical protection from cutting or piercing implements or from firearms, spiritual protection from evil spirits, and the like.

Hunting and Fishing

Some hunters and fishermen do use talisman for protection or for success in the enterprise. In N Nigeria, a charm worn against the power of the hyena, *Hyaena hyaena,* was found to be composed of the root and other parts of *Kosteletzkya grantii,* an erect course malvaceous herb, pulverized and wrapped in a leather purse (2). In the Democratic Republic of the Congo (Zaire), *Aerva lanata,* a straggling amaranth herb, is a good luck talisman for hunters. In Rwanda, it is reported to safeguard the well-being of widows (6). In Gabon, *Scleria boivinii (barteri),* a scandent perennial sedge, is deemed to be a good talisman when the backs of the hands are beaten with the leaves, together with the leaves of *Staudia stipitata,* a forest tree, for catching crustacean and shell fish naked-handed (15). In this country, the bark of *Plagiostyles africana,* a shrub or forest tree, is made into talisman for trappers (or by an interesting extension of sophistry) to attract suitors (15). (For more of such plants, see LOVE).

Fertility Charms

Some women might wear a talisman as charms for fertility or fecundity. For instance, in Gabon, the invasive habit of *Ipomoea intrapilosa,* an introduced tree with large white flowers, is considered to be a good talisman for fecundity, so that pregnant women sometimes wear a liane around the waist (15). In this country the great floriferousness of *Combretum racemosum,* a scandent shrub or forest liane, is a fecundity talisman (15).

Fighting Charms

Plants enter various charms for protection against adversaries. There are also charms for the sustenance of energy and for protection in games involving physical exertion and possible injury. In N Nigeria the bast fibre of *Sida linifolia,* an erect malvaceous weed popularly believed to be fetish, when twisted, dried and worn as a waist band is used by the Hausa as charms for stability and balancing in mock fights called *kokua,* in which combatants always land on their feet like a cat, *Felis domestica,* irrespective of how they are thrown down (11). An important ritual requirement for the effectiveness of the charm is that the band must be woven by a left-handed person (10). (For plant prescriptions used for physical protection against sharp weapons, see PROTECTION: CUTTING OR PIERCING WEAPONS).

Firearms

Special garments call *batakari* or *fugu* and studded with talisman may be worn by medicine-men as insignia of office or by warriors for spiritual and physical protection. The head gear may also be studded with talisman. The bark of *Securidaca longepedunculata* Rhode's violet, a shrub or small tree in savanna vegetation, called 'kyirituo' in Akan, meaning 'against firearms', for instance, enters into the preparation of the talisman for such garments (12). A necessary condition is that the shrub must be debarked with the necessary rites (see HUNTING: THE HUNTED PROTECTED). Similarly, the bark of *Salacia debilis*, a large lofty undulating forest liane, called 'homa-kyereben' in Akan, meaning 'mamba-like vine', debarked with the necessary ceremony, is wrapped in leather as talisman against firearms (14). (For other plants believed to offer protection against firearms, see PROTECTION: FIREARMS). Commenting on the effects of talisman on Ashanti warriors, it has been observed that 'their faith in them is almost incredible. They will, when decked out with their leather-cased charms, rush recklessly into any danger, certain in their own minds of coming out safe and sound. They believe it makes them invulnerable and invincible in war, and they are supposed to arrest every evil except sickness and death' (3).

Against Evil Spirit

By far the most popular use of a talisman, on the whole, is for spiritual protection against evil forces. This practice is not exclusive to simple, rural dwellers only. For, many successful businessmen, teachers and university professors, among them young people as well, seek protection from *abosomerafo* shrines or buy *asuman* (amulets and good luck charms) and *bayie* (witchcraft) which could prevent them from achieving their individual goals in life (5). The male bulb of *Allium sativum* Garlic is believed to possess extra magical powers and is used, encased in a protective cover, as a talisman against the forces of the Devil (11). This use occurs also in C America (13). In some traditional areas the leaves of *Phyllanthus niruri* var. *amarus (amarus), Sida linifolia,* both common weeds, and *Bryophyllum pinnatum* Resurrection plant wrapped as talisman is believed to offer both physical and spiritual protection among many of the ethnic groups. The leaf of Resurrection plant may also be chewed and one kept in the pocket to supplement the talisman (10).

In Ghana and Ivory Coast, assorted items comprising a few

Bryophyllum pinnatum Resurrection plant (symbol of resurrection)

leaves of *Senna (Cassia) occidentalis* Negro coffee, a piece of stem of *Spiropetalum heterophyllum,* a lofty undulating forest liane with red 'blood' exudate believed to be fetish and locally called 'homa-kyem' in Akan; a piece of the stem of *Gouania longipetala,* another forest liane known as 'homa-biri' in Akan; the seeds of *Ricinus communis* Castor oil plant; a piece of wood cut from a tree across a path; and a bit of thunder stone - all collected with the necessary rituals and wrapped in a piece of cloth encased in a leather pendant, is prescribed by medicine-men to be worn as a talisman for spiritual protection against the forces of evil (4).

Ricinus communis Castor oil plant

Among the Hausa of N Nigeria, the Zarma of Mali and Niger, the Fula and some other tribes the crushed leaves of *Pimenta dioica* Allspice or Pimento and *P. racemosa* Bay-tree, both exotic trees, are wrapped in leather case as talisman for personal protection against adversaries (11). In Ivory Coast and Burkina Faso, *Aframomum melegueta* Guinea grains or Melegueta, believed to be a fetish plant, is used in making a number of juju, amulets and protective talisman. Normally seven seeds or a multiple of seven are used (7). In these countries, *Microglossa pyrifolia,* an erect or straggling composite shrub, is incorporated into a protective talisman against evil spirits and illness (7). In its distribution area the clawed seeds of *Martynia annua,* a foetid decorative annual herb around villages (often tended or sometimes cultivated), may be worn as a necklet - either naked (in which case the claws are trimmed) or enclosed in a leather like a talisman against evil spirits (1). The seeds are sold in medicine

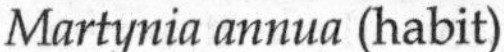
Martynia annua (habit)

Martynia annua (seeds)-1/4 natural size

markets for this and other purposes.

Fetish

In Mexico, the fruit of *Solandra* species, the god-plant of mythology which grows on rocky cliffs, are a very powerful fetish; especially, the oily or resinous ones are considered the best and are sewn into the belt or pouch by the Huichol Indians, and carried everywhere. It is very dangerous to lose a fruit of the god-plant because this would certainly anger the god greatly and cause him to do harm to an individual so careless (8). (See also EVIL).

References.

1. Abaye, (pers. comm.); 2. Dalziel, 1937; 3. Ellis, 1881; 4. Eshun, (pers. comm.); 5. Fink, 1989; 6. Hauman, 1951; 7. Kerharo & Bouquet, 1950; 8. Knab, 1977; 9. Meek, 1931; 10. Mensah, (pers. comm.); 11. Moro Ibrahim, (pers. comm.); 12. Noamesi, (pers. comm.); 13. Pitkanen & Prevost, 1970; 14. Sereboe, (pers. comm.); 15. Walker & Sillans, 1961.

48 TWINS

And when her days to be delivered were fulfilled
Behold, there were twins in her womb.
Genesis 25:24

In some cultures throughout Africa and other parts of the world, twins are believed to be special beings who require special attention and protection spiritually. For instance, in S Nigeria, the Yoruba believe twins are sacred, and a cult is made to them (19). In Benin Republic, the Ho-ho fetish (so called because the fetish is modelled in twos) protects twins, who are always named Ho-ho in the Efon language, as in Ghana and Ivory Coast they are called Ata (7a) - by the Akan-speaking tribe. (For the specific names of twins among some of the ethnic groups, see below under CEREMONIES: Birth: Names of Twins). Ceremonial rituals to protect twins spiritually are also recorded from Gabon (25), from Congo (Brazzaville) (2), from Ivory Coast (15) and from S Africa (9, 26). However, there were other cultures who believed that twin-birth was an abomination, and therefore killed the children. (See below under CEREMONIES : Taboo).

CEREMONIES

Various ceremonies and rites in which plants feature prominently characterise twin birth, anniversaries and also death. Among the Ewe-speaking tribe of Ghana, Togo and Benin Republic for example, *Uraria picta*, a perennial woody fibrous legume, traditionally believed preventive of all evil intentions, features in ceremonies connected with many twin rituals (12). Among the Ga-speaking people of Ghana, *Momordica balsamina* Balsam apple and *M. charantia* African cucumber, twining cucurbits traditionally believed to be fetish, have always been associated with twin rituals.

Spirit Possession

The Fanti, the Ga, the Ewe, the Nzima, the Yoruba and some other coastal tribes in the Region use African cucumber together with the assorted plants listed in Table 1 below, and other plant ingredients in water amid incantations and libation, in rituals to induce spirit possession in a subject during twin ceremonies, and on other occasions (18). The ingredients include thunder stone (8).

Species	Plant Family	Common Name	Remarks
Adenia rumicifolia var. miegei (lobata)	Passifloraceae	-	a tall woody climber
Hyptis suaveolens	Labiatae	Bush tea-bush	an aromatic labiate
Ocimum gratissimum	Labiatae	Fever plant	an aromatic labiate
Paullinia pinnata,	Sapindaceae	-	a climbing plant with tendrils
Phragmanthera species	Loranthaceae	Mistletoes	an epiphytic parasite
Rourea (Byrsocarpus) coccinea	Connaraceae	-	a savanna climbing shrub
Scoparia dulcis	Scrophulariaceae	Sweet broomweed	an annual weed
Securinega virosa,	Euphorbiaceae	-	a common shrub
Tapinanthus species	Loranthaceae	Mistletoe	an epiphytic parasite

Table 1. Assorted Plants Used with Momordica species to Induce Spirit Possession

Species	Plant Family	Common Name	Remarks
Adansonia digitata	Bombacaceae	Baobab	Tree
Ceiba pentandra	Bombacaceae	Silk cotton tree	Tree
Gouania longipetala	Rhamnaceae	-	Shrub or forest liane
Milicia (Chlorophora) excelsa	Moraceae	Iroko	Tree
Spiropetalum heterophyllum	Connaraceae	-	Liane, believed fetish

Table 2. Plants Assembled to Induce Spirit Possession

Another prescription assembled by medicine-men to induce spirit possession consists of pieces of the plants listed in Table 2 above. The items are all pounded together with rites in a mortar at midnight. Libation is offered using schnapps - some of this spirit being added to the pounded material. The whole preparation is placed in a brass pan, mixed with twenty one eggs laid by the local hen and then dried (23). With seven other assorted herbals added, the brass pan is carried by a chosen subject using either *Momordica balsamina* Balsam apple or *M. charantia* African cucumber, twining cucurbits, as a head pad. To

trigger spirit possession, the name of the 'medicine' is called out by the medicine-man (23).

In traditional religion, the possessed subject serves both as a medium and a channel of communicating with local gods and ancestral spirits for the purpose of revelation (see MEDICINE: INSIGHT: Revelation) and divination (see FORTUNE-TELLING AND DIVINATION) - among others. As to whether spirit possession is genuine or not, both (10) and (11) - writing over ten decades apart about the Ga-speaking tribe along the coastal savanna, and the Dormaa people in the forest interior, respectively - both in Ghana - record their personal observation of a possessed medium. Both writers were of the opinion that the spirit possession they witnessed was genuine. That does not rule out the possibility that there may be fake ones. A certain kind of communication takes place between the mediums and the spirit world, and the relationship between them and such forces is perceived as a genuine one (11). As a further support as to whether spirit possession is genuine or not, both authors give an account of an elderly woman whose face again assumed its senile expression after the end of spirit possession, whereas it had appeared very young while she had been dancing. Spirit possession seems to generate powers in mediums which make it possible for them to engage in physical and mental activities which under normal circumstances, would be difficult or dangerous (11). The source of energy required for the vigorous dancing for hours non-stop, needs an explanation.

Spirit possession may be cleansed or reversed by a wash with the leaves of *Newbouldia laevis*, a tree popularly believed to be fetish; or the subject is whipped with a broom (4); or the aerial roots of *Ficus* species are hung round the neck of the subject (18). (For fig trees with aerial roots system, see HUNTING: INVISIBILITY AND VANISHING CHARMS: Tables 1 & 2). The leaves of *Baphia nitida* Camwood, a shrub or small tree legume, or that of *Manihot esculenta* Cassava could be rubed on the head of a possessed subject to reverse spirit possession (8). As a final proof to distinguish between imposters and genuine subjects, the crushed leaves of *Ageratum conyzoides* Billy goat weed, an annual weedy composite, rubbed on the hands

Ageratum conyzoides Billy goat weed

prior to shaking hands with a possessed person reverses all genuine cases of spirit possession instantly (8). In Tanzania, the Zigua add *Achyranthes aspera* Devil's horsewhip, an annual weedy amaranth, to medicine to remove spirit possession (3a).

Birth

In an article THE 'MAGIC' YAM AND TWINS it is recorded that, 'yams make up the bulk of the diet of the Yoruba tribe of S Nigeria, and nowhere else in the world is the birth of twins so frequent'. The article continues, 'twins occur in about three per cent of births in the 18 million-strong tribe. This is double the percentage in the rest of the African population and three times that of white people.' The article concludes that, 'for those women who feel they would like to give birth to twins, perhaps they should take a leaf out of the Yoruba tribe's book, and include lots of yams in their diet'. Among some of the Ewe-speaking tribe, twins' and triplets' birth is also common; and with this tribal group, again yams are the staple food. The tribe regard twin-birth as a sign of good luck among the family, because the mother freely receives gifts - especially from market women and traders.

Twin-births are usually followed by special ceremonies to protect the babies spiritually among many tribes in Africa. In Sierra Leone, when twins are born reeds of *Costus afer* Ginger lily, a robust forest herb in the ginger family, are laid by an anthill (5). In addition, to counteract any intended evil associated with twin-birth and to bring good luck, a mat made from the leaves of either *Hypolytrum africanum* or *H. poecilolepis (heterophyllum),* or *H. heteromorphum* or *H. purpurascens,* all rhizomatous sedges usually by streams in forest, is used superstitiously in ceremonies (5). In Congo (Brazzaville), a decoction of the roots and leaves of *Strombosia zenkeri*, a forest tree, are used in a fanciful way to wash twins to make them strong (2).

In this country, the sap of *Musanga cecropioides* Umbrella tree is used to wash new-born twins as a protection from evil spirits; also the sap of *Tetracera alnifolia* and *T. potatoria* Sierra Leone water tree, scandent shrubs or lianes whose cut stem provides abundant clear drinking water, is given regularly to twins to strengthen them (2) - probably a superstitious usage. In Ivory Coast and Burkina Faso, *Blighia sapida* Akee apple is a fetish among the Gouro and reserved for twins born into Kweni families, who are washed once a week, at birth and during

the suckling period, with a bark decoction from the east and the west sides of the tree (15) - suggestive of sun-worship. In parts of traditional Africa trees are planted to commemorate twin-birth. For example, in Gabon two trees of *Ceiba pentandra* Silk cotton tree may be planted before a house where twins have been born (25). Similarly, the African peoples have sometimes endowed euphorbias with supernatural powers. In S Africa, the Xhosa tribesmen, for instance, on bearing twins, collect and plant a pair of tree euphorbias near the entrance of their hut to ward evil spirits (9).

In S Africa, after a Venda woman has given birth to twins, she and the husband have to be ceremoniously cleansed (26). A hole is tunnelled through an anthill and the leaves of *Scolopia ecklonii* var. *engleri,* a tree, in water are poured into the hole. The man then creeps into the hole followed by his wife. Each drinks some of the mixture and the remainder is thrown into the hole which is then closed up (26). In some cases a miniature hut is erected with two opposite entrances. A hole is dug in the ground in the hut and a fire made in it over which water with the leaves in it is boiled. The parents follow each other into the hut, step over the pot, leave by the other door and go home without looking back (26). In Gabon, a mother who has given birth to twins will wear on her body some of culms of *Cyperus articulatus,* a sedge with aromatic rhizomes, as protection to herself and the babies against evil influences (25).

Twin rituals

Names of Twins. Among some of the ethnic groups, there are specific names indicating whether twins are of two boys, or they are a boy and a girl, or they are of two girls. See Table 3 below:

Ethnic Group	Two Boys	Boy and Girl	Two Girls
Awuna	Edo and Dotse	Edo and Dofe	Heyi and Hetsa
Basari	Dana and Dewin	Dewin and Powin	Powin and Dana
Bulsa	Anaab and Anatie	Anaab and Anapo	Anapo and Anapobili
Dagarti	Naa and Ziama	Naa and Poziama	Ponaa and Poziama
Dagbanli	Dawuni and Danaa	Danaa and Pagnaa	Pagwuni and Pagnaa
Ewe	Ata and Kuma Atsu and Etse	Atsu and Atsufe	Atawa and Kuma Eyi and Etsa
Frafra	Awena and Awimbiri	Awena and Awimpoka	Awena and Azure
	Ayini and Atenga	Ayini and Atenga	Ayimima and Atenyema
	Yini and Ban	Yini and Zore	Zon and Bane
Fula	Hasan and Huseni	Adam and Hawa	Adama and Hawa
Ga	Oko and Akuete	Oko and Akwele	Akwele and Akuoko
Ga Adangbe	Ate and Lawei	Lawei and Ata	Akwele and Aako
Grunshie	Bekeri and Yuyua	Bisankana and Dibakere	Bisankan and Sile
Hausa	Fuseni and Alhasan	Fuseni and Hasana	Hasana and Fusena
Kabre	-	Kwatsa and Naka	Naka and Afia
Kante (Lama)	Ayepa and Anyaror	Aluanam and Alaju	Kpenda and Tora
Kasem	Atia and Ane	Katia and Amo	Katia and Kamo
Kasena	Awe and Atega	Awe and Katega	Katega and Kawe
Konkomba	Danaa and Dawuni	Danaa and Powun	Ponaa and Powun
Kotokoli	Fuseni and Alhasan	Alhasan and Fusena	Fusena and Asana
Krobo	Tawia and Lawei	Ata and Lawei	Ata and Atayo
Nzima	Ndakpen and Ndakyea	Ndanrena and Ndanale	Ndakpen and Ndakyea
Sissala	Baka and Bakawie	Baka and Abaka	Abaka and Abakawie
Wangara	Lasena and Lonsani	Lonsani and Sanata	Sanata and Afusatu
Zarma	Bajo and Alhasan	Bajo and Husena	Husena and Hasana

Table 3. Twin Names of Some Ethnic Groups in the Region

The tribal groups above are just a sample of the 200 plus covered in the text. Twin names in and around the Region is not limited to these 23 only.

Taboo

Twin rituals do not occur among the Ashanti of Ghana probably because like the Igbo, some of the Niger Delta tribes of S Nigeria and some tribes in Kenya twin-birth, was in the olden days - before the white man came - a bad omen, an abomination and a taboo. Thus the babies were killed, while the mother was banished to live alone in the forest. Where the birth occurred away from home - say on the farm, one of the infants was usually abandoned by the mother or buried with the after birth to simulate single birth. 'A woman who gave birth to twins was regarded with horror. The belief was that the father of the infants was an evil spirit, and that the mother had been guilty of a great sin; one at least of the children was believed to be a monster, and as they were never seen by outsiders or allowed to live, no one could disprove the fact. They were seized, their backs were broken, and they were crushed into a calabash or a water-pot and taken out and thrown into the bush, to be eaten by insects and wild beasts. Sometimes they would be placed alive into the pots. The mother was driven outside the society and compelled to live in the bush' (17).

For the same reasons, the Nzimas of Ghana killed the tenth born in former days. Until very recently it was customary in parts of Ahanta traditional area in Ghana for the tenth child born of the same mother to be buried alive (7b). In Sierra Leone, the fruits of *Maesobotrya barteri* var. *barteri* and var. *sparsiflora,* under-storey trees, are not given to twins for whom they are considered unlucky (22). In this country, perhaps also superstitiously, but the significance is not explained, the plant is used to make a cage in dwellings for twins (3b). In N Ghana, when a woman delivers twins - especially the first born - she is divorced by the husband and sacked from the household. The children may even be killed. On the twin-birth taboo, it has been observed that twins are not killed in Ashanti (with the single exception of those born into the royal family - the reason for this, it is alleged, being that such an event is 'hateful' to the Golden Stool) (21). Among the Dormaa people, neighbours to the north of the Ashantis, twins were of royal importance (11). 'Twins are of special importance in Dormaa belief. In former times they were put into a bronze pan shortly after their birth and presented to the king. Male twins are expected to become elephant-tail carriers, whereas female twins were regarded as potential wives of the king' (11).

Anniversary

Ceremonies in connection with twin-birth are repeated annually throughout the life of the twins. The occasions consist mainly of sacrificial offerings and libation to the gods and ancestral spirits for their protection, followed by a ceremonial meal usually of mashed yam with eggs (and sometimes with the meat of sheep, *Ovis* species; or goat, *Capra* species; in wealthy homes) consumed by the twins with the family and friends.

The black seeds of *Operculina macrocarpa (Merremia alata)*, known as 'ebia' or 'ayibiribi' in Akan, a stout climber with winged hollow stem, sometimes with the false fruits of *Coix lacryma-jobi* Job's tears, a perennial grass by river banks and moist places, are strung as bracelets called *abam* and worn on the right hand at such anniversaries by twins among the Fantis, the Gas, the Nzimas, the Ewes, some of the Akans and some of the other coastal tribes in the Region. Christians wear the *abam* on the leg under the cover of clothing during these anniversaries. The practice occurs among the Dormaa people as well - a forest tribe to the north (11). 'Friday is the weekday dedicated to twins. Once every year before the beginning of the new harvest, the *abam* ceremony, the ritual of twins, is held on a Friday. A ritual bath in gold and white clay dust forms part of the *abam.* Afterwards the twins are adorned with wristlets of red and yellow beads and cowry-shells' (11).

Besides the twins, *abam* is also worn by the mother, the father and the *tawia* (the one born immediately after the twins). A fetish priest can converse with the spirit of the *abam*, and by a trick of ventriloquism, which makes it appear that the words proceed from the *abam*, the spirit will tell him the future of the twins (14). This trick of ventriloquism is elaborated as further proof of a pre-Columbus contact with the New World by Malians (24). 'In both Mexico and Mali, the gourd rattle becomes a sort of ventriloquist's dummy for the voice of the god. This gourd rattle is the chief instrument of both the West African and American 'fetish-man'. In this gourd resides the speaking divinity or Devil' (24). (See also LIGHTNING AND THUNDER: RAINFALL: Rituals To Induce Rainfall: The Rain Dance).

In non-twin families, *abam* with gold (probably to drive away evil spirits) and blue or red beads, is worn by the *mensa* (third born),the *esuon* (seventh born), the *nkroma* (ninth born), the *badu* (tenth born) and the *odiko* (eleventh born). Among the Ga tribe of Ghana, the seventh born (like twins) is considered divine, and after the seventh all the odd numbers are venerated (10). Another instance when an *abam* is

worn is after washing a girl at puberty and presenting food offerings to the local gods (7b). 'After the washing, a bracelet, consisting of one white bead, one black, and one gold, treaded on white cord, is put on the girl's wrist. These three beads in conjunction are termed *abam,* and their being taken into use is a sign to the *sasa* that its protecting care is no longer required' (7b).

Death

Like twin-birth, the death of a twin is similarly associated with ceremonial rituals among some tribal groups in the Region both to protect the living partner and to ensure a peaceful rest for the departed - and also for a replacement of the lost child. Among some tribes in the Region when one of a pair of twin dies, the fruit of *Citrus aurantiifolia* Lime is cut into two halves - one half being placed in the coffin in a ritual to separate the soul of the dead twin from that of the living (6). With the Ga tribe, some of the plants listed above under **Spirit Possession**, but without *Momordica balsamina* Balsam apple or *M. charantia* African cucumber enter rituals for the spiritual separation of twins in the event of such casualty (4).

In addition, a wooden doll is used by the mother as representative of the dead one, and in the event of disease or any mishap, the leafy shoots of *Newbouldia laevis*, a medium-sized forest tree believed to be fetish, together with the doll, are used in ritual invocation by a fetish priest to foretell the future of the living twin (1). In S Nigeria, the practice occurs among the Yoruba tribe. 'If a twin dies the mother may carry about a wooden image representing the dead one, or even two images if both die; and offerings are made to ensure due birth of another child after the twins - for if this is not done some strange influence may get to work' (19). The wooden doll, perhaps in effect, serves the same function as the fertility doll (see FERTILITY:

Fertility dolls *(Akuaba)*

FERTILITY IN HUMANS: Fertility Dolls). In Ghana, the fertility doll, in fact, serves another purpose - spiritual protection - among the Twi-speaking people; 'for I found out that the image was not a doll at all but an image of the child's dead twin which was being kept near it as a habitation for the deceased twin's soul, so that it might not have to wonder about, and feeling lonely, call its companion after it' (16). In S Nigeria, the Yoruba includes a lot of beans in the diet of twins to encourage them to stay on earth because the tribe believes the practice prevents early infant death.

ASSOCIATED PLANTS

Some plants are closely associated with twin rituals - *Uraria picta*, a savanna legume, among the Ewe-speaking tribe, because of how two leaves adhere closely when the top surfaces are placed against each other; and *Momordica balsamina* Balsam apple or *M. charantia* African cucumber, twining cucurbits, among the Ga tribe. African cucumber features prominently in many ceremonies in general, as well as in twin rituals. *Drosera binata* an insectivorous plant in the Sundew family is known as Twin-leaved sundew (13); and *Jeffersonia diphylla* Twin-leaf, a plant in the Barbery family and native of Manohuria in S America, is aptly named because each leaf is neatly cleft down the centre so that there appear to be two leaves on each leaf stalk (20). Similarly, the National Flower of Sweden is *Linnaea borealis* Twin flower, a herbaceous annual named after Carolus Linnaeus (1707-1778), the father of modern plant taxonomy.

Uraria picta

References.

1. Avumatsodo, (pers. comm.); 2. Bouquet, 1969; 3a. Burkill, 1985; b. idem, 1994; 4. Cofie, (pers. comm.); 5. Dalziel, 1937; 6. Dogbey, (pers. comm.); 7a. Ellis, 1883; b. idem, 1887; 8. Eshun, (pers. comm.); 9. Everard & Morley, 1970; 10. Field, 1937; 11. Fink, 1989; 12. Gamadi, (pers. comm.); 13. Graf, 1986; 14. Heine, 1961; 15. Kerharo & Bouquet, 1950; 16. Kingsley, 1897; 17. Livingstone, 1914; 18. Mensah, (pers. comm.); 19. Parrinder, 1953; 20. Perry & Greenwood, 1973; 21. Rattray, 1927; 22. Savill & Fox, 1967; 23. Teiko Tackie, (pers. comm.); 24. Van Sertima, 1977; 25. Walker & Sillans, 1961; 26. Watt & Breyer-Brandwijk, 1962.

49 WALK EARLY, ENABLE CHILDREN TO

Step by step, O lead me onward, Upward into youth;
Wiser, stronger, still becoming In Thy truth.
Walter J. Mathams, 1853-1931.

Healthy children normally start to take their first steps when they are between nine to fifteen months old - on the average. Children who are unable to walk within this period, for some reason, are a concern and a problem to the nursing mother, because they need extra care and attention. In traditional medicine, some plants are used in prescriptions to assist or induce weekly children to walk - some probably based on superstition.

INDUCEMENT PLANTS

In Congo (Brazzaville), a length of stem of *Ipomoea involucrata,* a climbing twiner, is sometimes tied around a baby's loins to promote walking; or in rachitic or retarded children, the legs are rubbed with the leaf-sap of *Aneilema beninense,* a robust straggling commelinaceous herb, in palm-oil to help them to walk (2). In this country, a piece of liane of *Stephania laetificata,* a forest liane, is woven as a bracelet or a girdle worn by infants or tied as an anklet to infants to assist them to walk (2); and the stem and root-bark decoction of *Uapaca heudelotii* and *U. paludosa,* trees of riparian forest and swamp forest respectively, are used to massage the legs of children late in learning to walk (2). In Gabon, the natives also wash their children with an infusion of *Aneilema* species if they are backward in walking (3). In this country, Pygmy tribes pass the leaves of *Asplenium africanum,* an epiphytic fern, over fire to soften them for tying to the legs of suckling children to assist them to learn to walk (10).

In Ivory Coast, a decoction of the leaves of *Dissotis rotundifolia,* a herb, along with others, is used in a superstitious washing of young Akye child to hasten its beginning to walk (1). In Ghana, the limbs and joints of backward children are either rubbed with the leaves of *Launaea taraxacifolia* Wild lettuce (Bunting fide 5, 6); or rubbed with the leaves

of *Gongronema latifolium,* a climbing shrub, to induce or to help them to walk (Irvine fide 4). Among the Hausa of N Nigeria, the Zarma of Mali and Niger, the Fula and some other tribal groups in the Region, the whole plant of *Oldenlandia corymbosa,* a straggling herb, is used to massage the limbs of retarded children (on three consecutive days in boys and four in girls) to promote walking (7). In Bangui, Central African Republic, the leaves of *Indigofera simplicifolia,* a legume, are rubbed on the feet and behind the knees of an infant slow in learning to walk, or has difficulty in walking (9).

In N Nigeria, *Sphaeranthus senegalensis*, an annual composite herb, is an ingredient of a concoction of herbs which is given to young infants to give them strength and to make them walk quickly; and a decoction of the roots or leaf of *Stylosanthes erecta,* a herb legume, is used as a daily bath for an infant to enable it to walk early (5). In S Nigeria, *Canna indica* Indian shot, a perennial robust herb, enters into Yoruba invocation to help little children to stand; and an indeterminate plant locally called 'orum opewe' in Yoruba, meaning 'leaves do not grow in heaven' is invoked to help a little child to walk (8). In Ghana, *Thonningia sanguinea* Crown of the earth, a root parasite, is tied round the anklet of small children to teach them to walk, the pointed bracts preventing them from sitting down in comfort (Vigne fide 5). This is probably a superstitious use. Alternatively, a macerate of the whole plant is used in enemas.

Thonningia sanguinea Crown of the earth 1/2 natural size

Sympathetic Magic.

In traditional medicinal practice, instances of sympathetic magic to induce walking in retarded or backward children do occur. For example, in Sierra Leone, a paste of *Mikania cordata* var. *cordata* or var. *chevalieri* Climbing hemp-weed, scrambling composite herbs, with the fat of a goat, *Capra* species, is rubbed on the feet of infants slow to walk. It must be goat-fat and no other as the goat is an active animal (Boboh fide 4). In S Nigeria, the Efik tribe make a medicine of the fruits of *Cnestis ferruginea,* a shrub with conspicuous red fruits, and allied species such as *Agelaea, Rourea (Byrsocarpus)* and *Connarus*, which is given to weakly children to encourage them to walk (5). This is an instance of sympathetic magic to impart the agility of the dog, *Canis* species, since both the Efik and Igbo names 'utin-ebua' or 'usiere-ebua' and 'amonketa', 'okpunketa' or 'okpu iche nketa', respectively, have reference to the dog (5). In this country, the leaf of an indeterminate plant called 'erin' in Yoruba, meaning 'power to walk' is invoked to help a little child to walk (8).

References.

1. Adjanohoun & Aké Assi, 1972; 2. Bouquet, 1969; 3. Bowdich. 1819; 4. Burkill, 1985; 5. Dalziel, 1937; 6. Irvine, 1930; 7. Moro Ibrahim, (pers. comm.); 8. Verger, 1967; 9. Vergiat, 1970; 10. Walker & Sillans, 1961.

50 WAR

When comes the promised time
That war shall be no more-
Oppression, lust, and crime
Shall flee thy face before ?
Lewis Hensley, 1824-1905

Waring factions consulted the oracle in early times before going to war. In traditional Africa a fetish-, a medicine- or a juju-man or a possessed person served as medium through which the guardian and protective spirits spoke - to forecast victory over the enemy or defeat. During war, and in times of disturbance, the fetish is always consulted by the priests to ascertain the most suitable occasion for action. If the augury be unfavourable, even the strongest positions are abandoned without an effort being made to hold them (9a).

Plants were similarly used to forecast events . For instance, the bamboo-like stems of *Hypselodelphys poggeana* and *H. violacea,* marantaceous forest perennials, previously tied up in bundles, were trodden upon by the marching army in a ceremony to forecast the results of the impending war. Should the binding strings break up during the march, it was believed to be a sure sign of victory. In traditional medicine, herbal prescriptions collected and prepared with the necessary ingredients and rites, entered into charms for strength and bravery in war. Certain plants were also associated with war. Other plants served as emblem of victory, or signified a truce in war or surrender and settlement of peace terms; while yet other plants were associated with the war dead.

DEITIES

It is an open secret that chiefs and warrior chiefs spent several months making a strong fetish prior to any battle - the power of the fetish and the sacrificial offerings and time devoted to it, depending on the size and strength of the enemy. Among the Dormaa people of Ghana, *Ko-asuman* are war medicines (13). Their role is to protect warriors from being wounded in battle. These war medicines are in the possession of the king (Omanhene), his royals and the chiefs and are inherited

matrilineally (13). 'Gyabum' is the most famous and oldest *Ko-suman* among the people of Dormaa, and has knowledge of herb essence which protect warriors from being wounded. Another war god is 'Bentim' (the god that intercepts gun bullets) (13). ' 'Bentim' once led the Dormaa army in a war against the neighbouring state of Berekum in which the Dormaa gained victory. According to oral history, 'Bentim' protected the Dormaa warriors with a medicine against gun bullets and this made them invulnerable.' (13).

In Benin Republic, the Bo fetish is the special guardian of soldiers and preserves them from injury (9b). Among the Ga-speaking tribe of Ghana, the following are war-gods : 'Aflim', 'Dade', 'Dinkra', 'La Kpa', 'Obobonte' and 'Ogbame' (12). Among the Akropong Akwapim people of this country, 'Gyamfi' is a war *obosom,* while 'Amanfo' and 'Adade' are *asuman* for war (14). In this country, the Yendi chief brought the god 'Buruku' to help him in war, and now the Dagomba worship it (14). Other war gods in this country are 'Beyah', 'Bobowissi' and 'Fuan-fuan-fo' (9c). In S Nigeria, the war-god among the Yoruba is 'Ogun' (9d). Among the ritual uses of *Zea mays* Maize or Corn among the Aztecs, Temple virgins decked their heads with garlands of pop-corn for celebrations honouring the god of war (24).

COURAGE AND SUCCESS CHARMS

Among the Akan-speaking people of Ghana and Ivory Coast, it has been observed that, 'In time of war, before taking the field, each company assembles at its *ehsudu,* and the following ceremony is performed. A large bowl or brass basin, filled with water, is set down near the tree, and a priest, after bruising the leaves of *Baphia nitida* Camwood, a tree legume, locally called 'ardwin-haban', puts them into water, squeezes the sap out of them, and sprinkles the assembled men with the macerate. The operation is intended to render the warriors invulnerable, and, while performing it, the priest invokes the tutelary deity, calling upon him to protect his children who are going into battle. The sacrifice of a black fowl, *Gallus domestica,* then terminates the proceedings' (9c).

In Kenya, the Masai warrior is fond of drinking water in which the crushed bark of *Pappea capensis* Indaba tree or *P. capensis* var. *radlkoferi* Sand olive has been soaked, the blood red product being thought to produce a gain in courage; and in S Africa, the Southern Sotho use the ash of *Phygelius capensis*, a herb, or *Crassula rubicunda,* locally called 'feko', as one of the ingredients in the war horn, as a charm 'to give courage during war' (36). In Mozambique, a watery extract of the root of *Cyperus*

sexangularis, a sedge, is used by the Shangana to sprinkle over men about to go into battle (36). In the southern part of Senegal, *Combretum molle*, a tree, is ascribed with magical properties by the Fula and the Fouladou to promote courage in battle by inhaling smoke emitted from a fire on which bark and branches have been placed (19).

In Liberia, *Distemonanthus benthamianus* African satinwood, a forest tree with reddish bark, is imbued with a sense of invulnerability reflected in many medico-magical prescriptions (17). In Gabon, warriors dust themselves with the powdered bark to ensure that arrows and bullets striking the body glance off without penetrating; and soldiers cover themselves with *Cissus aralioides*, a forest liane, as a fetish shield against poisoned arrows and bullets (34). In S Nigeria, the stinging habit of *Tragia* species, slender twiners armed with stinging irritating hairs, appears to engender superstitious ideas towards them. For instance *T. tenuifolia* has been used by the Igbo to prepare a magical wash for use before going to war (4a, 7). In this country, an unidentified plant locally called 'ojiji orota' in Yoruba, meaning 'to wake on a stone', is invoked for strength. Anyone so doing is hard and healthy and can therefore beat an enemy wherever he goes (33a). In W Africa, the root-fibres of *Alchornea floribunda*, a small forest tree, macerated for some days in palm- or banana-wine produce a liquid drink to stimulate energy or excitement and was formerly taken when going to war (18).

In Europe, the leaves and flowers of *Borago officinalis* Borage, a garden herb from the Mediterranean region, is said to inspire courage (27). *Thymus vulgaris* Thyme may have come from the Greek word *thumus*, meaning courage (3); therefore, in Medieval times, it was regarded as a plant that could impart courage and vigour, and women often embroidered a sprig of Thyme on gifts for their favourite knight (3). Because Thyme was slightly intoxicating it came to be regarded as a symbol of courage and bravery (1). In Europe, *Solidago virgaurea* Solidago was once carried by soldiers as

Thymus vulgaris Thyme (symbol of courage and bravery)

they went into battle (1); and in Egypt, *Allium sativum* Garlic was used by soldiers as a way to increase their courage during battle (21). Garlic was rationed to the soldiers of the mighty Roman armies as well as their slaves and labourers (28). In Japan, soldiers were supplied with *Paullinia curana* Guarana to chew in order to keep up their stamina and aid in keeping courage (28). Similarly, in W Africa, a decoction of the whole plant of *Paullinia pinnata,* a woody or sub-woody twiner with tendrils, collected with traditional rites is used by fetish priests, medicine- and juju-men or other practitioners in its distribution area in ceremonious bath as charms to impart strength and courage in fighting. (5). Prior to World War II Germany had done research work with Yage (botanical identity unknown) and developed a formula for energy and stamina for military purposes which enabled their soldiers to make forced marches without food and rest for 48 hours periods; and the Indians used the root of *Sanguinaria canadensis* Bloodroot, which has a juice that is the colour of blood, for war and for the colouring of quills and grasses (28).

In Mozambique, the Tonga employ *Acridocarpus natalitius*, a climbing shrub, as a 'war' medicine (36). In S Africa, Zulu 'army doctors' in former times sprinkled each man with a switch made from the tail of a wildebeest, species of either *Connochaetes gnou* or *C. taurinus* (36). In addition, Zulu 'army doctors' applied a medicine made from the root of *A. natalitius* to their weapons - procedures which were supposed to protect warriors from danger (36). In this country, *Pelargonium pulveratum,* a herb, locally known as 'ikubalo likamlanjani', has been used as a 'war' medicine in a variety of ways by the Xhosa during the 1850-53 war - the root pointed towards the enemy was supposed to ward off their bullets and wet the powder in their guns. Also a piece of the root tied to the neck or the chewing of the root was thought to have the same effect (36). Other 'war' medicines used in this part of the world (36) include the plants listed in Table 1 below:

Species	Plant Family	Common name	Remarks
Annona chrysophylla	Annonaceae	Wild custard apple,	a shrub, by the Tonga of Mozambique
Cannabis sativa	Cannabinaceae	Indian Hemp or Marijuana	a shrub, by the African in Zimbabwe
Phytolacca dodecandra	Phytolaccaceae	Endod	a liane or climbing shrub
P. heptandra	Phytolaccaceae	Umbra tree	a liane or climbing shrub

Table 1. Some 'War' Medicines in Southern Africa

In W Africa, *Crotalaria pallida (falcata)*, a shrub legume, is much in repute for superstitious purposes, and for such uses the pale-leaved forms of this or other species - called 'fara' (white) 'birana' - are preferred. The plant forms an ingredient in a decoction used as wash by warriors before going to fight (7). In Nigeria, *Polycarpaea linearifolia* and *P. corymbosa*, small erect herbs, are used superstitiously as a charm to ensure success in war, hence the Hausa name 'mai-nasara' meaning 'luck bringer'; and the seeds of *Afzelia africana* African teak, a savanna and fringing forest tree legume, are sold in Hausa markets under the name of 'fasa daga' and 'fasa maza' (terms meaning 'scattering of men in war'), and used as a charm in fighting (7). In Gabon (34), and Senegal (2), a draught of the bark decoction of *Pentaclethra macrophylla* Oil bean tree gives warriors courage in battle. In S Nigeria, the Yoruba believe that *Entandrophragma candollei*, a timber tree, is able to cause the death of any nearby trees should that tree's roots grow to touch it; thus the tribe invoke the tree in an incantation 'to be victorious over an enemy' (Verger fide 4c, 33). The tribe also invoke the nut of *Cola acuminata* Commercial cola nut tree and that of *Cola nitida* Bitter cola in an Odu incantation to enable one to wage a successful fight (33). (In African tradition, kola-nuts are used both for and against war. See PEACE below). The Yoruba also invoke *Elytraria marginata*, a rosette annual acanth, and *Saccharum officinarum* Sugarcane in an Odu incantation calling for calm and to kill an enemy; and for the great satisfaction of being victorious over an enemy, respectively (33).

In Kenya, the Masai warrior often carries a bunch of the leaves of *Clausena anisata* Mosquito plant when he goes to a dance or to battle; and in S Africa, the Xhosa use *Capparis citrifolia* Cape caper as a war charm to render the warrior invisible or to enable him to escape detection (36). In this country, the Thembu, in the event of an impending fight, drinks a decoction of the twig of *Acokanthera venenata* Poison tree with the intention of causing his opponent to swell up so that he cannot see (36). In this country, the Southern Sotho use *Androcymbium melanthioides*, a lily, as a charm to stop the enemy in war; and the Swati use the root of *Schrebera saundersiae*, locally called 'sehlulamanye', as a war charm (36). In Botswana, *Urginea capitata*, a lily, is similarly used by the Sotho as a charm 'to inflict harm on enemies' (36). In Tanzania, a drink of the chopped-up leaves of *Mimosa pigra*, a scrambling sensitive shrub often on river banks and in fresh-water swamps, confers invisibility in war (4b).

ASSOCIATED PLANTS

Some plant names, or uses, or legends, myths and beliefs, associate them with war among many cultures worldwide. In W Africa, for example, the Ewe name of *Mimosa pudica* Sensitive plant, 'avadzo', meaning 'war is coming' (4b); and the Yoruba name of *Tribulus terrestris* Devil's thorn, a prostrate annual herb, 'da ogun duro' or 'daguro', meaning 'causing war to stop' (4d) - have opposing references to war. However, the reasons for both terms are not clear. There is also an instance when a plant is used to symbolize war. Writing about the Bubi tribe of Bioko (Fernando Po), it has been observed that when on the war-path an armlet of twisted grass (botanical identity indeterminate - probably any grass would do) is always worn by the men (20). The ancient Europeans called *Achillea millefolium* Yarrow, an annual herb in the composite family, *Herba Militaris,* the military herb - an ointment made from it was used as a vulnerary drug on battle wounds (3). In the eighteenth century, *Hypericum pertoratum* St. Johnswort, a shrub in the Guttiferae family, was known as the 'military plant' because it was valued by soldiers (26).

Lore

There is a folklore in Ghana which says that when the *Mussaenda* with red 'flag' – *M. erythrophylla* Ashanti blood, a climbing shrub with pink or red enlarged sepal (and probably also *M. elegans* and *M. nivea*) - disappears and only the yellowish-white flowered ones remain, there will be no more war in Ashanti. The light-coloured ones are *M. afzelii, M. arcuata, M. chippii, M. grandiflora* and *M. isertiana.* The rest are *M. landolphioides, M. linderi, M. tenuiflora* and *M. tristimatica.* The Ashantis (then a kingdom or an empire) were a war-like people, and fought many battles between 1550 and 1901 (35). There is a belief in the then Gold Coast, (now Ghana) that seven valiant giants, impervious to bullets were reported to have cut down the Ashantis in dozens with the fire from their enormous guns during the battle at Nkatamanso in 1826

Mussaenda erythrophylla Ashanti blood (legendary and mythical plant; symbol of war)

- also known as the battle of Dodowa. As the Ashanti's fled for their lives, the giants could no longer be seen; but the next day the Gas saw near the battle-field seven trees of *Diospyros mespiliformis* West African ebony tree where none had been before (16). The tree is believed to be a symbol of luck to some tribes in W Africa - especially orchids and Mistletoe growing on it. In The Gambia, there is a fanciful belief that wrestlers should not eat *Cucurbita pepo* Pumpkin, or they will be thrown down and defeated (31).

In S Africa, the Tlhaping believe that a person taking refuge in *Acacia giraffae* Giraffe thorn during war-time, will be protected from his enemies and from wild animals; and in former days a decoction of the tuber of *Bowiea volubilis*, a lily, known as 'igibiszilla' in Zulu, was used by the witch-doctor in the Transvaal to sprinkle on *impis.* This is supposed to cause the enemy to flee before them (36). In S Nigeria, *Termitomyces robustus,* an agaric mushroom, is revered by people in Ekiti, in the Western State, as it is held to have saved them from capture by the Yoruba people in an international warfare. The pathway by which the Yoruba thought the Ekiti had escaped had much mushrooms growing on it which had not been trampled upon. It was presumed therefore that the Ekiti had gone another way (30).

The lore about the use of *Onopordum acanthium* Scotch thistle, a prickly composite with striking sculptured leaves, as the national emblem of Scotland relates a parellel incident following an invasion of the Danes. In the early days of Scottish history armies only marched during the day (27). The Danes thinking to catch the enemy unaware, moved at night; but in order that they might not be heard, they went barefooted. Unfortunately, one warrior stepping on a particularly fine specimen of Thistle cried out in pain, thus arousing the Scots who drove them off with great slaughter (27). Similarly, *Rosa* species Rose has featured prominently in English heraldry; and since 1461 has been the emblem of Britain following the war of Roses. These wars were responsible for many other heraldic signs, and from these we got the 'Red Rose', the 'White Rose' and the 'Rose and Crown' (22). In Europe, *Ruta graveolens* Common rue or Herb of grace is also one of the few plants to feature in heraldry and is represented in the Collar of the British Order of the Thistle.

At the time of the Trojan War, in Ancient Greece, the hero Achilles, symbol of strength and virility, was injured with a poisoned arrow which his enemy Paris had shot at him (26). The wound was deep, and bled constantly. According to the myth, the goddess Aphrodite washed the wounded ankle of the famous warrior with *Achillea millefolium* Yarrow, a composite, which after this fact was named *achillea*, in honour of the hero (26).

VICTORY

In traditional medicinal practice in Africa and elsewhere, some plants enter into prescriptions to achieve victory in war. Some other plants signify victory in war, while others serve to decorate victorious warriors.

In S Nigeria, the Yoruba name of *Combretum constrictum*, a low shrub in damp places, 'ogan ibule' meaning finger lies on the ground, features in an Odu incantation to achieve victory over an enemy (33). In this country, 'eton', 'akakanikoko' and 'orikan jajata', three unidentified plants, meaning 'buffalo', *Syncerus caffer;* 'we twist with strength' and 'head fights more than pepper' respectively, are all invoked by the Yoruba to be victorious over an enemy (33). In Botswana, warriors formerly dipped their weapons into a decoction of *Asparagus rivalis* a climbing lily, locally called 'leungeli', as a charm to ensure victory (36); and in Lesotho, a decoction of *A. africanus*, a climbing lily with numerous wiry spiny branches, into which spears are plunged ensures victory in war (15). In S Africa, the Zulu in former times, garlanded the returning warrior with twigs of *Salix mucrinata* Cape willow and, if he returned with proof of having slaughtered any of the enemy, he wore a bracelet of small pieces of the wood. (36).

Laurus nobilis Sweet bay or Laurel (legendary and mythical plant; biblical plant; symbol of victory)

Similarly, in the palmy days of Rome, *Laurus nobilis* Sweet bay or Laurel, a S European shrub, was considered the emblem of victory - dispatches announcing victories were wrapped between Bay leaves, victorious generals were crowned with it, and in triumphal procession every private soldier carried a sprig in his hand (27). In ancient times, victorious armies after battle

would select a convenient tree or set up a tall trophy pole known as the *tropaeum,* on which they draped the armour and equipment of the vanquished foe as an emblem of victory (27). When Linnaeus saw the little Peruvian annual he called it *Tropaeolum minus* because the round peltate leaves resembled soldiers' shields, and the red and yellow flowers reminded him of the blood-stained helmet of the fallen (27).

PEACE

With some ethnic groups or nations or even with opposing waring factions in and outside Africa, some plants are traditionally believed to symbolize peace. This is duely acknowledged, respected and practised by the people. In S Nigeria, the Yoruba invoke *Ficus* species, fig trees, to make an enemy forget everything, calling the tree 'aba', meaning ankle irons, perhaps on the analogy of the clasping strangling roots and shackles clapped on a captured enemy (33). The tribe invoke *Sarcocephalus (Nauclea) latifolius* African peach to prevent war coming to a village, hence their name for the plant, 'abisi', meaning 'bend back' (Verger fide 4c, 33). The Yoruba also invoke an unidentified plant known as 'as unfunrun', meaning 'sleep and forget', to forget the memory of an enemy (33). In N Nigeria, when the leaves of *Lophira alata* Red iron-wood, a forest tree with very hard timber, are waved above the head, they are a signal for a truce in war (23). In the Region, messengers or envoys of peace in olden times were sent to the enemy with fronds of *Elaeis guineensis* Oil palm around their necks in order to settle terms of peace (18b). Victory was represented by the palm, and was often shown in connection with martyrs (6). In reference to the angels, the hymnist, Bishop Christopher Wordsworth writes:

> *Multitude, which none can number, like the stars in glory stands,*
> *Clothed in white apparel, holding palms of victory in their hands.*

However, in S Nigeria, the leaves or the branches of Oil palm are used to signify war among the Igbos (25).

All over W Africa, kola-nuts have great traditional significance. In Sierra Leone, for example, members of the Poro society among the Mende use two red nuts as a symbol of war (32). However, one white nut broken into two indicates peace (32). The nuts are the products from the fruits of *Cola acuminata* Commercial cola nut tree and *C. nitida* Bitter cola - both forest trees. In Ghana, it is the belief among the inhabitants of Akropong Akwapim that *Kalanchoe intergra* var. *crenata* Never die, a succulent herb often planted near the sacrificial stone in the area, has the power to soften things. Thus if a case is difficult, the plant will reduce its power and bring peace (14). In this country, *Portulaca oleracea* Pigweed or Purslane is used as an emblem of peace and of goodwill (18a); and in W Cameroon, *Dracaena arborea*, a palm-like tree believed to be fetish, is symbol of peace - for this reason it is used to mark property boundaries (8).

In W Africa, traditional practitioners use *Pupalia lappacea*, a herbaceous amaranth spiritually to unite parties who are always quarrelling (5) - likely based on the hooked burs. In Gabon, *Brillantaisia patula*, a stout acanth with violet-purple flower, is held to have the power to adjudicate in quarrels; and in this country, the Fang people claim that *Duboscia macrocarpa*, a forest tree, is able to bring litigants to an agreement (34). On the contrary, in Zambia, there is a superstition among the Wemba that if a part of *Rothmannia whitfieldii*, a shrub or small forest tree, is carried on a person, the bearer will become truculent and provoke other people to fight (4c). In Senegal, the Tenda belief that to burn a leaf or the wood of *Combretum molle*, a savanna tree, in a village will promote squabble (11).

In S Africa, a Zulu charm use is to chew leaves of *Oldenlandia caespitosa* var. *major*, a herbaceous weed, when passing the kraal of an enemy in order to keep him from coming to fight - after passing the kraal he spits the material from his mouth (36). In a description of *Aframomum melegueta* Guinea grains or Melegueta or Grains of paradise and other species of the genus with export potential, it has been observed that, 'the flower is very pretty, in some kinds a violet pink, but in the most common a violet purple, and they are worn as marks of submission by people in the Old River suing for peace' (20b). The occult property of *Nicotiana tabacum* Tobacco is the production of a state of peace (6).

In its growing area, *Olea europaea* Olive has traditionally been symbolic of freedom and purity; and in olden days, when seeking peace, warriors approached their enemies holding Olive branches in their hands (29). Olive is the National Flower of the United Nations - reflecting the objectives and aspirations of this world body. From the Scandinavian belief that *Viscum album* Mistletoe, an epiphytic parasite, was sacred, when enemies met beneath the plant in the forest, they laid down their arms and maintained a truce all day (27). From this the practice of hanging Mistletoe over a door way was adopted to denote peaceful intent (27). Among the Ancient Egyptians, *Nymphaea lotus* Water lily was a symbol of peaceful intent.

THE WAR DEAD

As by nature man finds use for plants, whatever the situation or occasion, so does he associate certain plants with those gallant men who lost their lives for their country in some famous wars. Floral blooms on battle sites in both World Wars and earlier battles were traditionally believed to have sprang from the blood of the gallant dead. An old legend relates that the red Poppies which followed the ploughing of the field at Waterloo after Wellington's victory over Napoleon, sprang from blood of soldiers killed in the battle (27) After the First World War, the battlefields at Flanders were scarlet with *Papaver rhoeas* Poppy flower (27). Similarly, *Chamaenerion (Epilobium) angustifolium* Fireweed or Great willowherb, a robust weed, became ubiquitous element in the townscapes of Britain and other parts of Europe during the Second World War, when it colonised bomb-sites, often in pure stand with its rosy-pink spikes (10). In many countries throughout the world, Poppy is symbolic of Remembrance Day or Poppy Day - a day to commemorate all the brave soldiers who died in World Wars I & II in defence of peace, freedom and justice. (See also PROTECTION, and TALISMAN).

Poppy field

References.

1. Back, 1987; 2. Berhaut, 1975; 3. Blumenthal & al., 2000; 4a. Burkill, 1994; b. idem, 1995; c. idem, 1997; d. idem, 2000; 5. Cofie, (pers. comm.); 6. Crow, 1969; 7. Dalziel, 1937; 8. Depommier, 1983; 9a. Ellis, 1881; b. idem, 1883; c. idem, 1887; d. idem, 1894; 10. Everard & Morley, 1970; 11. Ferry & al., 1974; 12. Field, 1937; 13. Fink, 1989; 14. Gilbert, 1989; 15. Guillarmod, 1971; 16. Hall, 1978; 17. Harley, 1941; 18a. Irvine, 1930; b. idem, 1961; 19. Kerharo & Adam, 1974; 20a. Kingsley, 1897; b. idem, 1901; 21. Kloss, 1939; 22. Matson, 1970; 23. Meek, 1931; 24. Moore, 1960; 25. Okigbo, 1980; 26. Pamplona-Roger, 2001; 27. Perry & Greenwood, 1973; 28. Pitkanen & Prevost, 1970; 29. Roberts, 1970; 30. Oso, 1975; 31. Tattersall, 1978; 32. Tremearne, 1913; 33a, Verger, 1967; b. idem, 1986; 34. Walker & Sillans, 1961; 35. Ward, 1955; 36. Watt & Breyer-Brandwijk, 1962.

51 WEALTH

A good name is rather to be chosen than great riches,
and loving favour rather than silver and gold.
Proverbs 22.1

Stories of 'sika duro' ('magic money' or 'spirit money') in Africa are many and varied. Regularly shaven hair, potency or fertility, the life of loved ones or of a partner or that of very close relatives or even one's own life or other cherished denials rank among the basic requirements believed to be spiritually exchanged or 'bartered' for wealth and earthly riches. The belief is that while, in some instances, this 'spirit money' is superficial, a temporal contract with only the individual, and spirited away on his or her demise, there are other instances where the family inherits the fortune indefinitely. This wide-spread gossip and belief that 'spirit money' may be obtained through fetish, 'medicine' witchcraft, the occult, 'mame-water' (mermaid), or some local deity is further strengthened and supported by occasional stories of ritual murder and grave exhumation for human parts, allegedly required among others, as vital and indispensable ingredients in prescriptions for getting rich quick - through mysterious and not easily explicable means. There is also the belief that some plants may be used, alone or with ingredients, as charms or in a form of witchcraft for acquiring money. In W Africa and elsewhere, some plants symbolize wealth or were at one time used as money, or are named as such.

CHARM PRESCRIPTIONS

In Nigeria, the seeds of *Leucas martinicensis,* a labiate, with butter enter into preparations as charms for riches. The seeds are rubbed along with butter in the palms of the hands while the person says 'arziki yaka' (which in Hausa means 'good fortune come'); and *Boswellia dalzielii* Frankincense tree and *B. odorata,* savanna trees, are often planted as a live-fence in this country to bring prosperity *(ba-samu)* (6). In Ivory Coast, *Kigelia africana* Sausage tree has superstitious uses as a charm to secure

riches and good fortune, but if taken too freely it is said to result in scrotal elephantiasis and oedema of the legs (3). In Senegal, the Bedik believe that a bath of the macerate leaves of *Guiera senegalensis* Moshi medicine is a charm for fortune, and the Serer believe that washing the body first thing in the morning with a leaf-macerate of *Cordia myxa* Assyrian plum ensures good fortune (15a, b). In S Nigeria, an unidentified plant locally called 'didikuridi' in Yoruba, meaning 'to tie and press together', is invoked by the tribe to have money always; and 'ogege', an unidentified tree, is the tree of wealth - one climbs it to get money (26). Among the Yoruba, 'Aje Shaluga' is the god of wealth (8b). Other plants invoked by this ethnic group in an Odu incantation to get money quickly or to acquire riches (26) are listed in Table 1 below:

Species	Plant Family	Common name	Habit	Part Used
Ampelocissus bombycina	Ampellidaceae	-	trailing savanna herb	whole plant
Boerhaavia repens	Nyctaginaceae	herbaceous annual		whole plant
Bryophyllum pinnatum	Crassulaceae	Resurrection plant	annual or perennial herb	whole plant
Chassalia kolly	Rubiaceae	-	forest shrub	leaves
Dialium guineense	Caesalpiniaceae	Velvet tamarind	savanna tree	whole plant
Echinops longifolius	Compositae	-	perennial savanna herb	leaves
Elaeis guineensis	Palmaceae	Oil palm	tree	fronds
Kalanchoe integra var. crenata	Crassulaceae	Never die	annual or perennial herb	whole plant
Manihot esculenta	Euphorbiaceae	Cassava	shrub	whole plant
Milicia excelsa	Moraceae	Iroko	forest timber tree	whole tree
M. regia	Moraceae	Iroko	forest timber tree	whole tree
Sesbania pachycarpa	Papilionaceae	-	herb of muddy swamps	whole plant

Table 1. Plants Invoked by the Yoruba to Acquire Money.

In S Nigeria, a Yoruba incantation invokes the protection of *Amblygonocapus andongensis,* a tree in savanna forest, against being cast under a spell, while a spell may be cast over others, so that one may gain wealth (26). The tribe also invokes *Dialium guineense* Velvet tamarind under the local name 'awin', meaning lender, to acquire money quickly (26). In N Nigeria, an unidentified species of *Aristida*, an annual grass called 'dacere ngala' in Fulfulde is credited with magical powers - 'men eat it, money appears' (14). In Gabon, *Mammea africana* African mammee- apple is held to have magical potency for acquiring riches; and *Pachyelasma tessmannii,* a forest tree, is regarded as a fetish for abundance and prosperity (27). In this country, people seeking good fortune or to obtain good employment, wash in water in which the leaves of *Mimosa pudica* Sensitive plant have been steeped, or beat their forehead with the leaves; and *Adenia rumicifolia* var. *miegei (lobata)*, a forest climber, is used as a talisman for acquiring fame and riches (27).

Mimosa pudica Sensitive plant

In S Nigeria, the dietetic value of *Amaranthus hybridus* ssp. *incurvatus* Love-lies-bleeding, a weedy annual - often cultivated as spinach, is obliquely recognised by the Yoruba in an incantation to obtain money (4a). In Ghana, a macerate of the dried leaves of *Lawsonia inermis* Henna with natron (native sesquicarbonate of soda) as ingredient in water used both in drink and for ritual washing is prescribed as charms for the acquisition of riches and fortune among some of the northern tribes (5). In the Region, to find aggry beads is traditionally considered a sure sign of continuance of good fortune (8a). (For the nature, value and possible origin of the beads, see INFANTS: PHYSICAL DEVELOPMENT: Other Products).

In W Africa, the seeds of *Garcinia kola* Bitter kola, a forest tree popularly used for chewing-sticks, and locally called 'minchinguro', meaning 'kola-nut for men' in Hausa, enters into juju charms for the acquisition of quick money; on condition that one swallows a whole seed, tied together with assorted ingredients including the red tail feathers of a parrot, *Psittacus erithacus*, attached to it into the stomach (see FETISH: FETISH PLANTS: Trees). Traditional medicine-men prescribe the hide of animals of prey ground with a few leaves of *Mallotus oppositifolius*, an erect branching shrub, collected with rituals, then mixed with palm oil as pomade to be applied as a charm to attract wealth (17). This charm may also be used by prostitutes to attract customers, with the invocation 'the hyena, *Hyaena hyaena*, never goes to hunt and returns empty handed' (17). On the contrary, the seasonal leaf-shedding of *Terminalia catappa* Indian almond, a shade tree with edible nuts introduced to the Region, is traditionally believed to be associated with 'getting broke' or with financial problems - especially when the tree is growing within commercial or industrial establishments. For this reason, the planting of the tree near the premises of factories and other business concerns tends to be avoided. On the contrary, a seed of *Blighia sapida* Akee apple in one's purse is a charm

against 'getting broke' (19). In S Nigeria, the Yoruba invoke *Sabicea calycina,* a scandent shrub, to make someone lose their property, naming the plant 'iso apare' (fart of a bushfowl, *Agelastes* species), thus likening the bird's wind being dissipated into thin air (26).

Among the Hausa of N Nigeria, the Zarma of Mali and Niger and the Fula *Phragmanthera* and *Tapinanthus* species Mistletoe, epiphytic parasites, growing on *Diospyros mespiliformis* West African ebony tree is believed to be a very lucky find indeed. The dried and ground epiphyte collected and prepared with rituals and taken in milk serves as a charm for wealth of no mean proportion that is said to persist within the family for generations (19). In these countries, the unopened leaves of *Piliostigma thonningii,* a shrub or small tree legume in savanna, collected with ceremonial rites enter into prescriptions among these tribal groups as a charm to prevent the creditor from demanding repayment from the debtor - an instance of sympathetic magic hinged on the unopened leaves (19). In Yoruba folklore, all the plants of the forest owed money except 'papasan' who paid his debts. Hence *Portulaca oleracea* Purslane or Pigweed features in an incantation for the recovery of owed money - the Yoruba name, 'papasan', meaning 'stick pays' (26). This is likely an instance of sympathetic magic hinged on the local name.

Among some of the people, particularly the coastal tribes in the Region, *Euphorbia hirta* Australian asthma herb collected early in the morning with the necessary traditional rituals and invoked for the purpose, then chewed, and one leaf placed under the tongue in advance of demanding repayment of a debt from a creditor, is believed to improve one's prospects of receiving the money promptly (2). In Ivory Coast, medicine-men use *Hybanthus enneaspermus,* an erect bush of savanna, magically to enforce the payment of debts (3). The nuts of *Elaeis guineensis* var. *idolatrica* King oil palm confer protection. They will promote the fortunes of a good man, and punish an evil one (4b). In order to receive gifts regularly from friendly people and from spirits, the scraped root-bark of *Evolvulus alsinoides,* a prostrate convolvulus; *Securidaca longepedunculata* Rhode's violet, a savanna shrub; and a young *Khaya senegalensis* Dry-Zone mahogany (that has never been debarked - the tree is very often debarked for medicinal purposes); are all collected with prayer by traditional medicine-men, ground together with ceremonial rites, and divided into four parts (17). One part is sprinkled in water for a bath on a Monday, and the remaining three parts are used for a bath on a Friday (17).

In Botswana, *Urginea capitata* Berg Slangkop, a bulbous perennial lily, is used by the Sotho as a charm 'to bring good fortune to friends'; and

the plant is believed to be capable of making the country flourish or otherwise (28). In its natural growing area in the Mediterranean, *Salvia officinalis* Sage, a decorative labiate, had magic powers attributed to it. Family fortunes were supposed to thrive with the plant if grown on their land (21). (Sage is an introduced garden plant to the Region). In Europe, the thick roots of *Mandragora officinarum* Mandrake, often shaped in human form, were valued by the Ancients for various purposes, and wrapped in a piece of sheet these small 'figures' were supposed to double the amount of money locked up in a box (20).

WITCHCRAFT

A mysterious method of getting rich quick is believed to be the infamous *'sika duro'* or *'Nzima beyiri'*, a form of witchcraft, in which one's life, or that of a loved one (or both) is spiritually exchanged for money supposedly vomited by a snake. 'The charms themselves appear to be harmless objects such as a piece of charcoal, or a ring, or a few beads, or a small wooden figure, but popular belief, however, is not satisfied with such a statement and it is said that the charm only appears harmless whereas in reality it is a snake - mostly invisible - but sometimes materialising' (7). This supports the general belief throughout traditional Africa that witches may materialise as snakes. In Ghana, several instances of snake-witch association have been cited from the witch's own confessions. The snakes are not only used as porters but identified with the witch itself (10).

'Witches can transform themselves into birds, chiefly owls, *Tyto* or *Alba* species; crows, *Corbus albus;* vultures, *Necrosyrtes monachus;* and parrots, *Psittacus erithacus* (23). Also into houseflies, *Musca domestica;* and fireflies, *Lampyris* or *Photinus* species; into hyenas, *Hyaena hyaena;* leopards, *Panthera pardus;* lions, *P. leo*; elephants, *Loxodonta africana;* bongo, *Tragelaphus euryceros;* into all *sasa* animals; and also into snakes. Some witches have a waist-belt of snakes (23). (For *sasa* animals - animals with vindictive spirits - see HUNTING: THE HUNTED PROTECTED: Animals with Vindictive Spirits). So strong is the conviction of this snake-witch association that when snakes are killed they are invariably burnt, with the inextricable belief coupled with jubilation that the human form will scorch and die as well. In S Africa, *Tulbaghia dieterlenii,* locally called 'sefothafotha', and *Pisosperma capense* Snake root or Snake soap enter into the composition of a decoction drunk by the Southern Sotho to rid the body of a 'snake' supposed to have been introduced by witchcraft (28). In this country, *Hermannia coccocarpa,* a tree, has magical

significance to the African. A root decoction is often used to drive out a little double-headed snake which may gain entry to the body (28). (For more on the topic, see WITCHCRAFT).

ASSOCIATED PLANTS

In Africa, as with other countries, some plants - either the seeds or the leaves - were used as money because they were valuable. Other plants were so named because of their colour, while others were symbolic of wealth. Years ago, the natives of W Africa employed the seeds of *Cola acuminata* Commercial cola nut tree and *C. nitida* Bitter cola as money in local trading (20). The seeds of *Theobroma cacao* Cocoa were held in high esteem by the Incas, Mayas and Aztecs; and at the beginning of the 16th century they were the basis for the monetary system in Mexico, and for a long time served as coinage within the distribution area of the plant (16). One hundred Cocoa beans could buy one slave.

Theobroma cacao Cocoa (Fruits) (associated with traditional religion; National Plant of Ecuador).

Theobroma cacao Seeds 1/5 natural size

A pre-Columbian poem on smoking in Africa documents the early use of *Nicotiana tabacum* Tobacco. 'At Kubacca' Captain Binger wrote, 'tobacco serves also as money' (25). And in Ancient Egypt, Rome and Greece, so dominant was grain in the national economy that taxes often were paid in measures of it (18). The grain referred here is *Triticum aestivum* Wheat. In W Africa, quantities or *Oryza sativa* Rice were supplied as part salary to the forces before the Second World War (1). In SE Senegal, the leaves of *Ficus ingens,* a fig tree, became a sort of currency during the dry season for paying Cheptel, a rent on cattle, *Ovis* species (9). In Mexico, bags of *Capsicum annuum* Pepper were part of tribute that conquered tribes were required to pay to the Aztec (12).

In W Africa, *Enantia chlorantha* and *E. polycarpa* African yellow wood, under-storey forest trees with deep yellow wood as the Common name suggests, are known as 'dua sika' in Twi, meaning 'tree of gold' in reference to this bright yellow wood (4a). In S Africa, *Gazania rigens*, a decorative composite with pink flowers which blooms throughout the year, is known as Gold nuggets or Treasure flower (11). In S Nigeria, the Yoruba name of *Aerva lanata*, a prostrate succulent herbaceous annual amaranth, is 'ewe owo' or 'ewe aje', both meaning 'money leaf" (4a). The reason for the names is, however, uncertain. In Guinea, the presence of *Berlinia grandiflora*, a forest tree, along watercourses in the Siguiri area is held to indicate the earth is auriferous (22). This unlikely premise probably rests on a Doctrine of Signatures for the bark contains a yellow honey-like resin of fraudulent usage. Prospectors panning for gold, dry and powder this bark, along with that of *Khaya senegalensis* Dry-zone mahogany and additions of laterite, for dusting over articles to give them the appearance of being gold (22). In Ghana, tradition has it that Pisemankama tribesmen, in the Nzima traditional area mistaking the reflection of the bright golden flowers of *Pterocarpus santalinoides*, a tree legume with brilliant yellow flowers on river banks, in a river for money, dived in and were all drowned (13).

In Japan, *Chloranthus glaber*, a low-growing evergreen forest shrub, is locally called 'senryo'; and *Ardisia crenata* Coral berry, a low evergreen pot plant, is locally called 'manryo' (24). The local names are equal to 1,000 'ryo' (an old gold coin) and 10,000 'ryo', respectively (24). The plants are considered to bring money and riches to households in the New Year (24). The Japanese name of *Adonis amurensis*, a dwarf native perennial with yellow flowers, locally called 'fukujuso', means 'wealth and happiness', and the bright flowers bursting the bare earth and snow really do bring joy at the coldest time of the year (24). According to Chinese mythology, *Narcissus pseudo-narcissus* Daffodil or Trumpet narcissus stands for good fortune, while *Paeonia* species 'King of flower', high-valued herbs and sub-shrubs of Asiatic origin mostly, symbolize wealth and honour (11). (See also MARKETING).

References.

1. Abbiw, 1990; 2. Atsu, (pers. Comm.); 3. Bouquet & Debray, 1974; 4a. Burkill, 1985; b. idem, 1997; 5. Cofie, (pers. comm.); 6. Dalziel, 1937; 7. Debrunner, 1959; 8a. Ellis, 1881; b. idem, 1894; 9. Ferry & al., 1974; 10. Field1937; 11a. Graf, 1978; b. idem, 1986; 12. Heiser, 1980; 13. Irvine, 1961; 14. Jackson, 1973; 15a. Kerharo & Adam, 1964; b. idem, 1974; 16. Lostchert & Beese, 1983; 17. Mensah, (pers. comm.); 18. Moore, 1960; 19. Moro Ibrahim, (pers. comm.); 20. Perry & Greenwood, 1973; 21. Pitkanen & Prevost, 1970; 22. Portères, s.d.; 23. Rattray, 1927; 24. Richards & Kaneko, 1988; 25. Van Sertima, 1976; 26. Verger, 1967; 27. Walker & Sillans, 1961; 28. Watt & Breyer- Brandwijk, 1962.

52 WITCHCRAFT

For we wrestle not against flesh and blood,
but against principalities, against powers, against
the rules of the darkness of this world,
against spiritual wickedness in high places.
Ephesians 6:11

Certainly, witchcraft is undoubtedly one of the most controversial of all the spiritual societies. All aspects of this secret art are highly debatable - from the belief, its nature, its acquisition, its transference, its manifestation and above all, to its activities.

BELIEFS AND ATTRIBUTES

In traditional Africa and elsewhere, the belief in the art and in the practice of witchcraft ranks high among several other possible causes of misfortune - such as the will of God or the gods, Kismet, Karma, Fate, Providence, the action of ancestral spirits, negligence, accident, chance, and so on. The belief in witchcraft explains why particular persons at particular times and places suffer particular misfortunes - accident, disease and so forth (32); and 'as long as the belief in the existence of witchcraft is not generally challenged, individual persons will attribute certain experiences and illnesses to the activities of *abayifo* (witches)' (29).

The causes of countless mishaps such as illness, failure in life, poverty, accidents, misfortune, erectal disorders, sterility (see FERTILITY), miscarriage, unemployment, alcoholism, drug addiction, mental disorders (see DISEASE: MENTAL DISORDERS), and even death (see DEATH AND DYING) are more often attributable among some people to witchcraft - a common, but embarrassing and shameful subject. The craft is believed to be second only to the Devil itself in evil and inhuman activities. It is the Devil's representative on earth (56). 'Witches of both sexes carried on illicit sexual relations with demons and practiced all sorts of capital crimes. They killed infants, and aimed to destroy men, women, animals, fields, fruits, vineyards, gardens and meadows. They made humans and animals ill, men impotent and women barren' (56).

The belief in witchcraft, as the cause of why misfortunes occur, was

officially banned in England in the reign of James II. In all former British colonies, accusations of witchcraft has not been entertained in law courts since the colonial era; but nevertheless the belief persists. Tradition dies hard. As in all Akan and most African societies, witchcraft forms part of the social reality of the Dormaa (29). 'I will assert categorically that there are witches in Africa; that they are as real as the murderers, prisoners, and other categories of evil workers, overt and surreptitious; this and not any imagination, is the basis of the strong belief in witchcraft' (29). This belief is neither among rural, illiterate pagan folks - nor among Africans only. In an article on the merciless persecution of witches in Europe, it has been observed that it is wrong to attribute the witch craze which occupied human beings for three centuries to dark superstition, to a lack of education, knowledge, or sense, to the mass hysteria of simple people who could neither read nor write, and could not be brought to reason. On the contrary, it was scientists, university professors, highly respected theologians, philosophers, philologists, and lawyers who came to believe what the Catholic church had originally disputed: that witches did indeed exist (56).

It is even feared that the practice has been increasing. According to several of my informants, the number of witches has risen dramatically since the turn of the century, and it is threatening the harmony of the community (29). 'They saw as the main reason for this development the spreading of Christianity and the concomitant weakening over the compliance with social norms' (29). Ironically, though the art and practice of witchcraft is generally feared by many among the various cultures, its true nature and concept is equally least understood by most - while yet others view its existence with sceptism.

NATURE AND CONCEPT

The specific concept of witchcraft is the idea of some supernatural power of which man can become possessed, and which is used exclusively for evil and antisocial purposes (50). A witch, generally speaking, is an antisocial person (13). However, others claim that it could also be used to the benefit of humanity. Witchcraft as a theory of causation embraces the theory of morals, for it says that witches are wicked people. It is the wicked feeling which causes their witchcraft to do harm; and it is the use of witchcraft, and not its mere possession, which is immoral (32). Witchcraft is an important problem for the various cultures in Cercle de Bondoukou in Ivory Coast and west central Ghana, as it is for so many

African societies (13).

Witchcraft is a substance in the bodies of witches, a belief which is found among many people of Central and W Africa (25). 'Witchcraft is a bad medicine directed destructively against other people, but its destructive feature is that there is no palpable apparatus connected with it. There are no rites, ceremonies, incantations, or invocations that the witch has to perform. It is simply projected at will from the mind of the witch' (28). *Bayie* (witchcraft) is nothing else but the evil thoughts of a person which can be directed at the person itself or at another person. These destructive thoughts slowly begin to work and materialize in the victim who believes in *bayie* (29). It has been defined by *The Oxford Dictionary* as the use of magic (30).

Witchcraft is worldwide and an ancient, highly secretive cult which manifests itself in various forms. For instance, voodoo, the local religion prevalent in Haiti and among W Indies and US Creoles and Negroes, is a form of witchcraft with an African root. Negroes in Africa and parts of America use *Aframomum melegueta* Melegueta or Grains of paradise seeds in voodoo rites; and whole roots of *Iris florentina* Orris root, resembling the Immam form, are highly valued in voodoo performances (54). In W Africa, leaflets of *Elaeis guineensis* Oil palm tied round the waist neutralizes any voodoo charm (11).

Witchcraft is believed to be practiced by women (witch) and men (wizard or man witch) alike. Among the Ashanti of Ghana, it has been observed that the majority of witches are women, but they need not necessarily be very old women (57). Among the Yoruba tribe of S Nigeria, the witch/wizard proportion is supported, but not the age group (23). Witches are more common than wizards, and here, as elsewhere in the world, it is the oldest and most hideous of their sex who are accused of the crime (23). Among the Dormaa people of Ghana, the predominance of this age group is corroborated. The majority of those suspected of witchcraft or sorcery are women in the period following the menopause (29). In northern Ghana, several places are known for witches, but those with position never sent their women to homes for witchcraft. It is mainly for those who are old, unlucky, deformed, barren, or very poor, and who thus have become isolated from society (Bob Loggah fide 48). The inmates at one such Witch Camp at Gambaga are all females. There is another Camp at Yendi. In Ashanti, there are two important limitations with regard to witches (57). Firstly, a non-adult cannot be a witch; and secondly, a witch is powerless to use her or his enchantment over anyone outside the witches' clan (57). (For explanation, see ACQUISITION). In

support of the latter limitation apparently, it has been observed among the Azande of S Sudan: 'witchcraft does not strike a man at a great distance, but only injures people in the vicinity. It is believed to act only at a short range so the further removed a man's homestead from his neighbour, the safer he is from witchcraft' (25).

The activities of witches are said to be mainly nocturnal. *Abayifo* (witches) are said to wander about naked at night, with their head down and surrounded by a red glow; or they gather in trees like *Milicia (Chlorophora) excelsa* or *M. regia* Iroko, locally called 'odum' in Akan, or *Ceiba pentandra* Silk cotton tree, locally called 'onyina' in Akan, and hold celebrations (29). Witchcraft materializes in various forms such as an antelope, *Adenota kob kob;* a crocodile, *Crocodylus* species; a leopard, *Panthera pardus;* a snake, an owl, species of *Alba* or *Tyto;* a bat, *Eidolon helvum* or *Epomophorus gambianus* or species of *Cynopterus, Glossophaga* or *Pteropus;* a firefly, species of *Lampyris* or *Photinus;* a cockroach, *Periplaneta americana;* or a ball of fire (29). This should only be glanced at cautiously and with the side of the eye to avoid repercussion (29).

An encounter with a couple such balls of fire while paddling alone at about 3.00 am on L Ncovi in Cameroon has been recorded: 'Down through the forest on the lake bank opposite came a violet ball the size of a small orange. When it reached the sand beach it hovered along it to and fro close to the ground. In a few minutes another ball of similar coloured light came towards it from behind one of the islets, and the two waver to and fro over the beach, sometimes circling round each other. I made off towards them in the canoe, thinking - as I still do - they were brand new kind of luminous insects. When I got on to their beach one of them went off into the bushes and the other away over the water. I followed in the canoe, for the water here is very deep, and when I almost thought I had got it, it went down into the water and I could see it glowing as it vanished in the depths' (39).

Another encounter with this ball of fire in S Sudan is recounted: 'I had only once seen witchcraft on its path. I had been siting late in my hut writing notes. About midnight, before retiring, I took a spear and went for the usual nocturnal stroll. I was walking in the garden at the back of a hut, among banana trees, when I noticed a bright light passing at the back of my servant's huts towards the homestead of a man called Tupoi. As this seemed worth investigation, I followed its passage until a grass screen obscured the view. I ran quickly through my hut to the other side in order to see where the light was going, but did not regain sight of it... Shortly afterwards, on the same morning, an old relative of

Tupoi and an inmate of the homestead died. This event fully explained the light I had seen' (25).

ACQUISITION

There are several means of acquiring the art of witchcraft. It could either be acquired at birth, or through breast feeding, or by inheritance from a dying relative, or by the administration of herbal preparation, or through ritual initiation, or by finding three beehives, or by a direct purchase.

At Birth

On the acquisition of the power of witchcraft among the Ga-folk of Ghana, it has been observed: 'the worst ones come when the witch is born (28). It is recorded among the Dormaa people of Ghana that '*bonsam-bayie* (witchcraft) can also be congenital. In this case, a witch takes possession of the *sunsum* (spirit) of the unborn child and takes it with her to witch gatherings at night in order to bathe it in herbal medicines and to teach it witch practices. Shortly before dawn, the *obayifo* (witch) returns the child's *sunsum*. At birth the child has already full knowledge of the art of *bayie* (witchcraft). Witches whose *bayie* is congenital are considered very powerful' (29). (See PLANTS AGAINST WITCHCRAFT: Children's Cases below for a plant prescription that is believed to counteract this evil).

Through Breast Feeding

The practice of witchcraft is believed to be an organic and hereditary phenomenon. It is thus said to be passed on by witch mothers to their children through breast feeding (47) - though not automatically since there must be the willingness on the part of the recipient for it. This supports the allegation and strong belief in folklore that the art of witchcraft is normally inherited from the grandmother. It has been observed that among the Azande of S Sudan 'witchcraft is transmitted by unilinear descent from the parent to the child. The sons of a male witch are all witches but his daughters are not, while the daughters of a female witch are all witches but her sons are not (25). On the contrary, it has been observed from southern Africa that 'the maternal line in some of these patriarchal societies is thus seen as carrying a mystical threat to the welfare and solidarity of the paternal

group. For they believe that it is in this line that the dreadful power of witchcraft may be inherited. For instance, among the Tsonga of Mozambique and the Tallensi of Gold Coast, witchcraft passes like haemophylia from woman to their children, so that men carry it, but do not transmit it. Stranger-women bring witchcraft into the group of men, and their daughters take it out again (32).

By Inheritance

It appears beneficiaries of witchcraft are deceived and tricked into accepting the craft, because it is usually baited or sugar-coated. An informant observes: 'witchcraft is of this world and cannot be taken to the one beyond' (7). It is traditionally believed that the forces responsible for witchcraft in the body have to be transferred by the dying - more usually to a relative. (See EXTERMINATION below for the consequences of non compliance). 'It is either left in clothing, beads or in food; and whosoever wears the clothing or eats the food acquires it' (7). Among the Ga tribe of Ghana, it has been recorded: I am told that a seed is a common vehicle for transferring witchcraft from one person to another, and that the recipient swallows the seed, but I have not learnt this from witches themselves (28). Like bad juju, witchcraft has to be vomited out - usually in the form of a small gold chain and given to a near relation to swallow before death (47). 'There are several ways of becoming a *bonsam-obayifo* (witch): usually the *bayie* (witchcraft) is inherited within the matrilineal family by an *obayifo* (witch) who spiritually transfers her *bonsam-bayie* to a close member of the family. This transfer takes place in dreams shortly before her death, either with or without the knowledge of the infant' (29).

Ritual Initiation.

The craft may also be transferred by ritual initiation if there is the willingness for it. 'Around mid-night I suddenly woke up and saw the landlady, an elderly woman, standing naked in the middle of the room which I had personally locked before retiring to bed. With one foot stepping on an empty shell of a snail, *Achatina achatina* or species of *Archachatina,* she ordered me to undress. My first reaction was to refuse her orders, but then I feared the consequences, knowing perfectly well that she could enter a locked room. During this split moment of indecision, I had a feeling of a state of vertigo. When I came to myself, I was among a group of other men and women - all naked with sac-covered heads - dancing around a big fire somewhere in a forest clearing. I shouted out the name of the Lord Jesus Christ for deliverance from these forces of evil. Instantly, I found myself back on the bed in my room. At about 5.00 in the morning there was a knock on the door. It was the landlady again - this time inquiring how I managed to return, when and why. She added that all the other dancers were prominent men and women in the society' (6).

By Administering Herbs

In N Nigeria, it has been observed among the Hausa: 'in the tales any woman may become a witch, and she is liable to do so by drinking a brew of the leaves of *Parkia biglobosa (clappertoniana)* West African locust bean' (61). Among the Ashanti of Ghana, if an old witch wishes her daughter to become a witch she will bathe her repeatedly with 'medicine' at the *suminaso* (the kitchen-midden) (57).

Beehives

The acquisition of the craft could also be linked with the number of beehives one finds. 'Among the Bemba of Northern Rhodesia (now Zambia) to find one beehive in the woods is luck, to find two is very good luck, to find three is witchcraft' (32). On the contrary, it is traditionally believed among many ethnic groups that honey protects one from witches and their activities. 'When a witch enters the room, she is invariably attracted to enjoy the honey first - with the result that she leaves you alone. If you keep honey in the house and notice that the level has been steadily decreasing, it could mean that witches have been visiting you' (45, 47).

By Direct Purchase

Finally, others maintain the art of witchcraft can be bought (45) - provided one already has a bit of it (24). Some demons are left as a legacy by a dying relative, some are set free, some are sold for a penny or less (28).

EXPOSURE

The agents responsible for exposing witches include ordeal trials or witch ordeals, deities, spiritual churches and also plants.

Ordeal Trials or Witch Ordeal

Witches are normally exposed in witch trials or witch ordeals (see ORDEAL: PLANT TESTS: Witch Trial).

Special Deities

Witches may also be exposed by special deities. It has been observed that in Ghana '*Abosomerafo* dwell in forests, waters, and stones. As divine executors, they sanction the moral and social order of the Dormaa. It is their main task to expose witches and to punish culprits. There used to be fire gods, who are now extinct. Their main task was to catch witches and sorcerers' (29). Other deities of the Dormaa are 'Apo' (exposes and treats); 'Dopo' (exposes); and 'Nanguro' (exposes and kills if necessary) (29).

Spiritual Churches

A number of spiritual churches and sects do include witch-hunting among their programme of activities. 'There have arisen in Africa movements designed to cleanse the country of witches, held responsible for social disintegration, for failing yields on over-cultivated lands, and for diseases. The philosophy of these movements against witchcraft is that if the Africans would cease to hate one another, and would love each other, misfortune would pass. These movements are short-lived, and tend to be replaced by religious movements involving messianic elements' (32). A similar movement has been recorded in Europe. 'In 1484 Pope Innocent VIII issued the famous Witches' Bull, in which he confirmed the two inquisitors' authority to extirpate all witches in large parts of Germany and Switzerland. The Witches' Bull was also important because it was a papal confirmation of the existence of witches and their crimes. For a long time, the church had denied the existence of witches, and even threatened to punish all those who believed in magic and witchcraft' (56).

Plants

In addition to the exposure through witch trials, deities, spiritual churches and other similar movements, it is traditionally believed among many cultures throughout the world that certain plants are also inimical to witchcraft and capable of exposing them or thwarting their activities. Among the Azande tribe of S Sudan for example, it has

been observed: 'it is not magic words nor ritual sequences which are stressed in initiation into the corporation of witch-doctors, but trees and herbs. An Azande witch-doctor is essentially a man who knows what plants and trees compose the medicine which, when eaten, will give him power to see witchcraft with his own eyes, to know where it resides, and to drive it away from its intended victim' (25).

Cleome gynandra

In W Africa, the Ewe-speaking tribe of Ghana, Togo and Benin Republic believe that *Croton zambesicus,* a small tree often planted in villages and towns, is a protective fetish. Accordingly, the tribe burn the leaves with those of *Cleome (Gynandropsis) gynandra,* a herbaceous annual often cultivated or tended around villages as spinach, into a black powder mixed with black (kotokoli) soap, and rubbed with rites on all four corners outside the house to entrap witchcraft. When the spirit arrives it will instantly materialize into the human form and drop down on its knees - in a plead for mercy (45). It is a general belief in Ghana and Ivory Coast that any witch who goes under the shade of *Okoubaka aubrevillei,* a rare forest tree popularly believed to be fetish, will be exposed of any demonic activities. In these countries, the root-bark of this tree, locally called 'odee' in Akan or 'okoubaka' in Anyi, collected and prepared with the necessary rituals, then burnt with *Hyptis* species, a herb labiate, and *Amaranthus hybridus* ssp. *incurvatus* Love-lies-bleeding in the house is against witchcraft and also exposes them (24).

Heliotropium indicum Indian heliotrope or Cock's comb

The male inflorescence of *Elaeis guineensis* Oil palm is burnt by many tribal groups in the Region to drive away witchcraft. The practice is said to be very effective, and often employed by husbands to drive away their witch wives from the house (21). The odour of burnt kernels of the Oil palm with naphthalene crystals is reported

inimical to the craft and exposes them (35). Similarly, to scare off a witch girl friend, the genitals are first washed with an infusion of *Heliotropium indicum* Indian heliotrope or Cock's comb, a herbaceous annual, before intercourse (45). The use of *Distemonanthus benthamianus* African satinwood, a tree legume with characteristic red bark, which is sacred to witchcraft, as fuel-wood also drives them away from the household - the vernacular name in Gabon meaning 'sorcerer's tree' (18). However, in Botswana the Southern Sotho use *Lasiosiphon anthylloides,* locally called 'moomang' as fuel only when no other is available as they believe that the smoke bewitches people (66).

In W Africa, *Ricinus communis* Castor oil plant is commonly used in discerning wizards. Sometimes the leaves are placed on a fire for the purpose. The dry seeds are also sometimes put on the fire as well, and when the wizard smells it and shows a sign of dislike, he is identified (36). The ground seeds of Castor oil plant may be burnt with the dry faeces of a dog, *Canis* species, for the same purpose (31). The bark of *Elaephorbia drupifera*, an euphorb in savanna, burnt as an incense is also inimical to witchcraft and exposes them (60). (The high forest equivalent is *E. grandifolia).* In S Nigeria, two unidentified plants locally called 'mopinaise' and 'ewe irin' in Yoruba, meaning 'I have no part in any offence' and 'press down', respectively, are used by the tribe to placate witches and to prevent a witch becoming angry, respectively (63c).

Calotropis procera Sodom apple

In N Nigeria, witches are believed to detest the smell of *Calotropis procera* Sodom apple, locally called 'bambami' in Fula - and a spell arising from a brew of some other plant may be countered with anti-dotal concoction made potent by being cooked over twigs of 'bambami' (McIntosh fide 14a). The Hausa of N Nigeria, the Zarma of Mali and Niger and the Fula believe that the fruit of Sodom apple (either fresh or dried) held in the hand or kept in one's pocket while greeting a witch results in a reaction to expose the victim (47). In many parts of W Africa, the fruit of *Citrus aurantiifolia* Lime or the whole plant is believed to be inimical to witchcraft and their activities. The plant is sometimes planted around the house for the purpose. Among many ethnic groups in its growing area, two leaves

of *Ageratum conyzoides* Billy goat weed, an annual composite, collected at night half naked, dried, powdered, and then mixed with palm oil and applied as a body ointment is protection against witchcraft; or in an incantation to expose witchcraft, the victim's name is mentioned while the leaves are burnt in a ceremonial ritual (15).

EXTERMINATION

Once a witch or a wizard has been exposed, society requires her or him to confess any past activities publicly. Among the Dormaa people of Ghana, it has been observed: '*Nokwabo* (stone of truth) - a seat made from piled-up stones - plays an important role as a confessional chair when *obosom* 'Nanguro' exposes witches. Those suspected of witchcraft must sit on it in order to confess. Only comprehensive confession can save the witch and guarantee the mercy of the *obosom* (deity). If they refuse to do so, 'Nanguro' can take possession of *nokwabo* and kill the victim' (29). It has further been observed among the Dormaa: 'Priests never actively killed those found guilty of witchcraft or witchcraft. Rather witches or sorcerers exposed by them were killed spiritually by the *obosom,* which meant that they were considered banished. The banished person accepts the punishment and loses his or her will to live so that the slightest physical illness may result in death' (29). Finally, it has been observed: 'Only in extreme cases *abosom* punish with death in spite of a confession of guilt. Once the death sentence is passed, the accused will die within a short period of time without any physical influence and without any visible signs of disease. The *abosom* never kill physically, but always spiritually' (29). (For the different forms of destruction or extermination of witches, see ORDEAL: PLANT TEST: Witch Trial).

The extermination of witches has also been recorded from Europe: 'Witches did indeed exist and were in league with the Devil to destroy Christendom if one did not track them down and exterminate them using every possible means' (56). This awareness: 'contributed greatly to the 'success' of *The Witches' Hammer,* that is the burning of hundreds of thousands - the number may be lies between 300,000 and 500,000 of witches, warlocks, and their children in Europe, many of them, perhaps most, in Germany' (56).

Witchcraft has to be transferred before death (see ACQUISITION: Inheritance above), and unless this transference is successfully accomplished, the dead body still remains activated by this demon which, as a result, continues its evil and antisocial activities - as if, and

when the body was alive. Under such a circumstance, the activated body does not decay, but still remains fresh. One such body of an old witch exhumed after several months in the grave, it is alleged (since the evil practices she was accused of still persisted) was found to be as fresh as if it had just been laid in state. On the successful performance of the necessary customary rites to exorcize the inherent demon, the body instantly began to bloat, followed by putrefaction and decay, dehydration, to marrowless white bones - all in a twinkle of the eye.

It is general knowledge throughout Africa that witches do normally confess their evil deeds just before death - recounting a catalogue of names of all the victims they have killed (9). The informant adds that to prevent the revelation of this gruesome act, the relatives at the bedside usually force cloth into the mouth to silence the dying witch. In W Africa, it is recorded among others that a person suspected of *bayie* (witchcraft) who dies as a result of its own mischief or is killed by the gods is buried on the same day without any ceremony (29). (See also GHOST AND SPIRIT WORLD: SACRIFICES). Among the Ewe-speaking tribe of Ghana, Togo and Benin Republic, in previous times witches were burnt when they died (see DEATH AND DYING: FUNERAL OBSEQUIES: Cremation). Among the Azande of S Sudan, it is observed that if maggots come out of the aperture of a dead man's body before burial, it is a sign that he or she was a wizard or a witch... and that when witches die they become evil ghosts (*agirisa*) (25).

COMPARED TO SORCERY

Sorcery is distinguished from witchcraft by many authorities - firstly, by the speed with which it acts, and secondly, by the method of execution. Though the end result and the motive might be same, the rapidity and the method by which this end result is achieved is different. The difference between a sorcerer and a witch is that the former uses the technique of magic and derives his power from medicine, while the latter acts without rites and spells and uses hereditary psycho-physical powers to attain his or her ends (25). For instance, when an Ojebway Indian desired to work evil on anyone, he made a little wooden image of his enemy and ran a needle into its head and heart. Or he shot an arrow into it, believing that wherever the needle pierced or the arrow struck the image, his foe will at the same instant be seized with a sharp pain in the corresponding part of the body (62). If he intended to kill the person outright, he burned or buried the image, uttering certain magic words as he did so (62).

In sorcery, therefore, 'bad medicine' is actually sprinkled, or buried, or hanged, or placed within the precinct of the victim. The leaves of the 'bombiri' tree, *Albizia aylmeri,* a legume, with thin sticks is reported (25) as a sample of 'bad medicine' planted in one's precinct with the purpose to kill the victim. On the contrary, in witchcraft there is no such physical contact, but rather spiritual projection. Suddenness and violence are some indications of sorcery (25). In witchcraft, the invalid wastes away gradually. (See also JUJU OR MAGIC: 'AFRICAN ELECTRONICS').

It has been observed in S Sudan: 'the Azande fear sorcery far more than they fear witchcraft which evolves anger rather than fear. This may be due partly to the absence of machinery for countering sorcery as adequate as that employed against witches. Indeed, today, apart from administering an antidote or making a counter-magic, nothing can be done to stop an act of sorcery' (25). Sorcery is associated with men generally, just as witchcraft tends to be associated with women (32). 'If Zulu men wish to harm others, they have deliberately to enter on the art of sorcery. It is important that among the Zulu, as in many other tribes, it is believed that men make this deliberate choice to do evil in these ways, while women are believed to be witches, who are possessed of innate evil power' (32). In W Africa, it is confirmed among the Dormaa of Ghana that *'bayie* (witchcraft) is almost exclusively found within one's own matrilineal family, and diseases attributed to *bayie* are also referred to as *abusua yadee* (family disease) or *abayifo yadee* (disease due to witchcraft). Conflicts within the matrilineage often result in accusations of witchcraft, but in the case of conflicts outside the matrilineage, *aduto* (evil magic) is often suspected to be involved (29).

SORCERY PLANTS

Traditionally, some plants are known, for their magic or evil or hallucinogenic properties, to feature in the art of sorcery; or as ingredients in the practice. Some other plants serve as counter-charms to sorcery.

Plants For The Craft.

In W Africa, *Cassytha filiformis* and *Cuscuta australis* Dodder, twining epiphytic perennial parasites, being rootless and of odd appearance, are mainly used superstitiously - this is supposed to be known only to sorcerers (18). In Senegal, *Annona senegalensis* var. *senegalensis* Wild custard apple has general reputation with sorcerers

for casting spells (38a). In Gabon, sorcerers employ *Tabernanthe iboga* Iboga, an apocynaceous shrub with hallucinogenic properties before demanding information from the spirits (55). In this country, the leaders of the Bwiti cult eat Iboga for an entire day before asking advice of the ancestors, topping it off with a root bark infusion, which is served in a manner similar to the Christian communion (55). In this country, the stem of *Musa sapientum* Banana and *M. sapientum* var. *paradisiaca* Plantain are reportedly used in many sorcery practices (65). In Congo (Brazzaville), *Olax latifolia,* a forest shrub, is able to heal wounded sorcerers; and *Barteria fistulosa* has a strong reputation in magical practices and sorcery (12); and in the Democratic Republic of the Congo (Zaire), *Microdesmis puberula,* a forest shrub, is put into various concoctions for the protection of sorcerers (12).

In S Nigeria, the 'sorcery in the path' which paralyzed the sight of even fierce warriors, causing them to make a long detour, consisted of a plantain sucker stuck in the ground, and lying about, a coconut shell, palm leaves and nuts (43). In N Nigeria, Fula sorcerers use *Aristida adscensionis,* an annual grass, in offensive magic locally known as 'lekki baatal' (37), but the circumstances and method are not recorded (14b); and the Igala use *Clerodendrum capitatum,* an erect or scrambling shrub of savanna and forest, in sorcery to harm named persons (14d). In this country, the powder of the fruits of *Polycarpaea linearifolia,* an erect annual or perennial herb, is said to have a great magical property and much used by sorcerers (14a). The Hausa name 'bakinsuda' or speech of 'suda', referring to a species of shrike (a bird) whose song only the initiated can interpret (18). The birds in the Region (58) are listed in Table 1 below:

Species	Common name	Identification
Corvinella corvina	Long-tailed shrike	Long tail, yellow bill, rounded dark-tipped rufous wings
Dryoscopus gambensis	Gambian puff-back shrike	Glossy blackupper parts, whitish rump
Laniarius atroflavus	Yellow-breasted shrike	Black upperparts, yellow underparts
L. barbarus	Barbary shrike	Olive-yellow crown, black upperparts, scarlet underparts
L. ferrugineus	Bell-shrike	Black upperparts, white underparts, usually conspicuous white wing-bar
L. leucorhynchus	Sooty boubou	Entirely blackish
Lanius collaris	Fiscal shrike	White shoulder and wing patches, long black white-tipped tail

L. excubitor	Great grey shrike	Mainly pale grey upperparts, white underparts, conspicuous black and white wings, white outer tail feathers
L. senator	Woodchat shrike	Chestnut crown and nape
Malaconotus cruentus	Fiery-breasted bush-shrike	Grey head, green mantle, yellow underparts, often washed crimson
M. multicolor	Many-coloured bush-shrike	Grey head, green mantle, scarlet- or orange- or black-breasted phase
Nilaus afer	Brubru shrike	Broad white stripe over eye, buff wing-bar
Prionops coniceps	Red-billed shrike	Red bill and eyelids, grey crown, black cheeks and throat
P. plumata	Long-crested helmet-shrike	White crest, white wing-bar and tail feather tips
Tchagra minuta	Little blackcap tchagra	Black head, no head stripes, black shoulder patch
T. senegala	Black-crowned tchagra	Black crown and stripe through eye, broad buff stripe above eye

Table 1. Types of Shrikes in the Region with their Identification.

Plants Against The Craft

Certain plants are, however, believed to be protective against sorcery. In Gabon, *Laggera alata*, a strongly aromatic composite herb, is planted around houses to counteract malign influences of sorcerers (65). In S Africa, the Xhosa put pegs of *Acridocarpus natalitius*, a shrub, around a homestead to protect it against sorcerers (66). In Botswana, the Southern Sotho use *Pentanisia prunelloides* Wild verbena as one of the ingredients of a preparation used to prevent a sorcerer from finding out the door of the hut; or use *Urginea capitata*, a perennial lily, in the belief that it keeps sorcerers away from huts (66). *Cynodon dactylon* Bermuda grass is used by the Sotho against sorcery (66). In Congo (Brazzaville), *Brenania brieyi*, a tree of evergreen forest, is reported to have magical usage in the preparation of the fetish *Nzobi*, and if a twist of the plant is kept in the house it will repel sorcerers and evil spirits (12). In this country, *Piptadeniastrum africanum* African greenheart, a tree legume, has a high reputation for the power of its magic, killing sorcerers and keeping away spirits. This power is equally reversible for criminal ends (12). Similarly, in S Cameroon, the 'oven' tree, *Didelotia africana*, a legume, is thought extremely powerful and used among the Beti in many traditional healing treatment, especially those involving sorcery (5).

In Congo (Brazzaville), *Smilax anceps (kraussiana)* West African sarsparilla, a prickly climbing perennial, together with *Erythrophleum suaveolens* Ordeal tree and *Strychnos icaja*, a large liane of evergreen

forest, is believed to be a suitable replacement for the lead bullet of a cartridge with which to shoot a sorcerer; and the sap of *Thomandersia hensii (laurifolia),* a composite shrub, is used as a drink to heal illnesses caused by sorcery (12). In this country, the sap from a piece of the bark of *Pseudospondias microcarpa* var. *microcarpa,* a tree in forest margins and fringing forest in savanna, cooked in a leaf is given to a girl suffering infatuation because she is under sorcery to make her vomit up the fetish (12). Also in this country, the tuber of *Icacina mannii,* a scandent shrub of evergreen forest, is hushed up and macerated for 2-3 days in water which is drunk for gastro-intestinal troubles and dysentery. If the patient does not respond, a leaf must be placed beneath the bed. If this leaf becomes rolled up or perforated, the patient is thus exposed as a sorcerer who, if a cure is desired, must confess. To effect a cure, a draught of the tuber-macerate is given (12).

In S Nigeria, *Croton zambesicus,* a small tree often planted in villages and towns as a fetish or juju, locally called 'ajekofole' or 'ajekobale' in Yoruba, is deterrent to sorcery, the local name literally meaning 'witch cannot alight upon' (18). The general belief in the Region is that the tree is inimical to witchcraft as well. In this country, the Yoruba put *Ouratea flava,* a forest shrub, to magical use: to obtain compassion of sorcerers (63d). In Senegal, Serer doctors hang a branch of *Morus mesozygia,* a forest tree, in the doorway of a house to warn off sorcerers (39b); and among the Pulaar, *Quassia undulata,* a soft-wooded tree of forest and savanna, has a place in superstition and magic in making application for charms and sorcery (38a, b). In Liberia, the leaves of *Cardiospermum grandiflorum* Balloon vine, a herbaceous climber, are used to evict sorcery from a village (Hardy fide 64). In Guinea, there is a superstitious invocation not to beat a friend with the stem of *Manihot esculenta* Cassava in the Loma country, for if he is a sorcerer he will die (8). In Sahara Morocco, the fruit of *Balanites aegyptiaca* Desert date or Soap berry hung round the neck is a protective amulet against sorcerers who drink blood (46).

The Influence of Mistletoe.

In W Africa Mistletoe, epiphytic parasites, growing on some specific hosts are used against sorcery. In traditional medicine, the properties of the parasite invariably depends on the host plant on which it is growing. For instance in SE Senegal, Mistletoe growing on *Psorospermum senegalense,* a small savanna bush, is tied to a post near a house by the Tenda to ward off sorcerers; and Mistletoe growing on *Euphorbia unispinosa*, an erect shrub, is held to repel and even frustrate the action of sorcerers. The host plant is used by Tenda singers at rituals in the belief that it improves the voice (27). In this country, should there be Mistletoe growing on a big *Afzelia africana* African teak, a tree legume of savanna and fringing forest, on a cultivated land, then the Mistletoe has power against sorcery; and a Mistletoe parasitizing *Dombeya quinqueseta*, a shrub or small tree of savanna forest, or parasitizing *Combretum tomentosum*, a savanna shrub, is deemed to confer protection from sorcerers (27). In this country, Mistletoe growing on *Pavetta cinereifolia*, a shrub of savanna woodland, is ascribed with protective property for which purpose it is tied round the neck as a guard against sorcery (27). Also in Tendaland (Senegal/Guinea), the leaf of *Erythrophleum suaveolens* Ordeal tree, a savanna tree legume with extremely poisonous properties, is given to convicted sorcerers to make them confess their sins before they die (27). (See DISEASE: LEPROSY: Table 1 for a list of Mistletoes in the Region).

PLANTS AGAINST WITCHCRAFT

As with sorcery, some plants also feature in witchcraft. In addition to the role believed to be played by the plant kingdom in the transmission of the craft or exposure of culprits (see ACQUISITION and EXPOSURE above), some plants are believed, in many of the folklore, to influence witchcraft - either negatively or positively, that is either against and preventive of the practice, or for their association and promotion of the craft. Other plants are named after witches. In Cameroon, the Mambila people use the fruit of *Gardenia ternifolia* ssp. *jovis-tonantis*, a savanna shrub, in a charm against witchcraft (44). In N Nigeria, the Jukun people use the bark of *Erythrophleum suaveolens* Ordeal tree or *Piliostigma thonningii*, a savanna shrub or small tree legume, since these are bitter and disliked by witches (44). Similarly, *Gardenia erubescens*, a savanna shrub with edible fruits, and the branches of *Moringa oleifera* Horse-radish tree or Oil of Ben tree enter into charms to protect against witchcraft (44). In S Nigeria, *Morus mesozygia*, a forest tree, is prescribed by Yoruba medicine-men as credited with strong juju properties to afford protection against witches (63a); or prescribe *Vitex thyrsiflora*, an under-shrub or small tree, so that witches may show one pity (63b). In this country, Yoruba medicine-men put *Euphorbia convolvuloides*, a savanna herb, in prescriptions designed to prevent witches beating anyone (63a).

Among the Ewe-speaking of Ghana, Togo and Benin Republic, and some of the coastal tribes in W Africa, *Momordica balsamina* Balsam apple

or *M. charantia* African cucumber, trailing cucurbits believed to be fetish, and *Heliotropium indicum* Indian heliotrope or Cock's comb, a common weedy annual around villages, are employed on the spiritual level to resist evil influences such as witchcraft (45). With the Akan-speaking tribe of Ghana and Ivory Coast, repeated ritual baths with a whole plant infusion of African cucumber at dawn is believed to strengthen one spiritually to withstand any evil influences, especially those attributed to witchcraft (24). In S Africa, the Ndebele grind the root of *Ximenia caffra* var. *natalensis* Natal plum and mix it with cow-dung for smearing on the floor of huts to keep witches away; or the Lobedu place a medicated pole of *Vangueria venosa* Wild medlar across a gate entrance to keep wizards out (66). In Botswana, *Hibiscus malacospermus*, locally called 'bohoyana', is used by the Southern Sotho as a charm to prevent villages being bewitched by the enemy (66).

In W Africa, the root of *Securidaca longepedunculata* Rhode's violet, a shrub or small savanna tree, is worn as a charm against witchcraft (18); and in N Nigeria, an unspecified part of *Securidaca* or the root of *Vetiveria zizanioides* Vetiver or *V. nigritana*, densely tufted perennial grasses, are used for purifying drinking water in Kano, in the superstitious belief that witches will then not be able to drink it (Samia Al Azharia Jahn fide 14c). In S Nigeria, *Croton lobatus*, a weedy herbaceous euphorb, has superstitious uses against witchcraft - the person who washes with a lotion prepared from it keeps evil spirits at bay (18). As a charm against witchcraft in Botswana, the ash of *Nolletia ciliaris*, a composite locally called 'moloka', mixed with the fat of a goat, *Capra* species, is burnt in the hut by the Southern Sotho, the smoke being credited with the capacity of countering this evil; or *Scirpus burkei*, a sedge, is an ingredient in the tribe's preparation used as a charm protection against witchcraft (66). In S Africa, a branch of *Combretum suluense* Zulu-willow is used in a magical sense, by erecting a pole to protrude perpendicularly from the top of the hut roof, to keep out witches who are sent in lightning (66).

In W Africa, a ritual bath on seven consecutive days with the crushed leaves of *Amaranthus hybridus* spp. *incurvatus* Love-lies-bleeding, an annual weedy herb sometimes cultivated as spinach; *Hyptis pectinata* or *H. suaveolens* Bush tea-bush, weedy aromatic labiates; and *Kalanchoe integra* var. *crenata* Never die, fleshy perennial herb; all collected with the necessary rites and invoked for the purpose, serves to protect one against witchcraft (45). In Tanzania, Bush tea-bush is also a witchcraft medicine at Mkalamo (14c). In W Africa it is believed that to see witches clearly in a vision, Love-lies-bleeding is collected from a clean environment, wrapped together with a dried frog, *Bufo* species, in a piece of white cloth and placed under one's pillow when going to bed (45). Alternatively, the leaves may be eaten as a spinach against witchcraft (45). In its distribution area, *Sida acuta* Broomweed, a common perennial malvaceous weed, offers protection against witchcraft and evil forces when used in ritual baths and in fumigations (45).

In S Africa, both the Tswana and the Southern Sotho believe that *Eucomis undulata,* a perennial lily, is a powerful charm against witchcraft and is used in inflicting harm on one's enemies. If a Tswana is poisoned by an ill-wisher, a decoction of the plant is taken along with meat and milk. The ash and partly burnt remnants of the root is then mixed with the fat of an ox, *Oryx* species; sheep, *Ovis* species; or goat, *Capra* species; and smeared over the body as a protection against contamination of any effects from witchcraft (66). In Botswana, an African medicine-man washes himself with a decoction of the plant to protect him from being bewitched (66). In Central African Republic, the leaves of *Antiaris toxicaria* Bark cloth tree are wrapped and stitched with the fibre from *Manniophyton fulvum,* a forest climbing shrub, to make a witch-doctor's magic wand; and in this country, the Pygmy ascribe magical potential to *Olax gambicola,* a forest shrub: witch-doctors instil a few drops of sap from the leaves into their eyes to enable them to see spirits better (49).

In W Africa, *Senna (Cassia) singueana,* a small tree legume, enters into charms against witchcraft and to bring an enemy into contempt (44); and *Calotropis procera* Sodom apple is placed over doorways as a preventive of witchcraft (18). Among some of the tribes in its distribution area, the root powder of Sodom apple rubbed between the joint of the fore- and hind-arm is believed to be a protection against a witch's spell and against all evil spirits generally (45). (But see PLANTS WITH DOUBLE SPIRITUAL NATURE below). In S Nigeria, it is reported that the chameleon, *Chamaeleon vulgaris,* a common prop of a wizard's paraphernalia, is said never to climb *Laggera alata,* a strongly aromatic composite herb, so it is invoked by the Yoruba to confer protection against witchcraft (63c) - hence the significance of the Yoruba vernacular name 'agemo-kogun', meaning 'chameleon does not climb it' (65). In this country, the tribe fumigate a house with *Hoslundia opposita,* an erect or scrambling shrub, to guard against witches (63b). In the Region, *Ipomoea argentaurata,* a woody climber, as a medicine against witchcraft is worn as an armlet, amulet or girdle; and clothing is fumigated with it not as a scent, but as a charm for the same purpose (18).

In S Africa, the Chagga use *Adhatoda engleriana,* an acanth locally called 'indungu', to counter the bewitching of food; and the Shambala rub the fresh juice of the ash of *Polygala gomesiana,* locally called 'kilemela ukuku',

into scarifications as a charm against witchcraft (66). In this country, the Lobedu of Northern Transvaal plant medicated posts of *Celtis africana* White stinkwood at the entrance to a village, and pegs of the wood are hammered into the ground in a magic circle to keep witches away; and the Shambala and the Bugu use the leaf ash of *Toddalia aculeata* Wild orange tree as charms against witchcraft (66). In Tanzania, the leaf of *Lannea amaniensis,* a tree, is used as a charm against witchcraft, especially for the renewal of spells (66). In E Africa, sometimes a stick with a rootstock of *Hypoxis villosa,* a lily, locally called 'thithigwani', stuck on it, is planted outside the kraal by the Lobedu to repel witches (66). In Zimbabwe, the Manyika use the leaf of *Myrsine africana* Cape myrtle as a charm to ward off thunder and witchcraft by washing the body with the leaf while standing in the river and allowing them to float away down stream (66). In Lesotho, an enema of *Withania somnifera* Winter cherry is taken to evict a snake from the body which gained entry by witchcraft (33, 66).

In Ghana, Togo and Benin Republic, it has been observed that the leaves of *Sida linifolia,* a small malvaceous weed said to be fetish, called 'wodoewogbugbo' in Ewe, meaning 'they planned but retreated' are added to any medication for diseases believed by the patient to be caused by witchcraft (51). Similarly, a little of the dried pulverized leaf of *Elaephorbia drupifera* or *E. grandifolia,* euphorb trees in savanna and forest respectively, is added to any medication to counteract the activities of witches and expose them. Any sign of a rise in the patient's body temperature after taking the medicine, is a positive indication that he or she is a wizard or a witch (60). In Liberia, the bark of *Isolona cooperi,* a small forest tree, is used as a charm against witchcraft by chiefs and important men (14a). The bark and ashes are mixed with palm-oil into a paste, and when in a strange country, or in danger of eating poisoned or bewitched food, one licks the paste, and the poison or spell is cast out (16). In this country, a snuff prepared from the pounded leaves of *Sabicea ferruginea,* a climbing forest shrub, is held to confer protection against enemy witchcraft, and furthermore to give one power to bewitch one's own enemies (16).

Tapinanthus bangwensis Mistletoe

In Europe in the past, *Antirrhinus majus,* a temperate plant, was valued as a specific against witchcraft; and the magical powers ascribed to *Viscum album* Mistletoe, an epiphytic parasite, included a charm against witches (53). In W Africa, *Phragmanthera* and *Tapinanthus* species Mistletoes have similar uses. The parasites, collected with the necessary rituals and used to scrub the shaven head of a confessed witch amid incantations dispossesses the victim of the craft (45). In Mexico, it has been observed that the Tarahumara shamans mix tobacco with the dry blood of a bat, either species of *Cynopterus, Eidolon, Epomophorus, Glossophaga* or *Pteropus;* and the dried meat of the turtle, either species of *Caretta, Chelonia* or *Eremochelys;* to purify medicine and protect patients against the evil of witchcraft (59).

In S Nigeria, *Tetrapleura tetraptera,* a tree legume, enters into a Yoruba Odu incantation to obtain the release of people tied by witches. The Yoruba name 'aidan' implies an imprecation to 'cast no spell' (63a, c). In Congo (Brazzaville), the fruit-pod has magico-medical properties and are used in certain ceremonies of exorcism (12). In S Nigeria, *Acanthus montanus* False thistle, a herbaceous plant in high forest, or *Dioscorea rotundata* Guinea yam or White yam is invoked by the Yoruba to save people from witches; and *Ageratum conyzoides* Billy goat weed enters into a Yoruba incantation on the strength of its smell 'to placate witches and kill bad medicine' (63c). In this country, *Canna indica* Indian shot, locally called 'ido' in Yoruba, enters into invocation by the tribe for protection against wizards and witches who are said not to eat 'ido' (63c). The Yoruba use *Ocimum gratissimum* Fever plant, a labiate, in an Odu incantation to imprecate witches already in a house to depart, and to fumigate a house as protection against witches (63c). In Trinidad, the leaves of Fever plant are put into a bath for bewitchment; or a bath with an infusion of *Cajanus cajan* Pigeon pea is against stroke and bewitchment (68). In N Nigeria, the Fulfulde take a leaf infusion of *Ocimum basilicum* Sweet basil as protection against witches (37).

In Botswana, *Halleria lucida* White olive is burned during spring each year and the ash, mixed with fat, is smeared on pegs made from the wood of *Rhamnus prinoides* Dogwood which are inserted in the ground around the villages and which are cultivated to protect them from wizards and bad weather (66). In this country, the Southern Sotho use the ash of *Phygelius capensis,* locally called 'mafifi-matso', against witchcraft; or use the root of *Asparagus declinatus,* a climbing lily, as a charm against witchcraft, by placing pegs which have been dipped in a decoction at the corner of the plot where a crop is growing (66). In Sierra Leone, *Palisota hirsuta,* a commelinaceous herb, is carried in the hands in witch divination (N.W. Thomas fide 18). In Ghana, the leaves of *Crescentia cujete*

Calabash tree or that of *Pancratium trianthum* or *P. hirtum* Pancratium lily, placed at the door, with a suitable invocation serve as preventive of witchcraft (47). This usage is also reported among the Hausa of N Nigeria, the Zarma of Mali and Niger and the Fula (47).

In many parts of W Africa, the ripe fruits of *Capsicum annuum* Pepper or *C. frutescens* Chillies with rock salt and a piece of charcoal are openly displayed by food sellers to 'keep away the evil eye' or to repel witchcraft - and probably also for marketing. A piece of charcoal and whole pepper may sometimes be added to the ingredients used in preparing soup or other dishes, especially in a household of more than one family, apparently for the same purpose. In S Sudan, the Azande also use charcoal against witchcraft. If a man sees the light of witchcraft he picks a piece of charcoal and throws it under his bed so that he may not suffer misfortune from the sight (25).

In Europe, *Helleborus niger* Christmas rose, a native of Central and S Europe has a romantic history and for centuries was thought to counteract witchcraft (53); and *Sambucus nigra* Elder growing near a house was believed to keep witches and bad spirits away (10). Another herb of supposedly magical properties was *Angelica archangelica* Angelica known as 'root of the Holy Ghost' which was used to counter the machinations of witches (54). In Mexico, the Kickappo Indians kept a bundle of *Rosmarinus officinalis* Rosemary, a temperate aromatic herbaceous labiate, in the house to ward off witchery (22).

The fumes of gum copal has been observed to be inimical to witches. Witches were kept away by the fumes of *Boswellia carteri* or *B. sacra* Frankincense and that of *Commiphora molmol* or *C. myrrha* Myrrh (17). (See EVIL: Table 1 for gum copal trees in the Region). People also protected themselves against witchcraft with holy water, or with consecrated salt, or with candles hallowed at Candlemas or with leaves blessed on Palm Sunday; and the horse-shoe over a doorway is a relic of protection against witchcraft (17). *Allium sativum* Garlic is mentioned in the *Egyptian Secrets* against witchcraft. 'Give to a human being or dumb animal, on St. Martin's Eve (10th November), three garlic bulbs to eat'.

Sympathetic Magic

Some of the prescriptions against witchcraft appear to be based on sympathetic magic. For instance, to the Chamba of N Nigeria, a decoction of *Echinops longifolius*, a perennial composite herb with spinose-margined leaves in savanna, added to bath water confers

Hygrophila auriculata

protection against suspected witchcraft (34). The practice is likely an instance of sympathetic magic based on the leaf spines. Among the Ga-speaking and other coastal tribes, the dried whole plant of *Hygrophila auriculata,* a thorny acanth in marshy places, kept in the house is a charm against evil forces in flight (15) - yet another instance of sympathetic based on the spines. A few leaves of this prickly acanth collected with ceremonial rituals and buried with seven fruits of *Capsicum frutescens* Chillies on the compound of a house before sunrise, is a charm against witchcraft (45). Some of the chillies may also be ground, mixed with water and poured over the hole. This is a further protection against these forces who are said never to visit the household again (45). This is probably an instance of sympathetic magic based both on the prickly acanth and the hot chillies. The clawed seeds of *Martynia annua,* a foetid erect herb, are also used against witchcraft by some of the tribal groups in its distribution area. The seeds are either worn like a pendant or encased in leather cover like a talisman (1); or placed in a pomade or in lavender which is regularly applied as a cosmetic (3). This application is likely an instance of sympathetic magic hinged on the clawed seeds.

Children's Cases

Children are particularly vulnerable to the evil influence of witchcraft, as such it is traditionally believed that some plant prescriptions are protective of this evil. Some coastal tribes in the Region believe that *Schweinkia americana,* an erect weedy herb of cultivation, is against witchcraft - especially its influence on children. The whole plant, collected with rites, is boiled with an egg which is eaten by one of the parents. The child only drinks some of the decoction and is bathed with the remainder for protection against witchcraft (45). Similarly, the child may be bathed with a decoction of *Evolvulus alsinoides,* a spreading hairy convolvulus herb, and fumigated with it to prevent his or her spirit from being taken away by witchcraft (45). The dried leaves of *Calotropis procera* Sodom apple

are burnt as incense or in fumigation against witchcraft - especially in cases involving children (47).

PLANTS WITH DOUBLE SPIRITUAL NATURE.

There are, however, a few plants with double spiritual nature - that is both for and against the craft. In Nigeria, *Calotropis procera* Sodom apple is endowed with special powers in witchcraft, both for and against the craft. The name of a person to be frightened is written with the blood of a cock, *Gallus domestica*, on the leaf (14a). Similarly, various superstitious uses of *Ricinus communis* Castor oil plant, *Milicia (Chlorophora) excelsa* and *M. regia* Iroko, forest timber trees, occur for purposes of witchcraft - both for and against. Among many tribes in W Africa, the stone in the heartwood of Iroko (excrescences of calcium malate) is believed to be inimical to witchcraft - for witches are said to detest water in which this stone has been immersed - though the tree itself is traditionally believed to promote the craft (45). Though the dried leaves of *Terminalia catappa* Indian almond are burnt into black powder and applied to the skin against witchcraft among some tribes in W Africa, the tree itself is traditionally believed to be a popular dining and meeting place for witches (45). In Mexico, *Solandra* species, trees on rocky cliffs believed to be the god-plant of mythology, are associated in the mind of the Huichol Indians with sorcery and witchcraft - the major causes of death and disease. However, the leaves are considered very powerful in curing cases of witchcraft and sorcery, or spells occasioned by them (40).

PLANTS FOR WITCHCRAFT.

On the contrary, some other plants are believed, among many cultures throughout the world, either to serve as a medium or a meeting point and venue of witches in furtherance of the craft and its evil activities; or are said to be positively associated with the practice as ingredients, or as accessories, or as vital constituents.

Plants Associated with Witch Meetings.

Ceiba pentandra Silk cotton tree, which occurs in both the high forest and the savanna country, and on both sides of the Atlantic, is reported to be an important feeding ground during the nocturnal activities of witches. Other forest trees on which witches are said to meet are *Milicia (Chlorophora) excelsa, M. regia* Iroko; *Triplochiton*

scleroxylon African whitewood; *Terminalia superba* Afara; and (rarely) *Khaya* species Mahogany (20). In S Sudan, it has been observed among the Azande that witches in the locality would beat their drums... and afterwards hold council in a baga tree (25). (The botanical name of the tree is *Khaya senegalensis* Dry-zone mahogany). *Cocos nucifera* Coconut palm is believed to be another feeding ground for witches. It has been confirmed that both Iroko and Coconut palm are associated with the nocturnal practices of witches (28).

In Ghana, *Vernonia cinerea* Little ironweed, an erect weedy composite herb, is reported to be used by witchcraft among the Ga-speaking and other tribes to aid and protect themselves during their nightly flights and travels (42). In Senegal, *Prosopis africana,* a savanna tree legume, is credited with soporific properties in native superstition. A witch who wishes (on Mondays and Fridays) to invoke the Devil without the knowledge of her sleeping spouse, places a stump or rod of the wood in contact with him in her absence (Perrot & Gerard fide 18). In Ghana, the leaves of *Baphia nitida* Camwood, locally called 'dwene' in Akan, or *B. pubescens*, small tree legumes, are used instead (20). It is the general belief among the tribes in the Guinea-savanna and the forest zone that the fruits of *Capsicum frutescens* Chillies used with rites and incantations to rub the 'vacated' body of a witch prevents the demon from re-entering the sleeping body (45). In a 'vacated' body, the breathing pace reduces considerably. The undue separation of this demon from the body to reactivate and restore it, causes the latter to die (45).

Other Plants For the Craft.

In Sierra Leone, the leaf petiole of *Carica papaya* Pawpaw is used in certain witchcraft to simulate a gun (Boboh fide 14a); and in Ghana, the branches of *Lantana camara* Lantana or Wild sage are used in witchcraft as firewood (41). In N Nigeria, the pods of *Cassia sieberiana* African laburnum are included in the Hausa *sandan mayu* or the Fula *chabbi-mai'be,* both meaning sorcerer's rod. The name is also applicable to the fronds of *Amblygonocarpus andongensis* and *Cassia arereh* (18). In Tanzania, *Amblygonocarpus* is used in witchcraft, and is said to have figured in a murder case (66). In N Nigeria, *Holarrhena floribunda* False rubber tree and other similar trees are used secretly by the Hausas in witchcraft - a twig of the tree is used by the wizard to rub his teeth, and whosoever he curses suffers evil or dies (18). In Nigeria, *Ageratum conyzoides* Billy goat weed, has a pleasant smell which is thought to

be pleasing to witches, so it is used in an incantation to make them happy (63a). In Liberia, *Leea guineensis*, a soft-wooded shrub, is important in witchcraft ceremonial; the bark of *Homalium letestui*, a forest tree, is used in native witchcraft; and *Cassipourea firestoneana*, a tree, is important in native superstition and for purposes of witchcraft (Cooper fide 18). In this country, the bark of *Scottellia coriacea*, a forest tree, is considered of great importance in witchcraft when juju amulets are losing their power (16).

In Lesotho, *Ilex mitis*, a mountainous tree usually by streams, has reputation in witchcraft (33, 66). A similar usage is reported in Madagascar (19). In Tanzania, the leaf and root of *Dombeya cincinnata;* a species of *Loranthus* Mistletoe, an epiphytic parasite; the fruit of *Kigelia africana* Sausage tree and *Mapronnea africana*, an euphorb, are used in witchcraft (66). In S Africa, the Pedi use the root of *Cyathula uncinulata*, an amaranth locally called 'maime', in witchcraft (66). In Gabon, the soft wood of *Spathodea campanulata* African tulip tree which smells of garlic when cut, is sometimes used as witch-doctor's wand, and the tree is associated with witchcraft. The blossoms are placed before huts of those who break tribal laws, and when the inmate dies (or is put to death) the flowers are burned with him to prevent his spirit returning (53). During the Middle Ages, *Hyoscyamus niger* Henbane joined the many ingredients used by witches and warlocks in their concoctions (52).

In W Africa, the extraordinary appearance of *Cissus quadrangularis* Edible-stemmed vine, a perennial climbing herb with square stems, naturally suggests superstitious uses, and it enters into magic prescriptions by wizards (18). In Liberia, *Impatiens irvingii* var. *irvingii* and var. *setifera*, semi-aquatic herbs, are generally avoided by some races as poisonous plants - but probably it is because of their association with witchcraft and sorcery (18). In this country, *Leea guineensis*, an erect or sub-erect shrub, plays an important part in local witchcraft to strengthen the oath medicine *(carfoo)*, and is believed to be impervious to the advances of even the most influential witch-doctors (16), thus giving assurance of purity of motive (18). In S Nigeria, the Yoruba name of *Crinum jagus*, a lily in forest swamps, 'edesuku', is given to a plaited hairstyle, *suku* style, which features in an invocation 'to send Sigidi to kill one's enemy' (63c). Though *Cnestis corniculata*, a woody forest climber, is not recorded for Nigeria, a plant said to resemble it, and likely to be *C. longiflora* or perhaps *C. grisea*, both Nigerian and both possessing the same pungent irritating hairs, is invoked in a Yoruba incantation under the name of 'witches bean cake' to make the vagina swell and so to harm a hostile woman, or to expel an intruder from one's house (63c).

ASSOCIATED PLANT NAMES

Some plants are, for various reasons - such as colour (usually bright or red colours), suspected or proven poisonous properties, destructive habits and evil or devilish attributes - named after witches. In Ghana, *Abrus precatorius* Prayer beads is called 'anyen enyiwa' by the Fanti tribe, meaning 'witches' eyes', with reference to the black and red seeds; and in Zanzibar, the whole plant and seeds are used in witchcraft (67). In S Nigeria, *Cnestis ferruginea,* a shrub or climber with bright red fruits, is locally called 'akara aje' by the Yoruba tribe, meaning 'witches' bread' or 'oyan oje', meaning 'witches' breast', inferring magical attributes - probably in reference to the bright red fruits (14a). In Liberia, *Dictyophleba leonensis,* a forest liane, is known by the Kru as 'waydoo', meaning 'witch vine' (14a). The twigs of *Ficus mucoso,* a fig tree, when dried and peeled glow at night and in SE Nigeria they are thus called 'pou buni', meaning 'witches fig tree' by the Ijo of Yenagoa (14d). In S Nigeria, *Mirabilis jalapa* Marvel of Peru has been superstitiously used as a medicine for witchcraft - the Yoruba name 'itanna pa osho' meaning 'the flower that kills the wizard' (18); or 'tanaposo', meaning 'light of fire kill wizard' (63c). The plant enters an incantation for protection against witches, and as a fetish and in medicine for witchcraft (63c).

Mirabilis jalapa Marvel of Peru

In Senegal, the Wolof name of *Stereospermum kunthianum,* a tree of savanna woodlands, is 'etudamo', 'etudemo' or 'yetudomo', meaning 'wand of a wizard or a sorcerer' (14e). In S Nigeria, the Yoruba name of *Uraria picta,* a woody legume herb in grassland, is 'alupayida', meaning 'wizardry'. In Sierra Leone, *Caladium bicolor* Heart of Jesus, a herbaceous perennial aroid cultivated for the decorative coloured leaves, is known by the contrary name 'hona gboji', meaning 'bewitched cocoyam', by the Mende tribe; and *Clerodendrum umbellatum,* a climbing shrub, is known by the tribe as 'hwana wulie', meaning 'bewitched' (!4a, e).

In Europe, *Vinca major* and *V. minor* Lesser periwinkle are known

Striga hermontheca Purple witchweed

as Sorcerer's violet, presumably because witches and sorcerers use it for making potions (53). In China, Japan and eastern N America occur *Hamamelis virginiana* Witch hazel and *H. mollis* Chinese witch hazel, useful garden shrubs. The plants were so named because of their popular use in former times as divining-rods (see FORTUNE-TELLING AND DIVINATION: THE FUTURE). The young twigs and roots bark of Witch hazel distilled in alcohol was used to make the astringent 'Witch Hazel' (53). The ripe woody capsules explode noisily, as such North American Indians believed the tree was bewitched (52).

The common name of *Striga* species Witchweed, locally called 'wumlim' among the Dagombas of Ghana, meaning 'evil' or 'poison', was given because the plants have a surprising, seemingly magical way of establishing themselves from seeds (26). Some of the species of *Striga* distinguished in the Region (4) are given in Table 2 below:

Species	Plant Family	Common Name	Local Name
Striga asiatica	Scrophulariaceae	Red witchweed	'wumlim'
S. gesneroides	Scrophulariaceae	Tobacco witchweed	'wumlim'
S. hermontheca	Scrophulariaceae	Purple witchweed	'wumlim'

Table 2. Some Species of Striga in the Region

Striga species are a threat to agriculture, especially in the woodland savanna and Sahel regions of W Africa as a serious parasitic weed of *Pennisetum glaucum (americanum)* Bulrush, Pearl or Spiked millet and *Sorghum bicolor* Guinea corn. Other crops attacked are *Oryza glaberrima* Upland rice, *Zea mays* Maize or Corn, *Vigna unguiculata* Cowpea, *Arachis hypogaea* Groundnut and *Saccharum officinarum* Sugarcane (2)

Other species of Striga in the Region are given in Table 3 below:

Species	Plant Family	Local Name
Striga aspera	Scrophulariaceae	'wumlim'
S. bilabiata ssp. rowlandii	Scrophulariaceae	'wumlim'
S. linearifolia	Scrophulariaceae	'wumlim'
S. macrantha	Scrophulariaceae	'wumlim'

Table 3. Other Species of Striga in the Region

In W Africa, the cultivation of *Senna (Cassia) occidentalis* Negro coffee for a year in fields affected by *Striga* causes the latter to disappear (2). In S Africa, Swati women sometimes cultivate *Kaempferia* species, a plant in the ginger family, in their garden in the belief that it will keep away Witchweed (66). (See also DEVIL; EVIL; JUJU OR MAGIC; MEDICINE; OCCULT; and WEALTH).

References.

1. Abaye, (pers. comm.); 2. Abbiw, 1990; 3. Afenu, (pers. comm.); 4. Akobundu & Agyakwa, 1987; 5. Amat & Cortadellas, 1972; 6. Ankoma-Ayew, (pers. comm.); 7. Apagyahene of Abiriw - Akwapim, (pers. comm.); 8. Appia, 1940; 9. Avumatsodo, (pers. comm.); 10. Back, 1987; 11. Beloved, (pers. comm.); 12. Bouquet, 1969; 13. Bravmann, 1974; 14a. Burkill, 1985; b. idem, 1994; c. idem, 1995; d. idem, 1997; e. idem, 2000; 15. Cofie, (pers. comm.); 16. Cooper & Record, 1931; 17. Crow, 1969; 18. Dalziel, 1937; 19. Debray & al., 1971; 20. Debrunner, 1959; 21. Dokosi, (pers. comm.); 22. Dolores & Latorre, 1977; 23. Ellis, 1894; 24. Eshun, (pers. comm.); 25. Evans-Pritchard, 1937; 26. Everard & Morley, 1970; 27. Ferry & al., 1974; 28. Field, 1937; 29. Fink. 1989; 30. Fowler & Fowler, 1953; 31. Gamadi, (pers. comm.); 32. Gluckman, 1963; 33. Guillarmod, 1971; 34. Hepper, 1965; 35. Holloway, (pers. comm.); 36a. Irvine, 1930; b. idem, 1961; 37. Jackson, 1973; 38a. Kerharo & Adam, 1964; b. idem, 1974; 39. Kingsley, 1897; 40. Knab, 1977; 41. Lamptey, (pers. comm.); 42. Lartey, (pers. comm.); 43. Livingstone, 1914; 44. Meek, 1931; 45. Mensah, (pers. comm.); 46. Monteil, 1953; 47. Moro Ibrahim, (pers. comm.); 48. Maja Naur, 1999; 49. Motte, 1980; 50. Nadel, 1953; 51. Noamesi, (pers. comm.); 52. Pamplona-Roger, 2001; 53. Perry & Greenwood, 1973; 54. Pitkanen & Prevost, 1970; 55. Pope, 1990; 56. Prause. 1988; 57. Rattray, 1927; 58. Serle, Morel & Hartwig, 1977; 59. Siegel, Collins & Diaz, 1977; 60. Thompson, (pers. comm.); 61. Tremearne, 1913; 62. Van Sertima, 1979; 63a. Verger, 1965; b. idem, 1966; c. idem, 1967; d. idem, 1972; 64. Visser, 1975; 65. Walker & Sillans, 1961; 66. Watt & Breyer-Brandwijk, 1962; 67. Williams, 1949; 68. Wong, 1976.

DISCUSSION

Finally, brethren, whatsoever things are true,
...honest,...just,...pure,...lovely,...of good report;
if there be any virtue, and if there be any praise,
think of these things.
Philippians 4:8

Evidence from all the fifty-two headings testifies that some traditional uses of plants in W Africa and elsewhere are based on beliefs, symbols, signs and traditional values. This appears to be in line with man's quest (since prehistoric times) to understand and interpret nature to unravel and utilize the hidden secrets of plants for his needs, and to his benefit and advantage. There are also instances of sympathetic magic or Doctrine of Signatures in achieving objectives - such as preventing an incident or protection from adverse intention. Some plants are symbolic of various concepts among the ethnic groups in the Region and beyond (see APPENDIX - Table 1). The spiritual values of plants evidently reflects as National, State, and Emblematic Plants in many countries worldwide (see APPENDIX - Table IV). Since traditional practices and beliefs are handed down from one generation to the next, these values are likely to be cherished and preserved indefinitely.

Traditional practices and beliefs together with the accompanying rituals, libation, ceremonies, prayers, incantations and invocations may be explainable or not easily so. The practices may be beneficial to mankind or may be primitive and cruel. Occasionally, there are some discrepancies.

EXPLAINABLE

Generally, many of the practices and beliefs are explainable. For instance, the practice of insisting that medicinal plants dug up for their roots should be covered up again with soil (see MEDICINE: RITUALS) as a necessary prerequisite for the efficacy of the medication, is in effect a conservation measure to preserve and conserve medicinal plants, though this has been shrouded in superstition. Another example is the belief that if the first fish that come up into the lake in the great dry season are killed, the rest of the shoal turn back (see TABOO: GROVES: Fishing). It is worth a note that in these two examples, and many other instances mentioned in the text, the practices and beliefs are not only beneficial to humanity and to the environment in the conservation of biodiversity, but also in ensuring sustainable yield.

NOT EASILY EXPLAINABLE

Some other traditional practices cannot be easily explained - especially those practices with a spiritual or supernatural touch. More often, this invites scepticism and wonder; and it is just beyond comprehension. In effect this aspect of our cultural heritage is the real crux of the subject. The immersion of *Mormordica charantia* African cucumber with other assorted plant ingredients in water, among incantations and libation, in rituals to induce spirit possession in a medium during ceremonies in connection with twins or on other occasions (see TWINS: CEREMONIES: Spirit Possession), is an instance. Similarly, the accuracy with which the creeper *benge*, a species of *Strychnos*, can foretell or predict future events - either by causing a fowl, *Gallus domestica*, to die or not to die, is equally spiritual (see FORTUNE-TELLING AND DIVINATION: PREDICTING THE FUTURE).

The traditional use of *Senecio (Crassocephalum) biafrae*, a climbing forest composite, for purification by two separated tribal groups in Africa, without any record of its use for this ritual by any of the tribes in between them, requires some explanation as well (see CEREMONIAL OCCASIONS: PURIFICATION CEREMONIES: Purification). It is even more puzzling that a tribal group in W Africa uses *Amorphophallus* species, perennial aroids with underground tuber, to symbolize death; and another tribal group in Sumatra, far away in Indonesia, also uses *Amorphophallus titanus*, for the same purpose (see DEATH AND DYING: PLANT NAMES AND DEATH). Similarly, the use of *Paullinia pinnata*, a climbing plant with tendrils, for courage during war in W Africa; and *Paullinia curana* Guarana also for courage during war in Japan, is a puzzle (see WAR: COURAGE AND SUCCESS CHARMS).

The invocation of certain plants as charms allegedly to effect invisibility or vanishing powers (see HUNTING: INVISIBILITY AND VANISHING CHARMS) certainly belongs to the supernatural realm. Similarly, bullet-proof prescriptions (see PROTECTION: FIRE ARMS); immunity from cutting instruments (see PROTECTION: CUTTING OR PIERCING WEAPONS); invoking or redirecting lightning from its

original course (see LIGHTNING AND THUNDER: CHARMS AGAINST LIGHTNING); are equally all spiritually initiated. The hidden secrets behind these traditional practices and beliefs still baffle mankind.

DISCREPANCY.

There are some recorded instances of opposing traditional practices and beliefs. For example, in W Africa, *Alstonia boonei* Pagoda tree, is known as 'Sky-God's tree - a symbol of light and perfection (see RELIGION: RELIGIOUS CEREMONIES: Rituals: *Nyame-dua*); but in Indomalaya, *Alstonia scholaris* is called 'Devil Tree' - a symbol of darkness (see DEVIL: PLANTS ASSOCIATED WITH THE DEVIL). In parts of W Africa, some tribal groups use the fronds of *Elaeis guineensis* Oil palm as symbol of peace, but in S Nigeria, the fronds are used by the Igbos to symbolize war (see WAR: PEACE). A controversial religious taboo surrounds the planting of *Cola acuminata* Commercial cola nut tree and *C. nitida* Bitter cola. To some tribal groups the event is feared to result in death when the tree flowers; but to another it commemorates a joyous event (see TABOO: GENERAL FOR ALL AND SUNDRY: Farming and Fishing). There is a similar traditional belief about cutting *Dialium guineense* Velvet tamarind. To the people of Yaoundé in Cameroon, it is certain death within the year. However, to the Aku of Sierra Leone who are of Yoruba descent, the tree must rather be cut else when it grew up and seeded around the village, it brought fever (see TABOO: FELLING TREES).

In the East, *Cynodon dactylon* Bermuda grass is 'Preserver of Nations' or the 'Shield of India' (see SACRED PLANTS: NATIONAL AND STATE PLANTS); but in the West Indies, Bermuda grass is Devil's grass (see DEVIL: PLANTS NAMED AFTER THE DEVIL). Similarly, though *Adansonia digitata* Baobab, a large savanna tree often planted in villages, is revered throughout Africa by many tribal groups in its distribution area as a sacred tree, in Kenya, the tree is believed to have links with the Devil.

There are also few recorded instances where some plants are used both for and against the practice of witchcraft. Table 1 lists some of these plants.

Species	Plant Family	Common Name	Remarks
Calotropis procera	Asclepiadaceae	Sodom apple	Savanna shrub
Cola acuminata	Sterculiaceae	Commercial cola nut tree	Tree
C. nitida	Sterculiaceae	Bitter cola	Tree
Milicia (Chlorophora) excelsa	Moraceae	Iroko	High forest tree
M. (Chlorophora) regia	Moraceae	Iroko	High forest tree
Pupalia lappacea	Amaranthaceae	-	Herb
Ricinus communis	Euphorbiaceae	Castor oil plant	Shrub
Sida linifolia	Malvaceae	-	Herb
Solandra species	Solanaceae	-	God-plant of Huichol Indians
Terminalia catappa	Combretaceae	Indian almond	Introduced cultivated tree

Table 1. Plants With Double Spiritual Power

Milicia (Chlorophora) excelsa and *M. (Chlorophora) regia* Iroko, with ingredients, is used by traders in the belief of preventing money snatching, however, money saved in Iroko chest is believed to be either mysteriously halved or spirited away altogether (see MARKETING: PREVENTING MONEY SNATCHING). Despite its popular use mainly among the Ewe-speaking tribe for love charms and against evil, witchcraft and bad medicine, *Sida linifolia*, an erect malvaceous weed of cultivated land, also enters charms to kill someone among the same tribe (see JUJU OR MAGIC: 'AFRICAN ELECTRONICS': Sinister Rituals). Among this tribe, *Pupalia lappacea*, a weedy amaranth with hooked fruits, is used both as a love charm and to cause dissension among lovers (see LOVE: CHARM PRESCRIPTIONS). Similarly, kola- nuts do have opposing traditional uses depending on whether the nuts are red- or white-coloured. The use of the wrong coloured type is traditionally believed to have the opposite effect (see CEREMONIAL OCCASIONS: RITUAL CEREMONIES; MARKETING: CHARM PRESCRIPTIONS; RELIGION: RELIGIOUS CEREMONIES: Rituals: Kola-nuts; and WAR: PEACE). On the whole however, many of the practices and beliefs are generally uniform throughout the Region.

BENEFICIAL

Some of the traditional practices have been, and are still, beneficial to mankind. In fact, some are of conservation importance since the ultimate objective is sustained yield. Such practices are worth support and encouragement. For instance, the fear and respect of the people for trees professed to have supernatural powers and the subsequent

associated taboos eventually tends to preserve these plants (see JUJU OR MAGIC: CONSERVATION). Other examples are trees that are either forbidden to be used as fuel-wood or to be felled (see TABOO: FUEL-WOOD and FELLING TREES); or plants revered and designated as National and State Plants (see SACRED PLANTS: NATIONAL AND STATE PLANTS; and APPENDIX - Table IV). In general, all sacred, fetish and juju plants appear to have been conserved and protected from over exploitation and extinction to this day because of appropriate religious prohibitions and sanctions.

PRIMITIVE AND CRUEL

However, some other traditional practices are simply primitive. An example is scraping and scarifying the lips with the leaves of *Ficus asperifolia* Sandpaper tree and *F. exasperata* as punishment for evil-speaking (see EVIL: PLANTS AGAINST EVIL). The traditional practice of administering an ordeal poison to suspected persons for offences like alleged witchcraft, murder or adultery (see ORDEAL: PLANT TESTS); or killing someone either for violating a sacred day (see TABOO: GENERAL FOR ALL AND SUNDRY: Farming and Fishing); or killing others (often in numbers) as a sacrifice to the dead (see FETISH: APPEASEMENT SACRIFICES: Human Sacrifice); or killing babies because they are twin born or tenth born (see TWINS: CEREMONIES: Taboo); appear to be all primitive and cruel.

THE JUSTIFICATION

Before condemning these practices outright, however, they must be viewed critically in their right perspective and in the context of the whole historic background - the period, the environmental influences, the belief systems under which our ancestors lived; and the factors which might have compelled the institution of such rather strict rules and regulations with an equally severe punishment for defaulters. The steady decline of both cultural and moral values with 'civilization' and urbanization, gives reason to believe that simple primitive societies might have been highly conscious of these values. They respected, cherished , protected and enforced them; and set up very high moral standards and regulations to enforce and maintain these values.

Where the tribal group was half naked, naked or almost so, for example, the temptation for promiscuous sexual relations or illicit love affairs was high. In order to maintain the required moral standards expected within the society during these times and under these conditions, the severest punishment was meted to any offenders as a deterrent to others. Adultery in a wife, for instance, was then punishable by death or the right hand was cut off at the wrist (se MARRIAGE: FAITHFULNESS IN MARRIAGE: Punitive and Preventive Measures: Death or Mutilation). The latter punishment could also be applied to her paramour.

INFLUENCE OF CIVILIZATION.

With the opening up of the African continent, light penetrated darkness, and primitive or barbaric life with its associated jungle law has now given way to developing people and constituted laws. There has since been a remarkable change in the social, economic, political, religious, commercial and judicial aspects of the society. All these changes have had a corresponding influence on the beliefs and practices of the people. Christianity and Islam have been added on to traditional religious beliefs; and western laws have now replaced some of the traditional ones. Education has reduced illiteracy levels. Orthodox medicine is now being practiced side by side with traditional medicine. Science and technology has made transportation and communication easier and faster; and made agriculture less laborious and less time consuming. Finally, scientific explanation of events is gradually replacing superstitious interpretations.

With civilization and the accompanying reforms, the necessary changes and amendments to some of the outmoded traditional laws and regulations - especially those adversely affecting the whole society or against human rites - have been accordingly made. Consequently, slavery, trial by ordeal, ritual murder, killing of twins, carrying the body of a dead person to determine the murderer, accusations of witchcraft, burning a dead body as a punishment because the dead was a witch or wizard or murderer are all presently illegal in law.

PERSISTENT PRACTICES AND BELIEFS

Nevertheless, some traditional practices and beliefs still remain an inseparable part of the social structure throughout the Region. So long as beliefs in, or accusations and allegations of evil influences, bad juju, sorcery and witchcraft occur; or farm produce and other personal valuables are pilfered by thieves; or unfaithfulness in marriage or adultery is suspected; the people will naturally seek the required

protective or counter charms, armlets, talisman and the like. Similarly, the fear of a curse or a spell or a snakebite; or the desire for success in political aspiration, marketing, business, examination, fishing, hunting or sports; or for favour in court may compel the individual to consult a medicine-man or a fetish priest for the necessary charms.

REFORMS

On the whole, however, the majority of the practices and beliefs that have clearly outlived their usefulness, or are a hindrance to progress and development, have either died away naturally or been modified or been banned altogether by law. It is for the present generation to cherish and preserve those traditional practices and beliefs that are still beneficial and useful to the people, and in turn hand them over to future generations.

APPENDIX

SYMBOLIC PLANTS

Species	Common Name	Symbol of
Acanthus montanus	Falsethistle	Purification
Adansonia digitata	Baobab	Fertility
Adenia rumicifolia var. miegei	-	Purification
Aloe saponaria	Soap aloe	Purification
Alstonia boonei	Pagoda tree	Dormaa's attachment to God
Amaranthus species	Amaranth	Immortality
Amorphophalus species	-	Death
Anastatica hierochuntica	Resurrection or Mary's flower	Resurrection
Asparagus crispus	-	Fertility and prosperity of domestic animals
Baphia nitida	Camwood	Intelligence and wise counseling
Blighia sapida	Akee apple	Fecundity among the Ando
Bougainvillea species	Bougainvillea	Fragrance and purity and beauty
Bryophyllum pinnatum	Resurrection plant	Resurrection
Buxus sempervirens	Box	Resurrection. Palm Sunday
Cassytha filiformis	Dodder	Love
Ceiba pentandra	Silk cotton tree	Maleness
Cercis siliquastrum	Judas tree	Evil
Cola acuminata	Commercial cola nut tree}	Welcome and hospitality
C. nitida	Bitter cola }	Love, friendship and reconciliation
two red kola nuts	Kola nuts	War among the Mende
one white nut broken into two	Kola nuts	Peace among the Mende
Commiphora africana	African bdellium	Immortality
C. molmol	- }	Continence
C. myrrha	Myrrh }	
Copaifera religiosa	'Oven tree'	Fecundity, wealth, power and fame
Costus afer	Ginger lily	Purification from sin
Croton zambesicus	-	Ekoi people of Nigeria
Cupressus sempervirens	Cypress	Death
Cuscuta australis	Dodder	Love
Dianthus superbus	-	The ideal woman - a combination of strength and grace
Diospyros mespiliformis	W African ebony tree	Luck
Dissotis rotundifolia	-	Purification
Dracaena arborea	-	Peace
Distemonanthus benthamianus	African satinwood	Evil

Elaeis guineensis (fronds)	Oil palm	Palm Sunday and Easter Sunday
Elettaria cardamomum	Cardamon	Hospitality
Eragrostis tremula	-	Purification
Erythrina crista-galli var. variegata	-	Evil
E. senegalensis	Coral flower or Beed tree	Reconciliation
Ficus iteophylla	- }	
F. kamerunensis	- }	Fertility and
F. sur (capensis)	Fig tree }	Fecundity among the Ando of Ivory Coast
F. vogeliana	}	
Gardenia species	-	Graceful charm
Ginkgo biloba	Ginkgo or Maidenhair tree	Longevity
Gomphocarpus physocarpus	-	Fertility
Hypselodelphys poggeana }	-	Strength and durability
H. violacea }	-	
Hyssopus officinalis	Hyssop	Purification from sin
Iris florentina	Orris root	Power and majesty
Kigelia africana	Sausage tree	Fertility
Laurus nobilis	Sweet bay or Laurel	Victory
Lonicera periclymenun	Honeysuckle	True devotion
Magnolia species	Magnolia	Fragrance of virtues, Love and happiness
Malva silvestris	High mallow	Sweetness and calmness
Marantochloa species	-	Maleness
Milicia (Chlorophora) excelsa	Iroko }	Refuge
M. regia	Iroko }	
Momordica charantia	African cucumber	Purification after burial
Mussaenda erythrophylla	Ashanti blood	War
Myrtus communis	Myrtle	Compassion
Nelumbium nelumbo (Nelumbo nucifera)	Nelumbo or Blue lotus	The Ganges
Newbouldia laevis	-	Deities, Authority, Purification after burial and Reconciliation
Nigella damascena	Love-in-a-mist	Love
Nymphaea lotus	Water lily	Fertility. Chastity. Truth and purity Good or evil, Light or darkness Male or female. Intellect Immortality, purity and resurrection Peaceful intent, Regeneration & purification
Olea europaea	Olive	Peace, Freedom and purity
(the oil)	Olive	The Holy Spirit
Paeonia species	'King of flowers'	Wealth and honour
Pandanus species	-	Longevity

Papaver rhoeus	Poppy flower	Remembrance Day, Poppy Day
Passiflora species	Passion flower	Our Saviour's Passion
Piliostigma thonningii	-	The bush cat, Viverra civetta
Pinus species	Pine	Integrity and dignity Long life and constancy
Piper guineense	West African black pepper	Ghosts among the Ekoi
Plumeria rubra var. acutifolia	Frangipani	Fragrance and purity and beauty Remembrance
Portulaca oleracea	Purslane or Pigweed	Purification, Goodwill, Peace
Prunus domestica	Plum }	Brotherliness and cordial relationship
P. persica	Peach }	Physical charm and loveliness of women as well as of the spirit
P. persica	Peach	Longevity
Punica granatum	Pomegranate	Babylonian and Syrian god of thunder and of storms Fertility, fecundity, longevity and love
Pupalia lappacea	-	Human emotions
Quercus species	Oak	England
Reseda odorata	Mignonette	Mildness
Rosa species	Rose	Love and charity, Discretional secrecy
Rosmarinus officinalis	Rosemary	Fidelity and remembrance
Salix babylonica	Weeping willow	Friendship and mercy as well as grace
S. woodii	Wild willow	Palm Sunday
Senecio biafrae	-	Purification
Solanum incanum	Egg plant or Garden egg	Fertility
Spondias mombin	Hog plum or Ashanti plum	Purification after burial
Tanacetum vulgare	Tansy	Bitter herbs eaten at Passover
Taxus baccata	Yew	Death
Thaumatococcus daniellii	Katemfe	Female fertility
Thymus vulgaris	Thyme	Courage and bravery
Tilia europaea	Linden	Family and homelife
Vitis vinifera	Grape vine	People of Israel (in biblical times)
A tree in African myth and stories	-	Wisdom, authority and custom Mediator and judge
A tree in glazed containers	-	Those qualities that enrich life
Bamboo in Chinese mythology	-	Virtues as purity of mind, fidelity, humility, wisdom and gentleness
Firewood	-	Submission
Orchids	-	Culture, refinement and nobility of character

LEGENDARY AND MYTHICAL PLANTS

Species	Common Name	Plant Family
Adansonia digitata	Baobab	Bombacaceae
Aloe vera	Aloe	'Liliaceae' (Aloaceae)
Amanita muscaria	Death cap	Basidiomycetes
Angelica archangelica	Angelica	Umbelliferae
Balanites aegyptiaca	Desert date or Soap berry	Zygophyllaceae
Borassus aethiopum	Fan palm	Palmae
Ceiba pentandra	Silk cotton tree	Bombacaceae
Daniellia oliveri	African copaiba balsam tree	Caesalpiniaceae
Diospyros mespiliformis	West African ebony tree	Ebenaceae
Distemonanthus benthamianus	African satinwood	Caesalpiniaceae
Ficus species	Fig trees	Moraceae
Gardenia ternifolia ssp. jovis-tonantis	-	Rubiaceae
Ginkgo biloba	Ginkgo or Maidenhair tree	Ginkgoaceae
Hyssopus officinalis	Hyssop	Labiatae
Laurus nobilis	Laurel or Sweet bay	Lauraceae
Lawsonia inermis	Henna	Lythraceae
Lodoicia maldivica	Coco-de-mer	Palmae
Mandragora officinarum	Mandrake	Solanaceae
Metasequoia glyptostroboides	The fossil tree	Taxodiaceae
Milicia (Chlorophora) excelsa	Iroko	Moraceae
M. regia	Iroko	Moraceae
Morus mesozygia	-	Moraceae
Mussaenda erythrophylla	Ashanti blood	Rubiaceae
Myrtus communis	Myrtle	Myrtaceae
Ocimum gratissimum	Fever plant	Labiatae
Okoubaka aubrevillei	-	Santalaceae
Olea europaea	Olive	Oleaceae
Origanum dictamnus	Dittany of Crete	Labiatae
Paeonia officinale	Common paeony	Paeoniaceae
Panax quinquefolium	Gingseng	Araliaceae
P. schinseng	Gingseng	Araliaceae
Papaver rhoes	Poppy flower	Papaveraceae
Phragmanthera species	Mistletoe	Loranthaceae
Punica granatum	Pomegranate	Punicaceae
Quercus species	Oak	Fagaceae
Rosmarinus officinalis	Rosemary	Labiatae
Ruta graveolens	Common rue	Rutaceae

Salix babylonica	Weeping willow	Salicaceae
Securidaca longepedunculata	Rhode's violet	Polygalaceae
Sempervivum officinale	Houseleek	Crassulaceae
Sequoiadendrum giganteum	Giant sequioa	Taxodiaceae
Solandra species	-	Solanaceae
Tabernanthe iboga	Iboga	Apocynaceae
Tamarindus indica	Indian tamarind	Caesalpiniaceae
Tapinanthus species	Mistletoe	Loranthaceae
Taxodium mexicana	Mexican cypress	Taxodiaceae
Verbena officinalis	Vervain	Verbenaceae
Viscum album	Mistletoe	Viscaceae
Vitis vinifera	Grape vine	Vitidaceae
Welwitschia mirabilis	-	Welwitschiaceae

RELIGIOUS PLANTS

Traditional Religion

Species	Common Name	Plant Family
Adansonia digitata	Baobab	Bombacaceae
Alstonia boonei	Pagoda tree	Apocynaceae
Blighia sapida	Akee apple	Sapindaceae
Ceiba pentandra	Silk cotton tree	Bombacaceae
Chrysophyllum albidum	White star apple	Sapotaceae
Detarium microcarpum	Tallow tree	Caesalpiniaceae
D. senegalense	Tallow tree	Caesalpiniaceae
Elaeis guineensis var. idolatrica	King oil palm	Palmae
Ficus dekdekena (thonningii)	-	Moraceae
F. lutea	-	Moraceae
F. lyrata	-	Moraceae
F. sur (capensis)	Fig tree	Moraceae
F. sycomorus (gnaphalocarpa)	Sycomore	Moraceae
F. vogeliana	-	Moraceae
Kigelia africana	Sausage tree	Bignoniaceae
Lodoicea maldivica	Coco-de-mer	Palmae
Milicia (Chlorophora) excelsa	Iroko	Moraceae
M. regia	Iroko	Moraceae
Okoubaka aubrevillei	-	Santalaceae
Pennisetum purpureum	Elephant grass	Gramineae
Sarcostemma viminale	-	Asclepiadaceae
Sclerocarya caffra	Cider tree	Anacardiaceae
Tabernanthe iboga	Iboga	Apocynaceae
Theobroma cacao	Cocoa	Sterculiaceae

Islam and Oriental Religion

Species	Common Name	Plant Family
Cola acuminata	Commercial cola nut tree	Sterculiaceae
C. nitida	Bitter cola	Sterculiaceae
Dictamnus albus	Gas plant or Burning bush or Candle plant	Rutaceae
Elaeocarpus angustifolius	-	Elaeocarpaceae
Ficus benghalensis	Banyan	Moraceae
F. religiosa	Pepul or Bo-tree	Moraceae
Ginkgo biloba	Ginkgo	Ginkgoaceae
Gloriosa superba	Climbing lily	'Liliaceae' (Asphodelaceae)
Hydrangea macrophylla var. serrata	-	Saxifragaceae
Ixora grandifolia	-	Rubiaceae
Nelumbium nelumbo (Nelumbo nucifera)	Nelumbo or Blue lotus	Nelumbonaceae
Nymphaea lotus	Water lily	Nymphaeaceae
Phoenix dactylifera	Date palm	Palmae
Sesamum orientale (indicum)	Sesame or Beniseed	Pedaliaceae
Thevetia peruviana	Exile oil plant	Apocynaceae

Other Religions

Species	Common Name	Plant Family
Amanita muscaria	Death cap	Basidiomycetes
Dianthus caryophyllus	Carnation	Caryophyllaceae
Erythroxylum coca	Coca	Erythroxylaceae
Nandina domestica	Heavenly bamboo	Berberidaceae
Nicotiana rustica	Wild tobacco	Solanaceae
Psilocybe mexicana	-	Basidiomycetes
Verbena officinalis	Vervain	Verbenaceae
Viscum album	Mistletoe	Viscaceae

The Christ

Species	Common Name	Plant Family
Acanthus spinosus	Spiny bear's breach	Acanthaceae
Ailanthus altissima	Chinese tree of heaven	Simaroubaceae
Anemone coronaria	De Caen or Giant flowered French or Blood drops of Christ	Ranunculaceae
Caladium bicolor	Heart of Jesus	Araceae
Canotia species	-	Canotiaceae
Crataegus oxyacantha	Hawthorn	Rosaceae
Euphorbia millii (splendens)	Crown-of-thorns or Christ's thorn	Euphorbiaceae

Holacantha emoryi	Desert crucifixion-thorn	Simaroubaceae
Hyssopus officinalis	Hyssop	Labiatae
Koeberlinia species	-	Capparaceae
Paliurus spina-christi	Christ's thorn	Rhamnaceae
Passiflora caerulea	Passion flower	Passifloraceae
P. edulis	Passion flower	Passifloraceae
P. glabra	Stinking passion flower	Passifloraceae
P. laurifolia	Water-lemon	Passifloraceae
P. quadrangularis	Giant granadilla	Passifloraceae
Poterium spinosum	Spiny burnet	Rosaceae
Porana paniculata	Christ plant or Christmas-vine	Convolvulaceae
Tragia species	-	Euphorbiaceae
Ziziphus spina-christi	Christ's thorn	Rhamnaceae

The Virgin Mary

Species	Common Name	Plant Family
Anastatica hierochuntica	Rose of Jericho or Mary's flower or Resurrection flower	Cruciferae
Juniperus communis	Juniper	Cupressaceae
Lilium candida	Madona lily	Liliaceae
Merremia discoidesperma	Mary's bean or Crucifixion-bean	Convolvulaceae
Silybum marianum	Blessed Mary thistle	Compositae

Biblical Plants

Species	Common Name	Plant Family
Artemisia absinthium	Wormwood	Composite
Boswellia carteri	Frankincense	Burseraceae
B. papyrifera	Frankincense	Burseraceae
B. sacra	Frankincense	Burseraceae
Brassica nigra	Black mustard	Cruciferae
Chlorella species	Chlorella	Chlorophyceae
Cinnamomum zeylanicum	Cinnamon	Lauraceae
Commiphora gileadensis	Balm of Gilead	Burseraceae
C. molmol	Myrrh	Burseraceae
C. myrrha	Myrrh	Burseraceae
Coriandrum sativum	Coriander	Umbelliferae
Crocus sativus	Saffron	Iridaceae
Cyperus papyrus	Pappyrus	Cyperaceae
Ficus carica	Common fig	Moraceae

F. sycomorus	Sycomore	Moraceae
Juglans regia	Walnut	Juglandaceae
Laurus nobilis	Bay tree	Lauraceae
Morus nigra	Black nigra	Moraceae
Myrtus communis	Myrtle	Myrtaceae
Nandostachys jatamonsi	Spikenard	Valerianaceae
Nerium oleander	Oleander	Apocynaceae
Nymphaea lotus	Water lily	Nymphaeaceae
Ornithogalum arabicum	Star of Bethlehem	Liliaceae
O. narbonense	Star of Bethlehem	Liliaceae
O. umbellatum	Star of Bethlehem	Liliaceae
Pimpinella anisum	Anise	Umbelliferae
Platanus orientalis	Oriental plane tree	Platanaceae
Punica granatum	Pomegranate	Punicaceae
Prunus armeniaca	Apricot	Rosaceae
Rubus sanquineus	Bramble	Rosaceae
Ruta graveolens	Rue	Rutaceae
Sternbergia lutea	Lilies of the field'	Amaryllidaceae
Tanacetum vulgare	Tansy	Composite

Temples (Plants used in building or planted around)

Species	**Common Name**	**Plant Family**
Cantua buxifolia	-	Polemoniaceae
Caralluma dalzielii	-	Asclepiadaceae
C. decaisneana	-	Asclepiadaceae
Cedrus libani	Cedar of Lebanon	Pinaceae
Cryptomeria japonica	Sacred 'Japanese cedar'	Taxodiaceae
Cupressus sempervirens	Cypress	Cupressaceae
Dipterocarpus trinervis	-	Dipterocarpaceae
Ficus religiosa	Pepul or Bo-tree	Moraceae
Ginkgo biloba	Ginkgo or Maidenhair-tree	Ginkgoaceae
Helianthus annuus	Sunflower	Compositae
Hyssopus officinalis	Hyssop	Labiatae
Illicium anisatum (religiosum)	-	Illiciaceae
Juniperus drupaccea	Juniper	Cupressaceae
J. excelsa	Juniper	Cupressaceae
Michelia champaca	-	Magnoliaceae
Nandina domestica	Heavenly bamboo	Nandinaceae
Nelumbium nelumbo (Nelumbo nucifera)	Nelumbo or Blue lotus	Nelumbonaceae

Paeonia suffruticosa	The tree paeony or The moutan	Paeoniaceae
Phoenix dactylifera	Date palm	Palmae
Pinus halepensis	Allepo pine	Pinaceae
Plumeria rubra var. acutifolia	Frangipani or Temple flower	Apocynaceae
Rosmarinus officinalis	Rosemary	Labiatae
Santalum album	Sandalwood	Santalaceae
Savastana odorata	Holy grass	Gramineae
Styrax officinalis	Storax	Styracaceae
Tagetes lucida	Marigold	Compositae
Thespesia populnea	Portia tree	Malvaceae

NATIONAL, STATE OR EMBLEMATIC PLANTS

Country	Species	Common Name	Plant Family
Alabama	*Camellia species*	Camellia	Theaceae
Alaska	*Myosotis alpestris*	Alpine Forget-me-not	Boraginaceae
Albania	*Quercus coccifera*	Kermes oak	Fagaceae
Andorra	*Leontopodium alpinum*	Edelweiss	Compositae
Argentina	*Erythrina crista-galli*	Ceibo or Coral-tree	Papilionaceae
Arizona	*Cereus giganteus*	Sagurao cactus	Cactaceae
Arkansas	*Pyrus malus*	Apple blossom	Rosaceae
Ascension	*Zantedeschia aethiopica*	Calla	Araceae
Australia	*Eucalyptus globulus*	Blue gum	Myrtaceae
Austria	*Lilium species*	White lily	Liliaceae
Barbados	*Poinciana pulcherrima*	Barbados-pride	Caesalpiniaceae
Belgium	*Papaver rhoeas*	Red poppy or Coca poppy	Papaveraceae
Bermuda	*Sisyrinchium bermudiana*	Blue-eyed grass	Iridaceae
Bolivia	*Cantua buxifolia*	Inca magic-flower or Khantua	Polemoniaceae
Borneo	*Nipa fruticans*	Nipa palm	Palmae
Brazil	*Caesalpinia echinata*	Pau Brazil	Caesalpiniaceae
Bulgaria	*Rosa species*	Red rose	Rosaceae
Burma	*Oryza sativa*	Rice	Gramineae
Caicos Islands	*Melocactus communis*	Turk's-cap cactus	Cactaceae
California	*Eschscholzia californica*	California poppy	Papaveraceae
Cambodia	*Oryza sativa*	Rice	Gramineae
Canada	*Acer plantanoids*	Maple	Aceraceae
Canary Islands	*Dracaena draco*	Dragon tree	'Liliaceae' Dracaenaceae
Ceylon	*Corypha umbracalifera*	Talipot palm	Palmae
Chile	*Lapageria rosea*	Chilean bell flower	Philesiaceae

China	*Narcissus tazetta var. orientalis*	Chinese sacred-lily	Amaryllidaceae
Colombia	*Cattleya trianae*	Mayflower orchid	Orchidaceae
Colorado	*Aquilegia caerulea*	Columbine	Ranunculaceae
Congo (Brazzaville)	*Adansonia digitata*	Baobab	Bombacaceae
Connecticut	*Kalmia latifolia*	Mountain laurel	Ericaceae
Costa Rica	*Cattleya trianae*	Mayflower orchid	Orchidaceae
Cuba	*Roystonea regia*	Royal palm	Palmae
Czechoslovakia	*Dianthus caryophyllus*	Carnation	Caryophyllaceae
Delaware	*Prunus persica*	Peach	Rosaceae
Denmark	*Fagus sylvatica*	Beech	Fagaceae
Dominican Republic	*Swietenia mahogani*	Caoba mahogany	Meliaceae
Ecuador	*Theobroma cacao*	Cocoa	Sterculiaceae
Egypt	*Nymphaea lotus*	Water lily	Nymphaeaceae
El Salvador	*Yucca elephantipes*	Spanish-bayonet	Agavaceae
England	*Rosea gallicca*	Red Tudor rose	Rosaceae
Eritrea	*Zantedeschia species*	Calla	Araceae
Ethiopia	*Coffea arabica*	Arabian coffee	Rubiaceae
Finland	*Convallaria majalis*	European lily of the valley	Liliaceae
Florida	*Citrus trifoliata*	Orange blossom	Rutaceae
France	*Iris pseudacorus*	Golden Fleur de Lis	Iridaceae
Georgia	*Rosa sinica*	Cherokee rose	Rosaceae
Germany	*Centaurea cyanus*	Cornflower or Keiserblume	Compositae
Ghana	*Milicia (Chlorophora) excelsa*	Iroko	Moraceae
Greece	*Viola odorata*	Violet	Violaceae
Guatemala	*Lycaste skinneri*	White-nun orchid	Orchidaceae
Guernsey Island	*Nerine sarniensis*	Guernsey-lily	Amaryllidaceae
Haiti	*Roystonea regia*	Royal palm	Palmae
Honduras	*Pinus rigida*	Pitch pine	Pinaceae
Hungary	*Tulipa species*	Tulip	Liliaceae
Idaho	*Philadelphus lewisii*	Mock orange	Philadelphaceae
Illinois	*Viola pedata*	Bird's-foot violet	Violaceae
India (star of)	*Nelumbium nelumbo*	Nelumbo or Blue lotus	Nelumbonaceae
Rosa species	*Rose*	Rosaceae	
	-	Two crossed palm branches	Palmae
(shield of)	*Cynodon dactylon*	Bermuda grass	Gramineae
Indiana	*Paeonia species*	Paeony	Paeoniaceae
Indonesia	*Jasminum sambac*	Melati jasmine	Oleaceae
Iowa	*Rosa virginiana*	Wild rose	Rosaceae
Iran	*Rosa species*	Rose	Rosaceae
Ireland	*Trifolium repens minus*	Irish shamlock	Papilionaceae

Israel	*Vitis vinifera*	Grape vine	Vitidaceae
Italy	*Pinus pinea*	Italian stone pine	Pinaceae
Japan	*Chrysanthemum species*	Chrysanthemum	Compositae
Paulownia tomemtosa	*Scrophulariaceae*		
Kansas	*Helianthus annuus*	Sunflower	Compositae
Kentucky	*Solidago species*	Goldenrod	Compositae
Korea	*Magnolia species*	Magnolia	Magnoliaceae
Laos	*Ficus religiosa*	Bo-tree or Pepul	Moraceae
Latvia	*Chrysanthemum leucantheum*	Ox-eye daisy	Compositae
Lebanon	*Cedrus libani*	Cedar of Lebanon	Pinaceae
Liberia	*Elaeis guineensis*	Oil palm	Palmae
Libya	*Punica granatum*	Pomegranate	Punicaceae
Liechtenstein	*Lilium bulbiferum var. croceum*	Orange lily	Liliaceae
Lithuania	*Ruta graveolens*	Rue	Rutaceae
Los Angeles	*Strelitzia reginae*	Bird of Paradise flower	Strelitziaceae
Louisiana	*Magnolia grandiflora*	Laurel magnolia	Magnoliaceae
Luxembourg	*Rosa species*	Rose	Rosaceae
Madagascar	*Ravenala madagascariensis*	Traveller's palm	Strelitziaceae
Maine	*Pinus strobus*	Pine cone	Pinaceae
Malaysia	*Cocos nucifera*	Coconut palm	Palmae
Maryland	*Rudbeckia hirta*	Black-eyed Susan	Compositae
Massachusetts	*Epigaea repens*	Mayflower	Ericaceae
Mexico	*Euphorbia pulcherrima*	Poinsettia	Euphorbiaceae
Michigan	*Pyrus malus*	Apple blossom	Rosaceae
Minnesota	*Cypripedium species*	Lady's slipper	Orchidaceae
Mississippi	*Magnolia grandiflora*	Laurel magnolia	Magnoliaceae
Missouri	*Crataegus coccinea*	Hawthorn	Rosaceae
Monaco	*Dianthus caryophyllus*	Carnation	Caryophyllaceae
Montana	*Lewisia rediviva*	Bittercup	Portulacaceae
Morocco	*Quercus suber*	Cork oak	Fagaceae
Nebraska	*Solidago serotina*	Goldenrod	Compositae
Nepal	*Nelumbium nelumbo*	Nelumbo or Blue lotus	Nelumbonaceae
Netherlands	*Citrus sinensis*	Orange	Rutaceae
Nevada	*Artemisia tridentata*	Sagebrush	Compositae
New Hampshire	*Syringa vulgaris*	Lilac	Oleaceae
New Jersey	*Viola species*	Violet	Violaceae
New Mexico	*Yucca species*	Yucca	Agavaceae
New South Wales	*Telopea speciosissima*	Waratah	Euphorbiaceae
New York	*Rosa species*	Rose	Rosaceae

New Zealand	*Agathis australis*	Kauri-pine	Araucariaceae
Nicaragua	*Ceiba pentandra*	Silk cotton tree	Bombacaceae
North Carolina	*Cornus florida*	Dogwood	Cornaceae
North Dakota	*Rosa arkansana*	Wild prairie rose	Rosaceae
Northern Ireland	*Lilium usitatissimum*	Flax	Liliaceae
Norway	*Calluna vulgaris*	Heather	Ericaceae
Ohio	*Phoradendron flavescens*	Mistletoe	Loranthaceae
Oregon	*Berberis aquifolium*	Oregon grape	Berberidaceae
Panama	*Sterculia apetala*	Panama-tree	Sterculiaceae
Paraguay	*Brunfelsia hopeana*	Jasmine-of-the-Paraguay	Solanaceae
Pennsylvania	*Kalmia latifolia*	Mountain laurel	Ericaceae
Peru	*Cantua buxifolia*	Cantua	Polemoniaceae
Philippines	*Jasminum sambac*	Sambagita jasmine	Oleaceae
Poland	*Viola tricolor*	Red pansy	Violaceae
Polynesia	*Cocos nucifera*	Coconut palm	Palmae
Portugal	*Lavandula stoechas*	Lavender	Labiatae
Prussia	*Tilia americana*	Linden	Tiliaceae
Puerto Rico	*Delonix regia*	Flamboyant or Flame tree	Caesalpiniaceae
Rhode island	*Viola species*	Violet	Violaceae
Rumania	*Rosa species*	Dog rose	Rosaceae
San Marino	*Cyclamen neopolitanum*	Cyclamen	Primulaceae
San Salvador	*Yucca elephantipes*	Yucca	Agavaceae
Saudi Arabia	*Phoenix dactylifera*	Date palm	Palmae
Saxony	*Reseda odorata*	Mignonette	Resedaceae
Scotland	*Onopordum acanthium*	Scotch thistle	Compositae
South Africa	*Protea species*	Protea	Proteaceae
South Carolina	*Gelsemium sempervirens*	Yellow jessamine	Loganiaceae
South Dakota	*Pulsatilla hirsutissima*	American pasque flower	Ranunculaceae
Spain	*Punica granatum*	Pomegranate	Punicaceae
Sudan	*Acacia species*	Acacia	Mimosaceae
Sweden	*Linnaea borealis*	Twin flower	Caprifoliaceae
Switzerland	*Leontopodium alpinum*	Edelweiss	Compositae
Syria	*Anemone caronaria*	Windflower	Ranunculaceae
Tennessee	*Iris species*	Iris	Iridaceae
Texas	*Lupinus subcarmosus*	Bluebonnet	Papilionaceae
Thailand	*Oryza sativa*	Rice	Gramineae
Tibet	*Eugenia jambos*	Rose-apple	Myrtaceae
Trinidad & Tobago	*Heliconia humilis*	Lobster-claw	Heliconiaceae
Tunisia	*Acacia species*	Acacia	Mimosaceae
Turkey	*Tulipa species*	Tulip	Liliaceae

Turk's Island	*Melocactus communis*	Turk's Island cactus	Cactaceae
United Kingdom	*Quercus species*	Oak	Fagaceae
United Nations	*Olea europaea*	Olive	Oleaceae
Uruguay	*Erythrina crista-galli*	Ceibo or Cockspur-coral	Papilionaceae
USSR	*Helianthus annuus*	Sun flower	Compositae
Utah	*Calochortus nuttallii*	Sego lily	Liliaceae
Vatican City	*Lilium longiflorum var. eximum*	Easter lily	Liliaceae
Venezuela	*Tabebuia pallida*	White-cedar	Bignoniaceae
Vermont	*Trifolium pratense*	Red clover	Papilionaceae
Victoria	*Epacris impressa*	Common heath	Epacridaceae
Virginia	*Cornus florida*	Flowering dogwood	Cornaceae
Wales	*Narcissus pseudo-narcissus*	Daffodil or Trumpet narcissus	Amaryllidaceae
Washington	*Rhododendron california*	Coast rhododendron	Ericaceae
West Africa	*Elaeis guineensis*	Oil palm	Palmae
West Indies	*Plumeria acutifolia var. glabra*	Frangipani or Forget-me-not	Apocynaceae
West Virginia	*Rhododendron maximum*	Rhododendron	Ericaceae
Wisconsin	*Viola species*	Violet	Violaceae
Wyoming	*Castilleja linariifolia*	Indian paintbrush	Scrophulariaceae
Yemen	*Coffea arabica*	Arabian coffee	Rubiaceae
Yugoslavia	*Convallaria majalis*	Lily-of-the-valley	Liliaceae

PLANTS FORBIDDEN TO BE USED AS FUEL-WOOD OR CUT

Plants Not Used As Fuel-wood

Species	Common Name	Plant Family
Afzelia africana	-	Caesalpiniaceae
Annona chrysophylla	Custard apple	Annonaceae
A. senegalensis var. deltoides	-	Annonaceae
A. senegalensis var. senegalensis	-	Annonaceae
Anthocleista kerstingii	-	Loganiaceae
A. nobilis	-	Loganiaceae
Barteria nigritana ssp. nigritana	-	Passifloraceae
Combretum molle	-	Combretaceae
Cymbopogon schoenanthus	-	Gramineae
Distemonanthus benthamianus	African satinwood	Caesalpiniaceae
Elaeis guineensis	Oil palm	Palmae
Indigofera conjugata	-	Papilionaceae

Lophira alata	Red ironwood	Ochnaceae
Newbouldia laevis	-	Bignoniaceae
Premna angolensis	-	Verbenaceae
P. quadrifolia	-	Verbenaceae
Sterculia setigera	Karaya gum tree	Sterculiaceae
Stereospermum kunthianum	-	Bignoniaceae

Plants Not Cut Down

Species	Common Name	Plant Family
Acacia mellifera var. detinens	Hookthorn	Mimosaceae
Adansonia digitata	Baobab	Bombacaceae
Afzelia africana	African teak	Caesalpiniaceae
Ceiba pentandra	Silk cotton tree	Bombacaceae
Dalbergiella welwitschii	West African blackwood	Papilionaceae
Dialium guineense	Velvet tamarind	Caesalpiniaceae
Distemonanthus benthamianus	African satinwood	Caesalpiniaceae
Gouania longipetala	-	Rhamnaceae
Khaya senegalensis	Dry-zone mahogany	Meliaceae
Loesenera kalantha	-	Celastraceae
Milicia (Chlorophora) excelsa	Iroko	Moraceae
M. regia	Iroko	Moraceae
Newbouldia laevis	-	Bignoniaceae
Ochna afzelii	-	Ochnaceae
O. membranacea	-	Ochnaceae
Okoubaka aubrevillei	-	Santalaceae
Piptadeniastrum africanum	-	Mimosaceae
Salacia debilis	-	Celastraceae
Sambucus nigra	Elder	Sambucaceae
Spiropetalum heterophyllum	-	Connaraceae
Spirostachys africanum	African sandalwood	Euphorbiaceae
Stereospermum kunthianum	-	Bignoniaceae
Withania somnifera	Winter cherry	Solanaceae
Ziziphus mucronata	Buffalo thorn	Rhamnaceae

BIBLIOGRAPHY AND REFERENCES

ABAYE, K. (1988-89) Personal Communication.

ABBAN, J.K. (1987-90) Personal Communication.

ABBIW, D.K. (1990) *Useful Plants of Ghana.* Intermediate Technology Publications. London; and Royal Botanic Gardens, Kew.

ABRUQUAH, J.W. (1971) *The Catechist.* Ghana Publishing Corporation. Accra.

ADAMS, C.D. (1957) Activities of Danish Botanist in Ghana. 1783-1850. Trans Ghana Historical. Sec. 3, 30-46.

ADAMS, R.F.G. (1943) Efik vocabulary of living things. Nigerian Field 11: 156-169.

ADANDÉ, A. (1953) Le Maïs et ses usages dans le Bas-Dahomey. Bull. Inst. Franc, Afr, Noire,
15: 220-282.

ADDAE-MENSAH, I. (1975) Herbal Medicine - Does It Have A Future In Ghana? Universitas, Legon No. 5:17-30.

ADDO, J.K. (1988-89) Personal Communication.

ADEGBOLA, E.A.A. (Editor) (1983) *Traditional Religion in West Africa.* Asempa Publishers. Accra.

ADIBUER, O. (1988-89) Personal Communication.

ADJANOHOUN, E. & AKÉ ASSI, L.(1972). Plantes Pharmaceutiques de Côte d'Ivoire (mimeographed).

ADOBOE, (1971) Ex DRUM MAGAZINE.

AFENU, L. (1988-89) Personal Communication.

AGBORDZI, K. (1987-89) Personal Communication.

AGYEMAN, O. (1987-89) Personal Communication.

AINSLIE, J.R. (1937) A list of plants used in native medicine in Nigeria. Imperial Forest Inst. Oxford. Inst. Paper 7 (mimeographed).

AKÉ ASSI, L. (1980) Les Plantes et la Therapie de la Sterilite des Femmes en Côte d'Ivoire. Miscellaneous Paper 19 Landbouwnogeschool Wageningen.
- , (1988) Personal Communication.

AKLIGO-ZOMADI, Y. (1986-88) Personal Communication.

AKOBUNDU, J.O. & AGYAKWA, C.W. (1987) *A Handbook of West African Weeds.* International Inst. of Trop. Agric. Ibadan, Nigeria.

AKOTO, J. (1987-90) Personal Communication.

AKPABLA, G.K. (1986-87) Personal Communication.

ALANDO, G. (1988-89) Personal Communication.

ALBERT-PULEO, M. (1978) Mythology, Pharamcology and Chemistry of Thuione-Containing Plants and Derivatives. Econ. Bot. 32:65-74. Jan-Mar.

AMAT, B. & CORTADELLAS, T. (1972) Ngovayangui: un village du sud Caméroun. Contribution a une étude de la santé en Afrique. Thesis, École Practique des Haute Études, Paris, France (unpub).

AMATO, T. (1987-90) Personal Communication.

AMEYAW, K. (1988-89) Personal Communication.

AMMAH-ATTOH, P. (1989-90) Personal Communication.

AMPOFO, OKU. (1983) *First Aid in Plant Medicine.* Gha. Rur. Recon. Move., Mampong Akwapim.

AMPOFO, J.K. (1987-90) Personal Communication.

AMUZU, T. (1988-89) Personal Communication.

ANKOMA-AYEW (1986) Personal Communication.

ANNAN, J. (1988-89) Personal Communication.

ANOBAH, D. (1989-90) Personal Communication.

ANON (1611) *The Holy Bible.* King James Version. American Bible Society. New York.

-, *(1963) The Methodist Hymn Book.* Novell & Company Ltd. London.

-, (1971) Doris in Dwarfland. Drum Magazine. Ghana Edition. (May).

-, (1971) Dwarf In Our Area, They Play Marbles. Drum Magazine. Ghana Edition. (August).

-, (1982) Breaking the Chains of Superstition. The Watchtower Magazine. Vol. 103, No. 17.

-, (1986) Are the Dead Alive? The Watchtower Magazine. Vol. 107, No. 5.

-, (1986) The Fascination of the Occult. Awake Magazine. Vol. 67, No. 16.

-, (1987) Fortune-Telling - Still in Fashion. The Watchtower Magazine. Vol. 108, No. 5.

-, (1987) The Supreme Being is Unique. The Watchtower Magazine. Vol. 108, No. 12.

-, (2004) An Amazing Survivor. Awake Magazine. March 8th

-, (2004) Methuselah on the Mountain. Awake Magazine. March 22nd.

ANSTEY, R. (1975) *The Atlantic Slave Trade and British Abolition.* 1760-1810. The MacMillan Press Ltd. London.

ANTUBAM, K. (1963) *Ghana's Heritage of Culture.* Koehler & Amelang. Leipzig.

APAGYAHENE OF ABIRIW - AKWAPIM (1987-88) Personal Communication.

APPIA, B. (1940) Superstitious Guinéennes et Sénégalaise. Bull. Inst. Franc. Afr. Noire 2:358-395.

ARDAYFIO-SCHANDORF, E. (1987-90) Personal Communication.

ARKELL, A.J. (1939) Throwing-sticks and Throwing-knives in Dafur. Sudan Notes and records
23:251-267.

ASAMOAH, R.F.K. (1985) Uses of fallow trees and farm practices in Ho Forest District (Ghana). Thesis, Inst. of Renewable Nat. Resources, Univ. of Science & Technology, Kumasi, Ghana (unpublished).

ASHIEBOYE-MENSAH, J. (1988-89) Personal Communication.

ASIEDU BINEY, ALHAJI (1986) Personal Communication.

ASUMADU-SAKYI, Y. (1988-89) Personal Communication.

ATOE, G. (1988-89) Personal Communication.

ATSU, R. (1989-90) Personal Communication.

ATWIE, G. (1987-89) Personal Communication.

AUBRÉVILLE, A. (1959) *La Flore Forestière de la Côte d'Ivoire* 3 Vol. Centre Technique Forestière Tropical. Nogent-Sur-Marne - France.

AVUMATSODO, S.K. (1986-88) Personal Communication.

AYENSU, E.S. (1965) Notes on the Anatomy of the Dioscoreaceae. Gha. Journ. of Sci. Vol. 5, No.1.

-, (1972) Comments on old and new world Dioscoreas of Commercial Importance. Instituto Nacional de

Investigaciones Forestales. Mexico.

-, (1974) Plant & Bat Interaction in W Africa. Annals of the Missouri Bot. Gdn. Vol. 12, No. 8.

-, (1981) A worldwide role for the healing power of plants. In: Smithsonian Mag. Vol. 12, No. 8.

-, and COURSEY, D.G. (1972) Guinea Yams. The Botany, Ethnobotany, Use and Possible Future of Yams in W Africa. Economic Bot. Vol. 26, No. 4.

AYIM, K. (1987-90) Personal Communication.

AYIVOR, B. (1988-89) Personal Communication.

BACK, P. (1987) *The Illustrated Herbal.* Guild Publishing. London.

BADU, J.K. (1988-90) Personal Communication.

BAILEY, L.H. (1954) *Manual of Cultivated Plants.* The MacMillan Company. New York.

BALLE, S. (1951) Phytolacaceae, In: R. BOUTIQUE, op. cit. II:92-99.

BALLY, P.R.O. (1937) Native Medicinal and Poisonous plants of East Africa. Bull. Misc. Inf.
1937: 19-26.

BASDEN, G.T. (1921) *Among the Ibos of Nigeria.* Oxford University Press. London.

BAUMER, M.C. (1975) Catalogue des Plantes utiles du Kordofon (République du Soudan) particuliérement du point de vue pastoral. J. Arg. trop. Bot. appl. 22: 81-119

BELOVED, K. (1988-90) Personal Communication.

BENSON, L. (1959) *Plant Classification.* Heath & Company. Massachusetts.

BERHAUT, J. (1967) *Flore Illustrée du Sénégal, Dicotylédones,* VI, Linicées à Nymphaeacées. Dakar.

-, (1988) *Flore Illustrée du Sénégal, IX, Monocotylédones: Agavacées à Orchidacées.* Dakar, Governement du Sénégal.

BETHEL, M. (1968) *The Healing Power of Herbs.* Wilshire Book Company. California.

BINET, J. (1974) Drogue et mystique: le Buiti de Fangs (Caméroun). Diogène No. 86.

BISACRE, M., CARLISLE, R., ROBERTSON,D, & RUCK, J. (Editors) (1984) *The Illustrated Encyclopedia of Plants.* Marshall Cavendish limited. London.

BISSET, N.G. (1970) The African species of *Strychnos,* I. The Ethnobotany, Lloydia 33: 201-243.

-, & LEEUWENBERG, A.J.M. (1968) The use of *Strychnos* species in Central African ordeal and arrow poisons. Lloydia 31: 208-222.

BLANKSON, K. (1988-89) Personal Communication.

BLUMENTHAL, M., GOLDBERG, A. & BRINCKMANN, J. (Ed.) (2000) *Herbal Medicine.* Integrative Medicine Communications. Newton, MA. USA.

BONATI, A. (Editor) (1992) FITOTERAPIA Vol. LXIII, No. 1. Milan. Italy.

BONSRA, E. MANTE (2010) Personal Communication

BOUQUET, A. (1969) Féticheurs et Médicines traditionnelles du Congo (Braz). Mém. ORSTOM 36.

-, (1972) Plantes médicinales du Congo-Brazzaville. *Uvariopsis, Pauridiantha, Diospyros.* Trav. Doc. ORSTOM 13.

-, & DEBRAY, M. (1974) Plantes médicinales de la Côte d'Ivoire. Trav. Doc. ORSTOM 32.

BOUSCAYROL, R. (1949) Notes sur le peuple ébrié. Bull. Inst. franç. Afr. Noire 11: 382-408.

BOWDICH, T.E. (1819) *Mission to Ashantee.* John Murray, Albernarie-Street. London.

BRAVMANN, R.A. (1974) *Islam and Tribal Art in West Africa.* Cambridge University Press.

BROWN, F.J. (1878) From England to Darkest Africa or A voyage in the Missionary Steamer 'Henry Venn'. Article from Jan. 1956 of Huron Church News (mimeographed).

BUBUWA TUBRAR, (1985) (ined) Ebina plant names and plant medicinal uses in Dumme village, Adamawa, Nigeria, R. BLENCH fide H.M. BURKILL.

BURKILL, I.H. (1935) *A dictionary of the economic products of the Malay Peninsula.* Crown Agents for the Colonies. London.

BURKILL, H.M. (1985) *The Useful Plants of West Tropical Africa.* Vol. 1. Royal Bot. Gardens, Kew.

-, (1994) *The Useful Plants of West Tropical Africa.* Vol. 2. Royal Bot. Gardens, Kew.

-, (1995) *The Useful Plants of West Tropical Africa.* Vol. 3. Royal Bot. Gardens, Kew.

-, (1997) *The Useful Plants of West Tropical Africa.* Vol. 4. Royal Bot. Gardens, Kew.

-, (2000) *The Useful Plants of West Tropical Africa.* Vol. 5. Royal Bot. Gardens, Kew.

BURTT DAVY, J. & HOYLE, A.C. (Ed.) (1937) *Check-list of the Forest Trees and Shrubs of the British Empire,* No. 3, Draft of First descriptive check-list of the Gold Coast. Imp. For. Inst. Oxford.

BUSSON, F. (1965) *Plantes alimentaires de l'Ouest Africain.* Leconte, Marseille.

BUTTERFIELD, H. (1954) *Christianity and History.* G. Bell and Sons Limited. London.

CANTOR, M. (1990) Ibogain: Miracle Cure? In: The Truth Seeker Vol. 117, No. 5.

CASEY, E. (1988-89) Personal Communication.

CHADHA, Y.R. (1985) (Ed.) *Wealth of India.* Raw Materials, Vol. 9 (Rh-So); New Delhi: C.S.I.R., India. Revised Ed.

CHANEY, W.R. & BASBOUS, M. (1978) The Cedars of Lebanon. Witnesses of History. Econ. Bot. 32:119-123. Apr-June.

CHÉVALIER, A. (1913) *Étude sur la Flore de l'Afrique centrale francaise.* Paris

- , (1932) *Resource Végétales du Sahara et de ses confins nord et sud.* Paris.

COFIE, R.C. (1986-90) Personal Communication.

COMMEH-SOWAH, A. (1987-89) Personal Communication.

COOKE, R.C. (1977) *Fungi, man and his environment.* Longman. London & New York.

COOPER, G.P. & RECORD, S.J. (1931) *The Evergreen Forests of Liberia.* Yale University, School of Forestry Bull. 31.

COPPO, P. (1978) Considérations Préliminaires sur l'État de l'assistance Psychiatrique dans un Pays en Voie de Dévelopemant (République du Mali) et sur Quelques aspects de la Médicine Traditionnelle. In: FITOTERAPIA Vol. XLIX. Milan Italy.

COUSTEIX, P. (1961) L'art et la Pharmacopée des Guérisseurs Ewondo (Région de Yaoundé). Recherches et Étude Caméroinaise. Numero Special 6.

CROW, W.B. (1969) *The Occult Properties of Herbs.* Samuel Weiser Inc. New York.

CUDJOE, K. (1988-89) Personal Communication.

DALZIEL, J.M. (1937) *Useful Plants of West Africa.* Crown Agents for the Colonies. London.

DE FEO, V. (1992) Medicinal and magical plants in the northern Peruvian andes. In: FITOTERAPIA Vol. LXIII, No. 5. Milan Italy.

DE WILDEMAN, E. (1906) *Notices sur les plantes utiles ou intérisantes de la Flore du Congo.* 2, 1:145-6. Brussels.

DEBRAY, M., JACQUEMIN, H. & RAZAF INDRAMBOA, R. (1971) Contribution à l'inventaire des plantes médicinales de Madagascar. Trav. Doc. ORSTOM, No. 8. Paris.

DEBRUNNER, H. (1959) *Witchcraft in Ghana.* Presbyterian Book Depot Limited. Accra.

DELANY, M.J. & HAPPOLD, D.C.D. (1979) *Ecology of African Mammals.* Longman. Lond. & N. Y.

DEPOMMIER, D. (1983) Aspects de la foresterie villageoise dans l'Ouest et Le Nord (Caméroun). Report of an internship. Institut de Recherche agronomique and Centre Tropicale Forestier Technique, 'Nkolbisson, Caméroun' (mimeographed).

DIARRA, N'G. (1977) Quelqe plantes vendue sur les marchés de Bamako; J. Agric. trad. Bot. appl. 24:41-49.

DOBKIN de RIOS, M. (1977) Plant Hallucinogens and the Religion of the Mochica - an Ancient Peruvian People. Econ. Bot. 31:189-203. Apr-June.

DODU, S.R.A. (1972) Our Heritage - The Traditional Medicine of Mankind. Ghana Univer. Press. Accra.

DOE-LAWSON, G. (1988-90) Personal Communication.

DOKOSI, O.B. (1969) Some Herbs Used in The traditional System of Healing Diseases in Ghana - 1. Ghana Journal of Science Vol. 9, No. 2.

-, (1986-88) Personal Communication.

DOKU KORLE, Y. (1989-90) Personal Communication.

DOLORES, L. & LATORRE, F.A. (1977) Plants used by the Mexican Kickapoo Indians. Econ. Bot. 31:340-357. July-Sept.

DONGMO, J. (1985) L'Evolution de Systéme agraire en pays Banen. Étude Geographique: Université Yaoundé, Faculté des Lettres et Sciences Humaines. Thése, Department of Geography. Yaoundé, Caméroun (mimeographed)

DUNDAS, K.R. (1910) Notes on the Tribes inhabiting the Baringo District, East Africa Protectorate. J.R. Antrop. Inst. 40:49-73.

ELLIS, A.B. (1881) *West African Sketches.* Samuel Tinsley & Co. London.

-, (1883) *The Land of Fetish.* Chapman & Hall Ltd. London.

-, (1885) *West African Islands.* Chapman and Hall Ltd. London.

-, (1887) *The Tshi-Speaking Peoples of the Gold Coast of West Africa.* Chapman and Hall Ltd. London.

-, (1894) *The Yoruba-Speaking Peoples of the Slave Coast of West Africa.* Chapman and Hall Ltd. London.

EMBODEN, W.A. (1972) Ritual Use of *Cannabis sativa* L. : a historical ethnographic survey, pp 214-236. In: P.T. FURST, Flesh of the Gods. *The ritual use of hallucinogens.* Praeger. New York.

ENTSUAH, K.A. (1988-89) Personal Communication.

ESHUN, J.K. (1989) Personal Communication.

EVANS-PRITCHARD, E.E. (1937) *Witchcraft, Oracles and Magic among the Azande.* The Clarendon Press. Oxford.

EVERARD, B. & MORLEY, B.D. (1970) *Wild Flowers of the World.* Ebury Press & Michael Joseph. Italy.

EWER, D.W. & HALL, J.B. (1972) Ecological Biology 1. Longman. London.

FABIANU, Y. (1978-90) Personal Communication.

FALCONER, J. (1990) *The major significance of 'minor' forest products.* FAO & UN. Rome.

- , (1992) *Non-timber Forest Products in Southern Ghana.* ODA For. Series No. 2. UK.

FERNANDEZ, J.W. (1972) *Tabernanthe iboga:* Narcotics Ecstasis and the work of the Ancestors. In:
PT FURST: *Flesh of the Gods: The Ritual Use of Hallucinogens,* Praeger, New York.

FERRY. M.P., GESSAIN, M. & GESSAIN, R. (1974) Ethnobotanique Tenda. Docums Centre Rech. anthrop., Mus. Homme, No. 1.

FIELD, M.J. (1937) *Religion and Medicine of the Ga People.* Oxford Univ. Press. London.

FINK, H.E. (1989) *Religion, Disease and healing in Ghana. A case Study of Traditional Dormaa Medicine.* Trister Wisseschaft. Germany.

FLORES, F.A. & LEWIS, W.H. (1978) Drinking the South American hallucinogenic Ayahuasca. Econ. Bot. 32:154-156. Apr-June.

FOWLER,F.B. & FOWLER, H.W. (1953) *The Pocket Oxford Dictionary.* The Clarendon Press. Oxford.

FREYBERG, H. (1935) (Trans. by K.S. SHELVANKAR) *Out of Africa.* Hurst & Blackett Ltd. London.

GALT, A.H. & GALT, J.W. (1978) Peasant Use of Some Wild Plants on the Islands of Pantelleria, Sicily. Econ. Bot. 32:20-26. Jan-Mar.

GAMADI, M. (1986-89) Personal Communication.

GARRETT, W.E. (Editor) (1986) *The World Map.* National Geographic Society. Washington DC.

GILBERT, G. & BOUTIQUE, R. (1952) Mimosaceae. In: R. BOUTIQUE: *Flore du Congo-Belge et du Ruanda-Urundi.* 3: 137-233.

GILBERT, M. (1986-88) Personal Communication.
- , (1989) In: *The Creativity of Power.* I. Karp (Ed.) Smithsonian Inst. Washington DC.

GLEDHILL, D. (1972) *West African Trees.* Longman. London.

GLUCKMAN, M. (1963) *Custom and Conflict in Africa.* Basil Blackwell. Oxford.

GOLDING, J. (Ed.) (2002) *Southern African Plant Red Data Lists.* Southern African Botanical Diversity Network. Pretoria. South Africa.

GOLLINHOFER, O. *et al* (1975) Art et artisant Tsogho, Gabon. Travaux et Documents de l'ORSTOM 42, Paris, France.

GOMES E SOUSA, A. de F. (1930) Subsidios para e conhecimento da Flora da Guiné Portugesa. Mem. Soc. Brot. 1.

GOROG-KARADY, V. (1970) L'arbre justicier. In: Calame-Griaule, G. (Ed.) (1980) Le théme de L'arbre dans Les contes Africaines, SELAF No. 20:23-62.

GRAF, A.B. (1978) *Exotica - Pictorial Cyclopedia of Exotic Plants from Tropical and Near-Tropical Regions.* Roehrs Company. USA.
- , (1980) *Tropica - Colour Cyclopedia of Exotic Plants and Trees.* Roehrs Company. USA.

GREEN, A.H. (1951) *Pararistolochia goldieana* (Hook, f.) Hutch. & Dalz. Kew Bull. 6:132.

GREENWAY, P.J. (1941) Dyeing and Tanning Plants in East Africa. Bull. Imp. Inst. 39:222-245.

GUILHEM, M. & HERBERT, R.P.G. (1965) Notes additive sur 'Les Divins en pays Toussain'. Notes Afr. 107:92-95.

GUILLARMOD, A.J. (1971) Flora of Lesotho (Basutoland). J. Cramer.

GUNN, C.R. (1977) *Merremia discoidesperma:* Its Taxonomy and Capacity of its Seeds for Ocean Drifting. Econ. Bot. 31:237-252. Apr-June.

- , & DENNIS, J.V. (1979) *World Guide to Tropical Drift Seeds and fruits.* Quadrangle/The New York Times Book Co. New York.

GYEDU, J.K. (1987-90) Personal Communication.

HALL, J.B. (1978) Sacred Groves. (mimeographed).

- , &LOCK, J.M (1972) Ghana Herbarium Specimen No.GC43461

- , & - , (1975) Use of vegetative Characters in the identification of species of Salacia (Celastraceae). Boissiera 24:331-338.

- , & SWAINE, M.D. (1981) *Distribution and ecology of vascular plants in a tropical rain forest. Forest Vegetation in Ghana.* Dr. W. Junk Publishers. The Hague.

HALLAM, G. (1979) *Medicinal uses of Flowering Plants in The Gambia.* Yundum, Dept. of Forests (mimeographed).

HAMMOND, R. (1988-89) Personal Communication.

HARGREAVES, B.J. (1978) Kill and Curing: Soc. Malawi J. 31:21-30.

HARLEY, G.W./W.T. (1941) *Native African Medicines.* Harvard Univ. Press.

HARRIS, B.J. & BAKER, H.G. (1959) Pollination of Flowers by Bats in Ghana. The Nigerian Field Vol. 24, No. 4.

HAUMAN, L. (1948) *Ulmaceae.* In: R. BOUTIQUE: *Flore du Congo Belge et Ruanda-Urundi, Spermatophytes I. 8. Ulmaceae,* 39-51; I.N.E.A.C., Brussels.

- , (1951) Fam. 24. Amaranthaceae. op. cit. 2:12-81. In: R. BOUTIQUE Flore du Congo-Belge et du Ruanda-Urundi. Spermatophytes 1, I.N.É.A.C. Brussels.

HAYNES, F.H. & COURSEY, D.C. (1969) Gigantism in the Yam. Tropical Science 11:93-96.

HEINE, H. (1961) *Operculina macrocarpa* (L.) Urban (Convolvulaceae) in West Africa. Kew Bull. Vol. 14, No. 3.

-, (1966) Revision du Genre *Thomandersia Baill.* (Acanthaceae). Bull. Jard. Bot. Nation. Belg. 36:207-248.

HEISER, C.B. (1980) Peppers of the Americas. National Arboretum. USA.

HEPPER, F.N. (1965) The Vegetation and Flora of the Vogel Peak. Massif, Northern Nigeria. Bull. Inst. Franc. Afr. Noire A 27:413-513.

-, (1976) *The West African Herbaria of Isert and Thonning.* Bentham-Moxon Trust. Royal Botanic Gardens, Kew.

-, (1982) *Bible Plants at Kew.* Her Majesty's Stationery Office. London.

-, (1987) *Planting a Bible Garden.* Royal Botanic Gardens. Kew.

-, (2004) Personal Communication.

HIVES, D.O.F. (1930) *Juju and Justice in Nigeria.* John Lane. London.

HOLLAND, J.H. (1922) *The Useful Plants of Nigeria.* His Majesty's Stationery Office. London.

HOLLOWAY, G. (1987-90) Personal Communication.

HUBER, J. (1910) Mattas e madeiros amazonicas. Boilet. Mus. Goeldi 6:91-225.

HUNTING TECHNICAL SURVEYS (1968) *Land and water resource surveys of the Jabel Marra area, Republic of the Sudan: reconnaissance vegetation survey,* LA:SF/SUD/17, F.A.O. Rome.

HUTCHINSON, J. & DALZIEL, J.M. (1927-36) *Flora of West Tropical Africa.* 3 Vols. Crown Agents for Overseas Governments and Administration. London.

HUYNH, K.N. (1988) Étude des *Pandanus* (Pandanaceae) d'Afrique Occidentale (11e partie): Espèces nouvelle de la Sierra Leone, Bot. Helv. 98: 171-194.

ILOGU, E. (1974) *Christianity and Ibo Culture.* E.J.BRILL. Leiden.

IRVINE, F.R. (1930) *Plants of the Gold Coast.* Oxford Univ. Press. London.

-, (1925) Ghana Herbarium Specimen No. GC 2053.

-, (1927) Ghana Herbarium Specimen No. GC 690.

-, (1957) Indigenous African Methods of Beekeeping. Bee World 38 (5):13-128.

-, (1961) *Woody Plants of Ghana.* Oxford Univ. Press. London.

JACKSON, G. (1973) Fulani in N Nigeria. Botany Department. Ibadan Univ. Nigeria.

JADREJ, M.C. (1986) Dan and Mende masks: A structural composition. African 5 (1) Afri. Inst. London.

JORU, (1973) Notes of a herbalists' meeting with H.M. BURKILL at Joru, SE Sierra Leone; msc. ined. Herb. Kew.

KEAY, R.W.J. (1961) Botanical Collectors in West Africa prior to 1800. *In: Comptes Rendus.* A. Fernandes. Lisbon.

-, (1989) *Trees of Nigeria.* Clarendon Press. Oxford.

KERHARO, J. (1966) La Pharmacopeé Sénégalaise: Note sur les Rosacées utilisées en médicin traditionnelle. J.W. Afri. Sci. Ass, 11:77-80.

-, (1967) A propos de la pharmacopée Sénégalaise apercu historique concernant les recherches sur la flore et des plantes médicinales du Sénegal. Inst. Fond. Afr. Noire. A 20. 131-1434.

-, (1974) *La Pharmacopée Sénégalaise Traditionnelle. Plantes Médicinales de Toxique.* Vigot Frères. Paris.

-, & BOUQUET, A. (1950) *Plantes Médicinales et Toxique de la Côte d'Ivoire-Haute Volta.* Vigot Frères. Paris.

-, & ADAM, J.G. (1962) Premier inventaire des plantes médicinales et toxiques de la Casamance (Sénégal). Ann. Pharm. Franc. 20:726-744, 823-841.

-, &- (1963) Deuxieme inventaire des plantes médicinales et toxiques de la Casamance (Sénégal). Ann. Pharm. Franc. 21:733-792.

-, &- (1964) Note sur quelques plantes médicinales des Basari et des Tandanké du Sénégal oriental. Inst. Franc. Afr. Noire A. 26:403-437.

-, &-(1974) *La Pharmacopée Sénégalaise traditionnelles. Plantes médicinales et toxiques.* Vigot Frères. Paris.

KINGSLEY, M.H. (1897) *Travels in West Africa.* MacMillan & Co. Ltd. London.

-, (1901) *West African Studies.* MacMillan & Co. Ltd. London.

KIRCH, P.V. (1978) Indigenous Agriculture in Uvea (W Polynesia). Econ. Bot. 32:157-181. Apr-June.

KLAH, D. (1987-90) Personal Communication.

KLOSS, J. (1939) *Back to Eden.* Back to Eden Bks. Pub. Co. Loma Linda, California.

KLUGAH, A. (1987-90) Personal Communication.

KNAB, T. (1977) Notes concerning use of *Solandra* among the Huichol. Econ. Bot. 31:80-86. Jan-Mar.

KOAGNE, H. (1986) La dynamique des plantes et dérivés alimentaires dans la Chéfferie Batoussam. Mémoire de Maitrise, Department of Geography, University of Yaoundé, Cameroon (unpublished).

KOKWARO, J.O. (1976) *Medicinal Plants of East Africa.* East African Literature Bureau. Nairobi.

KRÔGER. F. (1986-88) Personal Communication.

KULEFIANU, K. (1987-89) Personal Communication.

LABURTHE-TOLRA, P. (1981) Le seigneurs de la forêt: Essai sur le passé historique, l'organisation sociale et les normes éthiques des anciens Bétis de Caméroun. Publication de la Sorbonne. Paris.

LACK, A. (1978) The Ecology of the Flowers of the savanna tree *Maranthes polyandra* and their visitor with particular reference to bats. Journal of Ecology 66, 287-295.

LAMPTEY, N. (1988-89) Personal Communication.

LANS, C. (2001) *Creole Remedies. Case studies of ethnoveterinary medicine in Trinidad and Tobago.* Posen & Looijen, Wageningen.

LANZARA, P. & PIZZETTI, M. (1977) In: SCHULER, S. (US Ed.) *Guide to Trees.* Simon & Schuster. New York.

LARTEY, B. (1988-90) Personal Communication.
LAWLER, L.J. (1984) Ethnobotany of the Orchidaceae. Orchid Biology. Reviews and Perspectives. 3: 27-149.

LAWRENCE, G.H.M. (1963) *Taxonomy of Vascular Plants.* The MacMillan Company. New York.

LÉONARD, J. (1950) Étude botanique des Copalier du Congo Belge, Publ. De l'Institut National pour l'Étude Agronomique du Congo Belge (I.N.E.A.C.), sér. sci. no. 45.

-, (1952) Cynometreae et Amherstieae. In: R. BOUTIQUE Flore du Congo Belge et du Ruanda-Urundi, 3:279-376.

LEROUX, H. (1948) Animisme et Islam dans la subdivision de Maradi. Bull. Inst. Franç. Afr. noire 10: 595-697.

LIVINGSTONE, W.P. (1914) *Mary Slessor of Calabar. Pioneer Missionary.* Hodder & Stoughton. London. New York. Toronto.

LOCK, J.M. & MARSHALL, A.C. (1976) Possible Pollination of *Parinari polyandra* by Bats. The Nigerian Field Vol. XLI. No. 2.

LORDZISODE, A. (1988-89) Personal Communication.

LOTSCHERT, W. & BEESE, G. (1983) *Collins Guide to Tropical Plants.* Collins Grafton St. London.

LOWRY, P. (1988-90) Personal Communication.

LOWY, B. (1972) Mushrooms symbolism in Maya codices. Mycologia 66:816-821.

MACFOY, C.A. & SAMA, A.M. (1983) Medicinal Plants in Pujehun District of Sierra Leone. Journal of Ethnopharmacology 8:215-223. Elsevier Scientific Publishers. Ireland Ltd.

MACMUNN, G. (1933) *The Underworld of India.* London. Jarrolds

MAIRE, H. (1933) Études sur le Flore et la vegetation de Sahara Central. Mém. Soc. Hist. Nat. Afr. Nord. Algiers.

MALIKI, A.B. (1981) *Ngaynaaka - Herding according to the Wodaabe,* Min. Rural Develop., Niger Range and Livestock Project, Discusion Paper No. 2 (mimeographed).

MALLART GUIMERA, L. (1969) L'arbre oven. In: CALAME-GRIAULE, G. (Ed.) La théme de l'arbre dans les contes Africans. SELAF, No. 16.

MAGNUS, A. (?) *Egyptian Secrets or White and Black Art for man and Beast.* Guidance House. N. Y.

MAJA NAUR. (1999) *Medicinal Plants and Rural Development in the Savanna Region of Northern Ghana: The Role of Women in Conservation, Management and Utilization.* A Report Presented to the World Bank, AFTR

MARTIN, G. J. (1995) *Ethnobotany.* Chapman & Hall. London. Glasgow. Madras. New York. Tokyo.

MATSON, S.J. (1970) The Story of Inn Signs. In: The Gliksten Journal Vol. 10, No. 6.

MAUNDU, P., BERGER, D., SAITABAU, C., NASEKU, J., KIPELIAN, M., MATHENGE, S., MORIMOTO, Y. & HÖFT, R. (2001) *Ethnobotany of the Loita Maasai.* UNESCO. Paris.

McKENZIE, P.R. (1976) *Inter-religious Encounters in West Africa.* Blackfriars Press Ltd. Lescester, England.

McLEOD, M.D. (1981) *The Ashanti.* British Museum Publications Limited. London.

MEEK, C.K. (1925) *The Northern Tribes of Nigeria 2 Vols.* Oxford University Press. London.

-, (1931) *A sudanese Kingdom* (Ethnographical Study of the Jukun-speaking Peoples of Nigeria). Oxford Univ. Press. London.

-, (1931) *Tribal Studies in Northern Nigeria 2 Vols.* Oxford University Press. London.

MENSAH, A.N.K. (1987-88) Personal Communication.

MICHAUD, Max. (1966) Contribution à l'étude des Olacées d'Afrique tropicale. Mém. Inst fond Afr. noire, 75: 157-284.

MILLSON, A. (1891) Indigenous plants of Yorubaland. Bull. Misc. Inf. 1891:206-219.

MOLDENKE, H.N. & MOLDENKE, A.L. (1983) *Verbenaceae.* In: M.D. DASSANAYEKE & F.R. FOSBERG, *A revised handbook of the Flora of Ceylon.* Vol. IV: Washington, U.S.A., Smithsonian Inst. Nat. Sci. Found.

MONOD, TH. (1950) Vocabulaire Botanique Teda. In: R. MAIRE & TH. MONOD, Études sur la flore et la végétation du Tibesti. Mém. Inst. Fanc. Afr. Noire B.

MONTEIL, V. (1953) *Institut des Haute Études Marocaine. Notes and Documents. VI. Contribution a l'études de la Flôre du Sahara Occidental,11.* Larose, Paris.

MOORE, A.C. (1960) *The Grasses.* The MacMillan Company. New York.

MORO IBRAHIM (1986-90) Personal Communication.

MORTIMER, W.G. (1901) *Peru. History of Coca. The Divine Plant of the Incas'* .Vail & Co., New York

MOTTE, E. (1980) A propos des thérapeutes Pygmées Aka de la Région de la Lobaye (Centrafrique); J. Agr. Tra. Bot. appl. 27:113-132

MOUNTFIELD, D. (1976) *A History of African Exploration.* The Hamlyn Printing Group Ltd. London. New York. Sydney. Toronto.

MUIR, J. (1937) Seed-drift of South Africa. S African Dept. Agric and Forestry Bot. Sur. Mem. No. 16.

MURREY, F. (1987-89) Personal Communication.

NADEL, S.F. (1953) Witchcraft and Anti-witchcraft in Nupe Society. In: Africa VIII pp. 424.

NALINI, K., ARBOOR, A.R., KARANTH, S.K. & RAO, A. (1992) Effects of *Centella asiatica* fresh leaf aqueous extract to learning and memory and biogenic amine turnover in albino rats. In: FITORERAPIA Vol. LXIII. No. 3. Milan, Italy.

NEWBERRY, R. (1938) Some games and pastimes of Southern Nigeria. 1: Some Yoruba Games. Nigerian Field 7:85-90.

NICHOLAS, F.J. (1953) Onomastique personnelle des l'Éta de la Haute-Volta. Bull. Inst. Franc. Afr. Noire 15:818-847.

NKONGMENECK, B. (1985) Le genre cola au Caméroun. Rev. Sci. et Tech. Serie Science Agronomique, Vol. 1 (3):57-70.

NKWANTABISA, G. (1987-89) Personal Communication.
NOAMESI, G.K. (1986-88) Personal Communication.

NOVAK, F.A. (1965) *The pictorial Encyclopedia of Plants and Flowers.* Paul Hamlyn. London. Crown Pub., Inc. New York.

NTIM, J. (1990) Personal Communication.

NYAMWAYA, D. & SOBER, R. (1983) (Editors) MILA. A Biannual Newsletter of Cultural Research, Vol. 6, No.2. Institute of African Studies. University of Nairobi.

NYARKO, N. (1988-90) Personal Communication.

OKAFOR. J.C. (1979) Edible Indigenous Woody Plants in the Rural Economy of the Nigerian Forest Zone. In: The Nigerian Rainforest Ecosystem (Ed. D.U.U. OKALI) Man and the Biosphere National Committee. Ibadan, Nigeria.
- , (1988) Personal Communication.

OKIGBO, B.N. (1980) Plants and Food in Igbo Culture. Ahiajoku Lecture. Owerri Imo State, Nigeria.

OLIVER, B. (1960) *Medicinal plants in Nigeria.* Nigerian College of Arts, Sciences and Technology.

OLIVER-BEVER, B. (1983) Medicinal Plants in Tropical West Africa, II. Plants acting on the nervous system; loc. Cit. 7: 1-93.

O'MALLEY, R. & THOMPSON, D. (1957) *Rhyme and Reason.* Chatto & Windus. London.

OPOKU, A.A. (1970) *Festivals of Ghana.* Ghana Publishing Corporation. Accra.

OSBURN, W. (1865) Notes on the Chiroptera of Jamaica. Proc. Zool. Sec. 82:61-85.

OSEI, K. (1988-89) Personal Communication.

OSMANU, Y. (1988-90) Personal Communication.

OSO, B.A. (1977) Mushrooms in Yoruba Mythology and Medicinal Practices. Econ. Bot. 31:367-371.

OTEDOH, M.O. (1972) *The rediscovery of Raphia regalis in Nigeria.* N.I.F.O.R. seminar of 12/4/72 (mimeographed).

OTU, J.K. (1988-90) Personal Communication.

OUEDRAGO, J. (1950) Les funérailles en pays Mossi. Bull. Inst. Franc. Afr. Noire 12:441-445.

PAMPLONA-ROGER, G.D. (2001) *Encyclopedia of Medicinal Plants Vols. 1 & 2.* MARPA Artes Gráficas - E-50172 Alfajarin, Zaragoza, Spain.

PÂQUES, V. (1953) L'estrado royale des Niare. Bull. Inst. Franc. Afr. Noire 15:1642-1654.

PARKER, T.J. & HASWELL, W.A. (1967) *A Text-book of Zoology.* Vol.II. MacMillan & Co. Ltd. Lond.

PARRINDER, G. (1953) *Religion in an African City.* Oxford Univ. Press. London.

PERRY, F. & GREENWOOD, L. (1973) *Flowers of the World.* Hamlyn, London. New York. Sydney.Toronto.

PICHON, M. (1953) Monographie des Landolphieés (Classification des Apocynacées XXXV). Mém. Inst. Franc. Afri. Noire 35.

PIPER, C.V. & DUNN, S.T. (1922) A revision of *Canavalia,* Kew Bull. Misc. Inf. 1922: 129-145.

PITKANEN, A.L. & PREVOST, R. (1970) *Tropical Fruits, Herbs, Spices, etc.* R. Prevost, Publisher, Lemon Grove, Cal.

POBÉGUIN, H. (1912) *Plantes médicinales de la Guinée.* Paris.

POPE, H.G. (1990) *Tabernanthe iboga:* an African Narcotic. In: The Truth Seeker Vol. 117, No. 5.

PORTÈRES, R. (1935) Plantes toxiques utilisées par les peuplades Dan et Guéré de la Côte d'Ivoire. Bull. Comité d'Études Historiques et Scientifiques de l'AOF 18.

- , s.d. *Reliquiae.* Lab. Ethnobot., Paris.

PRAUSE, G. (1988) German annals 500 years ago: No mercy for a witch. In: Lufthansa Bordbuch. W Germany.

QUARCOO, A.K. (1968) The Visual Arts of Ghana. In: *Insight and Opinion.*

Vol. 3, No. 3. (Ed. J.P. SCHOLS). Accra Catholic Press.

QUISUMBING, E. (1951) Medicinal Plants of the Phillippines. Dept. Agric. Nat. Res. Tech, Bull. 16.

RASOANAIVO, P., PETITJEAN, A. & CONAN, J.Y. (1993) Toxiz and Poisonous Plants of Madagascar: an Ethnopharmacological Survey. In FITOTERAPIA Vol. LZIV, No. 2. pp 114-129. Italy.

RATTRAY, R.S. (1923) *Ashanti.* Oxford Univ. Press. London.

- , (1927) *Religion and Art in Ashanti.* Oxford Univ. Press. London.

RAVEN, P.H., EVERT, R.F. & CURTIS, H. (1976) *Biology of Plants.* Worth Publishers. Inc.New York.

RAYNER, E. (1977) Orchids as medicine. S African Orchids. J. 8: 120.

RICHARDS, B.W. & KANEKO, A. (1988) *Japanese Plants. Know Them and Use Them.* Shufunotomo Co. Ltd. Tokyo, Japan.

ROBERTS, J. (1970) The Olive Trees of Majorca. The Gliksten Journal Vol. 10, No. 6. J. Glicksten & Sons Ltd. London.

ROSE INNES, R. (1971) Fire in West African Vegetation. Proceedings. Annual Tall Timbers Fire Ecology Conference. April 22-23.

ROWLEY, G. (1972) Voyage into the impossible - I meet *Welwitschia.* The journal of the Royal Horticultural Society Vol. 97. Part 8.

SANAGO, D. (1983) Bois sacré: temple ou école? Revue Ivoirienne d'Anthropologie et de Sociologie 3:59-62.

SASU-YAWLULE, P. (1987-88) Personal Communication.

SAUNDERS, H.N. (1958) *A handbook of West African Flowers.* Oxford Univ. Press. London.

SAVILL, P.S. & FOX, J.E.D. (1967) *Trees of Sierra Leone.* MS in Forestry Department. Freetown.

SCHNELL, R. (1946) Sur quelques plantes à usages religieux de la

region forestière d'Afrique Occidentale. Journal de la Société Africaniste 16:29-37.

- , (1950) *Manuel Ouest-africans 1. La forêt dense. Introduction à l'étude botanique de la région forestière d'Afrique Occidentale.* Paris.

SCHULTES, R.E. (1972) *An overview of hallucinogens in the Western Hemisphere.* In: P.T. Furst, *Flesh of the Gods. The ritual use of hallucinogens*, pp. 3-54; Washington, Praeger Publishers.

SEREBOE, E.K. (1986-89) Personal Communication.

SERLE, W., MOREL, G.J. & HARTWIG, W. (1977). A *Field Guide to the Birds of West Africa.* Collins, Grafton Street, London.

SHEPPHERD, (1988) *A Leaf of Honey.* Bodhi Publications.

SIDIBE, M. (1939) Famille, vie sociale et vie religieuse chez les Birifer et les Oulé (region de Diébougou, Côte d'Ivoire). Bull. Inst. Franç. Afr. Noire 1:679-742.

SIEGEL, R.K., COLLINGS, P.R. & DIAZ, J.L. (1977) On the Use of *Tagetes lucida* and *Nicotiana rustica* as a Huichol Smoking Mixture: the Aztec 'Yahutli' with Suggestive Hallucinogenic Effects. Econ. Bot. 31:16-23. Jan-Mar.

SIKES, S.K. (1972) *Lake Chad.* Eyre Methuen. London.

SITI, J. (1988-89) Personal Communication.

SMITH, E.W. (Ed.) (1930) *African Ideas of God.* Edinburgh House Press. London.

SOFAR, Y. (1987-90) Personal Communication.
STAUCH, A. (1966) Le basin Caméroun ais de la Benoue et la peche. Mém. ORSTOM, Paris (15):152 p.

STEWART, J. & HENNESSY, E.F. (1981) *Orchids of Africa.* MacMillan Press Limited. London and Basingtoke.

STORRS, A.E.G. & PIEARCE, G.D. (1982) *Don't eat these.* (A guide to some

local poisonous plants). Forestry Department, Ndola, Zambia.

STUDSTILL, J. (1970) L'arbre ancéstral. In: CALAME-GRIAULE, G. (Ed.) (1969) Le théme de l'arbre dans les contes Africains. SELAF (Société d'études Linguistiques et Anthropologique de France) 20:119-137. Paris, France.

SUTHERLAND, D.A. (1956) *State Emblems of the Gold Coast.* Government Printing Dept. Accra.

SWART, E.R. (1963) Age of the Baobab Tree. Nature 198:708-709.

SWITHENBANK, M. (1969) *Ashanti Fetish Houses.* Ghana Univ. Press. Accra.

TALBOT, P.A. (1912) *In the Shadow of the Bush.* Oxford Univ. Press. London.
- -, (1926) *The peoples of Southern Nigeria.* Vols. 1-4. Oxford Univ. Press. London.
- -, (1927) *Some Nigerian Fertility Cults.* Oxford Univ. Press. London.
- -, (1932) *Tribes of the Niger Delta.* Oxford Univ. Press. London.

TATTERSALL, S.L. (1978) *The lesser-known food plants of the Gambia.* Dept. Agr., The Gambia (mimeographed).

TAYLOR, C.J. (1960) *Synecology and Sylviculture in Ghana.* Thomas Nelson & Sons Ltd. London.

TEIKO, T. (1988-89) Personal Communication.

THOMAS, N.W. (1910) *Report, Edo-Speaking Peoples of Nigeria.* (Edo-Bini) pt. 2. Oxford Univ. Press. London.
- -, (1913-14) *Report, Ibo-Speaking Peoples of Nigeria* 2 & 5. Oxford Univ. Press.London
- -, (1916) *Report, Timne-Speaking Peoples of Sierra Leone.* Oxford Univ. Press. London.

THOMAS, O.O. (1988) Perspectives on ethno-phytotherapy of 'Yoruba' medicinal herbs and preparations. In: FITOTERAPIA Vol. LX, No. 1:49-

60. Milan, Italy.

THOMSON, S. (1988-90) Personal Communication.

TOMPKINS, P. & BIRD, C. (1974) *The Secret Life of Plants.* Allen Lane. London.

TREMEARNE, A.J.N. (1913) *Hausa Superstitions and Customs.* John Bale, Sons & Danielsson Limited.London.

TRINCAZ, J. (1980) L'arbre garant de la perennité culturelle d'une société d'émigreée menacée. Cahiers d'ORSTOM Série Science Humaines 17 (3-4):285-289.

TROCHAIN, J. (1940) La végétation du Sénégal. Mém. Inst. Afr. Noire 2.

TWUMASI, P.A. (1975) *Medical Systems in Ghana.* Ghana Publishing Corporation. Accra.

TYLER, V.E. Jn. (1966) The physiological properties and chemical constituents of some habit-forming plants. Lloydia 29:275-292.

UNDERWOOD, P. (1971) *A Gazetteer of British Ghosts.* Pan Books Limited. London.

UNESCO (1975) *Cultural Policy in Ghana - Studies and Documents on Cultural Policy.* The Unesco Press. Paris.

VAN SERTIMA, I. (1976) *They came before Columbus.* Random House. New York.

VELDCAMP, J.F. (1971) *Oxalidaceae,* in Fl. Males. 7:151-178.

VERGER, P.F. (1967) *Awon ewe osanyin* (Yoruba medicinal leaves). University of Ife. Nigeria.

VERGIAT, A.M. (1969) Plantes magiques et médicinales des féticheurs de l'Oubangui (Central Afrique). Journal d'Agriculture Tropicale et Botanique Appliquées 16 (2-10), 17 (1-9):3 parts

VISSER, L. (1952) Plantes médicinales de la Côte d'Ivoire. Mfivojelaededelingen Landbouwhogeschool (No, 75-115), Wageningen, Netherlands.

VOORHOEVE, A.G. (1965) *Liberian High Forest Trees.* Centre for Publications and Documentation, Wageningen.

WALKER, A.R. (1952) Usage pharmaceutiques des plantes spontanées du Gabon. 1. Bull. Inst. Études Centrafr. n.s. 4:181-186.
- -, & SILLANS, R. (1961) *Les plantes utile du Gabon.* Paul Lechevalier. Paris.

WALU, L.D. (1987) (ined.) *Mwalami: an ancient food-plant of the Goemai (Ankwe) of Plateau State, Central Nigeria;* msc., K.

WARD, W.E. (1935) *A Short History of the Gold Coast.* Longman, Green & Co. London. New York.

WARREN, D.M. (1974) Disease, Medicine and Religion among the Techiman-Bono of Ghana: A study of culture change (mimeographed).
- -, BUCKLEY, A.D. & AYANDOKUN, J.A. (1973) *Yoruba Medicines.* Institute of African Studies, Univ. of Ghana. Legon.

WATT, J.M. & BREYER-BRANDWIJK, M.G. (1962) *The Medicinal and Poisonous Plants of Southern and Eastern Africa.* 2nd Edition . Livingstone: Edinburgh and London.

WEIKANG, Fu (1985) *Traditional Chinese Medicine and Pharmacology.* Foreign Lang. Press. Beijing.

WELCOMME, R.L. (1985) *River Fisheries.* FAO Fisheries Technical Paper 262.

WHITE, F. (1957) Notes on Ebenaceae, III, Bull. Jard. Bot. Nation. Belg. 27: 515-131.

WIAFE-ANNOR, K. (1979) *Bosomtwe, the Sacred Lake.* Wavelite Publications. Accra.

WICKENS, G.E. (1987) Ecosystems data Base and Geographic

Distribution of Molluscidal Plants. In: *Plant molluscides* (Edited K.E. MOTT). Wiley & Sons Ltd. Chichester. New York. Brisbane.Toronto. Singapore.

WILLIAMS, R.O. (Jn) (1949) *The useful and ornamental plants of Zanzibar and Pemba.* Zanzibar.

WILLIAMS, J.T. & FARIAS, R.M. (1972) Utilization and Taxonomy of the Desert Grass, *Panicum turgidum.* Econ. Bot. 26: 13-20.

WILLIAMSON, K. (1970) Some food-plant names in the Niger Delta. Int. J. Amer. Linguistics 36:136-167.

WILLIS, J.C. (1931) *A Dictionary of Flowering Plants and Ferns.* 6th Edition. Cambridge Univ. Press.

- , (1973) *A Dictionary of Flowering Plants and Ferns.* 8th Edition. Revised by H.K. AIRY SHAW. Cambridge Univ. Press.

WOFA YAW, K. (2001) Personal Communication.

WONG, W. (1976) Some folk medicinal plants from Trinidad. Econ. Bot. 30: 103-142.

ZEVEN, A.C. (1964) The Idolatrica palm. Baileya 12: 11-18.

- , (1967) *The semi-wild oil palm and its industry in Africa.* Agr. Res. Rept. 689, State Agr.Uni., Inst. Plant Breeding, Wageningen.

ZOBERI, H.M.(1973) Some Edible Mushrooms from Nigeria.The Nigerian Field. Vol. 38, No. 2.

[illegible]

[illegible]

[illegible]

[illegible]

[illegible]

[illegible]

[illegible]

[illegible]

[illegible]

[illegible]

GLOSSARY

Abortifacient – a drug or material that causes the expulsion of the foetus
Affirmation – solemn declaration made in court instead of an oath
Amenorrhoea – absence of menstruation
Anthelmintic – a drug which causes the expulsion of intestinal worms
Anthropology – study of mankind – especially its origin, development, customs and beliefs
Anthropomorphic – treating gods and animals as human in form and personality
Antimony – a brittle silvery-white metal used in alloys
Aphrodisiac – any substance or drug arousing sexual desire
Apoplexy – sudden inability to feel or move due to blockage or rupture of brain artery - stroke
Apothecary – person who prepares and sells medicines and medical goods
Astringent - an agent that causes diminution of discharges or bleeding from the body
Auriferous – yielding gold – usually applied to rocks
Biodiversity – the variability among all living things and the ecological complexes of which they are part – this includes the diversity within species, between species and of ecosystems
Bioluminescence - the ability of plants and animals to emit a florescent light
Bitters – liquor flavoured with bitter herbs
Blennorrhoea- an excessive mucous discharge from inflammation of the mucous membrane
Cabalistic virtues – manifestations of emergent behaviour in society or governance on a part of community of person who have well established public affiliation or kingship
Cadaver – corpse or dead body of a person
Candelabriform branching - curved smoothly and ascending at right angles branching system
Chilblaines – painful swelling on hand or foot caused by exposure to cold
Cicatrisant - a scar which replaces a damaged tissue due to injury or skin disease
Circumcision – action or ceremony to cut off the foreskin of the male organ
Clairvoyance – ability of forecasting the future or happening out of sight
Climacteric - period of life when physical powers begin to decline such as menopause in women
Colostrum – the yellow fluid normally secreted from a woman's breast for a few days before and after childbirth
Comestibles – things to eat
Concoction – liquid obtained after boiling two or more plant species together
Congenital – disease which is present from or before birth
Coram publico - stark naked in public
Corneal opacity - an obstruction that prevents images from entering the eye or retina

Curandero - psychic healer
Curse - words spoken with the aim of punishing, injuring or destroying another person
Debilitate - make a person very weak - after illness
Decoction - liquid obtained from boiling only one plant species or plant part
Decongestant - drug that relieves blockage of the nostrils
Delirium - mental disturbance caused by illness, resulting in restlessness and often wild talk
Dementia - madness with loss of powers of thinking due to brain disease or injury
Difficult bodies - corpses of witches, wizards, lepers, decayed bodies, and those who died childless, committed suicide, or drowned
Dieresis - increased secretion of urine
Diuretic - a drug that causes an increase in the flow of urine
Djinns - See Genies
Dysmenorrheal - pains during the menstrual period
Emblem - an object that represents something or a symbol
Emetic - a drug that induces vomiting
Emmenagogue - a drug that promotes the menstrual flow
Enema - any liquid preparation injected into the rectum to empty the bowels instantly
Epidemic - disease that spreads quickly in the same place for a period
Erythrophleine - an alkaloid from the bark of Erythrophleum suaveolens used as cardiac tonic
Eserine - see physostigmine
Ethnophytotherapy - see phytotherapist
Etymology - study of the origin of the history of words and their meaning
Excision ceremony - rituals surrounding the removal by cutting of a tissue, structure or an organ
Excitant - the removed tissue or organ - (see excision ceremony)
Exorcize - drive out or expel an evil spirit by means of prayers or magic
Fecundity - productive or fertile
Fingerlings - immature or very young fish
Galactogenic - capable of or tending to increase milk secretion in breast-feeding women
Galactogogue - any agent that induces or increases milk secretion
Galena - cosmetic or antimony
Genies - spirit or goblin in Arabian stories with strange powers
Gibbet - an upright pole on which the bodies of executed criminals were hung in former days
Goblin - a small ugly mischievous manlike creature
Gynaecology - scientific study and treatment of diseases of the female reproductive system
Gynaeco-medico-magical - a supernatural treatment of female reproductive system diseases

Haematuria - discharge of blood in the urine and often associated with kidney disease
Haemophthysis - the expectoration (coughing up) of blood or of blood-stained sputum from the bronchi, larynx, trachea or lungs
Haemorrhoids - enlarged blood vessels of the anal canal -synonymous with piles
Haemostatic - a drug or substance that arrests bleeding
Hallucination - the illusion of hearing or seeing something when no such thing is present
Hallucinogenic - drug causing hallucination
Hallucogenic - see hallucinogenic
Herbal elixir - an imaginary plant medicine alleged to cure all ailments
Herpes - a virus disease that causes blisters on the skin
Homeopathy - healing system which applies to diseases a minimum dose of the same substance that in higher amounts would provoke in a healthy person equal or similar symptoms to those which are being fought
Hydrocephalus - an accumulation of fluid in the brain and spinal canal caused by blockage
Hysteria - wild uncontrollable emotion or excitement with laughter, crying or screaming
Incantation - series of words recited as a magic spell or charm
Inebriant effect - drunkenness
Infusion - liquid obtained from immersing plant or plant part in water -hot or cold
Interdiction - the period of forbidding a church member from taking part in service
Intoxicant - a drug or drink or gas that causes one to lose self-control
Invigorating -- making one feel more lively and healthy
Iridescence - that which changes colour as its position changes
Kaolin - fine white clay used in making porcelain and in medicine
Karité butter - the Wolof name for shea butter - prepared from the seeds of *Vitellaria paradoxa*
Lactogenic - promoting the secretion of milk
Lassitude - tiredness of mind or body
Magico-medical - a supernatural, psychic or mystical touch to treating diseases
Malediction - prayer that somebody or something may be destroyed or hurt - a curse
Marauder - person or animal looking for either something to steal or people to attack
Medianistic trance - theatrical séances simulate spiritualistic or mediumistic phenomena
Medico-magical - a supernatural, psychic or mystical touch to treating diseases
Mendicant - person making a living from begging
Messianic - of the messiah - a physiological state of mind

Mestizos - a term traditionally used in Latin America and Spain for people of mixed European and Native American heritage or descent
Mimetic magic - clever at imitating or mimicking supernatural forces to change things
Mummification - method of preserving corpse with oils and wrapping it with cloth
Mythological - collection of ancient, fictional stories dealing with gods, heroes and the like
Natron - native sesquicarbonate of soda
Novitiate - period or state when one is new or inexperienced in job or situation
Originator - the source of an evil intention
Palanquin - a boat-like receptacle in which chiefs are carried during festive occasions
Panacea - medicine for several diseases and ailments
Pantheon - temple dedicated to many gods, or where famous men are buried
Paramour - any man other than the husband of a married woman who has liaison or sex with her
Paraphernalia - numerous small articles or personal belongings for a hobby
Paraxysms - sudden fit or outburst of an ailment
Parturient - a woman in labour, relating to parturition
Patrilineal - pertaining to or derived from the father. Descended through the male line
Patrimony - property inherited from one's father or ancestors
Patronymic - derived from the name of one's father or some other male ancestor
Phallic imagery - relating to the penis
Pharmacological - concerning the scientific study of drugs and their use in medicine
Philologist - expert or student of the science or the study of the development of languages
Phosphorescent - glowing in the dark or giving out light without heat
Phytopharmacognosy - an established field of drug research where the active substances come from plants
Phytostigmine - an alkaloid extracted from *Physostigma venenosum* Calabar bean
Phytotherapist - one who specializes in the treatment of diseases with plants and their extracts
Phytotherapy - the treatment of diseases with herbal medicine
Portent - sign or warning of a future, often unpleasant, event; an omen
Poultice - previously ground, moistened, moulded and dried plant material; applied topically by rubbing on any rough surface and adding water - sometimes also licked
Prophylactic - tending to prevent a disease or misfortune
Psilocin - triptamine compound from the fungus *Psilocybe mexicana,* which is reputed to cause psychotropic or hallucinogenic conditions when ingested

Psilocybin – phosphorelated tryptamine compound produced by *P. mexicana.* This compound and psilocin are reputed to cause psychotropic or hallucinogenic conditions following ingestion of the mushroom fruiting bodies

Psychedelic – of drugs - producing hallucination

Psychic healer – one who uses supernatural powers in administering herbal medicine

Psychic recitations - an ability to perceive information hidden from the normal senses. Extra Sensory Perception (ESP)

Psycho-physical – the relationship between the mind and the body

Pudenda – the external genital – especially of a woman

Puerperium – of or related to childbirth

Purification – the act of making something pure by removing dirty or harmful substances

Pythoness – the serpent-like dragon that twined round the tree in the Garden of the Hesperides and guarded the golden apples

Rachitic – relating to rickets

Reincarnation – bring back a soul after death in another body

Rogation ceremony – four days traditionally set apart for solemn possessions to invoke God's mercy – they are April 25th, the Major Rogation, coinciding with St. Marks day and the three days preceding Ascension Day, the Minor Rogations.

Sacrilege – treating a sacred thing or place with disrespect

Sacrilegious – adjective - see sacrilege

Samurai – the military caste in feudal Japan

Sarcophagi – stone coffins, especially one with carvings and used in ancient times

Séance – meeting of spiritualists –at which people try to talk with the spirits of the dead

Shamanism – traditional healing system

Simianpromorphic – the higher primate: the monkeys and the apes, including humans

Sophistry – using clever but false arguments to deceive others

Soporific properties – sleep-inducing

Spell – words which when spoken are thought to have magical power – a charm

Strychnine – poisonous substance extracted from *Strychnos nux-vomica* and other *Strychnos* species

Subservient – giving too much respect, obedience, submissive

Suppository – a piece of medicinal substance inserted in the rectum or vagina to dissolve

Theological – institution studying the nature of God and the foundations of religious beliefs

Therapeutic – of the act of healing or the curing of diseases

Tisane – an aqueous plant extract – (see also decoction , concoction and infusion)

Tonic – any food drink

Tranquilizer - drug for making an anxious person feel calm - a sedative
Tropical drift dissemule - or sea-beans are fruits or seeds that are carried ashore by ocean currents - often in distant lands
Umbilical suppuration - a thick yellow liquid or pus forming in the organ due to infection
Vertigo - sensation that the world is whirling round an individual or vice versa
Vindictive - having or showing a desire for revenge - unforgiving
Votive offering - (presented especially in church) to fulfill a promise made to God
Vulnerary - capable of healing wounds and bruises
Winter solstice - about 22nd December when the sun is furthest North
Xeroids - plants that grow in arid conditions
Xerophytic - of plant that grows in a dry or arid habitat, such as the desert

COMMON NAMES (ENGLISH)

Aardvark — *Orycteropus afer*
Abraham's book — *Massonia bowkeri*
Abyssinian tea — *Catha edulis*
Acacia — *Acacia species*
Adonis — *Adonis amurensis*
Afara — *Terminalia superba*
African arrowroot — *Tacca leontopetaloides*
  bdellium — *Commiphora africana*
  blackwood — *Peltophorum africanum*
  bowstring hemp — *Sansevieria liberica*
    S. senegambica
    S. trifasciata
  breadfruit — *Treculia africana var. africana*
  cherry orange — *Citropsis articulata*
  copaiba balsam tree — *Daniellia oliveri*
  coral wood — *Pterocarpus soyauxii*
  cucumber — *Momordica charantia*
  greenheart — *Cylicodiscus gabunensis*
    Piptadeniastrum africanum
  kino — *Pterocarpus erinaceus*
  laburnum — *Acacia sieberiana*
    Cassia sieberiana
  linden — *Hallea stipulosa*
  locust bean — *Parkia biglobosa*
  mammee -apple — *Mammea africana*
  peach — *Sarcocephalus latifolius*
  pearwood — *Baillonella toxisperma*
  rhinoceros — *Diceros bicornis*
  sandalwood — *Osyris compressa*
    O. wightiana
    Spirostachys africanus
  satinwood — *Distemonanthus benthamianus*
  sensitive plant — *Biophytum petersianum*
  star apple — *Chrysophyllum delevoyii*
  teak — *Afzelia africana*
  tragacanth — *Sterculia tragacantha*
  tulip tree — *Spathodea campanulata*
  walnut — *Coula edulis*
  whitewood — *Triplochiton scleroxylon*
  wood-oil-nut tree — *Ricinodendron heudelotii*
  yellow wood — *Enantia chlorantha*
    E. polycarpa
Agrimony — *Agrimonia eupatoria*
Akee apple — *Blighia sapida*
Aleppo pine — *Pinus halepensis*
Alexander laurel — *Calophyllum inophyllum*
Allspice — *Pimenta dioica*
Alpine forget-me-not — *Myosotis alpestris*
Amanita — *Amanita muscaria*
Amaranth — *Amaranthus species*
Amaryllis — *Hippeastrum puniceum*
American basil — *Ocimum canum*
  pasque flower — *Pulsatilla hirsutissima*
Anatto — *Bixa orellana*
Angel's trumpet — *Brugmansia suaveolens*
Angelica — *Angelica archangelica*
Aningeria — *Pouteria altissima*
Anise — *Pimpinella anisum*
Ant, driver — *Dorylus species*
  fierce black — *Crematogaster species*
  white — *Macrotermes species*
Antelope — *Adenota kob kob*
  roam — *Hippotragus equinus*
  royal — *Neotragus pygmaeus*
Ape — *Pan troglodytes*
Apple — *Pyrus malus*
  blossom — *P. malus*
  of Sodom — *Solanum linnaeanum (sodomeum)*
Apricot — *Prunus armeniaca*
Arabian coffee — *Coffea arabica*
Arbor-vitae — *Thuja species*
Arnica — *Arnica chamissonis ssp. foliosa*
Arrow poison — *Strophanthus hispidus*
Ashanti blood — *Mussaenda erythrophylla*
  plum — *Spondias mombin*

Assegei wood	*Terminalia sericea*
Assyrian plum	*Cordia myxa*
Atta bean	*Pentaclethra macrophylla*
Australian asthma herb	*Euphorbia hirta*
Avocado pear	*Persea americana*
Avodire	*Turraeanthus africanus*
Bachelor's button	*Gomphrena globosa*
Bald crow	*Picathartes gymnocephalus*
Balloon vine	*Cardiospermum grandiflorum*
	C. halicacabum
Balm-of-Gilead	*Commiphora gileadensis*
Balsam	*Bulbine asphodeloides*
apple	*Momordica balsamina*
spurge	*Euphorbia balsamifera*
Bambara groundnut	*Vigna subterranea*
Bamboo	*Arundinaria species*
	Bambusa species
	Dendrocalamus species
	Oxytenanthera species
	Phyllostachys species
	Sasa species
	Shibataea species
Banana	*Musa sapientum*
Banyan	*Ficus bengalensis*
Baobab	*Adansonia digitata*
Barbados-pride	*Poinciana pulcherrima*
Barbary shrike	*Laniarus barbarus*
Bare-headed rock fowl	*Picathartes gymnocephalus*
Bark cloth tree	*Antiaris toxicaria*
Barley	*Hordeum vulgare*
Barnyard millet	*Echinochloa crusgalli var. frumentacea*
Barrel cactus	*Echinocactus simpsonii*
Bat	*Cynopterus species*
	Eidolon helvum
	Epomophorus gambianus
Bat	*Glossophaga species*
	Pteropus species
Bay-tree	*Pimenta racemosa*
Beach convolvulus	*Ipomoea pes-caprae*
Bead tree	*Adenanthera pavonina*
	Erythrina senegalensis
Bee	*Apis mellifera*
Beech	*Fagus sylvaticus*
Beetle, boring	*Apate monachus*
	Minthea obsita
dung	*Scarabaeus sacer*
Bell-shrike	*Laniarus ferruginea*
Beniseed or Sesame	*Sesamum orientale (indicum)*
Berg slangkop	*Urgenia capitata*
Bermuda grass	*Cynodon dactylon*
Betel	*Areca catechu*
Betony	*Stachys officinalis*
Billy goat weed	*Ageratum conyzoides*
'Bird of paradise flower'	*Strelitzia reginae*
Bird's-foot violet	*Viola pedata*
Birthworth	*Aristolochia clematis*
Bitter apple	*Cucumis myriocarpus*
bark tree	*Sacoglottis gabonensis*
cola	*Cola nitida*
	Garcinia kola
gourd	*Citrullus colocynthis*
leaf	*Vernonia amygdalina*
yam	*Dioscorea dumetorum*
Bittercup	*Lewisia rediviva*
Biwa	*Eriobotrya japonica*
Black afara	*Terminalia ivorensis*
cobra	*Naja nigricollis*
-crowned tchagra	*Tchagra senegala*
duiker	*Cephalophus niger*
-eyed Susan	*Rudberckia hirta*
magic	*Begonia species*
nightshade	*Solanum americanum (nigrum)*
pepper	*Piper nigrum*
plum	*Vitex doniana*
sesame	*Hyptis spicigera*

'Blackboys'	*Xanthorrhoea preissii*
Blackberry	*Rubus ludwigii*
Bladder-wort	*Utricularia species*
Blessed Mary thistle	*Silybum marianum*
thistle	*Cnictus benedictus*
Blombos	*Metalasia muricata*
Blood drops of Christ	*Anemone coronaria*
flower	*Asclepias curassavica*
	Scadoxus cinnabarinus
	S. multiflorus
	S. multiflorus ssp. katerinae
Bloodroot	*Sanguinaria canadensis*
Bloubekkie	*Heliophila suavissima*
Blue-eyed grass	*Sisyrinchium bermudiana*
gum	*Eucalyptus globulus*
lotus	*Nelumbium nelumbo*
pea	*Clitoria ternatea*
plumbago	*Plumbago capensis*
Bluebonnet	*Lupinus subcarmosus*
Blushing bride	*Serruria florida*
Boar	*Potamochoerus porcus*
Bonduc	*Caesalpinia bonduc*
Bongo	*Tragelaphus euryceros*
Bo-tree	*Ficus religiosa*
Borage	*Borago officinalis*
Bottle gourd	*Lagenaria siceraria*
Bougainvillaea	*Bougainvillaea glabra*
	B. spectabilis
Bowle's variety	*Vinca minor*
Box	*Buxus sempervirens*
Brab tree	*Borassus flabellifer var. aethiopum*
Bramble	*Rubus rigidus*
Brazil cress	*Spilanthes filicaulis*
Bread fruit	*Artocarpus altilis*
Bridal wreath	*Spiraea x arguta*
Brimstone tree	*Morinda lucida*
Bristle-cone pine	*Pinus longaeva*
Broomweed	*Sida acuta*
Brubru shrike	*Nilaus afer*

Bubinga	*Guibourtia ehie*
Buck, bush	*Tragelaphus scriptus*
reed	*Redunca redunca*
Buffalo	*Synceros caffer*
bean	*Mucuna poggei*
thorn	*Ziziphus mucronata*
Bug	*Clarigralla shadabi*
	Nezera viridula
Bugloss	*Anchusa officinalis*
Bulblet fern	*Cystopteris bulbifera*
Bull	*Bos indicus*
	B. taurus
Bulrush	*Typha domingensis*
millet	*Pennisetum glaucum*
Bur weed	*Triumfetta cordifolia*
	T. rhomboidea
Burdock	*Arctium lappa*
Burning bush	*Dictamnus albus*
Bush buck	*Tragelaphus scriptus*
butter tree	*Dacryodes edulis*
cat	*Viverra civetta*
clover	*Lespedeza bicolor*
cow	*Syncerus caffer*
fowl	*Agelastes species*
mallow	*Abutilon mauritianum*
pig	*Potamochoerus porcus*
tea-bush	*Hyptis suaveolens*
tick berry	*Chrysanthemoides monilifera*
willow	*Combretum salicifolium*
Butter tree	*Dacryodes edulis*
Cabbage palm	*Anthocleista kerstingii*
	A. nobilis
tree	*Cussonia spicata*
	C. paniculata
	C. spicata
Caeba mahogany	*Swietenia mahogani*
Calabar bean	*Physostigma venenosum*
Calabash	*Lagenaria siceraria*
nutmeg	*Monodora myristica*

tree	*Crescentia cujete*
Calendula	*Calendula officinalis*
California poppy	*Eschscholzia californica*
California swamp plant	*Darlingtonia californica*
Calla	*Zantedeschia aethiopica*
Camellia	*Camellia japonica*
Camphor	*Cinnamomum camphora*
tree or wood	*Tarchonanthus camphoratus*
Camwood	*Baphia nitida*
Candelabra flower	*Boophone disticha*
tree	*Euphorbia candelabra*
Candle nut	*Aleurites moluccana*
plant	*Dictamnus albus*
wood	*Zanthoxylum senegalense*
Canerat	*Thryonomys swinderianus*
Cape caper	*Capparis citrifolia*
clover	*Trifolium africanum*
holly	*Ilex mitis*
lilac	*Ehretia rigida*
mahogany	*Pteroxylon utile*
myrtle	*Myrsine africana*
valerian	*Valeriana capensis*
willow	*Salix mucrinata*
Cardamon	*Elettaria cardamomum*
Carline thistle	*Carlina acaulis*
Carnation	*Dianthus caryophyllus*
Carpet viper	*Echis carinatus*
Carrion flower	*Stapelia gigantea*
Carrisse	*Carissa edulis*
Cashew nut	*Anacardium occidentale*
Cassava	*Manihot esculenta*
Castor oil plant	*Ricinus communis*
Cat	*Felis domestica*
bush	*Viverra civetta*
Cat's tail	*Typha domingensis*
grass	*Sporobolus pyramidalis*
Cat-tail millet	*Pennisetum glaucum*
Catfish	*Chrysichthys species*
	Clarias species
Catfish	*Schilbe species*
	Synodontis species
electric	*Malapterurus electricus*
Cattle	*Bos species*
Ceara rubber	*Manihot glaziovii*
Cedar of Lebanon	*Cedrus libani*
Cedrela	*Cedrela odorata*
Cedròn	*Quassia (Simaba) cedron*
Ceibo	*Erythrina crista-galli*
Celanese cow plant	*Gymnema lactiferum*
Ceylon iron wood	*Mesua ferrea*
leadwort	*Plumbago zeylanica*
Chain of love	*Antigonon leptopus*
Chameleon	*Chamaeleon vulgaris*
Chamomile	*Chamomile matricaria*
Chaste tree	*Vitex agnus castus*
Cherokee rose	*Rosa sinica*
Chilean bell flower	*Lapageria rosea*
Chillies	*Capsicum frutescens*
Chinese cassia	*Cinnamomum cassia*
sacred-lily	*Narcissus tazetta var. orientalis*
tree of heaven	*Ailanthus altissima*
witch hazel	*Hamamelis mollis*
Chirinda redwood	*Catha edulis*
Chlorella	*Chlorella species*
Cholla	*Opuntia species*
Christ plant	*Porana paniculata*
	Euphorbia millii
	Paliurus spina-christi
Christ' thorn	*Ziziphus spina-christi*
Christmas bush	*Alchornia cordifolia*
	Ceratopetalum gummiferum
cactus	*Schlumbergera truncata*
	Zygocactus truncatus
candle	*Senna alata*
flower	*Euphorbia pulcherrima*
heather	*Erica melanthera*
jewels	*Aechmea racinae*
orchid	*Cattleya trianae*
palm	*Veitchia merrillii*

Common name	Scientific name
plant	*Euphorbia millii*
pride of Jamaica	*Ruellia paniculata*
rose	*Combretum racemosum*
	Helleborus niger
star	*Euphorbia pulcherrima*
tree	*Hildegardia barteri*
	Metrosideros tomentosa
	Morinda lucida
vine	*Porana paniculata*
Christmas-bells	*Blandfordia punicea*
	Saundersonia aurantiaca
-fern	*Polystichum acrostichoides*
Chrysanthemum	*Chrysanthemum species*
Cider tree	*Sclerocarya caffra*
Cimora	*Sanchezia species*
leon	*Acalypha macrostachya*
senorita	*Iresine herbstii*
Cimorilla	*Coleus blumei*
dominatora	*C. species*
Citron	*Citrus medica*
Civet cat	*Viverra civetta*
Climbing hemp-weed	*Mikania cordata*
	var. cordata
	var. chevalieri
lily	*Gloriosa simplex*
	G. superba
nettle	*Tragia benthamii*
Clove	*Eugenia caryophyllus*
Coast rhododendron	*Rhododendron californica*
Coat buttons	*Tridax procumbens*
Cobra	*Naja nigricollis*
plant	*Darlingtonia californica*
Coca	*Erythroxylum coca*
poppy	*Papaver rhoeas*
Cocaine	*Erythroxylum coca*
Cockroach	*Periplaneta americana*
Cock's comb	*Heliotropium indicum*
Cockspur coral-tree	*Erythrina crista-galli var. variegata*
Cocoa	*Theobroma cacao*
Coco-de-mer	*Lodoicea maldivica*
Coconut palm	*Cocos nucifera*
Cocuto	*Wormskioldia longepedunculata*
Coffee	*Coffea arabica*
Colocynth	*Citrullus colocynthis*
Columbine	*Aquilegia caerulea*
Columbus bean	*Entada gigas*
	E. rheedei (pursaetha)
Comfrey	*Symphytum officinale*
Commercial cola nut tree	*Cola acuminata*
Common ash tree	*Fraxinus excelsior*
canthium	*Psydrax parviflora*
cress	*Lepidium sativum*
fig tree	*Ficus carica*
foxglove	*Digitalis purpurea*
heath	*Epacris impressa*
milkwort	*Polygala serpyllifolia*
millet	*Panicum miliaceum*
paeony	*Paeonia officinalis*
reed	*Phragmites communis*
rue	*Ruta graveolens*
sowbread	*Cyclamen repandum*
Coral berry	*Ardisia crenata*
flower	*Erythrina senegalensis*
-tree	*E. crista-galli*
Corallita	*Antigonon leptopus*
Cornflower	*Centaurea cyanus*
Cotton	*Gossypium barbadense*
	G. herbaceum
Cow	*Bos species*
itch	*Mucuna pruriens var. pruriens*
-foot leaf	*Lepianthes peltata*
Crab	*Cardiosoma species*
mangrove	*Goniopsis cruentata*
	Sersama species
Crane, Crowned	*Balearica pavonina*
Cowpea	*Vigna unguiculata*
Crazy oil palm	*Elaeis guineensis*

	var. communis
	var. repandum
Crocodile	*Crocodylus species*
Cross vine	*Bignonia capreolata*
Crow	*Corbus albus*
Crown imperial	*Fritillaria imperialis*
of the earth	*Thonningia sanguinea*
-of-thorns	*Euphorbia millii*
Crowned crane	*Balearica pavonina*
hawk-eagle	*Stephanoaetus coronatus*
Crucifixion-bean	*Merremia discoidesperma*
Cuckoo flower	*Cardamine pratensis*
Cucuto	*Wormskioldia longepedunculata*
Cudweed	*Gnaphalium species*
Cupid's love	*Catananche coerulea*
Cycas	*Cycas circinalis*
Cyclamen	*Cyclamen neopolitanum*
Cypress	*Cupressus sempervirens*
Daffodil	*Narcissus pseudo-narcissus*
Damask rose	*Rosa damascena*
Date palm	*Phoenix dactylifera*
De caen	*Anemone coronaria*
Deadly nightshade	*Atropa bella-donna*
Death cap	*Amanita muscaria*
Desert crucifixion-thorn	*Canotia species*
	Holacantha emeryi
	Koeberlinia species
Desert date	*Balanites aegyptiaca*
rose	*Adenium obesum*
Devil bean	*Crotalaria retusa*
in the pulpit	*Tradescantia virginiana*
Devil's apple	*Datura stramonium*
	Mandragora officinarum
	Solanum aculeastrum
bit scabious	*Succisa pratensis*
bush	*Smeathmannia laevigata*
coach whip	*Starchytarpheta angustifolia*
	S. cayennensis
	S. indica
Devil's cotton	*Abroma angusta*
fig	*Argemone mexicana*
	Solanum hispidum
grass	*Afrotrilepis pilosa*
	Cynodon dactylon
hat	*Platycerium elephantotis*
	P. stemaria
herb	*Plumbago scandens*
horsewhip	*Achyranthes aspera*
ivy	*Epipremnum aureum*
plant	*Tacca species*
thorn	*Dicerocaryum zanguebaricum*
	Harpagophytum procumbens
	Emex australis
	Tribulus terrestris
tongue	*Amorphophalus species*
	Sansevieria liberica
	S. senegambica
	S. trifasciata
tree	*Alstonia scholaris*
tresses	*Cassytha ciliolata*
trumpet	*Datura metel*
stramonium	*D. stramonium*
Dittany of Crete	*Origanum dictamnus*
Divine flower	*Dianthus caryophyllus*
Djave nut	*Baillonella toxisperma*
Dodder	*Cassytha filiformis*
	Cuscuta australis
Dog	*Canis familiaris*
almond	*Andira inermis*
plum	*Ekebergia meyeri*
rose	*Rosa species*
Dogwood	*Rhamnus florida*
	R. prinoides
Dolphin	*Delphinium species*
Donkey	*Equus species*
Dove, Wild	*Streptopelia semitorquata*
Dragon tree	*Dracaena draco*
Dragon's blood tree	*Harungana madagascariensis*

Common name	Scientific name
Driver ant	*Dorylus species*
Drum tree	*Cordia millenii*
	C. platythyrsa
Dry-zone cedar	*Pseudocedrela kotschyi*
mahogany	*Khaya senegalensis*
Duck	*Anas species*
Duiker	*Sylvicarpa grimmia*
black	*Cephalophus niger*
yellow-backed	*C. sylvicultor*
Dum palm	*Hyphaene thebaica*
Dung beetle	*Scarabaeus sacer*
Dutchman's pipe	*Aristolochia albida*
	A. bracteolata
Eagle (martial)	*Polemaetus bellicosus*
Earthworm	*Lumbricus species*
East African copal	*Hymenaea verrucosa*
Indian rosebay	*Ervatamia coronaria*
Easter lily	*Lilium longiflorum var. eximum*
Ebony	*Diospyros species*
Eddoes	*Colocasia esculenta*
Edelweiss	*Leontopodium alpinum*
Edible-stemmed vine	*Cissus quadrangularis*
Egg plant	*Solanum incanum*
Egyptian sesban	*Sesbania sesban*
thorn	*Acacia nilotica ssp. tomentosa*
Elder	*Sambucus nigra*
Elecampane	*Inula helenium*
Electric catfish	*Malapterurus electricus*
	Thalapterurus electricus
Elephant	*Loxodonta africana*
creeper	*Argyreia nervosa*
grass	*Pennisetum purpureum*
Empress tree	*Paulownia tomentosa*
Endod	*Phytolacca dodecandra*
English holly	*Ilex aquifolium*
ivy	*Hedera helix*
Equinox flower	*Lycoris radiata*
Ethiopian pepper	*Xylopia aethiopica*

Common name	Scientific name
European lily of the valley	*Convallaria majalis*
Exile oil plant	*Thevetia peruviana*
False abura	*Mitragyna inermis*
rubber tree	*Funtumia africana*
	Holarrhena floribunda
thistle	*Acanthus montanus*
yam	*Icacina oliviformis*
Fan palm	*Borassus aethiopum*
Feather love grass	*Eragrostis amabilis*
Fern-palm	*Cycas circinalis*
Fever plant	*Ocimum gratissimum*
Fiery-breasted bush-shrike	*Malaconotus cruentus*
Finger-root	*Uvaria chamae*
Fire-ball lily	*Scadoxus cinnabarinus*
	S. multiflorus
	S. multiflorus ssp. katerinae
-fly	*Lampyris species*
	Photinus species
Fireweed	*Chamaenerion angustifolium*
Fiscal shrike	*Lanius collaris*
Fish poison	*Tephrosia vogelii*
bush	*Mundulea sericea*
Flamboyant }	*Delonix regia*
Flame tree }	*D. regia*
Flat-crowned tree	*Albizia gummifera*
Flax	*Linum usitatissimum*
'Flor de muerto'	*Lasianthus nigrescens*
Flower of death	*Vinca minor*
Flowering dogwood	*Cornus florida*
Foetid cassia	*Senna (Cassia) obtusifolia*
Forget-me-not	*Myosotis afropalustris*
	Plumeria rubra var. acutifolia
Fossil tree	*Metasequoia glyptostroboides*
Fowl	*Gallus domestica*
bare-headed rock	*Picathartes gymnocephalus*
bush	*Agelastes species*

guinea	*Numida meleagris*
Foxglove	*Digitalis purpurea*
Foxtail millet	*Setaria italica*
Frangipani	*Plumeria rubra* var. *acutifolia*
Frankincense	*Boswellia carteri*
	B. papyrifera
	B. sacra
tree	*B. dalzielii*
Frog	*Bufo species*
Fucus	*Fucus vesiculosus*
Funeral flower of Mexico	*Lasianthus nigrescens*
Gaboon nut	*Coula edulis*
viper	*Bitis gabonica*
Gambian puff-back shrike	*Dryoscopus gambensis*
tea-bush	*Lippia multiflora*
Garden croton	*Codiaeum variegatum*
egg	*Solanum incanum*
Garlic	*Allium sativum*
Gas plant	*Dictamnus albus*
Gazelle	*Gazella species*
Geneesblaar	*Withania somnifera*
Gentian	*Gentiana cruciata*
Germander speedwell	*Veronica chamaedrys*
Ghost orchid	*Epipogium aphyllum*
tree	*Davidia involucrata*
Ghost's palm	*Encephalartos barteri*
Giant flowered French	*Anemone coronaria*
granadilla	*Passiflora quadrangularis*
protea	*Protea cynaroides*
rat	*Thryonomys swinderianus*
sequoia	*Sequoiadendron giganteum*
Ginger	*Zingiber officinale*
lily	*Costus afer*
plum	*Neocarya macrophylla*
Gingerbread palm	*Hyphaene thebaica*
Ginkgo	*Ginkgo biloba*
Gingseng	*Panax quinquefolium*
	P. schinseng
Giraffe thorn	*Acacia giraffae*
Goat	*Capra species*
Gold nuggets	*Gazania rigens*
Golden fleur de Lis	*Iris pseudacorus*
Goldenrod	*Solidago serotina*
Good-luck plant	*Cordyline fruticosa*
Gorilla	*Gorilla gorilla*
Gotu kola	*Centella asiatica*
Granaatbossie	*Anthospermum rigidum*
Grains of paradise	*Aframomum melegueta*
Grape vine	*Vitis vinifera*
Grass-trees	*Xanthorrhoea preissii*
Grasscutter	*Thryonomys swinderianus*
Great grey shrike	*Lanius excubitor*
sequoia	*Sequoiadendron giganteum*
willowherb	*Chamaenerion angustifolium*
Greater burnet	*Sanguisorba officinalis*
yam	*Dioscorea alata*
Green amaranth	*Amaranthus lividus*
	A. viridis
mamba	*Dendroaspis viridis*
Grey-budded snake bark maple	*Acer rufinerve*
nickernut	*Caesalpinia bonduc*
parrot	*Psittacus erithacus*
Gromwell	*Lithospermum officinale*
Groundnut	*Arachis hypogaea*
Guana	*Varanus niloticus*
Guarana	*Paullinia cupana*
Guava	*Psidium guajava*
Guernsey-lily	*Nerine sarniensis*
Guinea corn	*Sorghum bicolor*
fowl	*Numida meleagris*
grains	*Aframomum melegueta*
grass	*Panicum maximum*
plum	*Parinari excelsa*
yam	*Dioscorea cayennensis*
	D. rotundata
Gum arabic tree	*Acacia karoo*

copal tree	*Daniellia ogea*
Gutta-percha tree	*Ficus platyphylla*
Haemorrhage plant	*Aspilia africana*
Hairy indigo	*Indigofera hirsuta*
thorn-apple	*Datura innoxia*
	D. metel
Hand-flower tree	*Chiranthodendron pentadactylon*
Hawk-eagle, Crowned	*Stephanoaetus coronatus*
Hawthorn	*Crataegus oxyacantha*
Hayme-huayme	*Salmea scandens*
Heart of Jesus	*Caladium bicolor*
Heartsease	*Viola tricolor*
Heartseed	*Cardiospermum grandiflorum*
	C. halicacabum
Heather	*Calluna vulgaris*
Heavenly bamboo	*Nandina domestica*
Hemp-leaved hibiscus	*Hibiscus cannabinus*
	H. lunariifolius
Hen	*Gallus domestica*
Henbane	*Hyoscyamus niger*
Henna	*Lawsonia inermis*
Herb of grace	*Ruta graveolens*
Paris	*Paris quadrifolia*
Hippopotamus	*Hippopotamus amphibius*
Hisakaki	*Eurya japonica*
Hog	*Potamochoerus porcus*
plum	*Spondias mombin*
red river	*Potamochoerus porcus*
wart	*Potamochoerus porcus*
Hogweed	*Boerhaavia coccinea*
	B. diffusa
	B. erecta
	B. repens
Holly	*Cnictus benedictus*
Holy grass	*Savastana odorata*
Hondeoor	*Cotyledon orbiculata*
Honesty	*Lunaria annua*
Honeysuckle	*Lonicera periclymenum*
Hookthorn	*Acacia mellifera var. detinens*
Hoopoe bird	*Upupa epops*
Hops	*Humulus lupulus*
Hornamo amarillo	*Senecio elatus*
morado	*Valeriana adscendens*
Hornbill	*Ceratogymna species*
	Tochus species
Horse	*Equus equus*
purslane	*Trianthema portulacastrum*
	Zaleya pentandra
Horse-eye bean	*Mucuna sloanei*
-radish tree	*Moringa oleifera*
-tail	*Equisetum arvense*
	E. ramosissimum
Horsewood	*Hippobromus pauciflorus*
Hound's tongue	*Cynoglossum lanceolatum*
Housefly	*Musca domestica*
Houseleek	*Sempervivum tectorum*
Huaminga chica	*Huperzia species*
Hydrangea	*Hydrangea macrophylla*
Hyena	*Hyaena hyaena*
Hyssop	*Hyssopus officinalis*
Iboga	*Tabernanthe iboga*
Iguana	*Varanus niloticus*
Immortelle	*Helichrysum bracteatum*
Inca magic-flower	*Cantua buxifolia*
wheat	*Amaranthus hybridus ssp. incurvatus*
Incense tree	*Canarium schweinfurthii*
Indaba tree	*Pappea capensis*
India (n) almond	*Terminalia catappa*
corn	*Zea mays*
heliotrope	*Heliotropium indicum*
hemp	*Cannabis sativa*
jujube	*Ziziphus mauritiana*
mustard	*Brassica juncea*
navelwort	*Centella asiatica*
paintbrush	*Castilleia linariifolia*
- rubber fig	*Ficus elastica*

shot	*Canna indica*
tamarind	*Tamarindus indica*
wormseed	*Chenopodium ambrosioides*
Irish shamrock	*Trifolium repens minus*
Iroko	*Milicia excelsa*
	M. regia
Ironweed	*Vernonia galamensis*
Ironwood	*Lophira alata*
Italian stone pine	*Pinus pinea*
Ivy	*Hedera helix*
Ixora	*Ixora coccinea*
Jack fruit	*Artocarpus heterophyllus*
Jamestown weed	*Datura stramonium*
Japanese apricot	*Prunus mume*
cedar	*Cryptomeria japonica*
love grass	*Eragrostis tenella*
millet	*Echinochloa crusgalli*
	var. frumentacea
Jasmine-of-the-Paraquay	*Brunfelsia hopeana*
Jerusalem thorn	*Parkinsonia aculeata*
Jew-bush	*Pedilanthus tithymaloides*
Job's tears	*Coix lacryma-jobi*
Johnson's arum	*Amorphophallus johnsonii*
Judas tree	*Cercis siliquastrum*
Juju palm	*Elaeis guineensis* *var. idolatrica*
Jujube tree	*Ziziphus mauritiana*
Juniper bush	*Juniperus communis*
	J. drupacea
	J. excelsa
Kaffir-cherry	*Gardenia neubria*
orange	*Strychnos innocua*
	S. spinosa
Kamerun grass	*Sorghum arundinaceum*
Kanot grass	*Flagellaria guineensis*
Karaya gum tree	*Sterculia setigera*
Karmes oak	*Quercus coccifera*
Katemfe	*Thaumatococcus daniellii*
Kauri-pine	*Agathus australis*
Kava kava	*Piper methystichum*
Keiserblume	*Centaurea cyanus*
Kenaf	*Hibiscus cannabinus*
Kenilworth ivy	*Linaria cymbalaria*
Kersting's groundnut	*Macrotyloma (Kerstigiella)*
	geocarpum
Khaki bur, weed	*Alternanthera pungens*
Khantua	*Cantua buxifolia*
King bamboo-palm	*Raphia vinifera*
of flowers	*Paeonia species*
oil palm	*Elaeis guineensis*
	var. idolatrica
Protea	*Protea cynaroides*
Kingcups	*Caltha palustris*
Kite	*Elanus caeruleus*
	E. riocourii
Kraaibos	*Heteromorpha arborescens*
'La grave'	*Vinca minor*
Ladislai regis herba	*Gentiana cruciata*
Lady's slipper	*Cypripedium species*
tongue	*Albizia lebbeck*
Lalang grass	*Imperata cylindrica*
Lance tree	*Lonchocarpus capassa*
Lantana	*Lantana camara*
Laurel	*Laurus nobilis*
magnolia	*Magnolia grandiflora*
Lavendar	*Lavandula angustifolia*
	L. stoechas
Lebbeck	*Albizia lebbeck*
Lemon balm	*Melissa officinalis*
grass	*Cymbopogon citratus*
-scented grass	*Elyonurus argenteus*
Leopard	*Panthera pardus*
Leper's lily	*Fritillaria meleagris*
Lesser periwinkle	*Vinca minor*
Liberian coffee	*Coffea liberica*
Life of Man	*Convolvulus tricolor*
Lignum vitae	*Guaiacum officinale*
Lilac	*Syringia vulgaris*
Lily-of-the-valley	*Convullaria majalis*
Lima bean	*Phaseolus lunatus*

Lime	*Citrus aurantiifolia*
Linden	*Tilia americana*
	T. europaea
Lion	*Panthera leo*
Little blackcap tchagra	*Tchagra minuta*
ironweed	*Vernonia cinerea*
Lobster	*Panulirus species*
	Scyllarides species
-claw	*Heliconia humilis*
Logwood	*Haematoxylon campechianum*
Loofah	*Luffa cylindrica*
Long-crested helmet-shrike	*Prionops plumata*
-tailed shrike	*Corvinella corvina*
Lotus, blue	*Nelumbium nelumbo*
Louse	*Pediculus humanus*
'Love and innocence'	*Quisqualis indica*
apple	*Lycopersicon lycopersicum*
-in-the-mist	*Nigella damascens*
-lies-bleeding	*Amaranthus hybridus ssp. incurvatus*
grass	*Chrysopogon aciculatus*
-vine	*Antigonon leptopus*
Lungwort	*Pulmonaria officinalis*
Madagascar periwinkle	*Catharanthus roseus*
Madona lily	*Lilium candidum*
Madras thorn	*Pithecellobium dulce*
Magic flower of the Incas	*Cantua buxifolia*
Magnolia flower	*Magnolia species*
Mahogany	*Khaya species*
Maidenhair-tree	*Ginkgo biloba*
Maize	*Zea mays*
Makore	*Tieghemella heckelii*
Mallow	*Malva parviflora*
High	*M. silvestris*
Malpighian hair	*Sphedamnocarpus pruriens*
Mamba, green	*Dendroaspis viridis*
Manatee	*Trichechus senegalensis*
Mandrake	*Mandragora officinarum*
Mango	*Mangifera indica*
wild	*Irvingia gabonensis*
Mangrove crab	*Cardiosoma armata*
	Goniopsis cruentata
	Sersama species
Many-coloured bush-shrike	*Malaconotus multicolor*
Map tree	*Euphorbia tetragona*
Maple	*Acer platanoides*
Red snake bark	*A. capillipes*
Marabou thorn	*Dichrostachys cinerea*
Marble vine	*Dioclea reflexa*
Marigold	*Tagetes lucida*
Marijuana	*Cannabis sativa*
Marvel of Peru	*Mirabilis jalapa*
Mary's-bean	*Merremia discoidesperma*
flower	*Anastatica hierocuntica*
Mayflower	*Epigaea repens*
orchid	*Cattleya trianae*
Mbandu	*Lonchocarpus capassa*
Meadowsweet	*Filipendula ulmaria*
Melbossie	*Euphorbia inaequilatera*
Melati jasmine	*Jasminum sambac*
Melegueta	*Aframomum melegueta*
Metel	*Datura metel*
Methuselah tree	*Pinus longaeva*
Mexican cypress	*Taxodium mexicana*
poppy	*Argemone mexicana*
Mignonette	*Reseda odorata*
Millet	*Pennisetum glaucum*
Milk bush	*Thevetia peruviana*
weed	*Asclepias fruticosa*
	Sideroxylon inerme
Milkwort	*Polygala vulgaris*
Millipede	*Iulus terrestris*
Miraculous berry	*Synsepalum dulciferum*
Mistletoe	*Helixanthera mannii*
	Loranthus species
	Phoradendron flavescens
	Phragmanthera species

	Tapinanthus species
	Viscum album
Mock orange	*Philadelphus lewissi*
Monarch of the East	*Sauromatum venosum*
Monitor	*Varanus niloticus*
Monkey	*Cercocebus species*
	Cercopithecus species
	Colobus species
	Erythrocebus species
	Papio species
	Piliocolobus species
	Procolobus species
Monkey's hand	*Chiranthodendron pentadactylon*
Monrovian coffee	*Coffea liberica*
Moon flower	*Brugmansia suaveolens*
Moshi medicine	*Guiera senegalensis*
Mosquito plant	*Clausena anisata*
Mother of thousands	*Linaria cymbalaria*
Motherwort	*Leonurus cardiaca*
Mountain laurel	*Kalmia latifolia*
Mournful widow	*Scabiosa atropurpurea*
Mouse	*Mus musculus*
Moutan, The	*Paeonia suffruticosa*
Mud-fish	*Protepterus species*
Mule	*Equus species*
Mushroom	*Calvatia species*
	Termitomyces microcarpus
hallucinogenic	*Amanita muscaria*
	Psilocybe mexicana
Musk mallow	*Malva moschata*
Myrrh	*Commiphora molmol*
	C. myrrha
Myrtle	*Myrtus communis*
-leaved milkwort	*Polygala myrtifolia*
Narrow-leaved coffee	*Coffea stenophylla*
Natal mahogany	*Kiggelaria africana*
plum	*Ximenia caffra* var. natalensis
Native pear	*Dacryodes edulis*
Nigerian powder-flask fruit	*Afraegle paniculata*
Nelumbo	*Nelumbium nelumbo*
Niger copal tree	*Daniellia oliveri*
	D. thurifera
Night blooming cestrum	*Cestrum nocturnum*
Nightshade	*Solanum capensis*
Negro coffee	*Senna (Cassia) occidentalis*
Never die	*Kalanchoe integra* var. crenata
Nipa palm	*Nipa fruticans*
Nodeweed	*Synedrella nodiflora*
Nogàl	*Juglans neotropica*
'Number one'	*Mareya micrantha*
Nutmeg	*Myristica fragrans*
Oak	*Quercus alba*
	Q. petraea
	Q. robus
	Q. species
Oil-bean tree	*Pentaclethra macrophylla*
of Ben tree	*Moringa oleifera*
palm	*Elaeis guineensis*
	E. guineensis var. communis
Okra	*Abelmoschus esculentus*
Okro	*A. esculentus*
Oleander	*Nerium oleander*
Olive	*Olea europaea*
Onion	*Allium. cepa*
Opium poppy	*Papaver somniferum*
Orange, sweet	*Citrus sinensis*
blossom	*C. trifoliata*
lily	*Lilium bulbiferum* var. croceum
Ordeal bean	*Physostigma venenosum*
tree	*Erythrophleum suaveolens*
Oregon grape	*Berberis aquifolium*
Orris root	*Iris florentina*
'Oven' tree	*Copaifera religiosa*
Owl	*Alba species*
	Tyto species

Ox	*Oryx species*
-eye daisy	*Chrysanthemum leucantheum*
Paeony	*Paeonia species*
Pagoda tree	*Alstonia boonei*
	Sophora japonica
Palmier fétiche	*Elaeis guineensis var. idolatrica*
Panama-tree	*Sterculia apetala*
Panax	*Panax species*
Pancratium lily	*Pancratium hirtum*
	P. trianthum
Panther	*Panthera species*
Papyrus	*Cyperus papyrus*
Parrot, grey	*Psittacus erithacus*
Parsley	*Carum petroselinum*
Pasque flower	*Anemone pulsatilla*
American	*Pulsatilla hirsutissima*
Passion flower, stinking	*Passiflora glabra*
fruit	*P. edulis*
Pau Brasil	*Caesalpinia echinata*
Paulownia	*Paulownia tomemtosa*
Pawpaw	*Carica papaya*
Peach	*Prunus persica*
Pear	*Pyrus communis*
Pearl millet	*Pennisetum glaucum*
Pepper	*Capsicum annuum*
	C. frutescens
West African black	*Piper guineense*
Pepul	*Ficus religiosa*
Periwinkle	*Vinca species*
Persian lilac	*Melia azedarach*
Peyote	*Lophophora williamsii*
Physic nut	*Jatropha curcas*
Pig	*Sus species*
bush	*Potamochoerus porcus*
Pigeon	*Columba guinea*
	C. unicincta
	C. species
Pigeonpea	*Cajanus cajan*
Pigweed	*Portulaca oleracea*
Pile root	*Sansevieria thyrsiflora*
Pimenta	*Pimenta dioica*
Pine	*Pinus halepensis*
cone	*P. strobus*
Pineapple	*Ananas comosus*
Pink poui	*Tabebuia rosea*
Piri piri	*Cyperus articulatus*
Pitanga cherry	*Eugenia uniflora*
Pitch pine	*Pinus rigida*
Plantain	*Musa sapientum var. paradisiaca*
	Plantago major
'Plena'	*Spiraea prunifolia*
Plum	*Prunus domestica*
Plumbago	*Plumbago capensis*
Poinsettia	*Euphorbia pulcherrima*
Poison tree	*Acokanthera venenata*
Pomegranate	*Punica granatum*
Poppy (flower)	*Meconopsis regia*
	Papaver rhoeas
Porcupine (brush-tailed)	*Artherurus africanus*
Portia tree	*Thespesia populnea*
Potato	*Solanum tuberosum*
yam	*Dioscorea bulbifera*
Prayer beads	*Abrus precatorius*
plant	*Marantha leuconeura*
Prickly amaranth	*Amaranthus spinosus*
poppy	*Argemone mexicana*
Primrose-willow	*Ludwigia octovalvis ssp. brevisepala*
Prophet flower	*Aipyanthus echioides*
Protea	*Protea species*
Pulmonaria	*Pulmonaria officinalis*
Pumpkin	*Cucurbita pepo*
Purple heath	*Setcreasea purpurea*
witchweed	*Striga hermontheca*
Purslane	*Portulaca oleracea*
Python	*Python regius*
Queen crape-myrtle	*Lagerstroemia speciosa*
of flowers	*Rosa damascena*

of the night	*Epiphyllum oxypetalum*
	Nyctocereus serpentinus
	Selenicereus grandiflorus
sago	*Cycas circinalis*
Queen's wreath	*Petrea volubilis*
'Ragwort'	*Senecio baberka*
Raisin tree	*Grewia flava*
Ram	*Ovis species*
Rama fibre	*Hibiscus lunarifolius*
Rangoon creeper	*Quisqualis indica*
Rat	*Cricetomys gambianus*
giant	*Thryonomys swinderianus*
tail grass	*Sporobolus pyramidalis*
tail verveine	*Starchytarpheta cayennensis*
Red-billed shrike	*Prionops coniceps*
clover	*Trifolium pratense*
ebony	*Rhamnus zeyheri*
-flowered silk cotton tree	*Bombax buonopozense*
head	*Asclepias curassavica*
ironwood	*Lophira alata*
pansy	*Viola tricolor*
pepper	*Capsicum annuum*
physic nut	*Jatropha gossypiifolia*
poppy	*Papaver rhoeas*
river hog	*Potamochoerus porcus*
rose	*Rosa species*
snake bark maple	*Acer capillipes*
thorn	*Acacia gerrardi*
Tudor rose	*Rosa gallica*
witchweed	*Striga asiatica*
Redbird cactus	*Pedilanthus tithymaloides*
Redwood	*Pterocarpus soyauxii*
	Nesogordonia kabingaensis
Reed buck	*Redunca redunca*
Resurrection flower	*Anastatica hierochuntica*
plant	*Bryophyllum pinnatum*
Rhinoceros, African	*Diceros bicornis*
Rhode's violet	*Securidaca longepedunculata*
Rhodesian holly	*Psorospermum febrifugum*
Rhododendron	*Rhododendron maximum*
Rice	*Oryza sativa*
Ringworm shrub	*Senna (Cassia) alata*
Rio Nunez coffee	*Coffea canephora*
Roan antelope	*Hippotragus equinus*
Roof iris	*Iris tectorum*
Rooi opslag	*Hermannia depressa*
Rose	*Rosa species*
apple	*Eugenia jambos*
moss	*Portulaca grandiflora*
of Jericho	*Anastatica hierochuntica*
	Nerium oleander
Rosemary	*Rosmarinus officinalis*
Rosewood	*Guibourtia tessmannii*
Royal antelope	*Neotragus pygmaeus*
palm	*Roystonia regia*
Rue	*Ruta graveolens*
Sabal	*Sabal mexicana*
Sacred Japanese cedar	*Cryptomeria japonica*
-lotus	*Nymphaea lotus*
Saffran	*Cassine croceum*
Sage	*Salvia officinalis*
Sagebrush	*Artemisia tridentata*
Sago lily	*Calochortus nuttallii*
-palm	*Cycas revoluta*
Salt and oil tree	*Cleistopholis patens*
Sampagita jasmine	*Jasminum sembac*
San Pedro	*Trichocereus pachanoi*
	T. peruvianus
Sand olive	*Pappea capens var. radlkoferi*
spurry	*Spurgularia rubra*
Sandalwood	*Santalum album*
Sandbox tree	*Hura crepitans*
Sandpaper tree	*Ficus asperifolia*
	F. exasperata
Satinwood	*Pericopsis laxiflora*
Sausage tree	*Kigelia africana*
Sasaki	*Cleyera japonica*
Sassafras	*Sassafras officinale*

Sasswood tree	*Erythrophleum ivorense*
Saxifrage	*Saxifraga granulata*
Scorpion	*Buthus species*
grass	*Myosotis species*
Scotch thistle	*Onopordum acanthium*
Screw pine	*Pandanus abbiwii*
	P. aggregate
Sea bean	*Entada gigas*
	E. rheedei (pursaetha)
bream	*Pagrus pagrus*
heart	*Entada gigas*
	E. rheedei (pursaetha)
hollies	*Eryngium maritimum*
moonflower	*Ipomoea tuba*
purse	*Dioclea reflexa*
Senegal lilac	*Lonchocarpus sericeus*
pendoring	*Gymnosporia senegalensis*
rose wood tree	*Pterocarpus erinaceus*
Sensitive plant	*Mimosa invisa*
	M. pigra
	M. pudica
	M. rubicaulis
	Schrankia leptocarpa
Sesame or Beniseed	*Sesamum orientale*
Sessile joyweed	*Alternanthera sessilis*
Shallot	*Allium cepa var. aggregatum*
Shamrock, Irish	*Trifolium repens minus*
Shark	*Carcharhinus amboinensis*
	C. leucas
Shea butter tree	*Vitellaria paradoxa*
Sheep	*Ovis species*
Shittim wood	*Acacia hockii*
Shoe flower	*Hibiscus rosa-sinensis*
Showers of gold	*Cassia fistula*
Shrike	*Corvinella corvina*
	Dryoscopus gambensis
	Laniarius atroflavus
	L. barbarus
	L. ferrugineus
Shrike	*L. leucorhynchus*
	Lanius collaris
	L. excubitor
	L. senator
	Malaconotus cruentus
	M. multicolor
	Nilaus afer
	Prionops coniceps
	P. plumata
	Tchagra minuta
	T. senegala
Sierra Leone (upland) coffee	*Coffea stenophylla*
gum copal	*Guibourtia copallifera*
water tree	*Tetracera alnifolia*
	T. potatoria
Silk cotton tree	*Ceiba pentandra*
Slipper flower	*Pedilanthus tithymaloides*
Slug	*Limax species*
Snail	*Achatina achatina*
	Archachatina species
Snake bean	*Swartzia madagascariensis*
gourd	*Trichosanthes cucumerina*
root	*Pisosperma capense*
soap	*P. capense*
Snake's tail	*Achyranthes aspera*
Sneezewood	*Ptaeroxylon utile*
Snuff-box tree	*Oncoba spinosa*
Soap aloe	*Aloe saponaria*
berry	*Balanites aegyptiaca*
Sour sop	*Annona muricata*
Sodom apple	*Calotropis procera*
Solidago	*Solidago virgaurea*
Sooty boubou	*Laniarus leucorhynchus*
Sorcerer's violet	*Vinca major*
	V. minor
Sorrel	*Hibiscus subdariffa var. subdariffa*
Sour plum	*Ximenia cafra*
Soybean	*Glycine max*

Spanish needles	*Bidens bipinnata*
-bayonet	*Yucca elephantipes*
Spider, house	*Nephilengys species*
Spiderwort	*Tradescantia species*
Spiked millet	*Pennisetum glaucum*
Spindle tree	*Euonymus europaeus*
Spiny bear's breach	*Acanthus spinosus*
burnet	*Poterium spinosum*
Spitting cobra	*Naja nigricollis*
Squirrel	*Protocerus species*
St. Johnswort	*Hypericum pertoratum*
Staff tree	*Gymnosporia buxifolius*
Stag-horn fern	*Platycerium elephantotis*
	P. stemaria
Star anise	*Illicium verum*
burr	*Acanthospermum hispidum*
of Bethlehem	*Ornithogalum arabicum*
	O. narbonense
	O. umbellatum
thistle	*Centaurea perrottetii*
	C. praecox
Starwart	*Synedrella nodiflora*
Stink grass	*Melinis minutiflora*
Stinkbas	*Pittosporum viridiflorum*
Stinking cassia	*Senna (Cassia) hirsuta*
passion flower	*Passiflora glabra (foetida)*
Stonecrop	*Sedum acre*
Storax	*Liquidambar orientalis*
	Styrax officinalis
Sugar cane	*Saccharum officinarum*
Sugarbush	*Protea roupelliae*
Sugurao cactus	*Cereus giganteus*
Sun plant of Brazil	*Portulaca grandiflora*
Sunflower	*Helianthus annuus*
Swamp arum	*Lasimorpha senegalensis*
Swartstorm	*Dicoma anomala*
Sweet basil	*Ocimum basilicum*
bay	*Laurus nobilis*
broomweed	*Scoparia dulcis*
flag	*Acorus calamus*
orange	*Citrus sinensis*
pigweed	*Chenopodium ambrosioides*
scabious	*Scabiosa atropurpurea*
sop	*Annona squamosa*
violet	*Viola odorata*
Switch sorrel	*Dodonaea viscosa*
Swizzle-stick	*Rauvolfia vomitoria*
Sword bean	*Canavalia ensiformis*
Talipot palm	*Corypha umbracalifera*
Tallow tree	*Allanblackia floribunda*
	Detarium senegalense
	Pentadesma butyraceum
Tamarisk	*Tamarisk gallica*
	T. hispida
	T. ramossisima
	T. tetandra
Tambootie	*Spirostachys africanus*
Tea bush	*Lippia chevalieri*
senna	*Chamaecrista mimosoides*
Temple flower	*Plumeria rubra*
	var. acutifolia
Ten o'clock plant	*Portulaca quadrifida*
Thistle	*Echinops species*
Thyme	*Thymus vulgaris*
Tiger	*Felis tigris*
flower	*Tigridia pavonia*
lily	*Lilium lancifolium*
nut	*Cyperus esculentus*
Toadstool	*Bolbitis species*
	Calvatia species
	Chlorophyllum species
	Hygrocybe species
	Lentinus tuber-regium
	Pleurotus species
	Pycnoporus species
	Schizophyllum species
	Termitomyces species
	Volvariella species
Tobacco	*Nicotiana tabacum*

witchweed	*Striga gesneroides*
Tomato	*Lycopersicon lycopersicum*
Tortoise	*Kinixys species*
Traveller's joy	*Clematis species*
palm or tree	*Ravenala madagascariensis*
Treasure flower	*Gazania rigens*
Tree of heaven, Chinese	*Ailanthus altissima*
-of-kings	*Cordyline fruticosa*
of life	*Guaiacum officinale*
paeony, The	*Paeonia suffruticosa*
True love	*Paris quadrifolia*
sea-bean	*Mucuna sloanei*
Trumpet narcissus	*Narcissus pseudo-narcissus*
stramonium	*Datura stramonium*
Tulip	*Tulipa species*
African	*Spathodea campanulata*
Turk's island cactus	*Melocactus communis*
-cap cactus	*M. communis*
Turtle	*Cavetta species*
	Chelonia species
	Eremochelys species
Twin flower	*Linnaea borealis*
-leaf	*Jeffersonia diphylla*
-leaved sundew	*Drosera binata*
Umbra tree	*Phytolacca heptandra*
Umbrella tree	*Musanga cecropioides*
Upland rice	*Oryza glaberrima*
Vegetable sponge	*Luffa cylindrica*
Velvet tamarind	*Dialium guineense*
Venus's hair	*Adiantum capillus-veneris*
Vervain	*Verbena officinalis*
Vetiver	*Vetiveria zizanioides*
Victorian Christmas bush	*Prostanthera lasianthos*
Vine rubber	*Landolphia owariensis*
Violets	*Viola odorata*
Viper	*Bitis species*
	Vipera species
Gaboon	*Bitis gabonicum*
Viper's bugloss	*Echium vulgare*
Voodoo lily	*Sauromatum venosum*
Vulture	*Necrosyrtes monachus*
Walking fern	*Camptosorus rhizophyllus*
Walnut	*Juglans regia*
Wandering Jew	*Cyanotis nodiflora*
	Tradescantia species
	Zebrina pendula
Waratah	*Telopea speciosissima*
Wart hog	*Potamochoerus porcus*
Wasp	*Sphex species*
Water forget-me-not	*Myosotis scorpioides*
leaf	*Talinum triangulare*
-lemon	*Passiflora laurifolia*
lettuce	*Pistia stratiotes*
lily	*Nymphaea caerulea*
	N. guineensis
	N. lotus
	N. maculata
	N. micrantha
	N. rufescens
melon	*Citrullus lanatus*
yam	*Dioscorea alata*
Waterberry	*Syzygium cordatum*
Waterbuck	*Kobus ellipsiprymnus*
Wax flax	*Linum thunbergii*
Weaver bird	*Ploceus species*
Wedding flowers	*Antigonon leptopus*
	Dombeya natalensis
Weeping cherry	*Prunus pendula*
love grass	*Eragrostis curvula*
Weeping willow	*Salix babylonica*
Wellingtonia	*Seqouiadendron giganteum*
Welwitschia	*Welwitschia mirabilis*
West African albizia	*Albizia adianthifolia*
black pepper	*Piper guineense*
blackwood	*Dalbergiella welwitschii*
ebony tree	*Diospyros mespiliformis*
locust bean	*Parkia biglobosa*
rubber tree	*Funtumia elastica*
sarsparilla	*Smilax anceps (kraussiana)*

serenpidity berry	*Dioscoreophyllum cumminsii*
Wheat	*Triticum aestivum*
Whistling pine	*Casuarina equisetifolia*
White-cedar	*Tabebuia pallida*
ironwood	*Vepris lanceolata*
lady	*Kalanchoe thyrsiflora*
lily	*Lilium species*
moonflower	*Ipomoea alba*
-nun orchid	*Lycaste skinner*
oak	*Anopyxis klaineana*
olive	*Halleria lucida*
seringa	*Kirkia acuminata*
star apple	*Chrysophyllum albidum*
stinkwood	*Celtis africana*
yam	*Dioscorea rotundata*
Widow's tears	*Tradescantia virginiana*
Wild amaranth	*Amaranthus lividus*
	A. viridis
asparagus	*Asparagus africanus*
banana	*Ensete gilletii*
cat	*Profelis aurata*
custard apple	*Annona chrysophylla*
	A. senegalensis
dove	*Streptopelia semitorquata*
flax	*Linum africanum*
garlic	*Tulbaghia violacea*
geranium	*Pelargonium species*
gourd	*Citrullus colocynthis*
grape	*Rhoicissus caneifolia*
hemp	*Leonotis leonurus*
lettuce	*Launaea taraxacifolia*
lime	*Ximenia americana*
mango	*Irvingia gabonensis*
medlar	*Vangueria venosa*
	Vangueriopsis lanciflora
orange tree	*Toddalia aculeata*
patata	*Ipomoea crassipes*
pink	*Dianthus crenatus*
pomegranate	*Burchellia bubalina*
prairie rose	*Rosa arkansana*
rice	*Oryza barthii*
rose	*Rosa virginiana*
sage	*Lantana camara*
soursop	*Annona chrysophylla*
squill	*Scilla lanceifolia*
tobacco	*Nicotiana rustica*
verbena	*Pentanisia prunelloides*
willow	*Salix woodii*
yam	*Dioscorea praehensilis*
Wildcape gooseberry	*Physalis angulata*
Wildebeest	*Connochaetes gnou*
	C. taurinus
Willow herb	*Epilobium hirsutum*
weeping	*Salix babylonica*
Windflower	*Anemone caronaria*
	A. hepatica
Wine palm	*Raphia farinifera*
	R. hookeri
	R. hookeri var. planifolia
	R. humilis
	R. palma-pinus
	R. regalis
	R. sudanica
	R. vinifera
	R. species
Winged yam	*Dioscorea alata*
Winter cherry	*Withania somnifera*
Wireweed	*Sida rhombifolia*
Witch hazel	*Hamamelis virginiana*
Witchweed	*Striga asiatica*
	S. aspera
	S. bilabiata
	S. elegans
	S. gesneroides
	S. hermontheca
	S. linearifolia
	S. macrantha
	S. rowlandii
Wolf	*Lupus species*
Wolf's grass	*Cymbopogon citratus*

	C. giganteus
Woodchat shrike	*Laniarus senator*
Woolly morning glory	*Argyreia nervosa*
Wormwood	*Artemisia absinthiun*
Yade	*indetermined*
Yam	*Dioscorea species*
-bean	*Sphenostylis stenocarpa*
Yarrow	*Achillea millefolium*
Yellow-backed duiker	*Cephalophus sylvicultor*
-breasted shrike	*Laniarus atroflavus*
chrysanthemum	*Chrysanthemum species*
guinea yam	*Dioscorea cayenensis*
jessamine	*Gelsemium sempervirens*
oleander	*Thevetia peruviana*
tassel flower	*Emilia coccinea*
Yew	*Taxus baccata*
Yoruba soft cane	*Marantochloa congensis*
	M. filipes
	M. leucantha
	M. mannii
	M. purpurea
	M. ramosissima
	Megaphrynium macrostachyum
	Sarcophrynium brachystachys
	S. prionogonium
Yucca	*Yucca elephantipes*
Yum yum	*Carissa bispinosa*
Yumbi	*Phytolacca weberbaueri*
Zulu (bush) -willow	*Combretum suluense*

LOCAL NAMES (AFRICAN)

aba (Yoruba)	*Ficus species*
abehene (Twi)	*Elaeis guineensis var. idolatrica*
abisi (Yoruba)	indeterminate
	Sarcocephalus latifolius
abiwéré (Yoruba)	*Hybanthus enneaspermus*
aboa-ngatsee (Nzima)	*Lagenaria breviflora*
abonsam aburow (Fante)	*Zea mays*
achineku (Tiv.)	indeterminate
adade oko (Yoruba)	*Rungia grandis*
adam (Akan)	*Palthothyrsus species*
ade-ile (Yoruba)	*Thonningia sanguinea*
adelamanyi (Awuna)	*Lantana camara*
adele (Yoruba)	*Thonningia sanguinea*
adosusu (Yoruba)	indeterminate
afagbara (Yoruba)	indeterminate
afema (Akan)	*Justicia flava*
agbu (Tiv)	*indeterminate*
agemo-kogun (Yoruba)	*Laggera alata*
	L. heudelotii
agidi magbayin (Yoruba)	*Sida acuta*
agogo-igun (Yoruba)	*Heliotropium indicum*
ahenegeen trso (Ga)	*Bombax buonopozense*
ahensaw (Akan)	*Momordica angustisepala*
aidan (Yoruba)	*Tetrapleura tetraptera*
ajade (Yoruba)	*Stereospermum kunthianum*
ajekobale (Yoruba)	*Croton zambesicus*
ajekofole (Yoruba)	*Croton zambesicus*
akakanikoko (Yoruba)	indeterminate
akara-aje (Yoruba)	*Cnestis ferruginea*
akidimmo (Igbo)	*Cleome rutidosperma*
akpakpa enyi miri (Igbo)	*Sorghum arundinaceum*
	S. lanceolatum
	S. vogeliana
akuakuaanisuo (Akan)	*Spathodea campanulata*
alhaji (Hausa)	*Aerva javanica*
alupayida (Yoruba)	*Uraria picta*
amonketa (Igbo, Owerri)	*Cnestis ferruginea*
amuje (Yoruba)	*Harungana madagascariensis*
amunimuye (Yoruba)	*Senecio abyssinica*
anafonanaeku (Akan)	*Hilleria latifolia*
anderabai (Temne)	*Adansonia digitata*
anon ekun (Yoruba)	*Acanthus montanus*
	Hibiscus asper
anon ekundudu (Yoruba)	*Acanthus montanus*
anyan (Yoruba)	*Distemonanthus benthamianus*
anyen-enyiwa (Fante)	*Abrus precatorius*
apako (Yoruba)	*Cleistopholis patens*
apara (Yoruba)	*Hexalobus crispiflorus*
apebentutu (Fanti)	*Lagenaria siceraria*
apem (Akan)	*Musa sapientum var. paradisiaca*
apese (Akan)	*Atherurus africanus*
araka (Azande)	indeterminate
asaman-akyekyea (Akan)	*Coccinea barteri*
	Legenaria breviflorus
	Ruthalicia longipes
ase (Yoruba)	indeterminate
ashorin (Yoruba)	indeterminate
asimawu (Yoruba)	indeterminate
asirisiri (Akan)	*Platostoma africanum*
asogbosato (Yoruba)	indeterminate
asuani (Akan)	*Cardiospermum grandiflorum*
	C. halicacabum
as unfunrun (Yoruba)	indeterminate
asurin (Yoruba)	indeterminate
atabese (Yoruba)	indeterminate
atwere nantem Akan)	*Diospyros monbuttensis*
avadzo (Ewe)	*Mimosa pudica*
awe biyemi (Yoruba)	*Euphorbia prostrata*
awendade (Akan)	*Rourea coccinea*
	Dialium dinklagei

awin (Yoruba)	*Dialium guineense*
awuku (Akan)	*Dioscorea alata*
ayada (Yoruba)	*Stereospermum kunthianum*
ayibiribi (Fante)	*Operculina macrocarpa*
ayinre ogo (Yoruba)	*Albizia adianthifolia*
babadua (Akan)	*Hypselodelphys species*
ba-fi latana (Hausa)	*Heliotropium bacciferum*
bafuafu (Azande)	indeterminate
baga (Azande)	*Khaya senegalensis*
bakin suda (Hausa)	*Polycarpaea linearifolia*
	P. corymbosa
baku (Akan)	*Tieghemella heckelii*
bambami (Fula)	*Calotropis procera*
bamoru (Azande)	indeterminate
batafoia kani (Ga)	*Anchomanes difformis*
begeyi (Hausa)	indeterminate
bene-fing dion (Bambara/Malinke)	*Hyptis spicigera*
benge (Azande)	*Strychnos species*
bi ta saisai (Hausa)	*Acanthaceae indet*
bi ta swai swai (Hausa)	*Acanthaceae indet*
bingha (Azande)	*Imperata cylindrica*
bingi (Nupe)	*Lagenaria siceraria*
bloe (Basa)	*Eugenia whytei*
bohoyana (Sotho)	*Hibiscus malacospermus*
bombiri (Azande)	*Albizia aylmeri*
bonsamdua (Akan)	*Distemonanthus benthamianus*
	Pericopsis laxiflora
bore daso (Twi)	*Dissotis rotundifolia*
bore kete (Twi)	*Dissotis rotundifolia*
brezinle nyema (Nzima)	*Cassytha filiformis*
	Cuscuta australis
bronyadua (Akan)	*Hildegardia barteri*
	Morinda lucida
bubeben mil (Diola)	*Secamone afzelii*
bubum enab (Diola)	*Microglossa afzelii*
	M. pyrifolia
bunga bangkai (Sumatra)	*Amorphophalus titanum*

chabbi-mai be (Fula)	*Amblygonocarpus andongensis*
	Cassia arereh
	C. sieberiana
che'diya (Hausa)	*Ficus dekdekena*
daashi mai-tawan rai (Hausa)	*Commiphora africana var.africana*
da ogun duro (Yoruba)	*Tribulus terrestris*
dacere ngala (Fulfulde)	*Aristida species*
daguro (Yoruba)	*Tribulus terrestris*
daudar maaguzaawaa (Hausa)	*Blepharis linariifolia*
daudawa (Hausa)	*Parkia species*
de-fia (Ewe, Krepi)	*Elaeis guineensis var. idolatrica*
didikuridi (Yoruba)	indeterminate
die die (Kru Bete,Guere)	*Milicia excelsa*
	M. regia
dikabi-awotsho (Krobo)	*Sarcocephalus latifolius*
dome atre (Akan)	*Cassytha filiformis*
	Cuscuta australis
dua sika (Twi)	*Enantia chlorantha*
	E. polycarpa
duapompo (Akan)	*Omphalocarpum ahia*
	O. elatum
	O. procerum
dundu (Hausa)	*Dichrostachys cinerea*
duwei iyau (Kolokuma)	*Calopogonium mucunoides*
dwaba (Xhosa)	*Popowia caffra*
dwene (Akan)	*Baphia nitida*
	B. pubescens
ebia (Fante)	*Operculina macrocarpa*
ebube-ago (Igbo)	*Sansevieria liberica*
	S. senegambica
	S. trifasciata
ede-chukwu (Igbo)	*Crinum jagus*
ede hinde (Diola)	*Momordica balsamina*
ede-obasi (Igbo)	*Crinum natans*
edesuku (Igbo)	*C. distichum*

	C. jagus
edule imemein (Ijo)	*Acanthus montanus*
efon (Yoruba)	indeterminate
egungun-ekun (Yoruba)	*Balanites wilsoniana*
enwomei-ate (Aowin)	*Cathormion altissimum*
ere otore (Ijo)	*Amorphophalus dracontioides*
erikesi (Yoruba)	*Diospyros monbuttensis*
erin (Yoruba)	indeterminate
erinmi oke (Yoruba)	indeterminate
esere (Ibibio)	*Physostigma venenosum*
esho (Yoruba)	*Bombax buonopozense*
esono-ababaa (Akan)	*Microglossa afzelii*
	M. pyrifolia
esono-tokwa-kofo (Akan)	*Stereospermum acuminatissimum*
	S. kunthianum
etichoro (Yoruba)	indeterminate
eto-mkpa (Efik)	*Jatropha curcas*
eton (Yoruba)	*indeterminate*
etudamo (Wolof)	*Stereospermum kunthianum*
etudemo (Wolof)	*Stereospermum kunthianum*
ewe-aje (Yoruba)	*Aerva lanata*
	Gloriosa simplex
	G. superba
ewe biyemi (Yoruba)	*Euphorbia prostrata*
ewe irin (Yoruba)	indeterminate
ewé loko lepon (Yoruba)	*Hybanthus enneaspermus*
ewe obi (Yoruba)	*Hallea stipulosa*
ewe-owu (Yoruba)	*Aerva lanata*
ewire (Akan)	*Acacia kamerunensis*
fa-de (Benin)	*Elaeis guineensis*
	var. idolatrica
farin gamo (Hausa)	*Ipomoea argentaurata*
fasa daga (Hausa)	*Afzelia africana*
fasa maza (Hausa)	*Afzelia africana*
faskara tooyi (Hausa)	*Blepharis linariifolia*
feko (Sotho)	*Crassula rubicunda*
fuladafo (Diola)	*Strophanthus sarmentosus*
funmo'mi (Yoruba)	*Clerodendrum capitatum*
furai nyame (Mandinka)	*Celosia leptostachya*
furen yan sarki (Hausa)	*Lonchocarpus laxiflorus*
gadon machi ji (Hausa)	*Merremia tridentata*
	ssp. angustifolia
	Zaleya pentandra
gaiza (Hausa)	*Combretum micranthum*
gaman sauwa (Hausa)	*Holarrhena floribunda*
gammon bawa (Hausa)	*Merremia tridentata*
	ssp. angustifolia
gantimi (Chamba)	*Adenia venenata*
garkuwar-wuta (Hausa)	*Commiphora kerstingii*
gbetenga (Krobo)	*Cymbopogon giganteus*
gboro ayaba (Yoruba)	*Ipomoea asarifolia*
ge-an (Basa)	*Vismia guineensis*
ghei (Mano)	*Milicia excelsa*
	M. regia
goga masu (Hausa)	*Mitracarpus hirtus*
gorgo baigore (Fula)	*Vernonia nigritiana*
gué (Guere, Kono)	*Milicia excelsa*
	M. regia
gu-ekura (Akan)	*Ageratum conyzoides*
gunar kuuraa (Hausa)	*Cucumis figarei*
	C. metuliferus
gwaska (Hausa)	*Andira inermis*
hak'orin machiji (Hausa)	*Achyranthes aspera*
hale gbokpo (Mende)	*Ipomoea asarifolia*
	I. mauritiana
hano (Hausa)	*Boswellia dalzielii*
hantsar gada (Hausa)	*Ipomoea nil*
	I. aitonii
hare gujje (Fula)	*Leonotis species*
hloko (Sotho)	*Elyonurus argenteus*
homa-biri (Akan)	*Gouania longipetala*

homa-kyem (Akan)	*Spiropetalum heterophyllum*
homa-kyereben (Akan)	*Salacia debilis*
hona gboli (Mende)	*Caladium bicolor*
hwana wulie (Mende)	*Clerodendrum umbellatum*
iboti (Yoruba)	indeterminate
ido (Yoruba)	*Canna indica*
igibiszilla (Zulu)	*Bowiea volubilis*
ikehegh (Tiv)	indeterminate
ikubalo likamlanjani (Xhosa)	*Pelargonium pulveratum*
ilamwadibi (Yoruba)	indeterminate
ile ago (Igbo)	*Hibiscus surattensis*
inaeyinfun (Yoruba)	indeterminate
indungu (Chagga)	*Adhatoda engleriana*
indungulu (Zulu)	*Kaempferia species*
inunurin (Yoruba)	indeterminate
iromi (Yoruba)	indeterminate
irugbefou (Yoruba)	*Eragrostis ciliaris*
	E. gangetica
	E. pilosa
	E. tremula
isidwaba (Zulu)	*Popowia caffra*
iso apare (Yoruba)	*Sabicea calycina*
issa (Akan)	*Celtis species*
isupeyinkota (Yoruba)	indeterminate
itanna pa osho (Yoruba)	*Mirabilis jalapa*
ite-oka (Yoruba)	*Oplimenus burmannii*
	O. hirtellus
iviromila (Benin)	*Elaeis guineensis var. idolatrica*
iyu (Yoruba)	*Clerodendrum capitatum*
jallo (Gure)	*Lagenaria siceraria*
je-ra-kpar (Basa)	*Bersama abyssinica ssp. paullinioides*
ji abana mwo (Yoruba)	*Smilax anceps (kraussiana)*
ka lonk (Diola)	*Cercestis afzelii*
kaajiijii (Hausa)	*Cyperus articulatus*
	C. maculatus
	Kylinga erecta
	K. pumila
	K. squamulata
	K. tenuifolia
kabubuwar makafi (Hausa)	*Centaurea perrottetii*
	C. praecox
ka-fi-malan (Hausa)	*Evolvulus alsinoides*
	Scoparia dulcis
kafulugay (Diola)	*Hyptis spicigera*
kalankuwa (Hausa)	*Aspilia africana*
kali-kuli (Mende)	*Bertiera racemosa var. racemosa*
kam barawo (Hausa)	*Leucas martinicensis*
kanrinkan-ayaba (Yoruba)	*Luffa cylindrica*
kantu (Guere)	*Inhambanella guereensis*
karan masallaachii (Hausa)	*Caralluma dalzielii*
karan sarki (Hausa)	*Saccharum officinarum*
kauchin aduwa (Hausa)	*Balanites aegyptiaca*
kheoha (Sotho)	*Psydrax ciliata*
kilemela (Shambala)	*Polygala gomesiana*
kimbar mahalba (Hausa)	*Rourea (Byrsocarpus) coccinea*
	Lantana rhodesiensis
kiobaobe (Lobedu)	*Clematis oweniae*
kiroroki (Chagga)	*Polygala usambarensis*
koditsana (Sotho)	*Rhus divaricata*
koko oba (Yoruba)	*Cymbopogon citratus*
	C. giganteus var. giganteus
kokora (Akan)	*Smilax anceps (kraussiana)*
kokote (Akan)	*Anopyxis klaineana*
	Potamochoerus porcus
kolawotso (Adangbe)	*Ochna afzelii*
	O. membranacea
koli-goie (Krio)	*Ruthalicia eglandulosa*
koli-ngengowi (Mende)	*Machaerium lunatum*
koliawatso (Krobo)	*Ochna afzelii*
	O. membranacea

kooba (Fula)	*Hippotragus equinus*
koron mboddi (Fula)	*Citrullus colocynthis*
kpledzoo ahi tso (Ga)	*Adansonia digitata*
kuakuaanisuo (Akan)	*Spathodea campanulata*
kubaol (Mpondo)	*Asparagus species*
kukulu (Grusi-Lyela)	*Adansonia digitata*
kumakuafo (Akan)	*Maytenus senegalensis*
kumenini (Akan)	*Lannea welwitschii*
kunga (Azande)	indeterminate
kwadu-pa Akan)	*Musa sapientum*
kwaduo (Akan)	*Cephalophus sylvicultor*
kyirituo (Akan)	*Securidaca longepedunculata*
layre ngabbu (Fula)	*Ipomoea asarifolia*
	I. mauritiana
lebatjana (Sotho)	*Cymbopogon dieterlenii*
lekki baatal (Fula)	*Aristida adscensionis*
lekxolela-la-basoth (Sotho)	*Harveya speciosa*
lematlana (Sotho)	*Nerine angustifolia*
lemelanthufe (Tswana)	*Delosperma herbeum*
lesibo (Sotho)	*Stephania umbellata*
letomokoane (Sotho)	*Silene undulata*
leungeli (Sotho)	*Asparagus rivalis*
loko (Sotho)	*Elyonurus argenteus*
maalamin maataa (Hausa)	*Mukia maderaspatana*
mafifi-matso (Sotho)	*Phygelius capensis*
maganin kunama (Hausa)	*Merremia tridentata*
	ssp. angustifolia
magoori (Hausa)	*Mitracarpus hirtus (scaber)*
mai-nasara (Hausa)	*Polycarpaea linearifolia*
	P. corymbosa
maime (Pedi)	*Cyathula uncinulata*
mama (Hausa)	*Glossonema boveanum*
	ssp. nubicum
manta uwa (Hausa)	*Angraecum species*
	Ansellia species
manta uwa (Hausa)	*Bulbophyllum species*
	Crotalaria arenaria
	Listrostachys species
	Polystachys species
	Stolzia species
maasaran machiji (Hausa)	*Gloriosa simplex*
	G. superba
masa (Yoruba)	indeterminate
masopolohane (Sotho)	*Anthospermum pumilum*
mbokok ekpo (Efik)	*Pennisetum purpureum*
mbomdo (Ubangi)	*Strychnos icaja*
mbondo (Ubangi)	*S. icaja*
mboundu (Ubangi)	*S. icaja*
mere (Azande)	*Dioscorea species*
mfutu (Nyanja)	*Vitex mombassae*
minan-sa-bine (Kono)	*Melinis minutiflora*
minchingoro (Hausa)	*Garcinia kola*
minta (Pabir)	*Afzelia africana*
mkpon ekpo (Efik)	*Caladium bicolor*
mohlatsisa (Sotho)	*Asclepias stellifera*
moloka (Sotho)	*Nolletia ciliaris*
monwokuro (Yoruba)	indeterminate
moomang (Sotho)	*Lasiosiphon anthylloides*
mopinaise (Yoruba)	indeterminate
moretio (Sotho)	*Lycium acutifolium*
moriri-wa-lehala (Sotho)	*Galium rotundifolium*
mothokxo (Sotho)	*Turbina oblongata*
mpemba (Luvale)	*Vangueriopsis lanciflora*
musaya (Yoruba)	indeterminate
ndhlamnhloshane (Zulu)	*Vernonia natalensis*
neh-mle-chu (Basa)	*Massularia acuminata*
ngoli foyi (Ewe, Awuna)	*Scaevola plumieri*
nkanfo (Twi)	*Dioscorea dumetorum*
nkwu-kamanu (Igbo)	*Elaeis guineensis*
	var. idolatrica
noonon karse (Hausa)	*Cucumis figarei*
noonon kuuraa (Hausa)	*C. metuliferus*
notsigbe (Ewe)	*Euphorbia hirta*

nsurogya (Akan)	*Adenia rumicifolia var. miegei*
	Gongronema latifolium
	Parquetina nigrescens
	Telosma africanum
ntew (Akan)	*Dioclea reflexa*
ntsebele (Sotho)	*Polygala gymnocladia*
nufuten (Akan)	*Kigelia africana*
nwomele-ate (Nzima)	*Leptoderris brachyptera*
nyame-dua (Akan)	*Alstonia boonei*
nyame kyim (Fante)	*Anchomanes difformis*
	A. welwitschii
nyamele-kukwe (Nzima)	*Vitex grandifolia*
nyankoma (Akan)	*Myrianthus arboreus*
	M. libericus
	M. serratus
nyini-fakata (Mende)	*Carica papaya*
nyotfo (Adangbe)	*Uvaria ovata*
ode doona (Igala)	*Striga asiatica*
odee (Akan)	*Okoubaka aubrevillei*
odelamanyi (Awuna)	*Lantana camara*
odubrafo (Twi)	*Mareya micrantha*
odum (Akan)	*Milicia excelsa*
	M. regia
odwen (Akan)	*Baphia nitida*
	B. pubescens
ogan ibule (Yoruba)	*Combretum constrictum*
ogege (Yoruba)	indeterminate
ogodo (Yoruba)	indeterminate
oja ikoko (Yoruba)	*Sansevieria liberica (fibre)*
oja koriko (Yoruba)	*S. senegambica (fibre)*
	S. trifasciata (fibre)
ojiji orota (Yoruba)	indeterminate
oko aja (Yoruba)	*Hedranthera barteri*
okotitaku (Ijo-Izon)	*Hedranthera barteri*
okoubaka (Anyi)	*Okoubaka aubrevillei*
okpu iche nketa (Igbo, Owerri)	*Cnestis ferruginea*

okpunketa (Igbo, Owerri)	*Cnestis ferruginea*
okumu (Igbo)	*Plukenetia conophora*
ologbogin (Yoruba)	indeterminate
ol-okora (Masai)	*Cordia ovalis*
olorofo (Yoruba)	indeterminate
omonigedegede (Oyo)	*Cuscuta australis*
oniyeniye (Yoruba)	*Hydrolea palustris*
onyina (Akan)	*Ceiba pentandra*
ope (Ga)	*Hibiscus rosa-sinensis*
ope-ifa (Yoruba)	*Elaeis guineensis var. idolatrica*
opongbe	indeterminate
orikan jajata (Yoruba)	indeterminate
oro-adele (Yoruba)	*Euphorbia poisonii*
	E. unispina
oru angi (Ijo)	*Mariscus alternifolius*
orum opewe (Yoruba)	indeterminate
orunmila (Yoruba)	*Elaeis guineensis var. idolatrica*
otabese (Yoruba)	indeterminate
otakiti (Yoruba)	indeterminate
owoduane (Akan)	*Momordica foetida*
owudako (Akan)	*Coix lacryma-jobi*
oyan oje (Yoruba)	*Cnestis ferruginea*
oye (Yoruba)	*Maranthes robusta*
pa ori omo da (Yoruba)	*Pleioceras barteri var. barteri*
padula (Fula)	*Redunca redunca*
pa-kin-kin (Sherbro)	*Oryza barthii*
papaisan (Yoruba)	*Portulaca oleracea*
pari omoda (Yoruba)	*Pleioceras barteri var. barteri*
pashan koriko (Yoruba)	*Sansevieria liberica (fibre)*
	S. senegambica (fibre)
	S. trifasciata (fibre)
petekunsoe (Fante)	*Acanthospermum hispidum*
poma magbe (Mende)	*Newbouldia laevis*
poponla (Yoruba)	*Bombax buonopozense*

pou buni (Ijo-Izon)	*Ficus mucoso*
pulumpun (Choggu)	*Sterculia setigera*
rauma fada (Hausa) }	*Evolvulus alsinoides*
rima fada (Hausa) }	*Heliotropium bacciferum*
roma fada (Hausa) }	*H. ovalifolium*
ruma fada (Hausa) }	*H. subulatum*
runhun zaki (Hausa)	*Dalbergia saxatilis*
sa furfura (Hausa)	*Crotalaria microcarpa*
saa tesagu (Bimoba)	*Adenium obesum*
sadii (Yoruba)	indeterminate
sadjamba (Mandinka)	*Trianthema portulacastrum*
	Zaleya pentandra
sama jewo (Mandinka)	*Ipomoea asarifolia*
	I. mauritiana
saman-anka (Akan)	*Citropsis articulata*
saman-awuram (Asante)	*Megaphrynium macrostachyum*
saman-kube (Akan)	*Rothmannia longiflora*
saman-ntew (Akan)	*Mucuna sloanei*
	Physostigma venenosum
saman-ntini (Akan)	*Rourea (Byrsocarpus) coccinea*
samandua (Akan)	*Clausena anisata*
	Sophora occidentalis
samangya (Akan)	*Bersama abyssinica*
	ssp. paullinioides
samanta (Akan) }	*Anthonotha macrophylla*
samantawa (Akan) }	*Berlinia species*
	Bussea occidentalis
	Calpocalyx brevibracteatus
	Chidlowia sanguinea
	Xylia evansii
samanyobli (Ga)	*Clausena anisata*
samataga-na (Susu)	*Sophora occidentalis*
sandan mayu (Hausa)	*Amblygonocarpuus andongensis*
	Cassia arereh
	C. sieberiana
sanya (Akan)	*Daniellia oliveri*
sapowpa (Akan)	*Combretum fragrans*
	Momordica angustisepala
sapu pora (Senoufo)	*Merremia hederacea*
sark'a (Hausa)	*Asparagus species*
sasabonsam-kyew (Akan)	*Platyserium elephantotis*
	P. stemaria
sawa (Mende)	*Gouania longipetala*
	Strophanthus gratus
sekiseki (Yoruba)	indeterminate
soharane (Sotho)	*Galium dregeanum*
sehlulamanye (Swati)	*Schrebera saundersiae*
sede (Awuna)	*Elaeis guineensis*
	var. idolatrica
sefothafontha (Sotho)	*Tulbarghia dieterlenii*
serelile (Sotho)	*Anacampseros arachnoides*
sha shatan (Hausa)	*Ludwigia octovalvis*
	ssp. brevisepala
shadjo tso (Ga)	*Adansonia digitata*
shafa (Kilba)	*Combretum collinum*
	ssp. binderianum
	C. collinum
	ssp. geitonophyllum
	C. colinum
	ssp. hypopilinum
snek-tamatis (Krio)	*Trichosanthes cucumerina*
solonoringo (Mandinka)	*Machaerium lunatum*
sunya (Ewe)	*Dracaena arborea*
	D. fragrans
tafo ka shamamark (Hausa)	*Glossonema boveana*
	ssp. nubicum
tanaposo (Yoruba)	*Mirabilis jalapa*
thithigwani (Lobedu)	*Hypoxis villosa*
thswene (Sotho)	*Cephalaria zeyheriana*
tsilabelo (Sotho)	*Rhus erosa*

tsire (Ga)	*Pagrus pagrus*
turare-wuta (Hausa)	*Cyperus articulatus*
	C. maculatus
	Kylinga erecta
	K. pumila
	K. squamulata
	K. tenuifolia
tutar 'yan sarki (Hausa)	*Leonotis nepetifolia var. africana*
	L. nepetifolia var. nepetifolia
tweapea (Akan)	*Garcinia kola*
tweeblaar-kanniedood (Afrikaans)	*Welwitschia mirabilis*
ubani (Swahili)	*Commiphora charteri*
ugbogielimi (Edo)	
	Pararistolochia goldieana
ugbogiorimmwin (Edo)	*Pararistolochia goldieana*
ugugukile (Zulu)	*Hibiscus pusillus*
umvusankunzi (Swati)	*Carissa bispinosa*
unoba (Yoruba)	*Cocos nucifera*
uqamamawene (Zulu)	*Begonia sutheriandii*
urdi loho'be (Fula)	*Rourea (Byrsocarpus) coccinea*
	Lantana rhodesiensis
usiere-ebua (Efik)	*Cnestis ferruginea*
utin-ebua (Efik)	*Cnestis ferruginea*
utommo (Yoruba)	*Hoslundia opposita*
uwar yara (Hausa)	*Euphorbia balsamifera*
	Ficus sur (capensis)
vuruma (Azande)	indeterminate
walkin machiji (Hausa)	*Chrozophora senegalensis*
wamma (Akan)	*Ricinodendron heudelotii*
way-doo (Kru-Bassa)	*Dityophleba leonensis*
wodoewogbugbe (Ewe)	*Sida linifolia*
wumlim (Dagomba)	*Striga species*
yetudomo (Wolof)	*Stereospermum kunthianum*
yun yun (Yoruba)	*Aspilia africana*
zugubetia (Choggu)	*Stereospermum kunthianum*

SCIENTIFIC NAMES

GENERAL INDEX

A

B

C

G

I

J

K

M

S

U

V

W

X

Y

Z